THE SCIENCE AND APPLICATIONS OF ACOUSTICS

AIP Series in
Modern Acoustics and Signal Processing

BOOKS IN SERIES

THE SCIENCE AND APPLICATIONS OF ACOUSTICS

Daniel R. Raichel
CUNY Graduate School
and
The School of Architecture and Environmental Studies
The City College of the City University of New York

With 216 Illustrations

Daniel R. Raichel
Department of Mechanical Engineering
The City College of the City University of New York
New York, NY 10031
USA
draichel@worldnet.att.net

Series Editor:
Robert T. Beyer
Physics Department
Brown University
Providence, RI 02912
USA

Library of Congress Cataloging-in-Publication Data
Raichel, Daniel R.
    The science and applications of acoustics / Daniel R. Raichel.
        p.   cm.—(Modern acoustics and signal processing)
    Includes bibliographical references and index.
    ISBN 0-387-98907-2 (hc.: alk paper)
    1. Sound.   I. Title.   II. Series: AIP series in modern acoustics
and signal processing.
    QC225.15.R35   2000
    534—dc21   99-39801

Printed on acid-free paper.

Production managed by Steven Pisano; manufacturing supervised by Joe Quatela.
Typeset by TechBooks, Fairfax, VA.
Printed and bound by Maple-Vail Book Manufacturing Group, York, PA.
Printed in the United States of America.

9 8 7 6 5 4 3 2 1

ISBN 0-387-98907-2 Springer-Verlag New York Berlin Heidelberg   SPIN 10736904

In Memoriam:
Regina and Israel Raichel
Rose and Sidney Wahrman
Richard Charles Jordan and James Conner Clapp

Series Preface

Soun is noght but air y-broke
—Geoffrey Chaucer
end of the 14th century

Traditionally, acoustics has formed one of the fundamental branches of physics. In the twentieth century, the field has broadened considerably and become increasingly interdisciplinary. At the present time, specialists in modern acoustics can be encountered not only in physics departments, but also in electrical and mechanical engineering departments, as well as in mathematics, oceanography, and even psychology departments. They work in areas spanning from musical instruments to architecture to problems related to speech perception. Today, six hundred years after Chaucer made his brilliant remark, we recognize that sound and acoustics is a discipline extremely broad in scope, literally covering waves and vibrations in all media at all frequencies and at all intensities.

This series of scientific literature, entitled Modern Acoustics and Signal Processing (MASP), covers all areas of today's acoustics as an interdisciplinary field. It offers scientific monographs, graduate-level textbooks, and reference materials in such areas as architectural acoustics, structural sound and vibration, musical acoustics, noise, bioacoustics, physiological and psychological acoustics, speech, ocean acoustics, underwater sound, and acoustical signal processing.

Acoustics is primarily a matter of communication. Whether it be speech or music, listening spaces or hearing, signaling in sonar or in ultrasonography, we seek to maximize our ability to convey information and, at the same time, to minimize the effects of noise. Signaling has itself given birth to the field of signal processing, the analysis of all received acoustic information or, indeed, all information in any electronic form. With the extreme importance of acoustics for both modern science and industry in mind, AIP press, now an imprint of Springer-Verlag, initiated this series as a new and promising publishing venture. We hope that this venture will be beneficial to the entire international acoustical community, as represented by the Acoustical Society of America, a founding member of the American Institute of Physics, and other related societies and professional interest groups.

It is our hope that scientists and graduate students will find the books in this series useful in their research, teaching, and studies. As James Russell Lowell once wrote, "In creating, the only hard thing's to begin." This is such a beginning.

Robert T. Beyer
Series Editor-in-Chief

Preface

The science of acoustics deals with sound transmission through solids and fluids. As a mechanical effect, sound is essentially the passage of pressure fluctuations through matter as the result of vibrational forces acting on that medium. Sound has the attributes of a wave phenomenon, as do light or radio signals. But unlike its electromagnetic counterparts, sound cannot travel through a vacuum. In *Sylva Sylvarum*, written in the early seventeenth century, Sir Francis Bacon deemed sound to be "one of the subtlest pieces of nature," but he complained that "the nature of sound in general hath been superficially observed." His accusation of superficiality, from the perspective of the modern viewpoint, was justified for his time, not only for acoustics, but for nearly all branches of physical science. Frederick V. Hunt (1905–67), one of America's greatest acoustical pioneers, pointed out that "the seeds of analytical self-consciousness were already there, however, and Bacon's libel against acoustics was eventually discharged through the flowering of a clearer comprehension of the physical nature of sound."

Modern acoustics is vastly different from the field that existed in Bacon's time and even twenty years ago. It has grown to encompass the realm of ultrasonics and infrasonics in addition to the audio range, as the result of applications in materials science, medicine, dentistry, oceanology, marine navigation, communications, petroleum and mineral prospecting, industrial processes, music and voice synthesis, animal bioacoustics, and noise cancellation. Improvements are still being made in the older domains of music and voice reproduction, audiometry, psychoacoustics, speech analysis, and environmental noise control.

This text—aimed at science and engineering majors in colleges and universities, principally undergraduates in the last year or two of their programs and graduate students, as well as practitioners in the field—was written with the assumption that users DD this text are sufficiently versed in mathematics up to and including the level of differential and partial differential equations, and that they have taken the sequence of undergraduate physics courses that satisfy engineering accreditation criteria. It is my hope that a degree of mathematical elegance has been sustained here, even with the emphasis on engineering and scientific applications. While the use of SI units is stressed, occasional references are made to physical parameters in English units. It is also strenuously urged that laboratory experience be included in the course (or courses) in which this text is being used. The student of acoustics will thus obtain a far keener appreciation of the topics covered in "recitation" classes when he or she gains "hands-on" experience in the use of sound-level meters, signal generators, frequency analyzers, and other measurement devices.

Many of the later chapters in the text are self-contained in the sense that an instructor may skip certain segments in order to concentrate on the agenda most appropriate to the class. However, mastery of the materials in the earlier chapters, namely, Chapter 1–6, are obviously requisite to understanding of the later chapters. Chapters such as those dealing with musical instruments or underwater sound propagation or the legal aspects of environmental noise can be skipped in order to either accommodate academic schedules or to allow concentration on certain topics of greater interest to the instructor (and, hopefully, his or her class) such as ultrasound, architectural acoustics, or other topics. Problems of different levels of difficulty are included at the end of most of the chapters. Many of the problems entail the theoretical aspects of acoustics, but a number of "practical" questions have also been included.

As an author, I hope that I have successfully met the challenge of providing a modern, fairly comprehensive text in the field for the benefit of both students and practitioners, whether they are scientists or engineers. In using parts of this book in prepublication editions in teaching acoustics classes, I have benefited from feedback and suggestions from my students. A number of them have proved to be quite eagle-eyed as they have supplied a continuous stream of suggestions and corrections. It is impossible to acknowledge them all, but Gregory Miller and José Sinibadi come to mind as being among the most assiduous. A number of my colleagues and friends also went through most of the chapters. In particular, I must acknowledge Paul Arveson of the Naval Surface Warfare Center, Carderock of Bethesda, Maryland, who went through the first three chapters with a fine-toothed comb, M. G. Prasad of Stevens Institute of Technology who made a number of extremely valuable suggestions for Chapter 9 on instrumentation, and Edith Corliss who greatly encouraged me on Chapter 10 dealing with the mechanism of hearing. Dr. Zouhair Lazreq, who did his postdoctorate under my tutelage, also looked over some of the chapters. Martin Alexander has been helpful in obtaining illustrations for Chapter 9 from Brüel and Kjær; Dr. Volker Irmer of Germany's Federal Environmental Agency supplied updates on international noise codes, and Armand Lerner arranged to have materials forwarded from Eckel Corporation of Cambridge, Massachusetts. James E. West of Lucent Bell Laboratories, the past president (1998–99) of the Acoustical Society of America, was instrumental in providing photographs of the anechoic chamber for use in this text. I am also indebted to Caleb Cochran of the Boston Symphony Orchestra, Steve Lowe of the Seattle Symphony Orchestra, Elizabeth Canada of the Kennedy Center, Sandi Brown of the Minnesota Orchestral Association, Thomas D. Rossing of Northern Illinois University, Ann C. Perlman of the American Institute of Physics, Karen Welty of Abbott Laboratories, Toni Radler of Hohner, Inc., and others, too many to list here, for their help in providing photographs, certain figures, and/or permission to reproduce the figures. Medical instrument companies, a number more than I can list here, provided information on state-of-the-art (and even beyond the current state-of-the art) equipment. Martin Alexander, Professor Mauro J. Caputi of Hofstra University, Michael Dallal, and John F. Wehner provided the additional pairs of eyes in going over the galley proofs.

I would be remiss if I did not mention here that it has been an extreme pleasure to work with a most courtly, knowledgeable editor in the person of Dr. Thomas von Foerster of Springer-Verlag's New York office. He did a superb job in providing guidance through the labyrinth of publishing procedures. Springer-Verlag is also fortunate in having an excellent production editor in the person of Steven Pisano. Dr. Robert Beyer, the editor of this AIP series dealing with acoustics, provided a great deal of encouragement. He has my unbounded admiration for the range of his knowledge and extraordinary wisdom. I deem it a rare privilege to know such a person.

Throughout this project, the chief source of inspiration and support came from my wife, Geri. My past and present works were stimulated by the radiance of her presence.

Daniel R. Raichel

References

1. Bacon, Sir Francis (Lord Veralum), *Sylva Sylvarum* (published posthumously, 1626), *The Works of Sir Francis Bacon*, vol. 2, Spedding, J, Ellis, R.L., Heath, D.D., et al. (eds.), London: Longman and Co., 1957.
2. Hunt, Frederick Vinton. *Origins in Acoustics*. Woodbury, NY: Acoustical Society of America, 1992.

Contents

xiv   Contents

1
A Capsule History of Acoustics

Of the five senses that we possess, hearing probably ranks second only to sight in regular usage. It is therefore with little wonder that human interest in acoustics would date to prehistoric times. Sound effects entailing loud clangorous noises were used to terrorize enemies in the course of heated battles; yet the gentler aspects of human nature became manifest through the evolution of music during primeval times, when it was discovered that the plucking of bow strings and the pounding of animal skins stretched taut made for rather interesting and pleasurable listening. Life in prehistoric society was fraught with emotion, so music became a medium of expression. Speech enhanced by musical inflection became song. Body motion following the rhythm of accompanying music evolved into dance. Animal horns were fashioned into musical instruments (the Bible describes the ancient Israelites' use of *shofarim*, made from horns of rams or gazelles, to sound alarms for the purpose of rousing warriors to battle). Ancient shepherds amused themselves during their lonely vigils playing on pipes and reeds, the precursors of modern woodwinds.

Possibly the first written set of acoustical specifications may be found in the Old Testament, Exodus XXVI:7:

And thou shalt make curtains of goats' hair for a tent over the tabernacle.... The length of each curtain shall be thirty cubits and the breadth of each curtain shall be four cubits...

Additional specifications are given in extreme detail for the construction and hanging of these curtains, which were to be draped over the tabernacle walls to ensure that the curtains would hang in generous sound-absorbing folds. More fine details on the construction of the tabernacle followed. Absolutely no substitution of materials nor deviation from prescribed methods was permitted.

With the advent of metal-forming skills, newer wind instruments were constructed of metals. The march generated from ceremonial processions on grand military and ceremonial occasions. Patriotic fervor often was elevated to a state of higher pitch by the blare of martial music, indeed to the point of sheer madness on the part of the citizenry, even in modern times, as epitomized during the 1930s by the grandiose thunder of Nazi goose-stepping marches through Berlin's boulevards to the accompaniment of the crowds' roar.

With sound as a major factor affecting human lives, it was only natural for interest in the science of sound, or *acoustics*, to emerge. In the 27th century BCE, Lin-lun, a minister of the Yellow Emperor Huangundi was commissioned to establish a standard pitch for music. He cut a bamboo stem between the nodes to make his fundamental note, resulting in the "Huang-zhong pipe"; the other notes took their place in a series of twelve standard pitch pipes. He also took on the task of casting twelve bells in order to harmonize the five notes, so as to enable the composing of music for royalty. Archeological studies of the unearthed musical instruments attested to the high level of instrument design and the art of metallurgy in ancient China. Approximately 2000 BCE, another Chinese, the philosopher Fohi, attempted to establish a relationship between the pitch of a sound and the five elements: earth, water, air, fire, and wind. The ancient Hindus systematized music by subdividing the octave into 22 steps, with a large whole tone containing four steps, a small tone assigned three, and a half tone containing two such steps. The Arabs carried matters further by partitioning the octave into 17 divisions. But the ancient Greeks developed musical concepts similar to those of the modern Western world. Three tonal genders—the diatonic, the chromatic, and the enharmonic—were attributed to the gods.

Observation of water waves may have influenced the ancient Greeks to surmise that sound is an oscillating perturbation emanating from a source over large distances of propagation. It cannot have failed to attract notice that the vibrations of plucked strings of a lute can be seen as well as felt. The honor of being the earliest acousticians probably falls to the Greek philosopher Chrysippus (*circa* 240 BCE), the Roman architect–engineer Vetruvius (also known as Marcus Vitruvius Pollio, *circa* 25 BCE) and the Roman philosopher Severinus Boethius (480–524). Aristotle (384–322 BCE) stated in his rather pedantic fashion that air motion is generated by a source "thrusting forward in like movement the adjoining air, so that sound travels unaltered in quality as far as the disturbance of the air manages to reach." Pythagoras (570–497 BCE) observed that "air motion generated by a vibrating body sounding a single musical note is also vibratory and of the same frequency as the body"; and it was he who successfully applied mathematics to the musical consonances described as the octave, the fifth and the fourth, and established the inverse proportionality of the length of a vibrating string with its pitch. The forerunner of the modern megaphone was used by Alexander the Great (400 BCE) to summon his troops from distances as far as 15 kilometers.

The principal laws of sound propagation and reflection were understood by the ancient Greeks, and the echo figured prominently in a number of classical tales. Quintillianus demonstrated the resonance of a string with the air of small straw segments. Vitruvius, after making use of the spread of circular waves on a water's surface as an example, went on to explain that true sound waves travel in a three-dimensional world not as circles, but rather as outwardly spreading spherical waves. He also described the placement of rows of large empty vases for the purpose of improving the acoustics of ancient theaters. While there may be some question of whether such vases were actually employed in these theaters (because archeological excavations have failed to disclose their shards), it does presage knowledge of room acoustics on Vitruvius's part. These vases would have had the effect of low-frequency absorption, similar to that of special panels that are used today as absorbers. As these amphitheaters were constructed in stony recesses which provide little or no low-frequency absorption, such vases would definitely improve the acoustics of the ancient theaters. There is evidence of Lucius Mummius, who, after destroying Corinth's theater, brought its bronze vessels to Rome and made a dedicatory offering from the proceeds of their sale to the goddess Luna in her temple at Rome.

Aristotle's eschewal of experiments (which he deemed unworthy of a scientist) to establish the validity of hypotheses essentially caused the stagnation of all natural sciences, including acoustics, such was the sway of his authority until the end of the Middle Ages.

Leonardo da Vinci (1452–1519) knew as the ancients did that "there cannot be any sound when there is no movement or percussion of the air." His observations led him to correlate the waves generated by a stone cast into water with propagation of sound waves as similar phenomena. He also discerned that wave motion of a sound have a definite value of velocity; and he noted that "the stroke of one bell is answered by a feeble quivering and ringing of another bell nearby; a string sounding on a lute, compels to sound on another lute, nearby, a string of the same note," thus anticipating by nearly a century Galileo Galilei's discovery of sympathetic resonance.

Almost no further progress in acoustics was made until the seventeenth century when a relationship was established between pitch and frequency. Marin Mersenne (1588–1648), a French natural philosopher and Franciscan friar, may be considered the 'father of modern acoustics." In *Harmonie universelle*, published in 1636, he rendered the first scientifically palpable description of an audible tone (84 Hz); and he demonstrated that the absolute frequency ratio of two vibrating strings, radiating a musical note and its octave, is of the frequency ratio 1:2. An analogue with water waves is drawn: the belief was registered that air motion generated by musical sounds is oscillatory in nature, and it was observed that sound travels with a finite speed. Sound is also known to bend around corners, suggestive of diffraction effects which are also commonly observed in water waves. The velocity of sound was measured by Mersenne counting the number of heart beats during the interval occurring between the flash of a shot and the perception of the sound.

Independently of Mersenne, Galileo Galilei (1564–1642), in his *Mathematical Discourses Concerning the New Sciences* (1638), supplied to date the most lucid statement and discussion of frequency equivalence. It is interesting to note that the wave viewpoint was not accorded unanimous acceptance among the early scientists. Pierre Gassendi (1582–1655), a contemporary of Galileo and Mersenne, argued for a ray theory whereby sound is attributed to a stream of atoms emitted by the sounding body; the velocity of sound is the speed of atoms in motion, the frequency is the number of atoms emitted per unit time. He also attempted to demonstrate that sound velocity was independent of pitch by comparing results for the crack of a rifle with those for the deep roar of a cannon.

Robert Boyle (1626–91) with the help of his assistant Robert Hooke (1635–1703) performed a classic experiment (1660) on sound by placing a ticking watch in a partially evacuated glass chamber. He proved that air is necessary, both for the production and emission of sound. In this respect he disproved Athenasius Kircher's (1602–80) negative experiment in which the latter enclosed a bell in a vacuum container and excited the bell magnetically from the exterior. Kirchner's results were erroneous because he did not take the precaution to prevent the conduction of sound through the bell's supports to the surroundings. Francis Hauksbee (1666–1713) repeated the Boyle experiment (in a modified form) before the Royal Society.

Mention should be made here of Joseph Sauveur (1653–1713) who suggested the term *acoustics* (from the Greek word for sound) for the science of sound. In describing his research on the physics of music at the College Royal in Paris he introduced terms such as *fundamental, harmonics, node,* and *ventral segment.*[1] It is also an interesting footnote to history that Sauveur may have been born with defective hearing and speaking mechanisms; he was reported to have been a deaf-mute until the age of seven. He took an immense interest in music even though he had to rely on the help of his assistants to compensate for his lack of keen musical acuity in conducting acoustic experiments.

Franciscus Mario Grimaldi (1613–63) published *Physicomathesis de lumine, coloribus et tride,* which dealt with experimental studies of diffraction, much of which was to apply to acoustics as well as to light; and in 1678 Hooke announced his law relating force to deformation, which established the foundations of vibration and elasticity theories.

Kircher's publication *Phonurgia, die neue Hall- und Tonkunst (The New Art of Sound and Tone),* issued in 1680, provides us a rather amusing insight into the world of misconception, nostrums, and plain scientific hokum that were prevalent at the time. While delving into the phenomena of echoes and whispering galleries, the text recommended music as the only remedy against tarantula bites and provided a discourse on wines. In the chapter on wines, Kircher claimed that old wine has purified itself and acquired a deep soul. If old wine is poured into a glass, which is then struck, a sound will emanate. On the other hand, new wine was deemed to be "jumpy" as a child and bereft of a sound. Another misconception widely believed at the time was that sound could be trapped in a little box and preserved indefinitely, the idea of attenuation or absorption of sound being completely alien then. It was even proposed by a Professor Hut of the music academy at Frankfurt that a communications tube be constructed to transmit speech over long distances.

Ernst F. F. Chladni (1756–1827), author of the highly acclaimed *Die Akustik,* is often credited for establishing the field of modern experimental acoustics through his discovery of torsional vibrations and measurements of the velocity of sound

---

[1] Nearly twenty years earlier, in 1683, Narcissus Marsh, then the Bishop of Ferns and Leighlin in the Protestant Church, published an article "An Introductory Essay to the Doctrine of Sounds, Containing Some Proposals for the Improvement of Acousticks" in the *Philosophical Transactions of the Royal Society of London.* He was using the term "acousticks" to denote direct sound as distinguished from reflected and diffracted sound.

with the aid of vibrating rods and resonating pipes. The dawn of the eighteenth century saw the birth of theoretical physics and applied mechanics, particularly under the impetus of archrivals Isaac Newton (1642–1726) and Gottfried Wilhelm Leibniz (1646–1716). Newton's theoretical derivation of the speed of sound (in the *Principia*) motivated a spate of experimental measurements by Royal Society members John Flamsteed (1646–1713), the Astronomer Royal, and his eventual successor (in 1720), Edmund Halley (1656–1742); also by Giovanni Domenico Cassini (1625–1712), Jean Picard (1620–82), and Olof Römer (1644–1710) of the French Acadêmie des Sciences, and nearly half-century later in 1738 by a team led by César François Cassini de Thury (1714–62), a grandson of the aforementioned G. D. Cassini who headed the earlier 1677 measurement team.

Newton's estimate was found to be incorrect, for in his observations he had erred by assuming an isothermal (rather than an isentropic) process as being the prevalent mode for acoustic vibrations.[2] Temperature was found to influence the speed of sound in independent separate experiments by Count Giovanni Lodovico Bianoni (1717–81) of Bologna and Charles Marie de la Condamine (1701–73). Other acoustic developments included the evolution of the exponential horn by Richard Helsham (1680–1758); this device loads the sound source heavily, thus causing the source to concentrate its energy more than it could without the horn and direct the output more effectively. Real understanding of this phenomenon did not come about until John William Strutt, Lord Rayleigh (1845–1919), treated the problem of source loading, and Arthur Gordon Webster (1863–1923) the theory of horns.

Each of the optical phenomena of refraction, diffraction, and interference was elucidated during the seventeenth century. But all of these phenomena were soon realized to apply to acoustics as well as to light. Willbrod Snell (or Snellius) (1591–1626) composed an essay in 1620 treating the refraction of light rays in a transparent medium such as water or glass; but he somehow neglected to publish his manuscript which was later unearthed and used by Christian Huygens (1629–95) in his own works, which secured posthumous fame for Snell in spite of a publication of the same law by the stellar René Descartes (1596–1650) who, it turned out, had made two erroneous assumptions, which were corrected by Pierre de Fermat (1601–65). Fermat's principle derives from the assumption that light always travels from a source point in one medium to a receptor point in the second medium by the path of least time. Diffraction was first observed by the Jesuit mathematician Francesco Maria Grimaldi (1618–63) of Bologna. His experiments were repeated by Newton, Hooke, and Huygens; and soon this phenomenon that light does not always travel in straight lines, but can diffuse slightly around corners, constituted a core issue in the controversy between the wave and corpuscular theories of light. But it took nearly 200 years following Newton's era to resolve the conflict by embracing elements of both theories. Newton essentially squelched the wave theory until its revival by Thomas Young (1773–1829) and Augustin Jean Fresnel (1788–1817), both of

---

[2] Actually, what Newton really did was to assume that the 'elastic force' of the fluid is proportional to its condensation, which is now realized, in the context of modern thermodynamics, to be the equivalence of the isothermal process.

whom, independently of each other, elucidated the principle of interference. On his analysis of diffraction, Fresnel drew heavily on Huygen's principle in which successive positions of a wavefront are established by the envelope of secondary wavelets.

Armed with the analytical tools afforded by the advent of calculus by Newton and Leibniz, the French mathematical school treated problems of theoretical mechanics. Among the major contributors were Joseph Louis Lagrange (1736–1813), the Bernoulli brothers James (1654–1705) and Johann (1667–1748), G. F. A. l'Hôpital (Marquis de St. Mesme) (1661–1704), Gabriel Cramer (1704–52), Leonhard Euler (1707–83), Jean Le Rond d'Alembert (1717–83), and Daniel Bernoulli (1700–83). And the next generation provided a further flowering of genius: Joseph Louis Lagrange (1736–1813), Pierre Simon Laplace (1749–1827), Adrian Marie Legendre (1736–1833), Jean Baptiste Joseph Fourier (1768–1830), and Siméon Denis Poisson (1781–1840). The nineteenth century was also dominated by discoveries in electricity and magnetism by Michael Faraday (1791–1867), James Clerk Maxwell (1831–79), Heinrich Rudolf Hertz (1857–94), and by the theory of elasticity, principally developed by Claude L. M. Navier (1785–1836), Augustin Louis Cauchy (1789–1857), Rudolf J. E. Clausius (1822–88), and George Gabriel Stokes (1890–1909).

These developments constituted the foundation for understanding the physical aspects of acoustics. In the attempt to grasp the nature of musical sound, Simon Ohm (1789–1854) advanced the hypothesis that the ear perceived only a single, pure sinusoidal vibration and that each complex sound is resolved by the ear into its fundamental frequency and its harmonics. Hermann F. L. von Helmholtz (1821–94) arguably deserves the credit for laying the foundations of spectral analysis in his classic *Lehre von den Tonempfindungen (Sensation of Sound)*. The monumental two-volume *Theory of Sound*, released in 1877 and 1878 by the future Nobel laureate, Lord Rayleigh, laid down in a dramatically bold fashion the theoretical foundations of acoustics.

When the newly constructed Fogg Lecture Hall was opened in 1894 at Harvard University, its acoustics were found to be so atrocious as to almost render that facility useless. This prompted Harvard's Board of Overseers to request of the physics department that something be done to rectify the situation. The task was assigned to a young Harvard researcher, Wallace Clement Sabine (1868–1919), and he discovered soon enough that excessive reverberations tend to mask the lecturer's words. In a series of papers (1900–15) evolving from his studies of the lecture hall, he almost single-handedly elevated architectural acoustics to scientific status. Sabine helped establish the Riverbank Acoustical Laboratories[3] at Geneva, Illinois. Just prior to his scheduled assumption of his duties at Riverbank, Sabine succumbed at the young age of 50 to cancer. His distant cousin, Paul Earl Sabine (1879–1958), also a Harvard physicist, took on the task of running the laboratory.

---

[3] Riverbank is possibly the first research facility set up specifically for study and research in acoustics.

The development of test procedures, methodology, and standardization in testing the acoustical nature of products arose from the pioneering efforts of the younger Sabine. A third member of the family, Paul Sabine's son Hale (d. 1981), began his career in architectural acoustics at the tender age of 10 by assisting his father at Riverbank, and his efforts centered on control of noise in industry and institutions. Both father and son served terms as president of the Acoustical Society of America.

The genesis of ultrasonics occurred in the nineteenth century with James Prescott Joule's (1818–99) discovery in 1847 of the magnetostriction effect, the alteration of the dimensions of a magnetic material under the influence of a magnetic field, and in 1880 with the finding by the brothers Paul-Jacques (1855–1941) and Pierre (1859–1906) Curie that electric charges result on the surfaces of certain crystals subjected to pressure or tension. The Curies' discovery of the *piezoelectric electric effect* provided the means of detecting ultrasonic signals. The inverse effect, whereby a voltage impressed across two surfaces of a crystals give rise to stresses in the materials, now constitutes the principal method of generating ultrasonic energy.

The study of underwater sound stemming from the necessity for ships to avoid dangerous obstacles in water supplied the impetus for ultrasonic applications. Until the early part of the twentieth century ships were warned of dangerous conditions by bells suspended from lightships. Specially trained crew members listened for these bells by pressing microphones or stethoscopes against the hulls. In the effort to counteract the German submarine threat during World War I, Robert Williams Wood (1868–1955) and Gerrard in England and Paul Langevin (1872–1946) in France were assigned the task of developing counter-surveillance methods.

The youthful Russian electrical engineer, Constantin Chilowsky (1880–1958), collaborated with Langevin in experiments with an electrostatic (condenser) projector and a carbon-button microphone placed at the focus of a concave mirror. In spite of troubles encountered with leakages and breakdowns due to the high voltages necessitated by the projectors, Langevin and Chilowsky by 1916 were able to obtain echoes from the ocean bottom and from a sheet of armor plate at a distance of 200 meters. A year later Langevin came up with the concept of using a piezoelectric receiver and employed one of the newly developed vacuum-tube amplifiers—the earliest application of electronics to underwater sound equipment—and Wood constructed the first directional hydrophone geared to locate hostile submarines. The first devices to generate directional beams of acoustic energy also constitute the first use of ultrasonics. In the course of their underwater sound investigations, Wood and his co-worker Alvin L. Loomis and Langevin observed that small water creatures could be stunned, maimed, or even destroyed by the effects of intense ultrasonic fields.

World War I ended before underwater echo-ranging could be fully deployed to meet the German U-boat threat. The years of peace following World War I witnessed a slow but nonetheless steady advance in applying underwater sound to depth-sounding by ships. Improvements in electronic amplification and processing, magnetostrictive projectors, and piezoelectric transducers provided refinements in echo-ranging. The advent of World War II heightened research activity on both sides of the Atlantic; and most of the present concepts and applications of underwater acoustics trace their origins to this period. The concept of target strength, noise output of various ships at different speeds and frequencies, reverberation in the sea, and evaluation of underwater sound through spectrum analysis were quantitatively established. It was during this period that underwater acoustics became a mature branch of science and engineering, backed by vast literature and history of achievement.

The invention of the triode vacuum tube and the advent of the telephone and radio broadcasting served to intensify interest in the field of acoustics. The development of vacuum tube amplifiers and signal generators rendered feasible the design and construction of sensitive and reliable measurement instruments. The evolution of the modern telephone system in the United States was facilitated by the progress of communication acoustics, mainly through the remarkable efforts of the Bell Telephone Laboratories.

The historic invention of the transistor (1948) at the Bell Laboratories in Murray Hill, New Jersey, gave rise to a whole slew of new devices in the field of electronics, including solid-state audio and video equipment, computers, spectrum analyzers, electric power conditioners, and other gear too numerous to mention here.

Experiments and development of theory in architectural acoustics were conducted during the 1930s and the 1940s at a number of major research centers, notably Harvard, MIT and UCLA. Vern O. Knudsen (1893–1974), eventually the chancellor of UCLA, carried on Sabine's work by conducting major research on sound absorption and transmission. The most notable of his younger associates was Cyril M. Harris (b. 1917), who was to became the principal consultant on the acoustics of the Metropolitan Opera House in New York, the John F. Kennedy Center in the District of Columbia, the Powell Symphony Hall in St. Louis, and a number of other notable edifices.

Sound decay, in terms of reverberation time, was discovered to be a decisive factor in gauging the suitability of enclosed areas for use as listening chambers. The impedance method of rating acoustical materials was established to predict the radiative patterns of sonic output; and prediction of sound attenuation in ducts was established on a scientific footing. The architectural acoustician now has a wide array of acoustical materials to choose from and to tailor the walls segmentwise in order to effect the proper acoustic environment. Acoustics also engendered the science of psychoacoustics.

Harvey Fletcher (1884–1990) led the Bell Telephone Laboratories in describing and quantifying the concepts of loudness and masking, and there many of the determinants of speech communication were also established (1920–40). Fletcher, now regarded as "the father of psychoacoustics," worked with the physicist Robert Millikan at the University of Chicago on the measurement of the electron charge. Fletcher indeed had performed much of the famed oil drop experiment, to the extent that many physicists still feel that the student should have shared the 1923 Nobel Prize in physics with his professor who received the award for this effort. At Bell Labs Fletcher also developed the first electronic hearing aid and invented stereophonic reproduction. Sound reproduction also constituted the domain of

Harry F. Olson (1902–82), who directed the Acoustical Laboratory at RCA and developed modern versions of loudspeakers. Warren P. Mason's (1900–86) major work in physical acoustics essentially laid down the modern foundations of ultrasonics, and Georg von Bekésy (1899–1972) earned the 1961 Nobel Prize in Medicine or Physiology for his research on the mechanics of human hearing. Acoustics penetrated the fields of medicine and chemistry through the medium of ultrasonics: ultrasonic diathermy became established and certain chemical reactions were found to become accelerated under acoustic conditions.

The outbreak of World War II served to greatly intensify acoustics research at major laboratories in Western Europe and in the United States, particularly in view of the demand both for sonar detection of stealthily moving submarines and for reliable speech communication in cacophonous environments such as those inside propeller aircraft and armored vehicles. This research not only has reached great proportions, it has continued unabated to this day at major universities and government institutions, among them the U.S. Naval Research Laboratory, Naval Surface Warfare Center, MIT, Purdue University, and Pennsylvania State University.

Prominent among the researchers were Richard H. Bolt (b. 1911) and Leo L. Beranek (b. 1914) who teamed up after World War II to found a major research corporation, Bolt, Beranek & Newman; Phillip M. Morse (1903–85) of MIT [who authored and co-authored with Karl Uno Ingard (b. 1921) major texts in physical acoustics]; R. Bruce Lindsay (1900–85) of Brown University; and Robert T. Beyer (b. 1920), also at Brown, who contributed to nonlinear acoustics. In 1947 Eugen Skudrzyk (1913–90) began research in nearly all aspects of acoustics at the Technical University of Vienna and went on to Pennsylvania State University in the United States; he wrote possibly the best comprehensive text on physical acoustics since Lord Rayleigh's *Theory of Sound*.

Karl D. Kryter (b. 1914) of California deals with the physiological effects of noise on humans; and Carleen Hutchins (b. 1911), is still providing in her 80s great insight into the design and construction of musical string instruments, in her dual role as investigating acoustician and craftsperson seeking to emulate the old Cremona masters in her hometown of Montclair, New Jersey. Laser interferometry was applied by Karl A Stetson (b. 1937) and by Lothar Cremer (b. 1905) to visualize vibrations of the violin body. Sir James Lighthill (1927–98), who held the Lucasian chair (once occupied by Newton) in mathematics at Cambridge University, laid down the foundations of modern aeroacoustics, building on the foundations of Lord Rayleigh's earlier research. UCLA's Isadore Rudnick (1917–97) performed major experiments in superfluid hydrodynamics, involving sound propagation in helium at cryogenic temperatures and also conducted studies of acoustically induced streaming modes of vibrations of elastic bodies and attenuation of sound in seawater. At the Applied Physics Laboratory at the University of Washington, Lawrence A. Crum (b. 1941) is directing major research on sonofluorescence as well as development of ultrasound diagnostic and therapeutic medical devices. Kenneth S. Suslick (b. 1952) and his co-workers at the University of Illinois are making major contributions in the field of sonochemistry. Whitlow W. L. Au at the University of Hawaii is conducting major studies on the characteristics of cetacean acoustics, including the target discrimination capabilities of dolphins and whales.

With acoustics research continuing apace, the number of great acousticians living surely exceeds that of deceased ones.

It can truly now be said that the U.S. Navy has done more (and is still doing more) than any other institution to further acoustics research at its widespread facilities, including the Naval Research Laboratory (NRL), the Office of Naval Research (ONR), and the Naval Surface Warfare Center (NSWC). Much magnificent work was done under the cloak of security classification during the days of the Cold War, with the consequence that many deserving researchers do not bask in the glory that have been publicly accorded professional societies' medal honorees and Nobel Prize laureates.

Robert J. Bobber (b. 1918) of NRL facility in Orlando paved the way in underwater electroacoustics measurements. Acoustics radiation constituted the domain of Sam Hanish, late of the NRL in the District of Columbia. At NSWC's David Taylor Basin in Bethesda, Maryland, Murray Strasberg (b. 1917) continues to make major contributions in the field of propeller noise, which entails the study of cavitation and hydroacoustics as he has done for the past three decades; David Feit (b. 1937) ranks as a leading expert in the field of structural acoustics; and William K. Blake reigns preeminent in the category of aero-hydroacoustics. Herman Medwin of the Navy Postgraduate School at Monterey, California, conducted major research in acoustical oceanography. As a senior research physicist at the U.S. Naval Surface Weapons Center Headquarters in Silver Spring, Maryland, Robert Joseph Urick (1915–96) elucidated the characteristics of underwater acoustical phenomena, including sonar effects. He later taught the principles of underwater sound at the Catholic University of America in Washington, D.C.

Acoustics is no longer the esoteric domain of interest to a few specialists in the telephone and broadcasting industries, the military, and university research centers. Legislation and subsequent action has been demanded internationally to provide quiet housing, safe, and comfortable work environments in the factory and the office, quieter airports and streets, and protection in general from excessive exposure to noisy appliance and equipment.

The wiser architects are increasingly using acoustical engineers to ensure environmental harmony with the aesthetic aspects of their designs. Acoustic instrumentation is being used in industry to facilitate manufacturing processes and to ensure quality control. Acoustics has even invaded the living room through the medium of high-fidelity reproduction, giving rise to a spate of new equipment such as Dolby processors, digital processors, compact disc (and more lately DVD) players, multispeaker "Surround Sound" environment conditioners, and music synthesizer circuit boards for personal computers. The escalating applications of ultrasound provide new diagnostic and therapeutic tools in the medical field, more reliable characterizations of materials, better surveillance methodologies, and improved manufacturing techniques.

And what does the future hold in acoustics? The continuing miniaturization of electronic circuitry is now resulting in digitized hearing aids that can circumvent the "cocktail party effect" (i.e., the prevalence of background noise which makes it difficult for the sensorneurally impaired listeners to focus on a conversation). Even newer diagnostic and therapeutic processes entailing acoustical signals are being developed and tested in major medical centers. More sensitive and versatile transducers that can withstand harsher environments lead to new acoustical devices such as sonic viscometers, undersea probes, and portable voice recognition devices. And if we can gain a greater understanding of how cetaceans make use of their natural sonars to assess the submarine environment and perhaps to communicate with one another, we might be well on the way to constructing far more sophisticated megachannel acoustical analyzers. The generation of acoustical waves in the gigahertz range can rival or exceed the optical microscope for resolution with greater penetrating power. The repertoire of what is to come should truly constitute an amazing cornucopia of beneficence to humanity.

References

Beranek, Leo J. 1995. Harvey Fletcher: Friend and scientific critic. *Journal of the Acoustical Society of America* 97, 5, Pt. 2: 3357.

Beyer, Robert T. 1995. Acoustic, acoustics. *Journal of the Acoustical Society of America* 98, 1: 33–34.

Beyer, Robert T. 1999. *Sounds of Our Times*. New York, NY: Springer-Verlag. [A fascinating history of acoustics over the past 200 years, with many allusions to even earlier history. This text picks up where Frederick Vinton Hunt left off in his unfinished, meticulously researched work which was published posthumously (see below).]

Blake, William K. 1964. *Aero-hydroacoustics for Ships*. 2 vols. Bethesda, MD: David Taylor Basin publication DTNSRDC-84/010, June 1964.

Bobber, Robert J. 1970. *Underwater Electroacoustic Measurements*. Washington D.C.: U.S. Naval Research Laboratory.

Chladni, E. F. F. 1802. *Die Akustik*. Leipzig: Breitkopf & Hartel.

Clay, Clarence S., and Medwin, Herman. 1977. *Acoustical Oceanography: Principles and Applications*. New York: John Wiley & Sons.

Fletcher, Steven Harvey. 1995. Harvey Fletcher: A son's reflections. *Journal of the Acoustical Society of America* 97, 5, Pt. 2: 3356–3357.

Galileo, Galilei. 1638 (translation published 1939). *Dialogues Concerning Two New Sciences* (1638). Translated by H. Crew H. and A. DeSalvio. Evanston, IL: Northwestern University Press, 1939.

Hanish, Sam. 1981. *A Treatise on Acoustic Radiation*. Washington, D.C.: U.S. Naval Research Laboratory, 1981.

Harris, Cyril M. 1995. Harvey Fletcher: Some personal recollections. *Journal of the Acoustical Society of America* 97, 5, Pt. 2: 3357.

Helmholtz. Hermann F. L. von. 1877. *Lehre von den Tonempfindungen*. Braunschweig: Vieweg.

Hertz, J. H. (ed.). 1987. *The Pentateuch and Haftorahs*. London: Soncino Press.

Hunt, Frederick Vinton. 1992 (reissue). *Origins in Acoustics*. Woodbury, NY: Acoustical Society of America. (Although left incomplete by the author at the time of his death, this text is one of the most definitive accounts by one of the great modern acoustical scientists of the history of acoustics leading up to the eighteenth century.)

Junger, Miguel C. and Feit, David. 1956. *Sound, Structure, and Their Interaction*. Cambridge, MA: The MIT Press.

Kopec, John W. 1994. The Sabines at Riverbank. *Proceedings, Wallace Clement Sabine Centennial Symposium*. Woodbury, NY: Acoustical Society of America, 1994: 25–28.

Lindsay, R. Bruce (ed.). 1972. *Acoustics: Historical and Philosophical Development*, Benchmark Papers in Acoustics. Stroudsburg, PA: Dowden, Hutchinson & Ross, Inc. (A most interesting compendium of selected papers by major contributors to acoustical science, ranging from Aristotle to Wallace Clement Sabine. A must-read for the serious student of the history of acoustics).

Lindsay, R. Bruce. 1966. The story of acoustics. *Journal of the Acoustical Society of America* 39(4): 629–644.

Lindsay, R. Bruce. 1980. Acoustics and the Acoustical Society of America in historical perspective. *Journal of the Acoustical Society of America* 68(1) (July 1980): 2–9.

Mersenne, Marin. 1636. *Harmonie universelle*. Paris: S. Cramoisy. English translation: J. Hawkins. *General History of the Practice and Science of Music*. London: J. A. Novello, 1853: 600–616, 650 ff.

Newton, Sir Isaac. 1687. *Philosophiae naturalis principia mathematica*. London: Joseph Streater for the Royal Society.

Pierce, Allan D. 1989 (reissue). *Acoustics, An Introduction to its Physical Principles and Application*. Woodbury, NY: Acoustical Society of America.

Raman, V. V. March 1973. Where credit is due: Sauveur, the forgotten founder of acoustics. *Physics Teacher*: 161–163.

Shaw, Neil A., Jesse Klapholz, and Gander, Mark R. 1994. Books and acoustics, especially Wallace Clement Sabine's "Collected Papers on Acoustics." *Proceedings, Wallace Clement Sabine Centennial Symposium*, Woodbury, NY: Acoustical Society of America, 1994: 41–44.

Skudrzyk, Eugen. 1971. *The Foundations of Acoustics: Basic Mathematics and Basic Acoustics*. New York: Springer-Verlag. (A text of classic proportions. Nearly one-quarter of this volume lays out the mathematical foundations requisite to analysis of acoustical phenomena.)

Strutt, J.W. (Lord Rayleigh). *Theory of Sound*. 1877. London: Macmillan & Co., Ltd., 2nd edition revised and enlarged 1894, reprinted 1926, 1929. Reprinted in two volumes, New York: Dover, 1945. (These volumes should be in every acoustician's library.)

Wang, Ji-qing. 1994. Architectural acoustics in China: past and present. *Proceedings, Wallace Clement Sabine Centennial Symposium*. Woodbury, NY: Acoustical Society of America: 21–24.

Webster, Arthur G. 1919. *Proceedings of the National Academy of Science* 5: 275.

2
Fundamentals of Acoustics

2.1 Wave Nature of Sound and the Importance of Acoustics

Acoustics refers to the study of sound, namely, its general transmission through solid and fluid media, and any other phenomenon engendered by its propagation through media. Sound may be described as the passage of pressure fluctuations through an elastic medium as the result of a vibrational impetus imparted to that medium. An acoustic signal can arise from a number of sources, e.g., turbulence of air or any other gas, the passage of a body through a fluid, or the impact of a solid against another solid.

Because it is a phenomenon incarnating the nature of waves, sound may contain only one frequency, as in the case of a pure steady-state sine wave, or many frequency components, as in the case of noise generated by construction machinery or a rocket engine. The purest type of sound wave can be represented by a sine function (Figure 2.1) where the abscissa represents elapsed time and the ordinate represents the displacement of the molecules of the propagation medium or the deviation of pressure, density, or the aggregate speed of the disturbed molecules from the quiescent (undisturbed) state of the propagation medium.

When the ordinate represents the pressure difference from the quiescent pressure, the upper portions of the sine wave then represent the compressive states and the lower portions the rarefaction phases of the propagation. A sine wave is generated in Figure 2.2 by the projection of the trace of a particle A traveling in a circular orbit. This projection assumes the pattern of an oscillation, in which the particle A's projection, or "shadow" A, onto an abscissa moves back and forth at a specified frequency. Frequency f is the number of times the sound pressure varies from its equilibrium value through a complete cycle per unit time. Frequency is also denoted by the radial frequency

$$\omega = 2\pi f = 2\pi / T \tag{2.1}$$

expressed in radians per second. The period T is the amount of time for a single cycle to occur, i.e., the length of the time it takes for a tracer point on the sine curve to reach a corresponding point on the next cycle. The reciprocal of

13

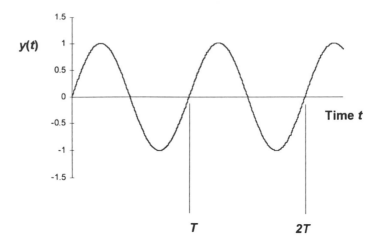

FIGURE 2.1. Plot of a sine wave $y(t) = \sin 2\pi f t$ over slightly more than two periods of $T = 1/f$, where f is the frequency of the sine wave. $y(t)$ may the displacement function x/x_0, velocity ration v/v_0, pressure variation p/p_0, or condensation variation s/s_0, where the subscript 0 denotes maximum values.

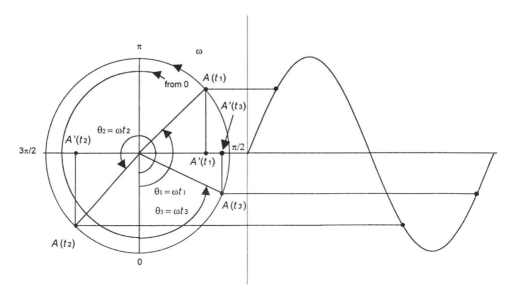

FIGURE 2.2. The oscillation of a particle A' in a sinusoidal fashion is generated by the circular motion of particle A moving in a circle with constant radial speed ω. A' is the projection of $A \cos \omega t = A \cos \theta$ onto the diameter of the circle which has a radius A. The projection of point A to the right traces a sine wave over an abscissa representing time t. The projections for three points at times t_1, t_2, and t_3 are shown here. The amplitude of the oscillation is equal to the radius of the circle, and the peak-to-peak amplitude is equal to the diameter of the circle.

period T is simply the frequency f. The most common unit of frequency used in acoustics (and electromagnetic theory) is the *hertz* (abbreviated Hz in the SI system), which is equal to one cycle per second. An acoustic signal may or may not be audible to the human ear, depending on its frequency content and intensity. If the frequencies are sufficiently high (>20 kilohertz, which can be expressed more briefly as 20 kHz), ultrasound will result, and this sound is inaudible to the human ear. This sound is then said to be *ultrasonic*. Below 20 Hz, the sound becomes too low (frequencywise) to be heard by a human. It is then considered to be *infrasonic*.

Sound in the *audio* frequency range of approximately 20 Hz to 20 kHz can be heard by humans. While a degree of subjectivity is certainly entailed here, *noise* conveys the definition of *unwanted sound*. Excessive levels of sound can cause permanent hearing loss; and continued exposure can be deleterious, both physiologically and psychologically, to one's well-being.

With the advent of modern technology, our aural senses are being increasingly assailed and benumbed by noise from high-speed road traffic, passing ambulances and fire engine sirens, industrial and agricultural machinery, excessively loud radio and television receivers, recreational vehicles such as snowmobiles and unmuffled motorcycles, elevated and underground trains, jet aircraft flying at low altitudes, domestic disputes penetrating flimsy walls, and so on.

Young men and women are losing their hearing acuity prematurely as the result of their sustained exposure to loud rock concerts, discotheques, use of personal cassette and compact disc players and megapowered automobile stereo systems. In the early 1980s, during the waning days of the Cold War, the Swedish navy reported considerable difficulty in recruiting young people with hearing sufficiently keen to qualify for operating surveillance sonar equipment for tracking Soviet submarines traveling beneath Sweden's coastal waters. Oral communication can be rendered difficult or made impossible by background noise; and life-threatening situations may arise when sound that conveys information becomes masked by noise. Thus, the adverse effects of noise falls into one or more of the following categories: (a) hearing loss, (b) annoyance, and (c) speech interference.

Modern acoustical technology also brings benefits: it is quite probable that the availability (and judicious use) of audiophile equipment has enabled many of us, if we are so inclined, to hear more musical performances than Beethoven, Mozart, or even the long-lived Haydn could have heard during their respective lifetimes. Ultrasonic devices are being used to dislodge dental plaque; overcome the effects of artereoclerosis by freeing up clogged blood vessels; provide non-invasive medical diagnoses; supply a means of nondestructive testing of materials; and clean nearly everything from precious stones to silted conduits. The relatively new technique of active noise cancellation utilizes computerized sensing to duplicate the histograms of offending sounds, but at 180 degrees out of phase, which effectively counteracts the noise. This technique can be applied to aircraft to lessen environmental impact and to automobiles to provide quieter interiors.

2.2   Sound Generation and Propagation

Sound is a mechanical disturbance that travels through an elastic medium at a speed characteristic of that medium. Sound propagation is essentially a wave phenomenon, as with the case of a light beam. But acoustical phenomena are mechanical in nature, while light, x-rays, and gamma rays occur as electromagnetic phenomena. Acoustic signals require a mechanically elastic medium for propagation and therefore cannot travel through a vacuum. On the other hand, the propagation of an electromagnetic wave can occur in empty space. Other types of wave phenomena include those of ocean movement, the oscillations of machinery, and the quantum-mechanical equivalence of momenta as propounded by de Broglie.[1]

Consider sound as generated by the vibration of a plane surface at $x = 0$ in Figure 2.3. The displacement of the surface to the right, in the $+x$ direction, causes a compression of a layer of air immediately adjacent to the surface, thereby resulting in an increase in the density of the air in that layer. Because the pressure of that layer is greater than the pressure of the undisturbed atmosphere, the air molecules in the layer tend to move in the $+x$ direction and compress the second layer which, in turn, transmits the pressure impulse to the third layer, and so on. But as the plane surface reverses its direction of vibration, an opposite effect occurs. A rarefaction of the first layer now occurs, and this rarefaction decreases the pressure to a value below that of the undisturbed atmosphere. The molecules from the second layer now tend to move leftward, in the $-x$ direction; and a rarefaction impulse now follows the previously generated compression impulse.

This succession of outwardly moving rarefactions and compressions constitutes a wave motion. At a given point in the space, an alternating increase and decrease in pressure occurs, with a corresponding increase and decrease in density. The spatial distance λ between one point on the cycle to the corresponding point on the next cycle is the *wavelength*. The vibrating molecules that transmit the waves do not, on the average, change their positions, but are merely moved back and forth under the influence of the transmitted waves. The distances these particles move about their respective equilibrium positions are referred to as *displacement amplitudes*. The velocity at which the molecules move back and forth is termed *particle velocity*, which is not to be confused with the *speed of sound*, the rate at which the acoustic waves travel through the medium.

The speed of sound is characteristic of the medium. Sound travels far more rapidly in solids than it does in gases. At a temperature of 20°C sound moves at

---

[1] The de Broglie theory assigns the nature of a wave to the momentum of a particle of matter in motion in the following way:

$$mv = \frac{h\nu}{c}$$

where mv represents the moment of the particle, h Planck's constant $= 6.625 \cdot 10^{-34}$ J-s, c the velocity of light $= 3 \cdot 10^{8}$ m/sec, and ν the radial frequency of the wave attributable to the particle.

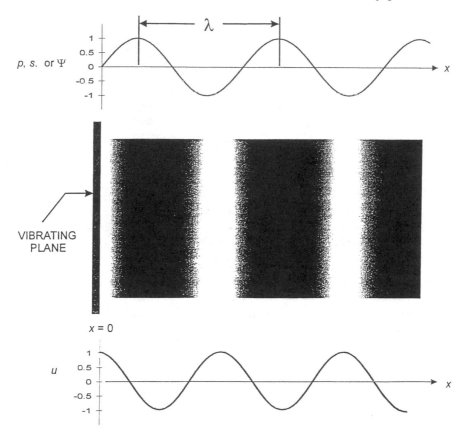

FIGURE 2.3. Depiction of rarefaction and condensation of air molecules subjected to the vibrational impact of a plane wall located at $x = 0$. The degree of darkness is proportional to the density of molecules. Lighter areas are those of rarefactions. Mini-plots of the local variations of molecular displacement Ψ, pressure p, condensation $s = (\rho - \rho_0)/\rho_0$, and particle displacement u are given as functions of x for a given instant of the sound propagation. Note that wavelength λ represents the distance between corresponding points of adjacent cycles.

the rate of 344 m/s (1127 ft/s) through air at the normal atmospheric pressure of 101 kPa (14.7 psia or 760 mm Hg). Sound velocities are also greater in liquids than in gases, but remain less in order of magnitude than those for solids. For an ideal gas the velocity c of a sound wave may be computed from

$$c = \sqrt{\frac{\gamma p}{\rho}} = \sqrt{\gamma R T} \qquad (2.2)$$

where γ is the gas constant defined as the thermodynamic ratio of specific heats, c_p/c_v, p is the quiescent gas pressure, ρ is the density of the gas, R is the thermodynamic constant characteristic of the gas, and T is the absolute temperature of the gas. For air at 20°C, the sound propagation speed c is found from

$$c = \sqrt{\gamma RT} = \sqrt{1.4[287 \text{ N} \cdot \text{m/(kg K)}](20 + 273.2) \text{ K}} = 343.2 \text{ m/s}$$

A simple relation such as equation (2.2) does not exist for acoustic velocity in liquids, but the propagation velocity does depend on the temperature of the liquid and, to a lesser degree, on the pressure. Sound velocity is approximately 1461 m/s in deaerated water. For a solid the propagation speed can be found from

$$c = \sqrt{\frac{E}{\rho}} \tag{2.3}$$

where E represents the Young's modulus (or modulus of elasticity) of the material and ρ is the material density. As an example, consider steel with a specific gravity of 7.85 and a modulus of elasticity of 207 GPa. Applying equation (2.3) and recalling that 1 newton is equal to 1 kg-m/s^2, we find that

$$c = \sqrt{\frac{207(10)^9 \text{ N/m}^2}{7850 \text{ kg/m}^2}} = 5135 \text{ m/s}$$

which does represent the propagation speed of sound in that material. Appendix A lists the speed of sound for a variety of materials.

The strength of an acoustic signal, as exemplified by *loudness* or *sound pressure level* (*SPL*), directly relates to the magnitudes of the displacement amplitudes and pressure and density variations, as we shall see later in Chapter 3.

When the procession of rarefactions and condensations occurs at a steady sinusoidal rate, a single constant frequency f occurs. If the sound pressure of a pure tone is plotted against distance for a given instant, the wavelength λ can be established as being the peak-to-peak distance between two successive waves. The wavelength λ is related to frequency f by:

$$\lambda = \frac{c}{f} \tag{2.4}$$

where c represents the propagation speed. From equation (2.4), it can be seen that higher frequencies will result in shorter wavelengths in a given propagation medium.

2.3   Thermodynamic States of Fluids

In the treatment that follows this section, we eschew the details of molecular motion and intermolecular forces by describing relevant effects in terms of macroscopic thermodynamic variables: pressure p, density ρ, and absolute temperature T. These variables relate to each other through an equation of state

$$p = p(\rho, T)$$

which is usually established experimentally. The implication of the equation of state is that only two of the variables are independent; this is to say that if the values of two of the independent thermodynamic variables are given for a fluid, the specific value of any other thermodynamic property is established automatically. The equation of state for an ideal gas,

$$\frac{p}{\rho} = RT \tag{2.5}$$

can be derived from simple kinetic theory. Here

$R =$ gas constant, energy per unit mass per degree

$R = \Re/\mathbf{M}$

$\Re =$ universal gas constant, energy per mole per degree

  $= 8.314$ kJ/kg-mol K $= 1545.5$ ft-lb$_f$/lb-mol $°$R

  $= 1.986$ Btu/lb$_m$-mol $°$R

$\mathbf{M} =$ molecular weight of gas, kg/kg-mol or lb$_m$/lb$_m$-mol

Each kilogram-mole of the gas contains $N_0 = 6.02(10)^{26}$ molecules. This value of N_0 constitutes Avogrado's number for the MKS system of dimensional units. With $\acute{\eta}$ representing the mass of a single gas molecule, $\mathbf{M} = N_0\acute{\eta}$, the number of molecules per unit volume is $N = \rho/\acute{\eta}$. The equation of state for the ideal gas can now be rewritten as:

$$p = N\frac{\Re}{N_0}T = NkT$$

where k is the Boltzmann constant $= \Re/N_0 = 1.38 \ 10^{-26}$ kJ/K

In liquids and gases under extreme pressures, the relationships between the thermodynamic variables p, T, ρ, χ (here χ is the quality or the fractional mass of gas comprising a saturated liquid–gas mixture, e.g., $\chi = 1.00$ represents a fully saturated gaseous state and $\chi = 0$ represents the fully saturated liquid state) are not so simple, but the fact remains that these parameters are fully dependent upon each other, and specifying two thermodynamic parameters (including enthalpy, entropy, etc.) will fully specify the thermodynamic state of the fluid.

2.4   Fluid Flow Equations

In the Eulerian description of fluid mechanics the field variables such as pressure, density, momenta and energy, are considered to be continuous functions of the spatial coordinates x, y, z and of time t. Because velocity has three components in three-dimensional space and only two independent thermodynamic variables need to be selected to fix the thermodynamic state of the fluid (we chose p and ρ), we have a total of five field variables for which we need five independent equations. We take advantage of conservation laws to establish these equations, namely, the conservation of mass which supplies one equation; the conservation of momentum along each of the three principal axes, which provides three equations; and the conservation of energy (or the equation of state, in the derivation of the actual wave equation)[2] which constitutes the fifth equation.

2.5   Conservation of Mass

In Figure 2.4 consider a parallelepiped serving as a control (or reference) volume $dV = dx\,dy\,dz$ through which fluid flows. Conservation of matter dictates that the net flow into this volume equals the gain or loss of fluid inside the volume, that is,

$$\dot{m}_{\text{exit}} - \dot{m}_{\text{enter}} = \left(\frac{\Delta m}{\Delta t_{\text{volume}}}\right)_{\Delta V \to dV}$$

Let the velocity $\mathbf{V}$ of the fluid resolve into u, v, w, the velocity components in the x, y, and z directions, respectively. In vector terminology

$$\mathbf{V} = u\mathbf{i} + v\mathbf{j} + w\mathbf{k}$$

where $\mathbf{i}, \mathbf{j}, \mathbf{k}$ represent the unit vectors along the x, y, z coordinates. The mass flux $Q(x, t)$ is defined as the flow of the mass of fluid per unit time per unit area, which

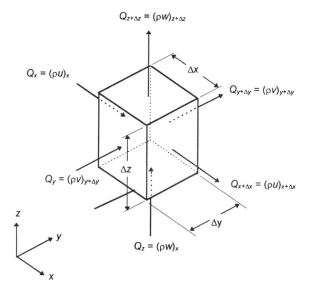

FIGURE 2.4 Flow $Q(\mathbf{v}, t)$ into and out of a control volume $\Delta V = \Delta x \Delta y \Delta z$ depicted for the derivation of the equation of continuity.

---

[2] It can be argued that because the equation of state derives from the principles of conservation of momentum and energy in classic kinetic theory, it effectively becomes the equivalent of the energy conservation principle in the extraction of the acoustic wave equations for a fluid, in conjunction with the equations of continuity and momentum.

is mathematically represented by

$$Q(\mathbf{x}, t) = \rho(\mathbf{x}, t)\, u(\mathbf{x}, t)$$

The rate of mass per unit time $\dot{m}_x$ flowing into the control volume dV in the x-direction is given by

$$\dot{m}_x = Q(x, t)\, dA_x = (\rho u)_{x,t}\, dA_x = (\rho u)_{x,t}\, dy\, dz \qquad (2.6)$$

at position x and the rate of mass per unit time $\dot{m}_{x+\Delta x}$ leaving dV at $x + \Delta x$ by

$$\dot{m}_{x+\Delta x} = Q(x + \Delta x, t)\, dA_x = (\rho u)_{x+\Delta x,t}\, dA_x = (\rho u)_{x+\Delta x,t}\, dy\, dz \qquad (2.7)$$

The subtracting of equation (2.6) from (2.7) yields the net flow in the x-direction

$$\dot{m}_{x+\Delta x} - \dot{m}_x = dA_x[(\rho u)_{x+\Delta x} - (\rho u)_x] = dA_x \frac{\partial(\rho u)}{\partial x}\, dx = \frac{\partial(\rho u)}{\partial x}\, dV \qquad (2.8)$$

Similarly, for mass flow in the y and z directions,

$$\dot{m}_{y+\Delta y} - \dot{m}_y = \frac{\partial(\rho v)}{\partial y}\, dV \qquad (2.9)$$

$$\dot{m}_{z+\Delta z} - \dot{m}_z = \frac{\partial(\rho w)}{\partial z}\, dv \qquad (2.10)$$

Summing the net mass flow equations (2.8)–(2.10) and equating them to the change of mass in the control volume results in:

$$\frac{\partial(\rho u)}{\partial x} + \frac{\partial(\rho v)}{\partial y} + \frac{\partial(\rho w)}{\partial z} = \frac{\partial \rho}{\partial t} \qquad (2.11)$$

Equation (2.11) is the *equation of continuity*, a general statement of the conservation of matter for compressible fluid[3] flow. In vector notation equation (2.11) may be written as

$$\frac{\partial \rho}{\partial t} + \nabla \cdot (\rho \mathbf{V}) = 0 \qquad (2.12)$$

wherein the gradient symbol represents:

$$\nabla = \mathbf{i} \frac{\partial}{\partial x} + \mathbf{j} \frac{\partial}{\partial y} + \mathbf{k} \frac{\partial}{\partial z}$$

in the rectangular coordinate system. In the cylindrical coordinate system the gradient operator ∇ appears as

$$\nabla = \frac{\partial}{\partial r} + \frac{1}{r}\frac{\partial}{\partial \phi} + \frac{\partial}{\partial z}$$

---

[3] If density ρ is constant, the fluid is said to be *incompressible*. As ρ is no longer a spatial or a time function, equation (211) simplifies to

$$\frac{\partial u}{\partial x} + \frac{\partial v}{\partial y} + \frac{\partial w}{\partial z} = 0$$

and in the spherical coordinate system as

$$\nabla \equiv \frac{\partial}{\partial r} + \frac{1}{r}\frac{\partial}{\partial \vartheta} + \frac{1}{r \sin \vartheta}\frac{\partial}{\partial \varphi}$$

2.6   Conservation of Momentum

In order to develop the equations of momentum for a fluid, let us consider in Figure 2.5 the motion of a fluid particle with a velocity field $\mathbf{V_p} = \mathbf{V}_t(x, y, z, t)$. At a later time $t + dt$, the velocity becomes $\mathbf{V_{p'}} = \mathbf{V}_{t+dt}(x + dx, \ y + dy, \ z + dz, \ t + dt)$. The change in velocity is given by

$$d\mathbf{V} = \Delta(\mathbf{V}_{p'} - \mathbf{V}_p) = \frac{\partial \mathbf{V}}{\partial x}\,dx + \frac{\partial \mathbf{V}}{\partial y}\,dy + \frac{\partial \mathbf{V}}{\partial z}\,dz + \frac{\partial \mathbf{V}}{\partial t}\,dt$$

and the total acceleration of the particle is therefore expressed as

$$\mathbf{a}_p = \frac{d\mathbf{V}}{dt} = \frac{\partial \mathbf{V}}{\partial x}\frac{dx}{dt} + \frac{\partial \mathbf{V}}{\partial y}\frac{dy}{dt} + \frac{\partial \mathbf{V}}{\partial z}\frac{dz}{dt} + \frac{\partial \mathbf{V}}{\partial t} \tag{2.13}$$

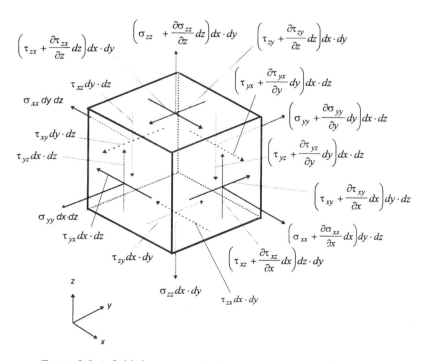

FIGURE 2.5. A fluid element acted on by normal and tangential stresses.

But $dx/dt = u$, $dy/dt = v$ and $dz/dt = w$. Equation (2.13) now can be written as:

$$\mathbf{a}_p = \frac{d\mathbf{V}}{dt} = u\frac{\partial \mathbf{V}}{\partial x} + v\frac{\partial \mathbf{V}}{\partial y} + w\frac{\partial \mathbf{V}}{\partial z} + \frac{\partial \mathbf{V}}{\partial t} = \frac{D\mathbf{V}}{Dt} \tag{2.14}$$

Here the operator

$$\frac{D}{Dt} = \frac{\partial}{\partial t} + u\frac{\partial}{\partial x} + v\frac{\partial}{\partial y} + w\frac{\partial}{\partial z}$$

represents the *total* or *convective* derivative of fluid mechanics. While equation (2.14) is a vector expression, we can rephrase it into scalar terms. With reference to a rectangular coordinate system the scalar components of equation (2.14) are written:

$$a_x = \frac{Du}{Dt} = \frac{\partial u}{\partial t} + u\frac{\partial u}{\partial x} + v\frac{\partial u}{\partial y} + w\frac{\partial u}{\partial z}$$

$$a_y = \frac{Dv}{Dt} = \frac{\partial v}{\partial t} + u\frac{\partial v}{\partial x} + v\frac{\partial v}{\partial y} + w\frac{\partial v}{\partial z}$$

$$a_z = \frac{Dw}{Dt} = \frac{\partial w}{\partial t} + u\frac{\partial w}{\partial x} + v\frac{\partial w}{\partial y} + w\frac{\partial w}{\partial z}$$

No acceleration or deceleration of the fluid will occur unless forces are acting upon it. Two types of forces act on the fluid element of Figure 2.5, namely, body forces and surface forces. Gravity constitutes a body force that pervades throughout the volume of the fluid. Surface forces include both normal forces (pressure) and tangential (shear) forces. A normal force is denoted by the symbol σ_{mm} where m denotes the direction of the normal. Because σ_{mm} is dimensionally expressed in force per unit area, it must be multiplied by the area normal to it in order to obtain the force.

A shear stress acts along the plane of the surface. It is represented by the symbol τ_{mn}, where the force produced by the shear is normal to coordinate m and parallel to coordinate n, and either m or n may represent the principal coordinate x, y, or z, provided that $m \neq n$. If $m = n$, then τ_{mm} really represents the normal force σ_{mm} and is thus no longer a tangential force. The shear stress is multiplied by area it is acting on to yield the shear force. For example, a shear τ_{xy} multiplied by area $(dx \cdot dy)$ represents the shear force normal to the x-axis and parallel to the y-axis, as shown in Figure 2.5 for a fluid element displayed in cartesian coordinates.

In order to determine the net force F_x in the x-direction for a fluid element of volume $V = dx\,dy\,dz$, all of the forces in the x-direction must be summed. From Figure 2.5 we can write

$$dF_x = \rho g_x\,dx\,dy\,dz + \left(-\sigma_{xx} + \sigma_{xx} + \frac{\partial \sigma_{xx}}{\partial x}dx\right)dy\,dz$$

$$+ \left(\tau_{yx} + \frac{\partial \tau_{yx}}{\partial y}dy - \tau_{yx}\right)dx\,dz + \left(\tau_{zx} + \frac{\partial \tau_{zx}}{\partial z}dz - \tau_{zx}\right)dx\,dy$$

which simplifies to:

$$dF_x = \left(\rho g_x + \frac{\partial \sigma_{xx}}{\partial x} + \frac{\partial \tau_{yx}}{\partial y} + \frac{\partial \tau_{zx}}{\partial z}\right) dx\, dy\, dz \qquad (2.15)$$

We can easily derive by extension the force summations for the other two principal axes:

$$dF_y = \left(\rho g_y + \frac{\partial \tau_{xy}}{\partial x} + \frac{\partial \sigma_{yy}}{\partial y} + \frac{\partial \tau_{zy}}{\partial z}\right) dx\, dy\, dz \qquad (2.16)$$

$$dF_z = \left(\rho g_z + \frac{\partial \tau_{xz}}{\partial x} + \frac{\partial \tau_{yz}}{\partial y} + \frac{\partial \sigma_{zz}}{\partial z}\right) dx\, dy\, dz \qquad (2.17)$$

From Newton's second law of motion

$$d\mathbf{F} = d(m\mathbf{a}) = \rho dV \frac{D\mathbf{V}}{Dt}$$

we can now formulate the differential momentum equations by combining the scalar components of equation (2.13) with equations (2.15)–(2.17) with the following results

$$\rho g_x + \frac{\partial \sigma_{xx}}{\partial x} + \frac{\partial \tau_{yx}}{\partial y} + \frac{\partial \tau_{zx}}{\partial z} = \rho \left(\frac{\partial u}{\partial t} + u\frac{\partial u}{\partial x} + v\frac{\partial u}{\partial y} + w\frac{\partial u}{\partial z}\right) \qquad (2.18a)$$

$$\rho g_y + \frac{\partial \tau_{xy}}{\partial x} + \frac{\partial \sigma_{yy}}{\partial y} + \frac{\partial \tau_{zy}}{\partial z} = \rho \left(\frac{\partial v}{\partial t} + u\frac{\partial v}{\partial x} + v\frac{\partial v}{\partial y} + w\frac{\partial v}{\partial z}\right) \qquad (2.18b)$$

$$\rho g_z + \frac{\partial \tau_{xz}}{\partial x} + \frac{\partial \tau_{yz}}{\partial y} + \frac{\partial \sigma_{zz}}{\partial z} = \rho \left(\frac{\partial w}{\partial t} + u\frac{\partial w}{\partial x} + v\frac{\partial w}{\partial y} + w\frac{\partial w}{\partial z}\right) \qquad (2.18c)$$

In order to use equations (2.18a)–(2.18c), the expressions for the stresses should be stated in terms of the velocity field. If a Newtonian fluid is assumed, the viscous stresses are proportional to the rate of shearing strain (i.e., the rate of angular deformation). Without delving into details, we express the stresses in terms of velocity gradients and viscosity coefficient μ as follows:

$$\tau_{xy} = \tau_{yx} = \mu \left(\frac{\partial v}{\partial x} + \frac{\partial u}{\partial y}\right) \qquad (2.19a)$$

$$\tau_{yz} = \tau_{zy} = \mu \left(\frac{\partial w}{\partial y} + \frac{\partial v}{\partial z}\right) \qquad (2.19b)$$

$$\tau_{zx} = \tau_{xz} = \mu \left(\frac{\partial u}{\partial z} + \frac{\partial w}{\partial x}\right) \qquad (2.19c)$$

$$\sigma_{xx} = -p - \frac{2}{3}\mu \nabla \cdot \mathbf{V} + 2\mu \frac{\partial u}{\partial x} \qquad (2.19d)$$

$$\sigma_{yy} = -p - \frac{2}{3}\mu \nabla \cdot \mathbf{V} + 2\mu \frac{\partial v}{\partial y} \qquad (2.19e)$$

$$\sigma_{zz} = -p - \frac{2}{3}\mu \nabla \cdot \mathbf{V} + 2\mu \frac{\partial w}{\partial z} \qquad (2.19f)$$

Here the term p is the local thermodynamic pressure, which is essentially an isotropic parameter at any given point in the fluid. If we assume the fluid to be frictionless, then $\mu = 0$, and we are left with equations (2.13d)–(2.13f) in the following format:

$$\sigma_{xx} = \sigma_{yy} = \sigma_{zz} = -p$$

and, neglecting the gravitational body force ρg_i (where $i = x, y, z$), we recast equations (2.19a)–(2.19c) as

$$-\frac{\partial p}{\partial x} = \rho \frac{Du}{Dt}$$

$$-\frac{\partial p}{\partial y} = \rho \frac{Dv}{Dt}$$

$$-\frac{\partial p}{\partial z} = \rho \frac{Dw}{Dt}$$

2.7   Conservation of Energy

The energy content W of a fluid is the sum of the macroscopic kinetic energy $\rho |\mathbf{V}|^2/2$ and the internal energy ρE of the fluid. In a gas, the microscopic kinetic energy (i.e., the thermal energy of the molecules) comprises the major portion of the internal energy, so the potential energy between molecules are negligible in comparison. Denoting the energy flux by S we write the equation for the conservation of energy as:

$$\frac{\partial W}{\partial t} + \frac{\partial S}{\partial x} = 0 \qquad (2.20)$$

The internal energy of a volumetric element can be increased through heat flow from the surrounding fluid or from external sources and by the work of compression $-\int p \, dV$ by the surrounding fluid pressure. This energy balance and the fact that the internal energy is a thermodynamic state that can be fully specified by two independent thermodynamic variables constitute the first law of thermodynamics.

With both the conservation equations discussed above and the equation of state, we have all the necessary equations to obtain solutions for the three components of velocity $\mathbf{V}$, ρ, p, and absolute temperature T. Because the fluid equations are nonlinear, solutions are not easy to come by, even with the aid of supercomputers to map the complex motions of atmospheric eddies, turbulent jet flows, capillary flow, and so on. Exact solutions exist principally for a few simple problems. Nevertheless, through the derivation of these equations, we have established the foundation for the derivation of acoustic field equations for fluids.

2.8   Derivation of the Acoustic Equations

We begin with the following assumptions:

(i) the unperturbed fluid has definite values of pressure, density, temperature, and velocity, all of which are assumed to be time independent and denoted by the subscript 0;

(ii) the passage of an acoustic signal through the fluid results in small perturbations of pressure, temperature, density, and velocity. These perturbations are expressed as $p_0 + p$, $\rho_0 + \rho$, u, and so on. The unperturbed velocity u_0 is set to zero; the unperturbed fluid does not undergo macroscopic motion, and u constitutes the perturbation velocity in the x-direction. Also, $p \ll p_0$, $\rho \ll \rho_0$, and $T \ll T_0$.

(iii) the transmission of the sound through the fluid is sufficiently transient that there is virtually no time for heat transfer to occur, and thus the ongoing thermodynamic action may be deemed an adiabatic process.

Under the above conditions we obtain an expansion of the continuity equation in the x-direction as follows:

$$\frac{\partial[(\rho_0 + \rho)u]}{\partial x} = \frac{\partial(\rho_0 + \rho)}{\partial t} = u\frac{\partial \rho}{\partial x} + \rho_0\frac{\partial u}{\partial x}$$

which, by discarding second-order terms, reduces to

$$\frac{\partial \rho}{\partial t} = \rho_0\frac{\partial u}{\partial x} \tag{2.21}$$

Here we consider $\rho_0 \approx \rho_0 + \rho$, also recalling that the quiescent density ρ_0 does not vary in time and space. Treating in the same fashion the one-dimensional momentum equation

$$\frac{\partial(p_0 + p)}{\partial x} = (\rho_0 + \rho)\left(\frac{\partial u}{\partial t} + u\frac{\partial u}{\partial x}\right)$$

yields

$$\frac{\partial p}{\partial x} = \rho_0\frac{\partial u}{\partial t} \tag{2.22}$$

In an adiabatic process involving an ideal gas:

$$p\rho^{-\gamma} = \text{const}$$

Here γ represents a thermodynamic constant, characteristic of the gas, equal to the ratio of the specific heats c_p/c_v. The numerator of this thermodynamic ratio is the specific heat at constant pressure, and the denominator, the specific heat at constant volume. By differentiation,

$$\rho^{-\gamma}dp - \gamma p\rho^{-\gamma-1}d\rho = 0$$

and rearranging

$$\frac{dp}{p} = \gamma \frac{d\rho}{\rho}$$

we have for this situation

$$\frac{dp}{p_0} = \gamma \frac{d\rho}{\rho_0}$$

The above expression can be differentiated with respect to time:

$$\frac{1}{p_0}\frac{\partial p}{\partial t} = \frac{\gamma}{\rho_0}\frac{\partial \rho}{\partial t} \qquad (2.23)$$

Combining equations (2.22) and (2.23),

$$\frac{\partial p}{\partial t} = \frac{\gamma p_0}{\rho_0}\frac{\partial \rho}{\partial t} = \gamma p_0 \frac{\partial u}{\partial x}$$

and then differentiating with respect to time t we obtain

$$\frac{\partial^2 p}{\partial t^2} = \gamma p_0 \frac{\partial^2 u}{\partial t \, \partial x}$$

Differentiating equation (2.21) with respect to x results in

$$\frac{\partial^2 p}{\partial x^2} = \rho_0 \frac{\partial^2 u}{\partial x \partial t}$$

Equating the above two cross-differential terms to each other, as we consider them to be equivalent regardless of their order of differentiation, we obtain the result

$$\frac{\partial^2 p}{\partial x^2} = \frac{\rho_0}{\gamma p_0}\frac{\partial^2 p}{\partial t^2} = \frac{1}{c^2}\frac{\partial^2 p}{\partial t^2} \qquad (2.24)$$

where

$$c^2 = \frac{\gamma p_0}{\rho_0} = \gamma R T$$

Here c, R, and T are, respectively, the propagation velocity of sound, the gas constant, and absolute temperature of the (ideal) gas. In three-dimensional form the wave equation (2.24) takes on the following appearance:

$$\nabla^2 p = \frac{1}{c^2}\frac{\partial^2 p}{\partial t^2} \qquad (2.25)$$

We can also eliminate p in favor of u by reversing the differentiation procedure between equations (2.22) and (2.23), in which situation we obtain:

$$\frac{\partial^2 u}{\partial x^2} = \frac{1}{c^2}\frac{\partial^2 u}{\partial t^2} \qquad (2.26)$$

for the one-dimensional situation, and

$$\nabla^2 \mathbf{V} = \frac{1}{c^2}\frac{\partial^2 \mathbf{V}}{\partial t^2} \qquad (2.27)$$

in the three-dimensional case. It is also a straightforward matter to derive the wave equation in terms of density, resulting in the following expressions:

$$\frac{\partial^2 \rho}{\partial x^2} = \frac{1}{c^2}\frac{\partial^2 \rho}{\partial t^2} \tag{2.28a}$$

for the one-dimensional case and

$$\nabla^2 \rho = \frac{1}{c^2}\frac{\partial^2 \rho}{\partial t^2} \tag{2.28b}$$

for three dimensions.

Equations (2.24)–(2.28) are second-order partial differential equations in x and t. Ordinarily we need two initial conditions and two boundary conditions for a fully defined solution for each of the equations, but we need not define these conditions in order to ascertain the nature of the general solutions. The general solution to equation (2.24) may be written as

$$p(x, t) = F(x - ct) + G(x + ct) \tag{2.29}$$

The function $F(x - ct)$ represents waves moving in the positive x direction and $G(x + ct)$ represents waves moving in the opposite direction. All solutions to equation (2.24) must be of the form represented in equation (2.29); otherwise any p that does not adhere to this form cannot constitute a solution. Because equations (2.26) and (2.28a) are functionally the same as equation (2.24), their respective general solutions take on the same cast as that of equation (2.29):

$$u(x, t) = \Phi(x - ct) + \Gamma(x + ct) \tag{2.30}$$

$$\rho(x, t) = \Theta(x - ct) + \Upsilon(x + ct) \tag{2.31}$$

The arbitrary functions F, G, Φ, Γ, Θ, Υ can be assumed to have continuous derivatives of the first and second order. Because of the manner in which the constant c appears in relation to x and t inside these functions, it must have the physical dimensions of x/t, so c must be a speed, which is indeed the experimentally determined rate at which the sound wave propagates through a medium. No matter how it is shaped, the propagating wave (or its counterpart, the backward traveling wave) moves without changing its form. To prove this, consider the sound pressure level at $x = 0$ and time $t = t_1$ for a wave moving in the positive x-direction. Thus, $p = f_\alpha(t_1)$ At time $= t_1 + t_2$, the sound wave will have traveled a distance $x = ct_2$. The sound pressure will now be

$$p = f_\alpha(t_1 + t_2) = f_\alpha\left(t_1 + t_2 - \frac{ct_2}{t_2}\right) = f_\alpha(t_1)$$

This means the sound pressure has propagated without change.

References

Beranek, Leo L. 1986. *Acoustics*. New York: American Institute of Physics. (An exceptionally clear text in the field.)

Crocker, Malcolm J. (ed.). 1997. *Encyclopedia of Acoustics*, vol. 1. New York: John Wiley
& Sons, Chapters 1 and 2. (Sir James Lighthill compared the significance of this four-
volume compilation with that of Lord Rayleigh's *The Theory of Sound*, which is not at all
far-fetched considering this encyclopedia contains contributions from an editorial board
whose members constitute a veritable *Who's Who in Acoustics*. *Handbook of Acoustics*
by the same editor and publisher (1998) is a truncated version of the *Encyclopedia*,
containing approximately 75 percent of the chapters. Chapters 1 and 2 are identical in
both publications.)

Shapiro, Ascher H. 1953. *The Dynamics and Thermodynamics of Compressible Fluid Flow*,
vol. I. New York: The Ronald Press Co; Chapter 1. (In spite of its venerable age, it is still
one of the best works on the topic of fluid dynamics.)

Problems

1. Write the expression for a simple sine wave having a frequency of 10 Hz and an
amplitude of 10^{-8} cm. What is the frequency expressed in radians per second?
Plot the expression on graph paper or, better yet, on a computer with the aid
of a program such as Mathcad®, MathLab®, etc. Repeat the process for a
frequency of 20 Hz and for 50 Hz.

2. If the frequency of a pure cosine wave is 100 Hz and the velocity of the wave
front is 330 m/s, what is the wave length of this signal? Express the frequency
in radians per second.

3. Air may be considered to be a nearly ideal gas with the ratio of specific heat
$\gamma = 1.402$. At 0°C its density is 1.293 kg/m³. Predict the speed of sound c for
the normal atmospheric pressure of 101.2 kPa (1 Pa = 1 N/m²).

4. Nitrogen is known to have a molecular weight of 28 kg/kg-mol. Predict the
speed of sound at 0°C, 20°C, and 50°C, with the assumption that nitrogen
behaves as an ideal gas. Repeat the problem for pure oxygen which has a
molecular weight of 32.

5. Compute the speed of sound (in ft/s) traveling through steel which has a Young's
modulus of 30×10^6 psi and a specific gravity of 7.7.

6. A solid material is known to have a density of 8.5 g/cm³. Sound velocity
traveling through this material was measured as being 4000 m/s. Determine
the Young's modulus in GPa for this material.

7. Find the speed of sound (in m/s) traveling through aluminum which typically
has a Young's modulus of 72.4 GPa and a specific gravity of 2.7.

8. For distilled water, the speed of sound c in m/s can be predicted within 0.05%
as a function of pressure P and temperature T from the experimentally determined formula

$$c(P,t) = 1402.7 + 488t - 482t^2 + 135t^3 + \left(15.9 + 2.8t + 2.4t^2\right)\left(P_a/100\right)$$

where P_a is the gauge pressure in bars (1 bar = 100 kPa) and $t = 0.01T$, with
temperature T in degrees Celsius. Find the speed of sound for the water at 20°C
and 1 bar. What will be the wavelength of a 200-Hz sine-wave signal traveling
through water? If the same signal travels through air at the speed of 343 m/s,
what will be the corresponding wavelength?

9. Explain why density and pressure are in phase and that both are out of phase with particle velocity.
10. When does the maximum amplitude of a pure sine wave occur with respect to the particle velocity and the instantaneous pressure?
11. A molecule exposed to a pure cosine sound wave undergoes a particle displacement y, with maximum amplitude A, according to

$$y = A \cos \omega t$$

Find the corresponding particle velocity and show that the expressions for both displacement and velocity constitute solutions to a wave equation.
12. Demonstrate that $y(x, t) = A_1 \cos(x - ct) + B_1 \sin(x + ct) + A_2 \cos^2 2(x - ct) + B_2 \sin^2 3(x + ct)$ constitute a solution to the wave equation. Which portion of the solution represent wave travel in the $+x$ direction and which portion denote propagation in the $-x$ direction?
13. If the density of a medium subject to wave propagation varies in the following manner

$$\rho = \rho_0 e^{i(x-ct)}$$

express the corresponding pressure $p(x, t)$ in terms of quiescent pressure p_0 and density ρ_0.

3
Sound Wave Propagation and Characteristics

3.1 The Nature of Sound Propagation

When energy passes through a medium resulting in a wave-type motion, several different types of waves may be generated, depending upon the motion of a particle in the medium. A *transverse* wave occurs when its amplitude varies in the direction normal to the direction of the propagation. This type of wave has been used to describe the transmission of light and alternating electric current. But the situation is almost completely different in the case of sound waves, which are principally *longitudinal*, in that the particles oscillate back and forth in the direction of the wave motion, with the result that the motion creates alternative compression and rarefaction of the medium particles as the sound passes a given point. The net fluid displacement over a cycle is zero, because it is the disturbance rather than the fluid that is moving at the speed of sound. The fluid molecules do not move far from their original positions.

Additionally, waves may also fall into the category of being rotational or torsional. The particles of a rotational waves rotate about a common center; the curl of an ocean wave roaring onto a beach provides a vivid example. The particles of torsional waves move in a helical fashion, that could be considered a vector combination of longitudinal and transverse motions. Such waves occur in solid substances, and shear patterns often result. These are referred to as *shear waves*, which all solids support.

3.2 Forward Propagating Plane Wave

In equation (2.29), which is the general solution to the one-dimensional wave equation, we consider only the wave moving in the $+x$ direction with the solution for a monofrequency wave represented by

$$p(x, t) = F(x - ct) = p_m \cos k(x - ct) \tag{3.1}$$

where p_m is the peak amplitude of the sound pressure, k is the wave number that equals $2\pi/\lambda$, and λ is the wavelength. Figure 3.1 shows the variation of sound

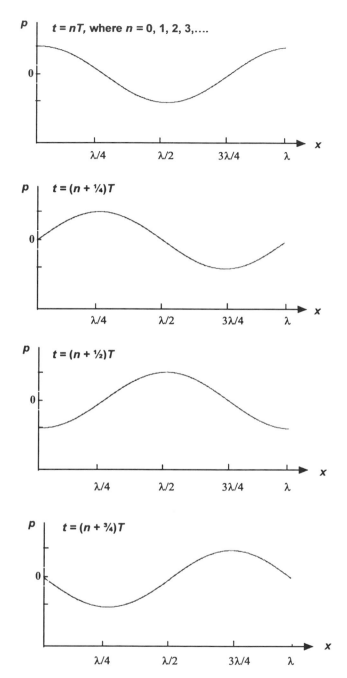

FIGURE 3.1. Variation of pressure from quiescent state along the x-axis for time instants $t = 0, t = T/4, t = T/2, t = 3T/4$, where $T = 1/f$ is the period for a complete cycle to occur.

pressure p for different time intervals $t = 0$, $t = T/4 = 1/(4f) = (1/2)\pi/\omega$, $t = T/2 = 1/(2f) = \pi/\omega$, $t = 3/(4f) = 3T/4 = (3/2)(\pi/\omega)$, and $t = T = 1/f = 2\pi/\omega$.

3.3 Complex Waves

The concept of simple sinusoidal waves lacks specificity to be of practical value in noise control, but complex periodic waveforms can be broken into two or more sinusoidal harmonically related waves. In Figure 3.2, a complex wave form resolves into a sum of harmonically related waves. The harmonic relationship in this example is such that the frequency of one harmonic is twice that of the other. The lowest-frequency sine term is the fundamental, and the next highest frequency the second harmonic, the next the third harmonic, and so on. Sound pressure waves radiating from pumps, gears and other rotating machinery are usually complex and periodic, with distinguishable discrete tones or pure tones. These sinusoidal waves can be broken down or synthesized into simple sinusoidal terms. In the analysis of the noise emanating from rotating machinery, there are often 8–10 harmonics present with frequencies that are integer-ordered multiples of the fundamental frequency. Even aperiodic sounds such as the hiss of a pressure valve of an autoclave, the broadband whine of a jet engine, or the pulsating sound of a jackhammer can be resolved and described in terms of sums of simple sinusoids. Integer harmonic relationships associated with periodic sound waves do not occur in these sounds, and the composition entails more than a simple series. But the principle of synthesis still applies.

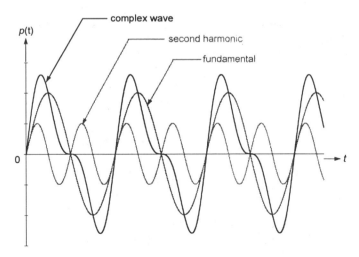

FIGURE 3.2. The resolution of a complex waveform into a set of harmonically related sinusoidal waves. The fundamental wave and the second harmonic sine wave add up algebraically to form the complex wave.

A general equation can be written to incorporate the elements of the sound pressure level associated with complex periodic noise sources:

$$p(t) = A_1 \sin(\omega t + \phi_1) + A_2 \sin(2\omega t + \phi_2)$$
$$+ A_3 \sin(3\omega t + \phi_3) + \cdots + A_n \sin(n\omega t + \phi_n)$$
$$= \sum_1^n A_n \sin(n\omega t + \phi_n) = \sum_1^n C_n e^{jn\omega t} \qquad (3.2)$$

where

$$A_n = \text{amplitude of the } n\text{th harmonic}$$
$$\phi_n = \text{phase angle of the } n\text{th harmonic}$$
$$C_n = \text{complex amplitude of the } n\text{th harmonic}$$

Equation (3.2) constitutes a form of the Fourier series, an analytical tool developed nearly two centuries ago by the French physicist Jean Baptiste Joseph Fourier to characterize complex functions and used by him to predict tides. Fourier's concept of complex wave synthesis constitutes one of the most powerful analytical and diagnostic tools available to the present-day acoustician.

When two or more sound waves become superimposed upon each other, they combine in a linear manner, that is, their amplitudes add algebraically at any point in space and time. The superposition generally results in a complex wave that can be synthesized into basic sinusoidal spectrum components. Two special phenomena resulting from superposition are of particular interest, namely, *beat frequency* and *standing waves*.

Consider the superposition of two sound waves of equal amplitudes but slightly differing frequencies. With A_0 denoting the amplitude of each wave and $\omega_1 \neq \omega_2$, the total superimposed pressure becomes

$$p(t) = A_0(\sin \omega_1 t + \sin \omega_2 t)$$

Applying the trigonometric identity

$$\sin \alpha + \sin \beta = 2 \cos \frac{(\alpha - \beta)}{2} \sin \frac{(\alpha + \beta)}{2}$$

the total pressure assumes the form

$$p(t) = 2A_0 \cos \frac{(\omega_1 - \omega_2)t}{2} \sin \frac{(\omega_1 + \omega_2)t}{2}$$
$$= 2A_0 \cos 2\pi \frac{(f_1 - f_2)t}{2} \sin 2\pi \frac{(f_1 + f_2)t}{2} \qquad (3.3)$$

where $\omega = 2\pi f$.

From equation (3.3) the resultant wave may be considered as a complex sound wave with a frequency of $(f_1 + f_2)/2$, as indicated by the sine factor and which is the average of the two superimposed waves. The amplitude is

$$p'(t) = 2A_0 \cos 2\pi \frac{(f_1 - f_2)t}{2}$$

When the argument of the cosine assumes integer values of π, the amplitude of the complex wave is a maximum that is equal to $2A_0$. Continuing the reasoning further, it is established that the amplitude of the complex wave vanishes when the argument of the cosine takes on integer odd values of $\pi/2$, that is,

$$2\pi \frac{(f_1 - f_2)t}{2} = \frac{(2n - 1)\pi}{2} \qquad (n = 1, 2, 3, \ldots) \qquad (3.4)$$

A graph of the envelope of this transient amplitude modulation is given in Figure 3.3. The modulation or *beat frequency* is simply the frequency difference $(f_1 - f_2)$ between the two superposed waves. To demonstrate this, let us solve equation (3.4) for those times t_n when the amplitude of the superimposed sound pressure is zero:

$$t_n = \frac{2n - 1}{2(f_1 - f_2)} \qquad (n = 1, 2, 3, \ldots)$$

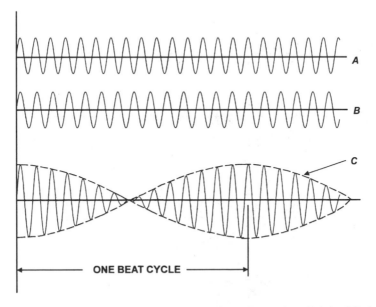

ONE BEAT CYCLE

FIGURE 3.3. Addition of waves A and B with equal amplitudes but slightly differing frequencies. The sum of the two sine waves yields an envelope C which has a beat frequency equal to the difference between the frequencies of the superimposed waves A and B.

Now consider in a general fashion the time difference between two consecutive beats, namely, the nth and the $(n + 1)$th:

$$t_{n+1} - t_n = \frac{2(n + 1) - 1}{2(f_1 - f_2)} - \frac{2n - 1}{2(f_1 - f_2)}$$

$$= \frac{1}{f_1 - f_2} \tag{3.5}$$

The time duration between beats is, by definition, the period T_b of the beat frequency; and the reciprocal of the period defined in equation (3.5) yields the beat frequency f_b

$$f_b = f_1 - f_2 \tag{3.6}$$

In the more general case when the amplitudes of the superposed equations are not equal, the amplitude of the superposed wave varies between the sum and difference of the component waves, as shown in Figure 3.4. The periodic variation in amplitude generates a rhythmic pulsating sound, and when the frequency difference is only a few hertz, say 4 or 5, the beat can readily be discerned by the human ear.

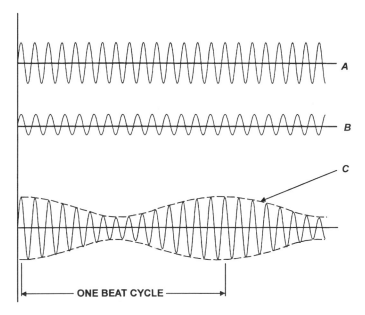

FIGURE 3.4. Addition of waves A and B with slightly different frequencies but unequal amplitudes. Envelope C that results from adding waves A and B shows beats in which the minimum strength is not zero.

3.4   Standing Waves

When a sound wave is superposed upon another wave of the same frequency, but traveling in a different direction, a standing-wave sound field is generated. As an illustration, consider the superimposition of two sound waves traveling in opposite directions as given by

$$p_1(t) + p_2(t) = A_1 \sin(2\pi ft - kx) + A_2 \sin(2\pi ft + kx) \qquad (3.7)$$

The first sine term in equation (3.7) represents a sound wave traveling in the positive x-direction with amplitude A_1 and frequency f. The second sine term represents a sine wave traveling in the negative x-direction with amplitude A_2 and identical frequency f. Using trigonometric identities for the sum and difference of angles, we rewrite equation (3.7) as follows:

$$p_1(t) + p_2(t) = A_1 \sin 2\pi ft \cos kx - A_1 \cos 2\pi ft \sin kx$$
$$+ A_2 \sin 2\pi ft \cos kx + A_2 \cos 2\pi ft \sin kx$$

For waves of equal amplitudes we obtain the following simplification:

$$p_1(t) = +p_2(t) = 2A_1 \cos kx \sin 2\pi ft \qquad (3.8)$$

Equation (3.8) may now be considered to be a simple sinusoidal function of time whose amplitude depends on the spatial location x of the observer. When the argument of the cosine assumes odd integer values of $\pi/2$, that is,

$$kx = \frac{\pi}{2}, \frac{3\pi}{2}, \frac{5\pi}{2}, \ldots, \frac{(2n-1)\pi}{2} \qquad (n = 1, 2, 3, \ldots)$$

the sound pressure vanishes, and there are nodal points in space where no sound occurs. Solution of the preceding equation for x_n yields the spatial locations of these nodes:

$$x_n = \frac{(2n-1)\pi}{2k} \qquad (n = 1, 2, 3, \ldots)$$

and because $k = \omega/c = 2\pi/\lambda$, we obtain

$$x_n = \frac{(2n-1)\lambda}{4} \qquad (n = 1, 2, 3, \ldots)$$

We thus note that the location of the nodes is simply related to the wavelength of the superimposed waves. By taking the difference between successive nodal locations, it can be demonstrated that the nodes occur every half wavelength, that is,

$$x_{n+1} - x_n = \frac{2[(n+1)-1]\lambda}{4} - \frac{(2n-1)\lambda}{4} = \frac{\lambda}{2}$$

We can also establish from equation (3.8) that the antinodes or points of maximum sound pressure in the standing wave occur when the argument of the cosine assumes integers values of π, that is,

$$kx = n\pi \qquad (n = 1, 2, 3, \ldots)$$

The amplitude of the antinodes is simply $2A_1$. These antinodes or points of maximum sound pressure are stationary, located halfway between the nodes and spaced one-half wavelength apart.

Example Problem 1

Consider a case where a hydraulic pump radiates a 600-Hz sound wave that is reflected back from a tile wall located 1 m from the wall. What is the spacing of the nodes or position of minimum sound for the fundamental tones and its second harmonic?

Solution

We first calculate the wavelength of the first harmonic from the relationship $\lambda = c/f$, with the speed of sound being taken at 344 m/s for the room temperature:

$$\lambda = \frac{c}{f} = \frac{344}{600} = 0.573 \text{ m}$$

The spacing between the nodes occurs every half-wavelength, or distance intervals of 0.287 m. With the hydraulic pump positioned in close proximity to the wall, the amplitude of the antinodes will be nearly twice as large, as the result of the summation of the radiated and the reflected waves. For the second harmonic at 1200 Hz, the spacing of the nodes will be 0.143 m, because its wavelength is half that of the fundamental. This example points out the necessity for caution in taking measurements in close proximity to highly reflecting surfaces, which can yield highly misleading results. In many situations reflective surfaces are not in close proximity and the amplitudes of reflected waves are relatively small compared to the original waves, so the variation in the amplitudes of the standing waves are correspondingly small and thus can be neglected.

3.5   Huygens' Principle

While originally conceived in the seventeenth century to explain optical phenomena, Huygens' principle applies equally to sound propagation. The principle states that advancing wavefronts can be considered to be point sources of secondary wavelets. Figure 3.5 illustrates the Huygens construction of a wave front at time $t + \Delta t$ from a wavefront at earlier time t. The new wavefront is the envelope of radii $c\Delta t$ centered at points on the preceding wavefront. Thus, a plane wave remains a plane wave and a spherical wave remains a spherical wave with ever-enlarging radius.

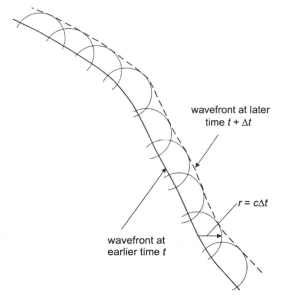

wavefront at later
time $t + \Delta t$

$r = c\Delta t$

wavefront at
earlier time t

FIGURE 3.5. Construction of a wavefront at time $t + \Delta t$ from its previous state at time t, according to Huygens's principle.

3.6   The Doppler Effect

When a sound source moves, the acoustic radiation pattern changes, thereby producing changes to the generated frequencies as perceived by a stationary observer. When an acoustic source generating a frequency f approaches an observer at velocity v, then during a single period $T = 1/f$ a signal emitted at the onset of the period travels a distance cT. But the signal is emitted at the end of the period from the source that is closer to the observer by a distance vT. The distance between the crests, that is, the wavelength λ, has been reduced to

$$\lambda = cT - vT = \frac{c - v}{f} \tag{3.9}$$

The resulting frequency heard by the observer is not the source's output frequency but that increased by the resulting drop in the wavelength, that is,

$$f_d = \frac{c}{\lambda} = \frac{fc}{c - v} = \frac{f}{1 - v/c} \tag{3.10}$$

Here f_d is the frequency perceived by the observer. When the source approaches at a velocity v, the observer hears a higher-frequency sound which represents the original frequency multiplied by a factor $(1 - v/c)^{-1}$. On the other hand, when v assumes a negative value, which means the source is moving away from the observer, f_d assumes a lower value than that of the source frequency f, because the velocity v in equation (3.10) assumes a negative value.

Example Problem 2

A train emits a 250-Hz signal while traveling at the rate of 200 km/h. What are the apparent frequencies in approaching the observer and retreating from the observer at the railroad crossing?

Solution

From equation (3.10)

$$v = (200,000 \text{ m/h})/(3600 \text{ s/h}) = 55.6 \text{ m/s}$$
$$f_d = 250 \text{ Hz}(1 - 55.6/344)^{-1} = 298 \text{ Hz for approach}$$
$$f_d = 250 \text{ Hz}(1 + 55.6/344)^{-1} = 215 \text{ Hz for retreat}$$

If an observer stands on a line making an angle θ with a source's direction of motion at speed V, the approach velocity of the source is $v = V \cos \theta$, and equation (3.10) modifies to

$$f_d = \frac{f}{1 - (V \cos \theta / c)} \qquad (3.11)$$

The relative frequency increases or diminishes when θ exists as an acute or obtuse angle, respectively.

3.7 Reflection

When sound impinges upon a surface, a portion of its energy is absorbed by the surface and the remainder bounces back or becomes reflected from the surface. A perfectly hard surface will reflect back all of the energy. A classic example of the reflection phenomenon is the *echo* which has intrigued and mystified humanity for centuries.

As waves impinge on a hard, smooth surface, the waves are reflected with shape and propagation characteristics unaltered, in accordance with Huygens' principle. Consider the impingement of a series of plane waves on reflecting surface $A - A'$ in Figure 3.6. The arrows normal to the wavefronts, or *rays*, which represent the directions of propagation, are drawn to represent the impingement and the consequent reflection of the wavefront. It follows from application of Huygens' principle and geometry that the angle of incidence is equal to the angle of reflection, where the angles are defined between a normal to the reflecting plane and the incident and the reflected rays, respectively. In Figure 3.7 the geometric ray construction is rendered for a diverging spherical wave incident upon a plane surface. The direction of the reflected sound can at least be determined qualitatively. It should be pointed out here that *standing wave interference patterns* will occur from these reflections.

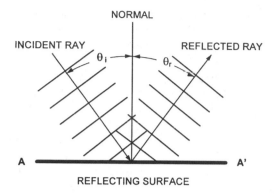

FIGURE 3.6. Geometric depiction of a plane wave reflection. Angle of reflection, θ_r, is equal to angle of incidence, θ_i.

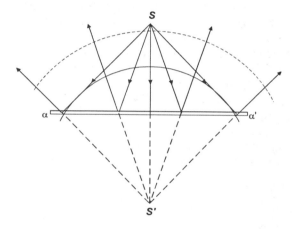

FIGURE 3.7. Construction for a spherical wave from point S incident upon plane surface α–α', Point S' is an imaginary point that is the mirror image of point S on the other side of plane α–α'.

It is of interest to consider the sound field resulting from reflection. Consider the sound waves in Figure 3.6 to be sinusoidal. As the incident or reflected wavefront intersects any normal to the wavefront, a scissorlike effect occurs, not unlike ocean waves breaking obliquely along a beach. The intersection of these waves along the normals constitutes a projection of the incident and reflected waves. From the concept of wave motion, the distance between crests along the normal may be established as a projected wavelength λ', which relates to the incident wave as follows

$$\lambda' = \frac{\lambda}{\cos \vartheta_1} = \lambda \sec \vartheta_1 \tag{3.12}$$

In obeying the laws of reflection, the reflected wave also scissors back along the normal in the opposite direction, producing a traveling wave with a projected wavelength also equal to λ'. Hence, there occurs along any normal line the superimposition of two waves traveling in opposite directions with wavelength λ'. From the concept of standing waves it can be inferred that nodes and antinodes occur along the normal line, and, moreover, the spacing between the nodes and antinodes needs only to be modified by the factor sec θ_1 in equation (3.12).

Consider a complex periodic wave that impinges upon a fully reflective plane surface. A standing wave sound field will exist. The distance d' between peaks along the normal ensues from equation (3.12) in the following manner:

$$d' = \frac{\lambda'}{2} = \lambda_n \sec \theta_i$$

where λ_n is the wavelength of the nth harmonic and θ_i is the angle of incidence of the propagating wavefronts. From the last equation it will be noted for the special case of $\theta_i = 0$ (normal incidence), the nodal spacing reduces to $\lambda/2$, according to equation (3.12). As the angle of incidence increases, the spacing between the nodes likewise increases, and in the limit $\theta_i = \pi/2$, there is no reflected wave, and thus the standing wave field vanishes.

The phenomenon of sound wave reflection finds many applications. The time it takes for a sound wave pulse to travel from a transducer at sea level to the ocean bottom and for the echo to travel back gives a measure of the depth of the water. Further, comparison of the spectral characteristics of the reflected wave with those of the generated waves provides an ample measure of the geological composition of the ocean bottom, for example, silt, rock, sand, coral, and so on. Reflected sound is also used in an analogous way by geologists to gauge the depth and composition of stratified layers in the earth crust, to locate oil, natural gas and mineral deposits.

3.8 Refraction

A phenomenon more familiar in optics than in acoustics is that of refraction, in which the direction of the advancing wavefront is bent away from the straight line of travel. Refraction occurs as the result of the difference in the propagation velocity as the wave travels from one medium to a different medium.

In the optical situation, refraction occurs suddenly when light waves cross the sharp interface between the atmosphere and glass at the surface of a lens, because light travels more slowly in glass than it does in air. At audible frequencies of sound waves, the wavelengths are so long that the apparatus would have to be extremely large in order to render observable acoustic refractions. However, at ultrasonic frequencies, which correspond to extremely short wavelengths, refraction constitutes the operating principle of the *acoustic microscope*. The device functions as indicated in Figure 3.8. A piezoelectric transducer P_z, under the impetus of an input voltage V to the opposite faces of the piezoelectric crystal, delivers a short

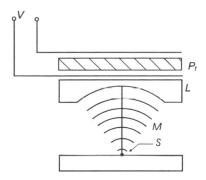

FIGURE 3.8. A schematic of an acoustic microscope. Voltage V causes the piezoelectric crystal P_t to launch a short train of waves into lens L. The propagation velocity of the waves slows down in liquid medium M, and the waves are focused toward point S on the surface of the specimen. Reflected waves follow the same paths in reverse, reaching the piezoelectric transducer which now has been switched into the detector mode.

train of ultrasonic waves into the lens L, which may consist of a small block of sapphire, which incorporates a spherical hollow in the face facing liquid medium M. The waves travel much faster in the crystal than in the adjoining liquid. As a result the central portion of each wave is retarded relative to the outer parts. All of the wavefront enter the liquid at the same time, but the refractive effect causes nearly all of the nearly spherical waves to focus at the central point of curvature. The strength of the reflection depends on the nature of the specimen surface at the focal point S. The operating mode of the lens and transducer now changes from the role of emitter to that of receptor. The lens gather the reflected signal and the transducer detects the signal. In this fashion this device resembles radar and sonar systems. As information is obtained from only one point at a time, the specimen must be moved in a raster pattern in the focal plane of the microscope while an image is progressively accumulated in a computer memory.

If water is used as a medium, a 3-MHz signal, with a propagation velocity of 1480 m/s, would have a wavelength of 500 μm = 0.5 mm, which would amount to a rather coarse resolution. Clearly, higher-frequency signals are called for, but such signals become strongly absorbed in water. A medium with a lower value of c and, more importantly, less absorption than water constitutes another possibility. An attractive choice turned out to be liquid helium, used with instruments that generate up into the 8-GHz frequency range. The wavelengths are small as 0.03 μm.

In Figure 3.9, a geometrical ray construction illustrates the refraction of sound passing from one medium to another. Application of Huygens' principle leads to the basic laws of refraction, the most useful of which is

$$\frac{\sin \theta_i}{c_1} = \frac{\sin \theta_r}{c_2} \qquad (3.13)$$

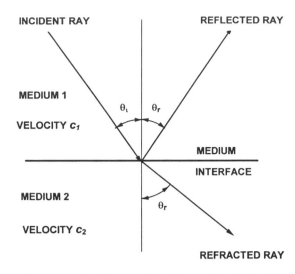

FIGURE 3.9. A sound wave passing from medium 1 to medium 2. In this case the speed of sound c_2 in medium 2 is greater than the speed of sound c_1 in medium 1.

where

$$\theta_i = \text{angle of incidence}$$

$$\theta_r = \text{angle of refraction}$$

$$c_1, c_2 = \text{speed of sound in medium 1 and medium 2, respectively}$$

Equation (3.13) should be recognized as the analogue to the Snell law for light refraction. While the analysis of refraction does not figure prominently in noise control, we cannot overlook the fact that zones of severe temperature differences do occur in the atmosphere and oceans. When sound travels from zone to zone, often across regions of severe temperature gradients, the direction of propagation changes measurably to an extent which cannot be ignored.

For example, the surface of the earth heats up more rapidly than the atmosphere on a sunny day. Due chiefly to conduction, the temperature of the air close to the ground rises correspondingly. Because the speed of sound is higher in the warmer lower layer, sound waves traveling horizontally are refracted upward. Similarly, on a clear night the earth's crust cools more quickly, and a layer of cooler air forms and bends the sound waves downward toward the surface. Thus, the noise from an industrial plant would be refracted downward at night and would seem louder to a homeowner residing near the plant than during the day (when upward refraction occurs), which is often the situation.

Nonuniform sound speed also constitutes a very important factor in underwater acoustics owing to the persistent presence of temperature and salinity and pressure gradients in the ocean. It is not unusual to find a minimum in c at some depth, usually in the order of one kilometer, with higher values above and below that strata. Interesting possibilities can occur, one of which is communication through *sound channels* in which trapped signals traveling horizontally retain their strength more effectively than if they had been able to spread in all directions. Another is the existence of *shadow zones*, where sound waves from a particular source never arrive, so they provide good places for submarines to hide.

3.9   Diffraction

In Figure 3.10, sound waves impinge upon a barrier. Some of the sound is reflected back, some continue onward unimpeded, and some of the sound bends or diffracts over the top. The barrier casts an *acoustical shadow* that is not sharply delineated. Another example of diffraction is bending of sound around a building corner. We usually can hear voices on the other side of a wall that is approximately 3 m high.

The analytical treatment of sound barriers is covered in Chapter 12, but it suffices for now to say qualitatively that sound at lower frequencies tends to diffract

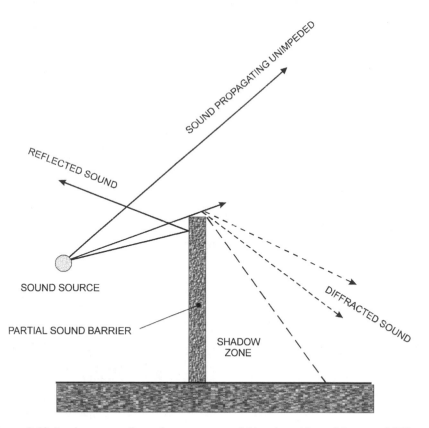

FIGURE 3.10. Impingement of sound waves on a partial barrier, with resulting sound diffraction and a shadow zone.

over partial barriers more easily than sound at higher frequencies. Moreover, the sharpness and extent of the shadow zone behind the barrier depends on the relative positions of the source and receiver. The closer the source is to the barrier, the longer is the shadow zone on the other side of the barrier, that is, the greater is the sound reduction.

3.10   Octave and One-Third Octave Bands in the Audio Range

For analytical purposes, the audio range of frequencies are divided into ten standard octave bands with center frequencies $f_C = 31.5, 63, 125, 250,$ and 500 Hz, and 1, 2, 4, 8, and 16 kHz. Each octave band center frequency f_C is double the preceding one and each bandwidth double the preceding one. The lower and upper limits of each octave band are given by

$$f_L = \frac{f_c}{\sqrt{2}}, \qquad f_U = \sqrt{2} f_C \qquad (3.14)$$

Because

$$f_C = \sqrt{f_L f_U} \qquad (3.15)$$

we see that the center frequency f_C constitutes the geometric mean of the upper band limit f_U and lower band limit f_L. The *bandwidth* **BW** for full octaves is defined by

$$\mathbf{BW} = f_U - f_L = f_C \left(\sqrt{2} - \frac{1}{\sqrt{2}} \right) = \frac{f_C}{\sqrt{2}} \qquad (3.16)$$

and thus the ratio $\mathbf{BW}/f_C$ is shown to constitute a constant. In order to avoid the use of irrational numbers the octave bands have been standardized in the field of acoustics, according to the Table 3.1.

TABLE 3.1. Octave Bands.

Lower band limit f_L	Center frequency f_C	Upper band limit f_U
22.4 Hz	31.5 Hz	45 Hz
45	63	90
90	125	180
180	250	355
355	500	710
710	1 kHz	1.4 kHz
1.4 kHz	2	2.8
2.8	4	5.6
5.6	8	11.2
11.2	16	22.4

TABLE 3.2. One-Third Octave Band.

Lower band limit f_L (Hz)	Center frequency f_C (Hz)	Upper band limit f_U (Hz)
18.0	20	24.4
22.4[a]	25	28.0
28.0	31.5[a]	35.5
35.5	40	45[a]
45[a]	50	56
56	63[a]	71
71	80	90[a]
90[a]	100	112
112	125[a]	140
140	160	180[a]
180[a]	200	224
224	250[a]	280
280	315	355[a]
355[a]	400	450
450	500[a]	560
560	630	710[a]
710[a]	800	900
900	1000[a]	1120
1120	1250	1400[a]
1400[a]	1600	1800
1800	2000[a]	2240
2240	2500	2800[a]
2800[a]	3150	3550
3550	4000[a]	4500
4500	5000	5600[a]
5600[a]	6300	7100
7100	8000[a]	9000
9000	10000	11200[a]
11200[a]	12500	14000
14000	16000[a]	18000
18000	20000	22400[a]

[a] Octave marking points.

One-third octave bands are formed by subdividing each octave band into three parts. The successive center frequencies increase in intervals by cube root of 2, and the upper and lower frequencies are related to the center frequency as follows:

$$f_L = \frac{f_C}{\sqrt[6]{2}}, \qquad f_U = \sqrt[6]{2}\,f_C, \qquad f_C = \sqrt{f_L f_U} \qquad (3.17)$$

From equations (3.16) and (3.17), the ratio $\mathbf{BW}/f_C$ is also a constant for the third-octave bands:

$$\frac{\mathbf{BW}}{f_C} = \sqrt[6]{2} - \sqrt[-6]{2}$$

Table 3.2 lists the standardized one-third octave limits and center frequencies.

While one-third octave bands generally suffice in providing adequate information, there are cases where one-tenth and even one-hundredth octaves are applied. For $1/n$th octaves, successive center frequencies are related as follows:

$$f_{n+1} = 2^{1/n} f_n \tag{3.19}$$

3.11   Root-Mean-Square Sound Pressure and the Decibel

Sound consists of small positive pressure disturbances (compression) and negative pressure disturbances (rarefaction) measured as deviations from the equilibrium or quiescent pressure value. The mean pressure deviation from equilibrium is always zero, because the mean rarefaction equals the mean compression. A simple way to measure the degree of disturbance is to square the values of the sound pressure disturbance over a period of time, thereby eliminating the countereffects of negative and positive disturbances by rendering them always positive. The mean-square sound pressure p_{rms} can be defined by

$$p_{rms} = \sqrt{\langle p \rangle^2} = \sqrt{\frac{\int_0^{\tau} p^2 \, dt}{\int_0^{\tau} dt}} \tag{3.20}$$

where τ is the time interval of measurement and p the instantaneous pressure. For a simple cosine wave over an interval of period $T = 2\pi/\omega$, there results

$$p_{rms} = \sqrt{\frac{\int_0^T p_m^2 \cos^2 k(x - ct) \, dt}{T}} = \frac{p_m}{\sqrt{2}} \tag{3.21}$$

The sound pressure as portrayed by the oscillation of the pressure above and below the atmospheric pressure is detected by normal human ear at levels as low as approximately 20 μPa (the SI unit of pressure is the *pascal*, abbreviated Pa, equivalent to 1.0 N/m²).[1] Because p_{rms} could vary over a wide range of orders of magnitude, it would be cumbersome to use it as the measure of loudness. At the threshold of pain, p_{rms} would reach approximately 40,000,000 μPa! The blastoff pressure in the vicinity of the launching pad of a Titan rocket can exceed a thousandfold the threshold of pain (i.e., 40 kPa). It is therefore more convenient to use the decibel as the folding-scale measure of loudness. This unit is defined by

$$L_p = 10 \log \left(\frac{p_{rms}}{p_0} \right)^2 = 20 \log \left(\frac{p_{rms}}{p_0} \right) \tag{3.22}$$

where

L_p = sound pressure level (dB)

log = common (base-ten) logarithm

$p_0 = 20 \times 10^{-6}$ Pa = the reference pressure

---

[1] One standard atmosphere equals 101.325 kPa.

From the context of equation (3.22) it can be established that the doubling of a mean square presure corresponds to approximately 6 dB increase in the sound pressure level. In order to determine the sound pressure level from a given value of L_p, equation (3.22) can be rewritten as

$$\frac{p_{\text{rms}}}{p_0} = 10^{L_p/20}$$

or

$$p_{\text{rms}} = 20 \times 10^{(L_p/20)-6}$$

Figure 3.11 illustrates the range of the decibel scale in terms of measured values of common sound sources. The audible range of sound, which encompasses music

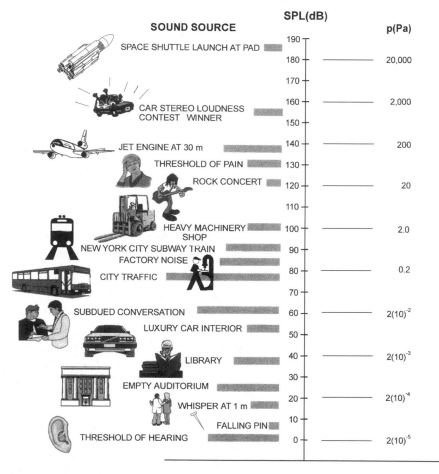

FIGURE 3.11. Sound pressure levels and corresponding pressures of various sound sources.

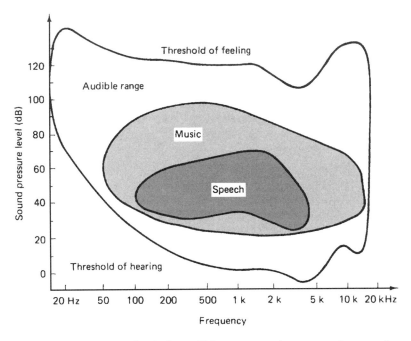

FIGURE 3.12. Sound pressure levels for audible range, music range, and range of speech versus frequency. (*Source:* Brüel & Kjær Instruments Division of Spectris Technologies, Inc.).

and speech, is shown delineated in Figure 3.12 in terms of sound pressure levels and frequencies.

3.12   Decibel Additions: Subtracting and Averaging

Most sound pressure levels do not arise from single sources, nor do they remain constant in time. Mathematical procedures must be used to add, substract and average decibels. From the definition of equation (3.22) it is apparent that decibels from single noise sources do not add or subtract directly. If we wish to add sound pressure levels (abbreviated to SPLs) $L_{p1}, L_{p2}, L_{p3}, \ldots L_{pn}$, we must obtain the antilogs of these SPLs to convert them into squares of rms pressures which can then be added directly to yield the square of total rms pressure, that in turn yields the total dB. Because

$$\left(\frac{p_i}{p_{\text{ref}}}\right)^2 = \log^{-1}\left(\frac{L_{p_i}}{10}\right)$$

the total sound pressure level L_{pt} becomes

$$L_{pt} = 10\log\left[\sum_{i=1}^{n}\left(\frac{p_i}{p_{\text{ref}}}\right)^2\right] \tag{3.23}$$

or, in terms of the sound pressure levels,

$$L_{pt} = 10 \log \left[\sum_{i=1}^{n} \log^{-1} \left(\frac{L_{pi}}{10} \right) \right] = 10 \log \left(\sum_{i=1}^{n} 10^{L_{pi}/10} \right) \qquad (3.24)$$

Example Problem 3

Find the total sound pressure level due to $L_{p1} = 96$ dB, $L_{p2} = 87$ dB and $L_{p3} = 90$ dB.

Solution

Applying equation (3.24) we have for this case

$$L_{pt} = 10 \log \left(10^{96/10} + 10^{87/10} + 10^{90/10} \right)$$
$$= 97 \text{ dB}$$

In certain situations, it is desirable to subtract an ambient or background noise from the total sound pressure level L_{pt} in order to establish the sound pressure level L_{ps} due to a particular source. Subtraction of decibels is analogous to the procedure for their addition. The total sound pressure level L_{pt} is converted into the mean-square pressure ratio, as in equation (3.23), and the background noise level L_{pB} is measured simply by turning off the noise source. The mean-square pressure ratio due to the background noise is obtained from

$$L_{pB} = 10 \log \left(\frac{p_B}{p_{\text{ref}}} \right)^2$$

or

$$\left(\frac{p_B}{p_{\text{ref}}} \right)^2 = \log^{-1} \left(\frac{L_{pB}}{10} \right) = 10^{L_{pB}/10}$$

The sound pressure level L_{ps} of the source is found from

$$L_{ps} = 10 \log \left(10^{L_{pt}/10} - 10^{L_{pB}/10} \right) \qquad (3.25)$$

Example Problem 4

Measurements indicated $L_{pt} = 93$ dB at a specific location with a lathe in operation. When the lathe is shut down the background noise measures at 85 dB. What is the sound level due to the lathe?

Solution

According to equation (3.25), the noise level due to the lathe is

$$L_{ps} = 10 \log \left(10^{93/10} - 10^{85/10} \right)$$
$$= 92 \text{ dB}$$

A requirement may arise occasionally to find the average decibels in order to determine the average sound pressure level L_p. In some situations we may wish to measure the SPL at a single location several times and determine an average value for engineering evaluation purposes. The procedure for averaging decibels is based on the same premise for the summation and subtraction of decibels, namely the application of equation (3.22). Equation (3.24) for the addition of decibels is modified by dividing the sum by n, the number of levels taken into consideration, in order to obtain the average value of L_p:

$$\bar{L}_p = 10\log\left(\frac{1}{n}\sum_{n=1}^{n}10^{\frac{L_{pi}}{10}}\right) \qquad (3.26)$$

Example Problem 5

Determine the average sound pressure level L_p for a series of measurements taken at different times: 96 dB, 88 dB, 94 dB, 102 dB, and 90 dB.

Solution

Using equation (3.26) we write

$$\bar{L}_p = 10\log\left[\frac{1}{5}\left(10^{96/10} + 10^{88/10} + 10^{94/10} + 10^{102/10} + 10^{90/10}\right)\right]$$

$$= 97 \text{ dB}$$

3.13   Weighting Curves and Associated Sound Levels

Human perception of loudness depends on the frequency of a sound. A noise having most of its energy concentrated in the middle of the audio spectrum (e.g., in the region of 1 kHz) is perceived as being louder than noise of equal energy but concentrated either in the low-frequency region (say, 40 Hz) or in the high-frequency region (near 15 kHz). This frequency effect becomes more apparent with soft sounds than the case with loud sounds, which provides the raison d'être for the presence of a *loudness control* on some audio amplifiers. This control supplies a loudness contour at low volumes, which applies greater amplification to the high- and low-frequency contents of the program material relative to the middle-frequency components.

Frequency-weighting takes typical human hearing response into account when the loudness generated by all of the audible frequency components present is to be represented by a single value. Rather than describing the sound level in each frequency band, we can use the A-weighted sound level to report the overall loudness. The A-weighted sound level is obtained from the conversion chart of Table 3.3, which also lists the B and C weightings in 1/3-octave bands. A-weighting

TABLE 3.3. Conversion of Sound Levels from Flat Response to A, B, and C Weightings.

Frequency (Hz)	A weighting (dB)	B weighting (dB)	C weighting (dB)
10	−70.4	−38.2	−14.3
12.5	−63.4	−33.2	−11.2
16	−56.7	−28.5	−8.5
20	−50.5	−24.2	−6.2
25	−44.7	−20.4	−4.4
31.5	−39.4	−17.1	−3.0
40	−34.6	−14.2	−2.0
50	−30.2	−11.6	−1.3
63	−26.2	−9.3	−0.8
80	−22.5	−7.4	−0.5
100	−19.1	−5.6	−0.3
125	−16.1	−4.2	−0.2
160	−13.4	−3.0	−0.1
200	−10.9	−2.0	0
250	−8.6	−1.3	0
315	−6.6	−0.8	0
400	−4.8	−0.5	0
500	−3.2	−0.3	0
630	−1.9	−0.1	0
800	−0.8	0	0
1000	0	0	0
1250	+0.6	0	0
1600	+1.0	0	−0.1
2000	+1.2	−0.1	−0.2
2500	+1.3	−0.2	−0.3
3150	+1.2	−0.4	−0.5
4000	+1.0	−0.7	−0.8
5000	+0.5	−1.2	−1.3
6300	−0.1	−1.9	−2.0
8000	−1.1	−2.9	−3.0
10000	−2.5	−4.3	−4.4
12500	−4.3	−6.1	−6.2
16000	−6.6	−8.4	−8.5
20000	−9.3	−11.1	−11.2

is almost exclusively used in measurements that entail human response to noise. Sound level that is measured with A-weighting is reported in terms of dB(A) or simply dBA rather than the generic decibel dB. Similarly, B-weighted and C-weighted measurements are designated in dB(B) (rarely used) and dB(C), respectively. In the conversion table, it will be noted that all of the weighting curves show an adjustment of 0 dB for the 1-kHz frequency band. The A-weighting was introduced for sound levels below 55 dB, B-weighting for levels between 55 dB and 85 dB, and C-weighting was for levels exceeding 85 dB, all of which corresponded, respectively, to human response to low, moderate, and loud sounds. C-weighting is relatively "flat" in the midrange frequencies, with less than 1 dB subtracted from actual dB levels measured in the frequency bands from 63 Hz through 4 kHz. The weighting curves are also shown graphically in Figure 3.13.

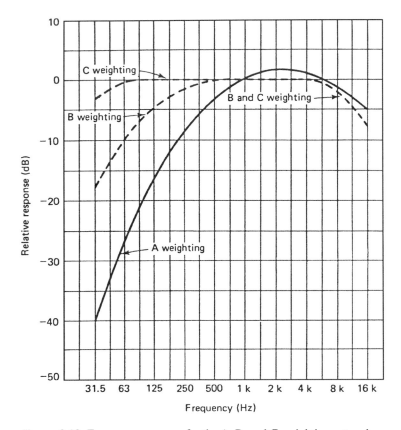

FIGURE 3.13. Frequency response for the A, B, and C weighting networks.

Example Problem 6

Find the total A-weight sound level L for the octave-band sound pressure levels given below:

Band Center Frequency (Hz)	Sound Pressure Level (dB)
31.5	73
63	68
125	72
250	68
500	80
1000	88
2000	95
4000	83
8000	97
16000	92

Solution

Use Table 3.3 to obtain the dB conversion from a flat response to dBA for each of the octave bands. This results in

$$73 \text{ dB at } 31.5 \text{ Hz} = 73 - 39.4 = 33.6 \text{ dB(A)}$$
$$68 \text{ dB at } 63 \text{ Hz} = 68 - 26.2 = 41.8 \text{ dB(A)}$$
$$72 \text{ dB at } 125 \text{ Hz} = 72 - 16.1 = 55.9 \text{ dB(A)}$$
$$68 \text{ dB at } 250 \text{ Hz} = 68 - 8.6 = 59.4 \text{ dB(A)}$$
$$80 \text{ dB at } 500 \text{ Hz} = 80 - 3.2 = 76.8 \text{ dB(A)}$$
$$88 \text{ dB at } 1 \text{ kHz} = 88 - 0 = 88 \text{ dB(A)}$$
$$95 \text{ dB at } 2 \text{ kHz} = 95 + 1.2 = 96.2 \text{ dB(A)}$$
$$83 \text{ dB at } 4 \text{ kHz} = 83 + 1.0 = 84 \text{ dB(A)}$$
$$97 \text{ dB at } 8 \text{ kHz} = 97 - 1.1 = 95.9 \text{ dB(A)}$$
$$92 \text{ dB at } 16 \text{ kHz} = 92 - 6.6 = 85.4 \text{ dB(A)}$$

The dB(A) values in each of the bands can be summed up for the total sound level L_p, through the use of equation (3.24).

3.14 Performance Indices for Environmental Noise

As the result of the passage of the Noise Control Act of 1972 by the U.S. Congress, the Environmental Protection Agency (EPA) issued two major documents published in April 1974, in accordance with Section 5 of the Act. One document deals principally with the criteria for time-varying community noise levels; and the other document is concerned with definitions of performance indices for noise levels. These indices are generally represented as single-number criteria, serving as an internationally recognized, simple means of assessing the noise environment. Three performance indices are described in this section. Because they utilize A-weighted measurements, these three statistically based methods of quantifying noise exposures tend to have good correlation with human response. These indices are L_N, which presents the levels exceeded N percent of the measurement time; L_{eq}, the equivalent continuous sound pressure level in dB(A); and L_{dn}, the day–night sound level average in dB(A).

L_N may be measured with the use of an amplitude distribution analyzer. An output of the device can provide a histogram, an example of which is shown in Figure 3.14. The time in any chosen band can be read as a percentage of the total observation time. The cumulative distribution curve in the figure indicates the probability of exceeding each range of decibel levels. In noise abatement planning, criteria are often specified in terms of sound levels that are exceeded 10%, 50%, and 90% of the time. These levels are customarily represented as L_{10}, L_{50}, and L_{90}, respectively.

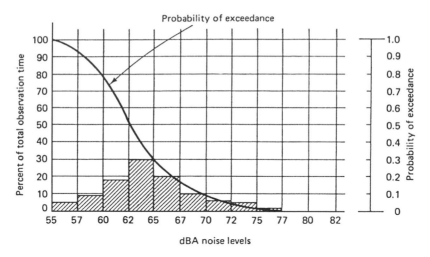

FIGURE 3.14. A histogram showing probability of exceedance for plant noise.

Example Problem 7

From examination of Figure 3.14, estimate the sound level that is exceeded 10%, 50%, and 90% of the time. Also establish the percentage of the total observation time that the sound was between 65 and 67 dB(A).

Solution

The probability-of-exceedance curve in Figure 3.14 is read to yield $\approx$69 dB(A) (exceeded 10% of the time), $L_{50} \approx 63$ dB(A) (exceeded 50% of the time), and $L_{90} \approx 58$ dB(A) (exceeded 90% of the time).

Equivalent sound level L_{eq} is the sound energy averaged over a given period of time T, that is, it is the rms or mean level of the time-varying noise. It is defined by

$$L_{eq} = 10 \log \left(\frac{1}{T} \int_0^T \frac{p^2}{p_{ref}^2} dt \right) \qquad (3.27)$$

where $p^2 = p^2(t)$ is the mean square (time-varying) sound pressure and $p_{ref} = 20 \ \mu$Pa. Equation (3.27) can be more conveniently rewritten in terms of sound level $L = L(t)$ using the relationship of equation (3.22):

$$L_{eq} = 10 \log \left(\frac{1}{T} \int_0^T 10^{L_p/10} dt \right) \qquad (3.28)$$

In order to facilitate digital processing in the measurement of L_{eq} through the use of an integrating sound level meter, the integral form of equation (3.28) is replaced by the equivalent summation:

$$L_{eq} = 10 \log \left(\frac{1}{N} \sum_{n=1}^{N} 10^{L_n/10} \right) \tag{3.29}$$

Here L_n are acquired instrumentally for each of N equal intervals to yield L_{eq}, in the course of digital processing of discrete samples.

Example Problem 8

Find L_{eq} in the case where $L_n = 90.5, 95, 103, 88,$ and 98 dB(A) are obtained as the respective average levels for five short, equal time intervals.

Solution

From equation (3.29)

$$L_{eq} = \log \left[\frac{1}{5} \left(10^{90.5/10} + 10^{95/10} + 10^{103/10} + 10^{88/10} + 10^{98/10} \right) \right]$$

$$= 97.9 \text{ dB(A)}$$

The *day–night equivalent sound pressure level*, L_{dn}, essentially a modification of L_{eq}, was conceived for the purpose of evaluating community noise problems. The modification consists of a nighttime penalty of 10 dB imposed on measurements between the hours from 10 P.M. to 7 A.M. With time t given in hours, equation (3.28) now becomes

$$L_{dn} = 10 \log \left[\frac{1}{24} \left(\int_{7\text{A.M.}}^{10\text{P.M.}} 10^{L/10} \, dt + \int_{10\text{P.M.}}^{7\text{A.M.}} 10^{(L+10)/10} \, dt \right) \right] \tag{3.30}$$

When the equivalent sound levels L_{eqd} and L_{eqn} are known for the day and night periods, respectively, the following version of equation (3.30) can be used to compute the day–night sound level:

$$L_{dn} = 10 \log \left[\frac{1}{24} \left(15 \times 10^{L_{eqd}/10} + 9 \times 10^{(L_{eqn}+10)/10} \right) \right] \tag{3.31}$$

Example Problem 9

Find L_{dn} for the situation where the daytime equivalent sound level is 82 dB(A) and the nighttime equivalent sound level is 76 dB(A).

Solution

Inserting the appropriate values in equation (3.31) we obtain

$$L_{dn} = 10 \log \left[\frac{1}{24} \left(15 \times 10^{82/10} + 9 \times 10^{(76+10)/10} \right) \right] = 84 \text{ dB(A)}$$

3.15   Particle Displacement and Velocity

Invoking equation (2.21) and inserting the wave equation solution (3.1), we obtain
the expression for the particle velocity u:

$$u = -\frac{1}{\rho} \int \frac{\partial \rho}{\partial x}\, dt = -\frac{p_m}{\rho c} \cos k(x - ct)$$

$$u = \frac{p}{\rho c} \tag{3.32}$$

where ρ = quiescent density of air = 1.18 kg/m^3 at a normal room temperature of
22°C and atmospheric pressure of 101.3 kPa, and c = speed of sound = 344 m/s.
The term ρc is the *characteristic* or *acoustic impedance* for a wave propagating in
air in a free-field condition. The value of ρc at standard conditions of temperature
and pressure is 40.7 rayls or 407 MKS rayls. The dimensional unit rayl is defined
as follows:

$$1\,\text{rayl} = 1.0\,\text{dyne-s/cm}^3$$

The particle displacement x for a cosine wave function can be found by simply
integrating equation (3.32) with respect to time:

$$x = \frac{p_m}{\rho c^2} \sin k(x - ct)$$

It is interesting to note that at 0 dB, the threshold of human hearing, the oscillation
of an air molecule covers an rms amplitude that is approximately only one-tenth
the diameter of a hydrogen atom.

The particle acceleration is obtained from the differentiation of equation (3.32)
with respect to time, and for a cosine wave function the acceleration is

$$\frac{du}{dt} = \frac{p}{\rho} \sin k(x - ct)$$

3.16   Correlated and Uncorrelated Sound

Correlated sound waves occur when they have a precise time and frequency relationship between them. An example of correlated sound waves is the output of
two identical loudspeakers located in the same plane, consisting of a pure tone
supplied by a single amplifier connected to both loudspeakers. Most of the sound
waves that we hear are generally *uncorrelated*.

Consider two sound waves that are detected at a point in space:

$$p_1 = P_1 \cos(\omega_1 t + \phi_1) \tag{3.33a}$$

$$p_2 = P_2 \cos(\omega_2 t + \phi_2) \tag{3.33b}$$

or in terms of complex exponential functions

$$p_1 = \text{Re}\left[P_1 e^{i(\omega_1 t + \phi_1)}\right] \tag{3.33c}$$

$$p_2 = \text{Re}\left[P_2 e^{i(\omega_2 t + \phi_2)}\right] \tag{3.33d}$$

where

$$p = \text{instantaneous sound pressure}$$
$$P = \text{amplitude of sound pressure}$$
$$\omega = \text{angular frequency}$$
$$\phi = \text{phase angle}$$

The instantaneous sound pressure resulting from the superimposition of the two waves is given by the sum of the two instantaneous sound pressures, and the mean-square sound pressure of the combined waves can be found from

$$p_{\text{rms}}^2 = \frac{1}{T}\int_0^T (p_1 + p_2)^2 \, dt \tag{3.34}$$

where T represents the averaging time, which should be an integer number of periods at both frequencies. In real measurements it suffices to have the averaging time cover many periods so that contributions from fractional periods become insignificant. This condition is met if $T \gg 1/f_{\text{lower}}$ where f_{lower} is the lower of the two frequencies. Inserting equations (3.33a) and (3.33b) or equations (3.33c) and (3.33d) into equation (3.34) and integrating, we obtain

$$p_{\text{rms}}^2 = \frac{P_1^2 + P_2^2}{2} = p_{\text{rms}1}^2 + p_{\text{rms}2}^2 \qquad \text{for } \omega_1 \neq \omega_2 \tag{3.35}$$

$$p_{\text{rms}}^2 = \frac{P_1^2 + P_2^2}{2} + P_1 P_2 \cos(\phi_1 - \phi_2) \qquad \text{for } \omega_1 = \omega_2 \tag{3.36}$$

Consider the case of two 4-kHz signals that are in phase at the receiving point. Each of the signals has a sound pressure level of 60 dB. From equation (3.36) for sound waves of the same frequency, the mean-square pressure is given by

$$p_{\text{rms}}^2 = \frac{P_1^2 + P_2^2}{2} + P_1 P_2 \cos 0 = 2P_1^2 = 4p_{\text{rms}}^2$$

The increase in sound pressure level as the result of adding an identical in-phase pure tone is

$$L_P - L_{P1} = 10 \log\left(\frac{p_{\text{rms}}^2}{p_{\text{ref}}^2}\right) - 10 \log\left(\frac{p_{\text{rms}1}^2}{p_{\text{ref}}^2}\right) = 10 \log\left(\frac{4 p_{\text{rms}1}^2}{p_{\text{rms}1}^2}\right)$$

$$= 10 \log 4 \approx 6 \text{ dB}$$

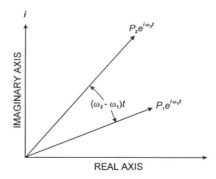

FIGURE 3.15. Complex exponentials portrayed as rotating vectors.

The combined signals result in a 4-Hz signal with an SPL of 66 dB. If the amplitudes would be out of phase by π radians (or 180°) and the frequencies are equal, the sound pressure would theoretically be zero at the observation point.

Let us now determine the effect of adding two 60-dB signals, which have frequencies of 1000 Hz and 1100 Hz, respectively. Because the frequencies are not equal, the mean-square sound pressure is double that of one wave. The sound pressure level increases by

$$10 \log \left(\frac{p_{rms}^2}{p_{rms1}^2} \right) = 10 \log 2$$

or approximately 3 dB. The combined SPL is 63 dB.

We gain a further insight into the phenomenon of beat frequency if we visualize the complex exponential functions (3.33c) and (3.33d) as rotating vectors in Figure 3.15. Without loss of generality we can define time $t = 0$ as the instant when both vectors lie along the positive real axis, resulting in the maximum sound pressure. The two vectors will be opposed along the real axis when $(\omega_2 - \omega_1)/t = 2\pi$. The envelope of the pressure–time curve yields a period $\tau = 2\pi/(\omega_2 - \omega_1) = 1/(f_2 - f_1)$. The term $(f_2 - f_1)$ is, of course, the beat frequency which has been previously discussed in Section 3.3.

3.17 Sound Intensity

An acoustic signal emanates from a point source in a spherical pattern over an increasingly larger area. When a closed surface completely surrounding the source is defined, the sound power **W** radiated by the source can be established from:

$$\mathbf{W} = \int_S I \cdot dS \qquad (3.37)$$

where

$$I = \text{sound intensity, W/m}^2$$

$$dS = \text{element of surface area, m}^2$$

$$S = \text{surface area surrounding source}$$

Here the surface integral in equation (3.37) is the integral of the sound intensity I normal to the element dS of the surface area. The integration can be executed over a spherical or hemispherical surface enclosing the source. If the source of power W is mounted on an acoustically hard surface (i.e., a surface that is totally reflective), the sound waves expand within a hemisphere. Other surfaces, such as those of a parallelopiped (representing, e.g., the walls of a room), are often used in practical applications. When the integration is performed over a spherical surface of radius r for a nondirectional source, sound intensity is related to sound power by

$$I(r) = \frac{W}{S} = \frac{W}{4\pi r^2} \tag{3.38}$$

where S denotes the area of a sphere having radius r. Equation (3.38) constitutes the inverse square law of sound propagation, which accounts for the fact that sound becomes weaker as it travels in open space away from the source, even if viscous effects of the medium are disregarded. For sound radiation within a hemisphere, with the sound source mounted at the origin above a totally reflective surface, equation (3.38) becomes:

$$I = \frac{W}{2\pi r^2}$$

Intensity, which represents the transfer of sound wave energy, equals the product of sound pressure and particle velocity,

$$I = p \cdot u \tag{3.39}$$

and for a simple cosine spherical wave, the pressure $p(r, t)$ is given as a solution to the spherical coordinate form of equation (2.25)

$$p(r, t) = \frac{A}{r} \cos k(r - ct)$$

A is a constant amplitude with its physical units in N/m. From equation (2.22) the velocity is

$$u(r, t) = -\frac{1}{\rho} \int \left[\frac{A}{r} k \sin k(r - ct) + \frac{A}{r^2} \cos k(r - ct) \right] dt$$

$$= -\frac{1}{\rho} \left[\frac{-kA}{kcr} \cos k(r - ct) - \frac{A}{r^2 kc} \sin k(r - ct) \right]$$

or

$$u(r, t) = \frac{A}{\rho c r} \cos k(r - ct) \left[1 + \frac{1}{kr} \tan k(r - ct) \right] \qquad (3.40)$$

At large values of kr equation (3.40) becomes:

$$u(r, t) \approx \frac{p(r, t)}{\rho c}, \qquad k^2 r^2 \gg 1 \qquad (3.41)$$

and for $k^2 r^2 \ll 1$:

$$u(r, t) \approx \frac{A}{k\rho c r^2} \sin k(r - ct) \approx \frac{p(r, t)}{\rho c k r} \qquad \angle 90° \text{ re } p(r, t) \qquad (3.42)$$

The difference between equations (3.41) and (3.42) connotes, respectively, the *far field* and *near field effects* of a spherical wave. As r approaches the center of the spherical source the sound pressure and particle velocity becomes progressively more out of phase, approaching 90° as the limit. In the near field the sound intensity is not simply related to the mean-square value of the sound pressure.

A sound source is generally directional, and the sound intensity does not have the same value at all points on the surface. In order to evaluate the integral of equation (3.37) it is necessary to execute an approximation by segmenting the surface into a finite number of subelements, each subtending an area S_i and to establish the sound intensity on each subelement in a direction normal to that element. A summation procedure over all of the surface subelements will yield the total sound power:

$$W = \sum_i I_i S_i \qquad (3.43)$$

where

I_i = sound intensity averaged over the ith element of area S_i, W/m^2

S_i = ith element of area, m^2

Equation (3.43) can be expressed logarithmically as

$$L_w = 10 \log \left(\frac{I}{I_0} \right) + 10 \log \left(\frac{S}{S_0} \right)$$

$$L_w = L_I + 10 \log \left(\frac{S}{S_0} \right)$$

where

L_w = sound power level, dB *re* 10^{-12} W

L_I = sound intensity level, dB *re* 10^{-12} W/m^2

S = area of surface, m^2

S_0 = reference area = 1.0 m^2

I_0 = reference sound intensity, internationally set at 10^{-12} W/m^2

3.18   The Monopole Source

A *monopole* can be described as an idealized point generating a spherical sound wave. A pulsating sphere can be considered a good approximation of a point source when its radius is small compared with the wavelength of the sound it generates. The three-dimensional equation (2.25) is expressed in spherical coordinates, without angular dependence, as

$$\frac{\partial^2 p}{\partial t^2} = c^2 \left(\frac{\partial^2 p}{\partial r^2} + \frac{2}{r} \frac{\partial p}{\partial r} \right) \tag{3.44}$$

The solution to equation (3.44) must be of the form

$$p = \frac{1}{r}[F_1(ct - r) + F_2(ct + r)]$$

with the term $F_1(ct - r)$ describing waves moving away from the source. We discard the term $F_2(ct + r)$ which describes waves traveling toward the source. With time $t = 0$ in order to eliminate a phase angle, a harmonic solution to the spherical wave equation (3.44) is given by

$$p = \frac{A}{r} \cos[k(ct - r)]$$

where A is a constant and k, the wave number equal to ω/c. The mean-square sound pressure p_{rms}^2 at a distance r from the source is given by

$$p_{rms}^2 = \frac{1}{T} \int_0^T p^2 \, dt = \frac{A^2}{r^2 T} \int_0^T \cos[k(ct - t)] \, dt = \frac{A^2}{2r^2} \tag{3.45}$$

Here $T = 1/f = 2\pi/\omega$ or the period needed to complete one cycle. If many frequencies are present, p_{rms}^2 and the root-mean-square pressure can be measured fairly accurately if the integration time is sufficiently large compared with the period of the lowest frequency. The sound power of the spherical source can be found by the use of equation (3.37) as follows:

$$W = \iint_s \mathbf{I} \cdot \mathbf{n} \, dS \tag{3.46}$$

Equation (3.46) represents the sound power of a source where

$$\mathbf{I} = \frac{1}{T} \int_0^T \rho \mathbf{u} \, dt \tag{3.47}$$

constitutes the vector sound intensity, $\mathbf{u}$ represents the particle velocity vector, S is any closed surface about the source, $\mathbf{n}$ is the unit normal to surface S, and T is the averaging time. The surface S for a spherical wave is defined by a sphere of radius r about the source, so that equation (3.46) becomes

$$W = \iint_S I_r \, dS \tag{3.48}$$

The magnitude of the sound intensity I_r is directed radially, that is, it runs parallel to unit normal **n**. When sufficiently far from the source, the sound pressure and particle velocity are in-phase. Applying equation (3.39) yields

$$I_r = p_{rms}u_{rms} = \frac{p_{rms}^2}{\rho c}$$

where ρ is the mass density of the propagation medium. With the sound source being isotropic (i.e., omnidirectional, with no angular-dependent variations), and integration of equation (3.46) over 4π steradians, the sound power is given by

$$W = I_r S = 4\pi r^2 I_r = \frac{4\pi r^2 p_{rms}^2}{\rho c} \tag{3.49}$$

If sound power is measured in half-space, that is, if the source lies on a reflective surface, then integration occurs over 2π steradians and

$$W = \frac{2\pi r^2 p_{rms}^2}{\rho c} \tag{3.50}$$

Here, W denotes the sound power of the source in watts, I is the sound intensity (W/m^2) in the direction of wave propagation, and r is the distance (m) from the center of the source.

For a spherical wave in full space, mean-square sound pressure and sound intensity in the direction of wave propagation are related to the sound power of the source by

$$p_{rms}^2 = \frac{\rho c W}{4\pi r^2} \tag{3.51a}$$

$$I = \frac{W}{4\pi r^2} \tag{3.51b}$$

The tendency of the sound intensity in equation (3.51b) to decrease with increasing distance from the source is called the *inverse square law*.

3.19   The Spherical Wave: Sound Pressure Level and Sound Intensity Level

Combining equations (3.51a) and (3.51b) with the definition of sound pressure level given by equation (3.22) we express the sound pressure L_p in terms of sound power W of the source and distance r from the source:

$$L_p = 10\log\left(\frac{p_{rms}^2}{p_{ref}^2}\right) = 10\log\left(\frac{\rho c W}{4\pi r^2 p_{ref}^2}\right)$$
$$= 10\log(\rho c W) - 20\log r + 83 \tag{3.52}$$

where $p_{ref} = 20$ μPa. The sound intensity level L_I in the direction of spherical propagation is found from

$$L_I = 10 \log \left(\frac{I}{I_{ref}} \right) = 10 \log \left(\frac{W}{4\pi r^2 p_{ref}^2} \right)$$

$$= 10 \log W - 20 \log r + 109$$

$$= L_W - 20 \log r - 11 \qquad (3.53)$$

where $I_{ref} = 10^{-12}$ W/m^2 and L_W is the sound power level (dB re 1 pW).

3.20   The Hemispherical Wave

If an omnidirectional sound power source, W is placed above an acoustically hard (i.e., totally reflective) surface, the sound waves expand within a half-space. For sound pressure and particle velocity in phase, sound intensity in the direction of propagation is given by

$$I = \frac{W}{2\pi r^2} \qquad (3.54)$$

and the sound intensity level by

$$L_I = L_W - 20 \log r - 8 \qquad (3.55)$$

A *hemianechoic chamber* is constructed by installing wedges of sound absorbing materials on the ceiling and walls of a room with an acoustically hard floor. In contrast, a full anechoic chamber is constructed with *all* of its surfaces, including the floor, lined with sound absorption wedges. A mesh floor or grating above the bottom surfacing provides structural support to equipment and laboratory personnel. With either type of chamber, a free-field condition is generated within, in which sound emitted by a source placed inside does not undergo reflection.

3.21   Energy Density

In the mathematical treatment of sound in enclosed spaces we need to know the amount of energy per unit volume being transported from a source to different parts of the room. Both kinetic energy and internal (potential) energy are involved in sound propagation. An interchange between these two forms of energy occurs from the compression/rarefaction process and the motion of the propagation medium particles. The kinetic energy density e_k is simply $\rho u^2/2$, and when the wave amplitude is small it will be fairly legitimate to assume the quiescent value of ρ_0 instead of ρ. In general, we may write an expression for kinetic energy e_k in terms of displacement vector $\mathbf{x}$:

$$e_k = \frac{\rho_0}{2} \left(\frac{\partial \mathbf{x}}{\partial t} \right)^2 \qquad (3.56)$$

The three-dimensional version of equation (2.24) is

$$\frac{1}{\rho_0}\nabla p = -\rho_0 \frac{\partial^2 \mathbf{x}}{\partial t^2}$$

which leads to

$$\frac{1}{\rho_0}\int \nabla p\, dt = \frac{\partial \mathbf{x}}{\partial t} \tag{3.57}$$

Inserting equation (3.57) into equation (3.56) yields

$$e_k = \frac{1}{2\rho_0}\left(\int \nabla p\, dt\right)^2$$

For a general case of a sine spherical wave described by

$$p = A\sin(r - ct)$$

we have

$$e_k = \frac{1}{2\rho_0 c^2}p^2 \tag{3.58}$$

From elementary thermodynamics, the change in energy per unit volume V_0 associated with the variation of density is given for a volume V of the fluid by

$$e_p = -\frac{1}{V_0}\int_V^{V_0} p\, dV \tag{3.59}$$

The negative sign indicates that the potential energy increases when compression occurs (i.e., when density increases) and decreases with rarefaction (when density decreases) under the impetus of an acoustic signal. In order to perform the integration we need to express all variables in terms of one variable, namely, instantaneous pressure p. From conservation of mass $p_0 V_0 = pV = \text{constant}$, and differentiating yields:

$$dV = -\frac{V}{\rho}d\rho \approx -\frac{V_0}{\rho_0}d\rho$$

From equation (2.23)

$$\frac{dp}{d\rho} = \gamma \frac{p_0}{\rho_0} = c^2$$

for an isentropic process in an ideal gas. Eliminating $d\rho$ between the preceding two equations gives

$$dV = -\frac{V_0}{\rho_0 c^2}dp$$

which now can be inserted into equation (3.59), which is then integrated from 0 to p to yield

$$e_p = \frac{1}{2}\frac{p^2}{\rho_0 c^2} \tag{3.60}$$

The sum of equations (3.58) and (3.60) constitutes the total *instantaneous energy density* denoted by e:

$$e = \frac{1}{2}\rho_0\left(u^2 + \frac{p^2}{\rho_0^2 c^2}\right) \tag{3.61}$$

Because the particle speed and acoustic pressure are functions of both time and space, the instantaneous energy density is not constant throughout the fluid medium. The time average E of e provides the energy density at any point in the fluid:

$$E = \frac{1}{T}\int_0^T e\, dt$$

where the time interval T represent one period of a harmonic wave. With the fact that $p = \pm\rho_0 cu$, as manifested in equation (3.32), equation (3.61) becomes

$$e = \rho_0 u^2 = pu/c$$

If we now let p and u represent the amplitude of the pressure and particle velocity, respectively, then the time-averaged energy E is written as

$$E = \frac{1}{2}\frac{pu}{c} = \frac{p^2}{2\rho_0 c^2} = \frac{1}{2}\rho_0 u^2 \tag{3.62}$$

For the cases of spherical or cylindrical waves or standing waves in a room, the pressure and particle velocity in equation (3.61) must be the real quantities derived from the superposition of all waves present. In these more complex cases, the pressure is not necessarily in phase with the particle speed, nor is the energy density given by $pu/2c$. But $E = pu/2c$ does constitute a good approximation for progressive waves if the surfaces of constant phase approach a radius of curvature much greater than a wavelength. This situation occurs for spherical and cylindrical waves at distances considerably far (i.e., many wavelengths) from their sources.

References

Federal Register, May 28, 1975. v. 40: 23105.
Federal Register, Oct. 29, 1974. v. 39: 38208.
Federal Register, Oct. 30, 1974. v. 39: p. 38338.
Harris, Cyril M. (ed.). 1991. *Handbook of Acoustical Measurements and Noise Control*, 3rd ed. New York: McGraw-Hill; Chapters 1 and 2. (A most valuable reference manual for the practicing acoustician.)
Lord, Harold, William S. Gatley, and H. A. Evensen. 1980. *Noise Control for Engineers*. New York: McGraw-Hill; pp. 7–30.
Morse, Philip M., and K. Uno Ingard, K. Uno. 1968. *Theoretical Acoustics*. New York: McGraw-Hill; Chapter 1. (A classic text which probably contains more details than necessary, with the result that the mathematics tends to obscure the physics of acoustics.)
Pierce, Alan D. 1981. *Acoustics: An Introduction to Its Physical Principles and Applications*. New York: McGraw-Hill; Chapters 1 and 2. (An excellent modern text by a contemporary acoustician.)

Problems

1. A signal consists of the following components:

$$y_1 = 10 \sin 4t, \qquad y_2 = 6 \sin 8t, \qquad y_3 = 4.3 \sin 10t$$

   Plot each component and add them up to obtain a composite wave.

2. A 300-Hz sound wave is propagating axially in a steel bar. Find the wavelength and the wave number.

3. Determine the wave number and the wavelength of a 30 Hz pure tone at 20°C.

4. Two sine waves have approximately the same amplitudes, but one is at a frequency of 135.3 Hz and the other at 136.0 Hz. What is the beat frequency? What is the time duration between the beats?

5. Show mathematically how noise cancellation can be effected by duplicating an offending signal and changing its phasing. In real-life situation, can the cancellation be a total one? If not, why not?

6. It is desired to place a worker at a "quieter" location near a machine that puts out a steady 400-Hz hum. Where would this location be, and where are the points where the noise would be greater?

7. A train whistle is measured with its frequency at 250 Hz when approaching the observer near the tracks at the rate of 125 km/h. Predict the frequency when the train is pulling away from the observer.

8. Train A traveling 60 km/h is approaching Train B traveling on a parallel track at 85 km per hour. Train A blows its whistle which has a fundamental of 255 Hz. What frequency will the engine man at Train B hear? What will be the perceived frequency after the whistle on Train B passes the locomotive of Train A? Neglect the distance between the tracks.

9. An observer stands 150 km from a railroad track. A train is 300 km from the normal from the track to the observer. It is traveling at 80 km/h and approaching this normal. Its whistle emits a fundamental of 300 Hz. What will be the frequency of this signal perceived by the observer?

10. Develop an equation for the rate of change of frequency with respect to an observer subtending an angle θ with a source's direction. The source is moving at a velocity V.

11. A boundary exists between two mediums, A and B, through which an acoustic signal travels. The signal traveling in Medium A impinges the boundary at an angle of 45°. In passing into Medium B on the other side of the boundary the signal refracts at an angle of 55°. The velocity of sound in Medium A is 450 m/s. Determine the velocity of sound in Medium B on the other side of the boundary.

12. Given a 150-Hz signal expressed as

$$p(x, t) = 35 \sin 2.5(x - 344t)$$

   determine

(a) the wave number

(b) the wave length

(c) the root mean square pressure.

13. Convert the following rms pressures into decibels:

(a) 20 μPa

(b) 150 μ

(c) 1 kPa

(d) 50 kPa

14. Convert the following values expressed in dB into rms pressure.

(a) 20 dB

(b) 60 dB

(c) 90 dB

(d) 130 dB

15. Two machines are running. One machine puts out 95 dB and the other 98 dB. What will be the combined sound pressure level in decibels?

16. Three machines are operating simultaneously. Their combined noise level is 115 dB. One machine is shut down and noise level drops to 110 dB. The remaining two machines have identical noise output. What is the noise output of these two machines?

17. An octave-band analysis of a machine yields the following results:

Band center frequency	SPL (Hz)
31.5	72
63	76
125	77
250	72
500	69
1000	84
2000	92
4000	83
8000	80
16000	78

Find the total A-weighted sound level, the total B-weighted sound level and the total C-weighted sound level.

18. The following SPL readings were taken of a noisy electric generator:

Amount of time, s	Unweighted SPL reading, dB
15	73.4
22	79.4
20	88.9
12	91.9

Find the equivalent sound pressure level L_{eq}.

19. At a property line the noise from a nearby machine shop was found over a 24-hr period to have the following averaged sound pressure levels.

Time	Noise level, dB(A)
7 A.M.–12 noon	87.5
12 noon–4 P.M.	84.6
4 P.M.–9 P.M.	78.5
9 P.M.–3 A.M.	76.5
3 A.M.–7 A.M.	77.4

Determine the L_{eq} and L_{dn}.

20. Determine the particle velocity for air at 1 atm and 22°C.

4
Vibrating Strings

4.1 Introduction

In dealing with vibrating systems it is commonly assumed that the entire mass of the system is concentrated in a single point and that the motion of the system can be described by giving the displacement as a function of time. This rather simplified approach yields approximations rather than accurate closed-form solutions. A spring, for example, certainly does not concentrate its mass at one end; nor can a loudspeaker be accurately depicted as being a massless piston engaged in an oscillating motion. The loudspeaker diaphragm consists of a considerable portion of its mass spread out over its surface, and each part of the diaphragm can vibrate with a motion that is different from those of other segments.

The vibrational modes of a loudspeaker constitute a complex affair, so it would behoove us to study simpler modes of vibration, say those of a vibrating string or bar, so that we can readily visualize the transverse vibrations. Even in the simplest of cases, certain simplifying assumptions have to be made which cannot be fully justified in the real physical world.

4.2 The Vibrating String:
Basic Assumptions

Consider a long, heavy string stretched to a moderate tension between two rigid supports. A momentary force is applied to the string which becomes displaced from its equilibrium position. The displacement does not remain in the initial position; it breaks up into two separate disturbances that propagate along the string apart from each other as shown in Figure 4.1. The propagation velocity of all *small* displacements depends only on the mass and tension of the string, not on the shape and amplitude of the initial displacement. The wave generated by such a transverse perturbation is generally known as a *transverse wave*.

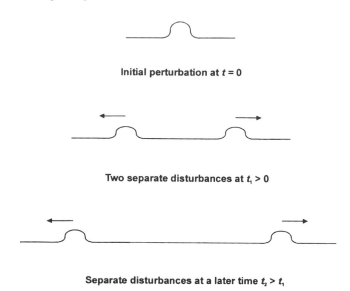

Initial perturbation at $t = 0$

Two separate disturbances at $t_1 > 0$

Separate disturbances at a later time $t_2 > t_1$

FIGURE 4.1. History of the propagation of a disturbance in a stretched string.

4.3   Derivation of the Transverse Wave Equation

In Figure 4.2, a portion of a string under tension T and rigidly clamped at its ends is shown. The string has negligible stiffness and a uniform linear density δ. Dissipation of vibrational energy is neglected. We let x represent the coordinate of a point along the horizontal distance with the origin at the left clamp of the string.

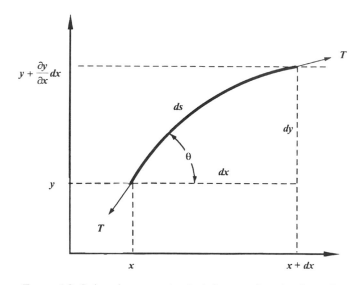

FIGURE 4.2. String element under the influence of tension force T.

The y-coordinate represents the transverse displacement from the equilibrium position. As the transverse displacements are defined as being small, tension T can be considered nearly constant ($T \cos \theta$ is even more so). Let θ denote the angle between a tangent to the string and the x-axis. In the segment of the string shown in Figure 4.2, the difference between the y-components of the tension at the two ends of element ds is the net transverse force given by

$$\Delta F_y = (T \sin \theta)_{x+\Delta x} - (T \sin \theta)_x \tag{4.1}$$

Here $(T \sin \theta)_{x+\Delta x}$ is the value of $T \sin \theta$ at $x + \Delta x$, and $(T \sin \theta)_x$ is the value at x. Letting $\Delta x \to dx$ and applying the Taylor's series expansion

$$f(x + dx) = f(x) + \frac{\partial f(x)}{\partial x} dx$$

equation (4.1) can be rewritten as

$$dF_y = (T \sin \theta)_x + \frac{\partial (T \sin \theta)}{\partial x} - (T \sin \theta)_x = \frac{\partial (T \sin \theta)}{\partial x} dx$$

As the displacement y is assumed to be small, θ will be correspondingly small and the relationship $\sin \theta \approx \tan \theta$ applies, with $\tan \theta$ equal to y/x. The net transverse force on the element ds then becomes

$$dF_y = \frac{\partial [T(\partial y/\partial x)]}{\partial x} dx = T \frac{\partial^2 y}{\partial x^2} dx$$

The mass of the string element is $\delta\, dx$. Applying Newton's law $F = ma$, we get:

$$T \frac{\partial^2 y}{\partial x^2} dx = \delta\, dx \frac{\partial^2 y}{\partial t^2}$$

Setting

$$c = \sqrt{\frac{T}{\delta}} \tag{4.2}$$

the equation of string motion becomes

$$\frac{\partial^2 y}{\partial t^2} = c^2 \frac{\partial^2 y}{\partial x^2} \tag{4.3}$$

The constant c defined in equation (4.2) represents the propagation velocity of the transverse wave. Equation (4.3) is the wave equation representing the wave disturbances propagated along the string. This equation was first derived by Leonhard Euler in 1748.

4.4   General Solution of the Wave Equation

The second-order partial differential equation (4.3) has the general solution

$$y = f(ct - x) + g(ct + x) \tag{4.4}$$

where the functions, $f(ct - x)$ and $g(ct + x)$, are arbitrary with arguments $(ct \pm x)$. The first term of the right side of equation (4.4) represents a wave moving to the right (in the positive x-direction) and the second term a wave moving to the left (in the negative x-direction). While each of the two wave shapes remains constant as the initial perturbation propagates along the string, the actual fact is quite the opposite because the simplifying assumptions are not fully realized in real strings. In relatively flexible strings with low damping, the rate of wave distortion is quite minimal as long as the initial perturbation is kept small. Large amplitudes, however, will result in a larger rate of change of the wave shapes.

The functions $f(ct - x)$ and $g(ct + x)$ cannot be freely arbitrary; they are constrained by initial and boundary conditions. The initial conditions, established at time $t = 0$, are dictated by the type and the location of application of the perturbing force applied to the string. To cite a musical example, the initial wave shape generated by plucking the string of a banjo or a harp will be quite different from the wave shape created by bowing a violin string. The boundary conditions extant at the ends of a string further limit the wave function. Real strings always have finite lengths and are fixed in some fashion at their ends. The displacement sum $y = f + g$ of equation (4.4) is constrained to have a zero value at all times at the clamping points. Also, when a string is sustained in a steady-state condition by periodic external driving forces, the functions f and g will also have the same frequency as the applied forces but the amplitudes of vibration are determined by the point of application of the force and by boundary condition at the ends of the string.

Example 1: String Clamped at Both Ends. Given a string of length l clamped rigidly at $x = 0$ and $x = L$. The solutions $y_1 = f(ct - x)$ and $y_2 = g(ct + x)$ are no longer arbitrary; and their sum must be zero at all times, that is,

$$f(ct - 0) + g(ct + 0) = 0$$

or

$$f(ct - 0) = -g(ct + 0) \tag{4.5}$$

The two functions must be of the same form, but with opposite signs. We can now rewrite equation (4.5) as

$$y(x, t) = f(ct - x) - f(ct + x) \tag{4.6}$$

The first term on the right of equation (4.6) represents a wave traveling to the right (in the positive x-direction) and the second term a wave moving leftward (in the negative x-direction).

4.5   Reflection of Waves at Boundaries

The reflection process at the boundary $x = 0$ can be viewed as one in which a second wave does not pass the boundary point, but is considered to reflect back, generating a similarly shaped wave of opposite displacement traveling in the positive x-direction. The presence of a fixed point at $x = L$ results in another reflection. In this case, the wave traveling in the positive x-direction reflects back as a similar wave of opposite displacement moving in the negative x-direction. The major result of these two reflections is that the motion of the free vibration becomes periodic. A pulse leaving $x = 0$ reaches $x = L$ after an interval of L/c seconds. There, it is reflected and returns to the origin where it again undergoes a reflection after a time lapse of $2L/c$ seconds. The shape of the pulse after its second reflection is identical to that of the original pulse. This periodicity has resulted from the specified boundary conditions, that is, fixed points at $x = 0$ and $x = L$.

4.6 Simple Harmonic Solutions of the Wave Equation

Simple harmonic vibrations frequently occur in nature, and we shall now consider a simple harmonic motion (SHM) propagating along a string. Any vibration of the string, however complex, can be resolved into an equivalent array of simple harmonic vibrations. This resolution of complex vibration into a series of SHMs is not a mere mathematical exercise, but constitutes the phenomenal principle of how the ear functions. The ear breaks down a complex sound into its simple harmonic components. This capability permits us to distinguish the differences between different voices and musical instruments. A piano sounding a note will sound differently from the same note played by an oboe. If all the frequencies present in the sound consist of a fundamental tone plus its harmonics, they will sound more harmonious than in the situation where the frequencies are not related so simply to each other.

The displacement of any point on the string exciting a SHM of angular frequency ω can be depicted by the special solution to equation (4.3):

$$y = a_1 \sin(\omega t - kx) + a_2 \sin(\omega t + kx) + b_1 \cos(\omega t - kx) + b_2 \cos(\omega t + kx)$$

$$(4.7)$$

where a_1, a_2, b_1, b_2 are arbitrary constants, and k is the wavelength constant given by

$$k \equiv \omega/c$$

Applying the boundary condition $y(0, t) = 0$ (i.e., $y = 0$ at $x = 0$), which describes a fixed point, equation (4.7) reduces to

$$(a_1 + a_2) \sin \omega t = -(b_1 + b_2) \cos \omega t \qquad (4.8)$$

Because this equation applies to all values of t, the following relations between the constants must exist

$$a_1 + a_2 = 0, \quad b_1 + b_2 = 0 \quad \text{or} \quad a_1 = -a_2, \quad b_1 = -b_2$$

The two limitations for the arbitrary constants of equation (4.7) are equivalent to the single restriction of equation (4.5) in the general solution of wave equation

(4.3). The two waves must be of equal and opposite displacements and must therefore differ in phase by π radians at $x = 0$. With these restrictions equation (4.7) becomes

$$y = a_1[\sin(\omega t - kx) - \sin(\omega t + kx)] + b_1[\cos(\omega t - kx)$$
$$- \cos(\omega t + kx)] \tag{4.9}$$

Making use of trigonometric transformations for the sine and cosine terms we simplify equation (4.9) to

$$y = [-2a_1 \cos \omega t + 2b_1 \sin \omega t] \sin kx$$

Thus y is expressed as a product of a time-dependent term and a coordinate dependent term.

Applying the boundary condition of $y(L, t) = 0$ (i.e., the displacement is zero at end point $x = L$) adds yet another restriction

$$\sin kL = n\pi$$

where $n = 1, 2, 3 \ldots$. Consequently, the string cannot vibrate freely at any random frequency; it can only vibrate with a discrete set of frequencies given by

$$\omega_n = n\pi c/L$$

where $n = 1, 2, 3, \ldots$, or, in terms of frequency,

$$f_n = nc/(2L) \tag{4.10}$$

4.7   Standing Waves

The boundary conditions at $x = 0$ and at $x = L$ reduce the general SHM solution equation (4.7) to a pattern of standing waves on the string. At the lowest or *fundamental* frequency, where $n = 1$, the displacement is given by

$$y_1 = (A_1 \cos \omega t + B_1 \sin \omega t) \sin k_1 x \tag{4.11}$$

Here $k_1 = \pi/L$, and A_1 and B_1 are arbitrary constants of which numerical values are established by the initial conditions, that is, the type of excitation imparted to the string at $t = 0$. This *fundamental mode* of vibration is associated with the fundamental (or *first harmonic*) frequency $f_1 = c/2L$. The nth mode of vibration corresponding to the nth harmonic frequency is represented by

$$y_n = (A_n \cos \omega_n t + B_n \sin \omega t) \sin k_n x \tag{4.12}$$

and the frequency is $f_n = nc/2L$, that is, n times the fundamental frequency. The constants A_n and B_n are determined by the initial excitation.

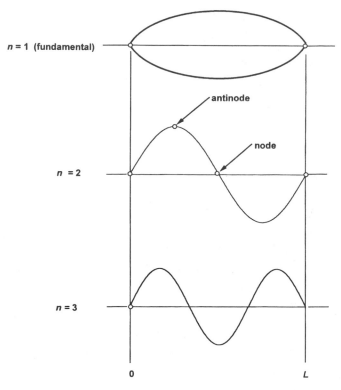

FIGURE 4.3. Different modes of vibration for a string for the fundamental and the first two harmonics.

In evaluating the term $\sin k_n x = \sin n\pi x/L$ we recognize that displacement $y_n = 0$ occurs for all values of x when $\sin n\pi x/L = 0$, that is,

$$n\pi x/L = m\pi$$

where $m = 0, 1, 2, 3 \ldots$. The cases for which $m = 0$ and $m = n$ corresponds to the boundary conditions at the fixed points at the two ends of the string. However, there are additional $(n - 1)$ locations, called *nodal points* or *nodes*, where the displacement produced by the nth harmonic mode of vibration remains at zero, as illustrated in Figure 4.3. This situation may be viewed as one in which the harmonic wave moving in the positive x-direction cancels precisely at the nodal points at all times t, the harmonic wave moving in the opposite direction. Because the points of zero displacement remain fixed, the resultant wave pattern constitutes what is known as *standing waves*. The distance between nodal points for the nth harmonic mode of operation is L/n, and the points of maximum vibrational amplitudes are referred to as *antinodes* or *loops*.

Let us take a "snapshot" of the vibrating string at a particular time for the sixth harmonic mode (Figure 4.4). The displacement of the string frozen in time occurs

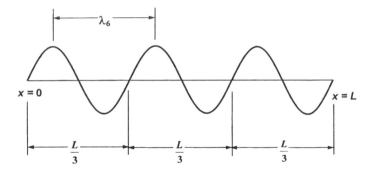

FIGURE 4.4. The sixth harmonic mode of a vibrating string stretched between $x = 0$ and $x = L$.

as a sinusoidal function of x. This function repeats itself every length $2L/n$ of the x-coordinate, which, in turn, is equal to the wavelength λ_n of the harmonic waves. In this case of the sixth harmonic mode the repetition occurs every $L/3$ of the string length. The wavelength is related to the velocity of propagation by

$$\lambda_n = c/f_n$$

where f_n is the frequency of vibration in the nth mode and the wavelength constant by $k_n = 2\pi/\lambda_n$. From equation (4.10) we derive

$$\lambda_n = 2L/n \tag{4.13}$$

that is, the wave length is twice the nodal distance of the associated wave pattern.

4.8   The Effect of Initial Conditions

The complete general solution to the general harmonic wave equation for a freely vibrating string rigidly clamped at its ends contains all the individual modes of vibration described by equation (4.12). It is expressed as

$$y_n = \sum_{n=1}^{\infty} (A_n \cos \omega_{nt} + B_n \sin \omega_n t) \sin k_n t \tag{4.14}$$

where A_n, B_n are amplitude coefficients dependent on the method of exciting the string to vibrate. The actual amplitude of the nth mode is

$$a_n = \sqrt{A_n^2 + B_n^2}$$

Consider the initial condition at $t = 0$ at the time when the string is displaced from its normal linear configuration so that the displacement $y(x, t)$ at each point of the string is given by the function

$$y(x, 0) = y_0(x)$$

The corresponding velocity $v(x, t)$ is given for $t = 0$ by

$$v(x, 0) = \frac{\partial y(x, 0)}{\partial t} = v_0(x)$$

In order that equation (4.14) represents the string at all times, it also must describe the displacement at $t = 0$ and therefore is written as

$$y_0(x) = y(x, 0) = \sum_{n=1}^{\infty} A_n \sin k_n x \qquad (4.15)$$

The derivative of y with respect to time must also represent the velocity at $t = 0$:

$$v_0(x) = v(x, 0) = \sum_{n=1}^{\infty} \omega_n B_n \sin k_n x \qquad (4.16)$$

We apply Fourier's theorem[1] to equation (4.14) in order to obtain

$$A_n = \frac{2}{t} \int_0^t y_0(x) \sin k_n \, dx$$

and then apply the theorem to equation (4.16) to get

$$B_n = \frac{2}{\omega_n t} \int_0^t v_0(x) \sin k_n x \, dx \qquad (4.17)$$

Example 2: String Pulled and Suddenly Released. Consider a string that is plucked by pulling it at its center a distance d and then is suddenly released at instant $t = 0$. In such a case, $v_0(x) = 0$ and all the coefficients B_n will be zero. The coefficients A_n are given by

$$A_n = \frac{2}{l} \left[\int_0^{\frac{L}{2}} \frac{2 \, dx}{l} \sin k_n x \, dx + \int_{\frac{L}{2}}^{L} 2\frac{d}{l}(l - x) \sin k_n \, dx \right]$$

$$= \frac{8d}{n^2 \pi^2} \sin \frac{n\pi}{2} \qquad (4.18)$$

---

[1] The theorem states that a complex vibration of period T can be represented by a displacement $x = f(t)$ written in terms of a harmonic series

$$x = f(t) + A_0 + A_1 \cos \omega t + A_2 \cos 2\omega t + \cdots + A_n \cos n\omega t + \cdots + B_1 \sin \omega t$$
$$+ B_2 \sin 2\omega t + \cdots B_n \sin n\omega t + \cdots$$

where $\omega = 2\pi/T$ and the constants are given by

$$A_0 = \frac{1}{T} \int_0^T f(t) \, dt$$

$$A_n = \frac{2}{T} \int_0^T f(t) \cos n\omega \, dt$$

$$B_n = \frac{2}{T} \int_0^T f(t) \sin n\omega \, dt$$

Therefore, all even modes $n = 0, 2, 4, \ldots$ have

$$A_2 = A_4 = A_6 = \cdots = 0$$

and the odd modes result in nonzero A_n coefficients:

$$A_1 = \frac{8d}{n^2}, \quad A_3 = -\frac{8d}{9\pi^2}, \quad A_5 = \frac{8d}{25\pi^2}, \quad \text{etc.} \tag{4.19}$$

The amplitudes of the various harmonic modes are given by the numerical values of A_n. In general, it may be observed that no harmonics are generated having a node at the point of the string initially plucked. As the nodal number n increases, the associated amplitudes decrease from the value of the fundamental amplitude, that is, the fundamental A_1 is 9 times larger than A_3 and 25 times larger than A_5, and so on.

Example 3: Sharp Blow Applied to String. If the string is struck a sharp blow (as opposed to being plucked, as described above), $v_0(x, 0)$ has nonzero values but no initial displacement exists. Then all the coefficients A_n are zero and the coefficients B_n are given by equation (4.17). A common example of a struck string is the impact of a piano hammer striking a string. It is interesting to note that pianos are designed so that the impact point of the hammer is one-seventh of the distance from one end of the string, thus eliminating the seventh harmonic (which would have produced a discordant sound).

4.9   Energy of Vibrating String

In any nondissipative system the total energy content remains constant, equal to the value of the maximum kinetic energy. For the nth mode of vibration the maximum value of the kinetic energy of a segment of length dx is

$$dE_n = \frac{\omega_n \delta}{2}\left(A_n^2 + B_n^2\right) \sin^2 k_n x \, dx \tag{4.20}$$

which is established by simply applying the relation $dE = (mv)\,dv/2$ in conjunction with equation (4.6) and the fact that the mass of the string element is given by

$$m = \delta \, dx$$

where δ is the linear density. With integration over the variable x from 0 to L, the maximum kinetic energy of the string is

$$E_n = \frac{\omega_n^2 \delta}{2}\left(A_n^2 + B_n^2\right)\frac{L}{2} = \frac{m}{4}\omega_n^2\left(A_n^2 + B_n^2\right)$$

Here m is the total mass of the string and $(A_n^2 + B_n^2)$ is the square of the maximum displacement of the nth harmonic. In a conservative system (which describes the dissipationless vibrating string) the maximum potential energy is also equal to the maximum kinetic energy of the system. From the relation of equation (4.20) the energy of the nth mode of vibration is

$$E_n = \frac{m}{4}\omega_n^2 A_n^2 = \frac{m}{4}\left(\frac{n\pi c}{L}\right)^2\left(\frac{8d}{n^2\pi^2}\right)^2 = \frac{16md^2c^2}{n^2\pi^2 L^2}$$

It becomes apparent that as n increases, the energy of the nth mode lessens. For example, the energy of the third harmonic is one-ninth that of the fundamental mode.

4.10   Forced Vibrations in an Infinite String

While this may appear to be a purely academic exercise, the simple case of a transverse sinusoidal force on an idealized string of infinite length can provide insight into the forced vibrations of finite strings and transmission of acoustic waves.

An ideal string of infinite length subject to a tension T receives a transverse driving force $T\cos\omega t$ at the string end $x = 0$. The end at $x = \infty$ is rigidly clamped but the point $x = 0$ is rigid only in the x-direction, being free to move in the y-direction, a support which can be approximated by a pivoted lever as shown in Figure 4.5. Thus, the driving force can move the lever as well as the string. We shall neglect the mechanical impedance of the pivoted lever (or hinge) which is deemed to have no friction and no stiffness. No waves are reflected from the far end $x = \infty$, and hence no waves travel in the negative x-direction. The displacement of the string can now be described by the general solution containing only the expression for a harmonic wave traveling in the positive x-direction:

$$y = a_1 \sin(\omega t - kx) + b_1 \cos(\omega t - kx)$$

or in complex format:

$$\mathbf{y} = \mathbf{A}e^{i(\omega t - kx)} \tag{4.21}$$

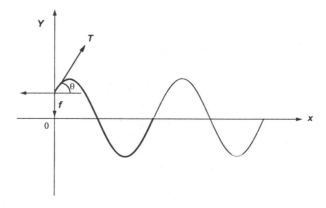

FIGURE 4.5. Forces acting at one end of an infinite string.

where $\mathbf{A}$ is a complex constant of which magnitude equals the displacement amplitude of the wave motion and whose phase angle renders the difference in phase between the motion of the string and the driving force.

In complex format the harmonic driving force can be written as

$$\mathbf{f} = \mathbf{D}e^{i\omega t} \qquad (4.22)$$

In Figure 4.5 the driving force is shown being applied to the string at an angle θ the string makes with the horizontal. This angle is given by

$$\tan\theta = \left(\frac{\partial y}{\partial x}\right)_{x=0}$$

The force exerted in the horizontal direction at the support at the end of the string is $-T\cos\theta$. Because the displacements are assumed small, $\cos\theta \approx 1$, and the magnitude of this force in the horizontal direction is essentially equal to tension T in the string. From similar considerations, the transverse force exerted by the support on the string is $-T\sin\theta$, approximated by

$$\mathbf{f} = -T\sin\theta = -T\left(\frac{\partial y}{\partial x}\right)_{x=0} \qquad (4.23)$$

Equation (4.23) indicates that for any applied transverse force the shape of the string at $\mathbf{x} = 0$ will vary. Inserting $\mathbf{f}$ and $\mathbf{y}$ from equations (4.21) and (4.22) into equation (4.23) yields for this boundary condition

$$\mathbf{F}e^{i\omega t} = -T(-ik)\mathbf{A}e^{i[\omega t - k(0)]}$$

or

$$\mathbf{A} = \frac{\mathbf{F}}{ikT} \qquad (4.24)$$

The term $\mathbf{F}/kT$ represents the magnitude of this complex amplitude $\mathbf{A}$. Inserting equation (4.24) into equation (4.21) and then differentiating with respect to time results in the complex velocity $\mathbf{v}$

$$\mathbf{v} = \mathbf{F}\left(\frac{c}{T}\right)e^{i(\omega t - kx)}$$

The mechanical (or wave) impedance $\mathbf{Z}_s$ of the string is defined as the ratio of the driving force to the transverse velocity of the string at $x = 0$:

$$\mathbf{Z}_s = \frac{T}{c} - \sqrt{T\delta} = \delta c$$

It turns out that $\mathbf{Z}_s$ is a real quantity, with no imaginary load. The mechanical load presented by the string to the driving force is purely resistance. The input impedance exists as a function of the linear density δ and the tension applied to the string, and it does *not* depend on the applied driving force; this means it is a property characteristic of the string, not the wave propagation in the string. The input, often termed *characteristic* or *mechanical impedance* (or *resistance*) is analogous to the characteristic electrical impedance of an infinite transmission line.

The average power input to the string is found from the average value of the instantaneous power $\mathbf{W} = fv$ evaluated at $x = 0$, or

$$W = \frac{F^2}{2\delta c} = \frac{\delta c V_0^2}{2}$$

where V_0 is the velocity amplitude at $x = 0$.

4.11   Strings of Finite Lengths: Forced Vibrations

In the case of a finite string, the reflections from the far end generate frequencies that cause the input impedance to change greatly with the frequency of the driving force. If the support at finite $x = L$ is fully rigid and no dissipative forces occur in the string, the input impedance becomes a pure reactance and no power is consumed in the string.

Because the complex expression for transverse waves on a finite string now needs a term descriptive of the reflected wave, equation (4.20) needs to be rewritten as:

$$\mathbf{y} = \mathbf{A}e^{i(\omega t - kx)} + \mathbf{B}e^{i(\omega t + kt)} \tag{4.25}$$

for all times t. The boundary condition for $x = 0$ is

$$Fe^{i\omega t} = -T\left(\frac{\partial y}{\partial x}\right)_{x=0} \tag{4.26}$$

for all values of t. Inserting equation (4.25) into equation (4.26) yields

$$F = -T(-ik\mathbf{A} + ik\mathbf{B}) \tag{4.27}$$

Applying $\mathbf{y}(L, t) = 0$ for the rigid clamp at $x = 0$, equation (4.25) becomes

$$0 = \mathbf{A}e^{-ikt} + \mathbf{B}e^{ikt} \tag{4.28}$$

Solving the equations (4.27) and (4.28) for $\mathbf{A}$ and $\mathbf{B}$ results in

$$\mathbf{A} = \frac{F}{ikT} \cdot \frac{e^{ikt}}{e^{ikt} + e^{-ikt}} = \frac{Fe^{ikt}}{2ikT \cos kL}$$

and

$$\mathbf{B} = -\frac{F}{ikT} \cdot \frac{e^{-ikT}}{e^{ikT} + e^{-ikT}} = -\frac{Fe^{ikT}}{2ikT \cos kL}$$

Substituting the above constants into equation (4.26) yields:

$$\mathbf{y} = \frac{Fe^{i\omega t}}{ikT} \cdot \frac{e^{ik(L-x)} - e^{-ik(L-x)}}{2 \cos kL} = \frac{Fe^{i\omega t}}{ikT} \cdot \frac{\sin kL(L - x)}{\cos kL} \tag{4.29}$$

The real portion of equation (4.29) graphs a pattern of standing waves on the string with nodes occurring at those points where $\sin(L - x) = 0$, in addition to

$y(0,\ L) = 0$. The displacement at $x = 0$, however, has an amplitude

$$y_0 = (F \tan kL)/kT \tag{4.30}$$

Singularities of equation (4.30) occur when $\cos kL = 0$, that is,

$$kL = \frac{\omega L}{c} = \frac{(2n-1)\pi}{2}, \qquad n = 1, 2, 3, \ldots$$

or

$$\omega_n = \frac{(2n-1)\pi c}{2L}$$

and

$$f_n = \frac{(2n-1)}{4L} c$$

These singularities connote infinite amplitudes which, of course, do not occur in real strings, because dissipative forces neglected in the foregoing analysis do actually exist. But the amplitudes do achieve maximum values at these frequencies. In a similar fashion we can ascertain the minimum amplitudes from the condition $\cos kL = \pm 1$, that is,

$$kL = \omega_n L/c, \qquad n = 1, 2, 3, \ldots$$

and

$$\omega_n = \frac{n\pi c}{L} \quad \text{or} \quad f_n = \frac{nc}{2L}$$

The minimum amplitudes decrease progressively with increasing frequencies. In fact, in comparing with equation (4.10) it is noted that the frequencies of minimum amplitudes are identical to those of free string vibration, and the term *antiresonance* has been applied to describe these frequencies. Differentiation of equation (4.29) with respect to time t yields the complex velocity v of the string

$$v = \frac{Fe^{i\omega t}}{T/c} \cdot \frac{\sin k(L-x)}{\cos kL}$$

The input mechanical impedance then becomes

$$\mathbf{Z}_s = \frac{fe^{i\omega t}}{v} = \frac{T \cos kL}{ic \sin kL} = -i\delta c \cot kL$$

which exists as a pure reactance, with no power absorbed by the string. The amplitude of vibration is a maximum at $\cot kL = 0$, which occurs at the frequency given by $f_n = nc/2L$. For extremely low frequencies input impedance has the limits

$$\mathbf{Z}_s = -\frac{i\delta c}{kL} = -\frac{iT}{\omega L}$$

4.12   Real Strings: Free Vibration

Real strings manifest some degree of stiffness, causing the observed frequencies to be higher than the theoretical values for idealized strings. This results from the presence of elastic boundary forces augmenting the action of tensile forces previously considered, with the net effect of increased restoring forces. The presence of stiffness exerts a greater influence with increasing frequency, and the overtones of a stiff string no longer constitute an exact harmonic series.

We must also be mindful of the fact that clamping at the ends of the string may not be exactly rigid and that yielding can occur at these points. The wave impedance at the ends will constitute the transverse mechanical impedance of the supports.

Consider a case where the left end of a finite string at $x = 0$ is attached to a pivot representing the slightly loose clamp. The transverse force f_0 exerted by the string on the hinge is

$$f_0 = T \sin \theta \approx T \left(\frac{\partial y}{\partial x} \right)_{x=0}$$

Here y is the complex expression for the transverse wave on the string, as given by equation (4.25). Because the motion of the swivel must match that of the end of the string, the velocity is

$$v_0 = \left(\frac{\partial y}{\partial t} \right)_{x=0} = i\omega y_0$$

Let z_0 denote the transverse mechanical impedance of the hinge at $x = 0$:

$$z_0 = \frac{f_0}{v_0} = \frac{T}{i\omega} \cdot \frac{(\partial y/\partial x)_{x=0}}{i\omega y_0}$$

and hence the boundary condition at $x = 0$ becomes

$$y_0 = \frac{T(\partial y/\partial x)_{x=L}}{i\omega Z_L}$$

where L_z is the mechanical impedance of the swivel located at $x = L$. If the supports were truly rigid, $Z_0 = Z_1 = \infty$, and the boundary conditions reduce to $y_0 = y_L = 0$.

References

Fletcher, Neville H., and Thomas D. Rossing. 1991. *The Physics of Musical Instruments.* New York: Springer-Verlag; Chapter 2. (This text provides an excellent exposition on the physical principles of musical instruments.)

Jean, Sir James. 1968. *Science and Music*, New York: Dover Publications; Chapter III. (Although rather skimpy on the mathematical details, the exposition provides the reader a good insight into the physics of a vibrating string.)

Kinsler, Lawrence E., Austin R. Frey, Alan B. Coppens, and James V. Sanders. 1982. *Fundamentals of Acoustics*, 3rd ed. New York: John Wiley & Sons; Chapter 2. (Still a good textbook which dates back to 1950.)

Morse, Philip M. 1982 (reprint from 1948 edition published by McGraw-Hill). *Vibration and Sound*. Woodbury, NY: Acoustical Society of America; Chapter 3.

Morse, Philip M., and K. Uno Ingard. 1968. *Theoretical Acoustics*, New York: McGraw-Hill; Chapter 4.

Reynolds, Douglas R. 1981. *Engineering Principles of Acoustics, Noise, and Vibration Control*, Boston: Allyn and Bacon; pp. 213–224.

Problems for Chapter 4

1. Show by direct substitution that each of the following expressions constitutes solutions of the wave equation:
   (a) $f(x - ct)$
   (b) $\ln [f(x - ct)\}$
   (c) $A(ct - x)^3$
   (d) $\sin[A(ct - x)]$

2. Show which of the following are solutions and *not* solutions to the wave equation:
   (a) $B(ct - x^2)$
   (b) $C(ct - c)t$
   (c) $A + B \sin(ct + x)$
   (d) $A \cos^2(ct - x) + B \sin(ct + x)$

3. Plot (by computer if possible) the expression $y = Ae^{-B(ct-x)}$ for times $t = 0$ and $t = 1.0$, with $A = 6$ cm, $B = 4$ cm^{-1}, and $c = 3$ cm/s. Discuss the physical significance of these curves.

4. Consider a string of density 0.05 g/cm, in which a wave form $y = 4 \cos(5t - 3x)$ is propagating. Both x and y are expressed in centimeters, and time t in seconds.
   (a) Determine the amplitude, phase speed, frequency, wavelength, and the wave number.
   (b) Find the particle speed of the string element at $x = 0$ at time $t = 0$.

5. A string is stretched with tension T between two rigid supports located at $x = 0$ and $x = L$. It is driven at its midpoint by a force $F \cos \omega t$.
   (a) Determine the mechanical impedance at the midpoint.
   (b) Establish that the amplitude of the midpoint is given by $F \tan (kL/2)/(2kT)$
   (c) Find the amplitude of displacement at the quarter point $x = L/4$.

6. Determine the mechanical impedance with respect to the applied force driving a semi-infinite string at a distance L from the rigid end. What is the significance of the individual terms in the expression for mechanical impedance?

7. Consider a string of density 0.02 kg/m that is stretched with a tension of 8 N from a rigid support to a device producing transverse periodic vibrations at the other end. The length of the string is 0.52 m. It is noted that for a specific driving frequency, the nodes are spaced 0.1 m apart and the maximum

amplitude is 0.022 m. What are the frequency and the amplitude of the driving force?

8. A device that has a constant speed amplitude $u(0, t) = U_0 e^{i\omega t}$, where U_0 is a constant, drives a forced, fixed string.

   (a) Find the frequencies of maximum amplitude of the standing wave.

   (b) Repeat the problem for a constant displacement amplitude $y(0, t) = Y_0 e^{i\omega t}$

   (c) Compare the results of (a) and (b) with the frequencies of mechanical resonances for the forced fixed string. Does the mechanical amplitude coincide with the maximum amplitude of the motion?

9. Consider a string fixed at both ends, with specified values of ρ_L, c, L, f, and T. Express the phase speed c' in terms of c and the fundamental resonance f' in terms of f if another string of the same materials is used, but

   (a) the length of the string is doubled;

   (b) the density per unit length is doubled;

   (c) the cross-sectional area is doubled;

   (d) the tension reduced by half;

   (e) the diameter of the string is doubled.

10. Consider a string of length L that is plucked at the location $L/3$ by producing an initial displacement δ and then suddenly releasing the string. Find the resultant amplitudes of the fundamental and the first three harmonic overtones. Draw (through computer techniques, if possible) the wave forms of these individual waves and the shape of the string occurring from the linear combination of these waves at $t = 0$. Redo this problem for time $t = L/c$, where c represents the transverse wave velocity of the string.

11. A string of length 1.0 m and weighing 0.03 kg has a mass of 0.15 kg hanging from it.

   (a) Find the speed of transverse waves in the string. (*Hint*: Neglect the weight of the string in establishing the tension in string.)

   (b) Determine the frequencies of the fundamental and the first overtone models of the transverse vibrations.

   (c) For the first overtone of the string, compare the relative amplitude of the string's displacement at the antinode with that of the mass.

12. A string having a linear density of 0.02 kg/m is stretched to a tension of 12 N between rigid supports 0.25 m apart. A mass of 0.002 kg is loaded on the string at its center.

   (a) Find the fundamental frequency of the system.

   (b) Find the first overtone frequency of the system.

13. A standing wave on a fixed-fixed string is given by $y = 3\cos(x/4)\sin 2t$. The length of the string is 36 cm and its linear density 0.1 gm/cm. The units of x and y are in centimeters, and t is given in seconds.

   (a) Find the frequency, phase speed, and wave number.

   (b) Determine the amplitude of the particle displacement the center of the string and at $x = L/4$ and $x = L/3$.

   (c) Find the energy density for those points, and determine how much energy there is in the entire string.

5
Vibrating Bars

5.1   Introduction

The theory underlying the physical operation of vibrating bars is of great interest to acousticians because a large number of acoustic devices employ longitudinal vibrations in bars, and frequency standards are established by producing sounds of specific pitches in circular rods of different lengths. The analysis of vibrating bars facilitates our understanding of acoustic waves through fluids, because the mathematical expressions governing the transmission of acoustic plane planes through fluid media are similar to those describing the travel of compression waves through a bar. Moreover, if the fluid is confined inside a rigid pipe, the boundary conditions bear a close correlation to those of a vibrating bar. An example of the sort of devices falling into the category of vibrating bars includes piezoelectric crystals which are cut so that the frequency of the longitudinal vibration in the direction of the major axis of the crystal may be used to monitor the frequency of an oscillating electric current or to drive an electroacoustic transducer.

The principal mode of sound transmission in bars is through the propagation of *longitudinal* waves. Here the displacement of the solid particles in the bar is parallel with the axis of the bar. The lateral dimensions of a bar are small compared with the length, so the cross-sectional plane can be pictured as moving as a unit. In reality, because of the Poisson effect which generally occurs in solid materials, the longitudinal expansion of the bar results in a lesser degree of lateral shrinkage and expansion; but this lateral motion can be disregarded in very thin bars.

5.2   Derivation of the Longitudinal Wave Equation for a Bar

In Figure 5.1, a bar of length L and uniform cross-sectional area $\hat{A}$ is subjected to longitudinal forces which produce a longitudinal displacement ξ of each of the molecules in the bar. This displacement in long thin bars will be the same at each point in any specific cross section. If the applied longitudinal forces varies in a wavelike perturbative manner, the displacement ξ is a function of both x and t and

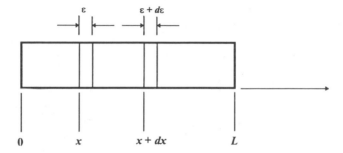

FIGURE 5.1. A bar undergoing longitudinal strain in the x-direction.

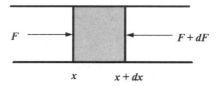

FIGURE 5.2. An element of the bar undergoing compression.

is fairly independent of lateral coordinates y and z. Thus,

$$\xi = \xi(x, t)$$

The x-coordinate of the bar is established by placing the left end of the bar at $x = 0$, with the right end terminating at $x = L$. Consider in Figure 5.2 an incremental element formed by dx of the unstrained bar positioned between x and $x + dx$. The application of a force in the positive x-direction causes a displacement of the plane at x by a distance ξ to the right and the plane at $x + dx$ by a distance $\xi + d\xi$ also to the right. A force acting in the opposite direction will likewise cause corresponding negatively valued displacements to the left. Because the element dx is small, we can represent the displacement at $x + dx$ by the first two terms of a Taylor series expansion of ξ about x:

$$\xi + d\xi = \xi + \left(\frac{\partial \xi}{\partial x}\right) dx$$

The left end of the element dx has been displaced a distance ξ and the right end a distance $\xi + d\xi$, thus yielding a net increase $d\xi$ in the length of the element given by

$$(\xi + d\xi) - \xi = d\xi = \left(\frac{\partial \xi}{\partial x}\right) dx$$

In solid mechanics the strain ϵ of an element is defined as the ratio of the change of its length to the original length, that is,

$$\epsilon = \frac{(\partial \xi / \partial x)}{dx} dx = \frac{\partial \xi}{\partial x} \tag{5.1}$$

In the situation in which static forces are applied to a uniform bar, the strain is the same for each point and is time-independent. But we are considering a dynamic case in which the strain in the varies with coordinate x and with time t. This type of variation generates a longitudinal wave motion in the bar in a manner analogous to the transverse waves in a string. When a bar undergoes strain, elastic forces are generated inside the bar. These forces act across each cross-sectional plane and essentially constitute reactions to longitudinally applied forces. We let $F_x = F_x(x, t)$ denote these longitudinal forces and adopt the convention that *compressive* forces are represented by *positive* values of F_x, and *tensile* forces by *negative* values of F_x. The stress σ in the bar is defined by

$$\sigma = F_x/\hat{A}$$

Here $\hat{A}$ is the cross-sectional area of the bar. We now apply Hooke's law

$$\sigma = \frac{F_x}{\hat{A}} = -E\epsilon = -E\frac{\partial \xi}{\partial x} \tag{5.2}$$

where E is the elastic constant, or *Young's modulus*, a property characteristic of the material used in the bar. Table A in Appendix A lists the values of Young's moduli for a number of commonly used solid materials. Because E must always have a positive value, a negative sign is introduced in equation (5.2) to accommodate the fact that a positive stress (compression) results in a negative strain, and a negative stress (tension) in a positive strain. We rewrite equation (5.2) to express force F_x at point x as follows:

$$F_x = -E\hat{A}\epsilon = -E\hat{A}\frac{\partial \xi}{\partial x} \tag{5.3}$$

Unlike the static case where the strain $\epsilon = \partial\xi/\partial x$ and hence the force F_x, remains constant throughout the bar, both the strain and F_x vary in the dynamic case, and a net force acts on element dx. F_x represents the internal force at x, and so $F_x + (\partial F_x/\partial x)\,dx$ constitutes the force at $x + dx$. The net force acting to the right becomes

$$dF_x = F_x - \left(F_x + \frac{\partial F_x}{\partial x}\,dx\right) = -\frac{\partial F_x}{\partial x}\,dx \tag{5.4}$$

Combining equations (5.3) and (5.4) results in

$$dF_x = E\hat{A}\frac{\partial^2 \xi}{\partial x^2}\,dx \tag{5.5}$$

The volume of the element dx is given by $\hat{A}dx$, and therefore the mass is $\rho\hat{A}\,dx$, where ρ denotes the density (kg/m^3) of the bar material. Applying Newton's equation of motion, with acceleration $\partial^2\xi/\partial t^2$ to equation (5.5), we obtain

$$\rho\hat{A}\,dx\frac{\partial^2 \xi}{\partial t^2} = E\hat{A}\frac{\partial^2 \xi}{\partial x^2}\,dx$$

Setting

$$c^2 = E/\rho \tag{5.6}$$

we now obtain the one-dimensional longitudinal wave equation:

$$\frac{\partial^2 \xi}{\partial t^2} = c^2 \frac{\partial^2 \xi}{\partial x^2} \tag{5.7}$$

Equation (5.6) corresponds to equation (4.3) for the transverse motion of the string, with the longitudinal displacement ξ assuming the role of the transverse displacement y. We note that equation (5.6) is identical to equation (2.3), and we have derived in this section the wave equation (5.7) which applies to acoustic propagation in a linearly elastic solid.

5.3 Solutions of the Longitudinal Wave Equation

The format of the general solution to equation (5.7) is identical with that of the solution to equation (4.3), that is,

$$\xi = f(ct - x) + g(ct + x) \tag{5.8}$$

The square root of equation (5.6) gives us the wave propagation velocity c.

$$c = \sqrt{\frac{E}{\rho}} \tag{5.9}$$

which indicates that c is a property of the bar material.

Let us write the solution (5.8) in the form of a complex harmonic solution

$$\xi = \mathbf{A}e^{i(\omega t - kx)} + \mathbf{B}e^{i(\omega t + kx)} \tag{5.10}$$

where $\mathbf{A}$ and $\mathbf{B}$ represent complex amplitude constants, and $k = \omega/c$ is the wave number. We now assume that the bar is rigidly fixed at both ends; the boundary conditions $\xi(x, t)$ becomes $\xi = 0$ at $x = 0$ and at $x = L$ at all times t. Applying the condition $\xi(0, t) = 0$ yields $\mathbf{A} = -\mathbf{B}$, and equation (5.10) revises to

$$\xi = \mathbf{A}e^{i\omega t}(e^{-ikx} - e^{ikx}) \tag{5.11}$$

The stipulation $\xi(L, t) = 0$ results in

$$e^{-ikL} - e^{ikL} = \frac{2 \sin kL}{i} = 0$$

or equivalently

$$\sin kL = 0$$

which means that

$$k_n L = n\pi, \quad n = 1, 2, 3, \ldots.$$

The allowed modes of vibration possess the radial frequencies ω_n and the corresponding cyclic frequencies f_n given by

$$\omega_n = \frac{n\pi c}{L} \quad \text{or} \quad f_n = \frac{nc}{2L}, \qquad n = 1, 2, 3, \ldots.$$

In simplifying equation (5.1), the complex displacement ξ for the nth mode of vibration is

$$\xi_n = -2i A_n e^{i\omega_n t} \sin k_n x \tag{5.12}$$

The real part of equation (5.12) is

$$\xi_n = \sin k_n x (A_n \cos \omega_n t + B_n \sin \omega_n t) \tag{5.13}$$

where the real amplitude constants A_n and B_n are related to the complex constant $\mathbf{A}_n$ as follows:

$$2\mathbf{A}_n = B_n + i A_n$$

The full solution to equation (5.7) consists of the sum of all of the individual harmonic solutions, i.e.,

$$\xi = \sum_{n=1}^{\infty} \sin k_n x \, (A_n \cos \omega_n t + \sin \omega_{nt}) \tag{5.14}$$

The constants A_n and B_n can be evaluated by using the Fourier analysis described in the last chapter, provided the initial conditions are known with respect to the displacement and the velocity of the bar.

5.4   Other Boundary Conditions

It should be understood that the boundary conditions corresponding to rigid supports are difficult to realize in practice. The free end condition, on the other hand, can be simulated by supporting the bar on extremely pliant supports placed some distance inward from the ends. The end of the bar can now move freely and no internal elastic force exists at that location. We now apply equation (5.3), setting $F_x = 0$; this gives rise to the condition $\partial \xi / \partial x = 0$ at the free end. If the bar is free to move at both ends (this is termed the *free–free bar*), the condition $\partial \xi / \partial x = 0$ applied to $x = 0$ in the wave equation solution (5.10) yields

$$\mathbf{A} = \mathbf{B}$$

with the result

$$\xi = \mathbf{A} e^{i\omega t} \left(e^{-ikx} + e^{ikx} \right) \tag{5.15}$$

Inserting the condition $\partial \xi / \partial x = 0$ into the above equation (5.15) for the location $x = L$ yields

$$-e^{-ikL} + e^{ikL} = 0 \quad \text{or} \quad \sin kL = 0$$

The allowable frequencies for a free–free bar are the same as those for the bar fixed at both ends (*fixed–fixed bar*). There will be, however, major differences in the respective wave patterns of the free–free bar and the fixed–fixed bar. Recasting equation (5.15) by making use of the relation

$$2\cos kx = e^{-ikx} + e^{ikx}$$

we can express the complex displacements corresponding to the nth mode of vibration as

$$\xi = 2A_n e^{i\omega_n t} \cos k_n x$$

The real part of the preceding equation gives the tangible vibrations described by

$$\xi_n = \cos k_n x (A_n \cos \omega_n t + B_n \sin \omega_n t) \tag{5.16}$$

By comparing equation (5.13) for the fixed–fixed bar with equation (5.14) for the free–free bar it will be seen that antinodes exist at the end points for the latter bar in contrast with the nodes that must exist at the end points of the fixed–fixed bar. The nodal patterns for both bars are shown in Figure 5.3. It is of interest to observe that when an antinode exists at the center of the bar the vibrations are symmetrical with respect to the center; otherwise a node in the center corresponds to asymmetrical vibration.

As with the vibrating string, we can rigidly clamp a bar at any one of its nodal positions without affecting the modes of vibrations which have a node at this

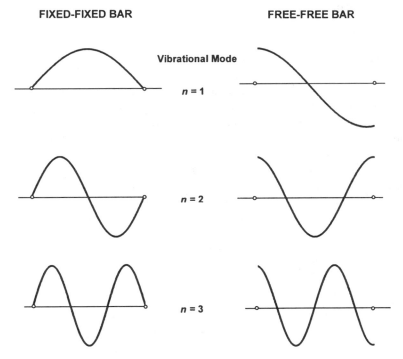

FIGURE 5.3. Typical standing waves for the first three modes of vibration in a fixed–fixed bar and in a free–free bar.

position. But the vibrations which do not have a node at this position will become suppressed. The nature of the vibration of the free–free bar is such that it is not possible to clamp it at any position that will not eliminate at least some of the allowed modes of vibrations.

The case of a free–fixed bar also makes for an interesting study. One end remains free at $x = 0$, and the other is rigidly clamped at $x = L$. The first condition $\partial \xi / \partial x = 0$ at $x = 0$ leads again to equation (5.15), while the second condition $\xi = 0$ at $x = L$ yields

$$e^{-ikL} + e^{ikL} = 0 \quad \text{or} \quad \cos kL = 0$$

which means that the allowable frequencies must satisfy

$$k_n L = \frac{\omega_n}{c} L = \frac{\pi}{2}(2n - 1), \qquad n = 1, 2, 3, \ldots$$

or

$$\omega_n = (2n - 1)\pi \frac{c}{2L}, \quad f_n = (2n - 1)\frac{c}{4L}$$

The fundamental frequency is half that for an otherwise identical free–free bar, and only odd-numbered harmonic overtones exist. The quality of the sound produced by an oscillating free–fixed bar will thus differ from that of a free–free bar, because of the absence of the even harmonics.

5.5   Mass Concentrated Bars

In practical situations a vibrating bar is not truly clamped totally, nor is it completely free to move at its ends. It may incorporate some type of mechanical impedance, most commonly as the result of concentrating a certain amount of mass at a certain location. An example is a diaphragm represented as a distributed mass located at one end of a vibrating tube inside a sonar transducer.

As an example let us consider a bar that is unfettered at $x = 0$ and has a loading consisting of a mass m concentrated at $x = L$. The mass is depicted as a point mass so that it does not move as a unit and thus merely sustain waves propagating through it. The boundary condition $\partial \xi / \partial x = 0$ at $x = 0$ again leads us to equation (5.15), as the result of $\mathbf{A} = \mathbf{B}$. For the boundary condition at $x = L$ we again invoke Newton's law of motion:

$$F_x(L, t) = m \left(\frac{\partial^2 \xi}{\partial t^2} \right)_{x=L} \tag{5.17}$$

A positive value of F_x, which compresses the bar, will result in acceleration of the mass in the positive x-direction. Because the mass is rigidly coupled to the bar, the accelerations of the mass and of the end of the bar should be identical. But if the mass had been concentrated at $x = 0$, a positive (compression) force would correspond to a reaction force to the left on the mass. The appropriate boundary condition for this case would be

$$-F_x(0,\ t) = m\left(\frac{\partial^2 \xi}{\partial t^2}\right)_{x=0} \tag{5.18}$$

Incorporating the boundary condition equation (5.18) into equation (5.16) results in

$$-E\hat{A}e^{i\omega t}(-ike^{-ikL} + ike^{ikL}) = mAe^{i\omega t}(-\omega^2)(e^{-ikL} + e^{ikL})$$

which rearranges to

$$kE\hat{A}\sin kL = -m\omega^2 \cos kL$$

or

$$\tan kL = -\frac{\omega mc}{E\hat{A}} \tag{5.19}$$

Because equation (5.19) is a transcendental equation, no explicit solution exists. However, if the mass m is very small, $m \approx 0$ and hence $\tan kL \approx 0$ and $kL \approx n\omega$, both of which constitute the allowed conditions for a free–free bar. This is a result that should occur, because light loadings render a bar virtually free at both ends. At the other extreme, for very heavy mass loadings, the mass behaves very nearly like a rigid support, and the allowed frequencies will approximate those of a free–fixed bar.

In the more general case of intermediate mass loading, it is rather cumbersome to solve by hand the transcendental equation (5.19) through graphic means. However, computer programs such as Mathcad®, MathLab®, Mathematica®, or even a professional-level spread sheet for IBM-compatible and Macintosh personal computers can be used to facilitate solutions. Eliminating Young's modulus in equation (5.19) by applying $E = \rho c^2$ from (5.9) and recognizing that the mass of the bar is given by $m_b = \rho\hat{A}L$, we can rewrite (5.19) as

$$\frac{\tan kL}{kL} = -\frac{m}{m_b} \tag{5.20}$$

The right-hand side of equation (5.20) is fixed by the amount of mass m_b in the bar and the loading mass m located at $x = L$. An example of the solution to (5.20) is given in Figure 5.4 for the case of a steel bar with a mass loading m/m_b of 20%. The longitudinal velocity of sound propagation of steel is taken at 5050 m/s. The fundamental frequency f_1 is found from $k_1 L$ through the relation

$$f_1 = \frac{k_1 L}{2\pi}\frac{c}{L}$$

and the higher frequencies are similarly established from

$$f_n = \frac{k_n L}{2\pi}\frac{c}{L}$$

But overtones corresponding to the higher frequencies are not harmonics of the fundamental. In the example of Figure 5.4, in which the plots of $(m/m_b)kL$ and

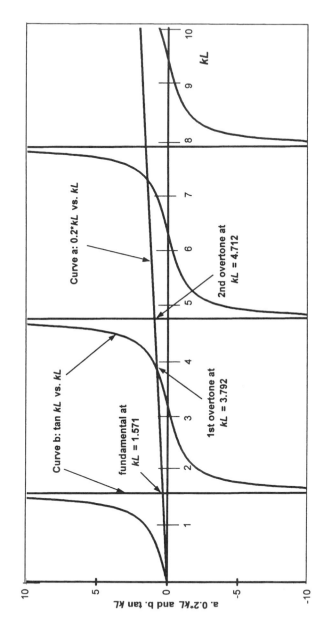

FIGURE 5.4. Plot of $0.2 \times kL$ (curve a) and $\tan kL$ (curve b) vs. kL. The intersections of the two sets of curves as shown above provide the first five solutions $k_n L$ ($kL = 0$ being trivial) of the transcendental equation (5.20) for the fundamental and overtones of a vibrating bar with 20% concentrated loading at location $x = L$. The fundamental occurs at $k_1 L = 1.571$, the first overtone at $k_2 L = 3.792$, the second overtone at $k_3 L = 4.712$; and neither of the latter two are harmonics of the fundamental.

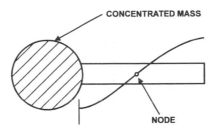

FIGURE 5.5. Location of node in a free mass-loaded bar and fundamental mode of vibration.

$\tan kL$ vs. kL yield intersections that constitute the solutions to equation (5.20), the ratio of the first overtone to the fundamental is $3.792/1.571 = 2.414$, *not* the value of 2.0 that would have specified the overtone to be a harmonic of the fundamental. Situations exist where the occurrence of nonharmonic overtones could be useful: for example, in a properly mass-loaded loudspeaker, a pure monofrequency input would not result in harmonics that could arise from the driving signal, if at all. The nodes of the vibration in the bar exist at locations where $\cos kx = 0$. The fundamental mode where $kL = 3.792$ engenders a node at $3.792\, x/L = \pi/2$, or $x = 0.414\, L$, not $x = 0.5\, L$ that would normally occur in a second harmonic. In contrast to the case of the free–free bar, the node in a free mass-loaded bar is no longer at the center—it shifts toward the loading mass, as shown in Figure 5.5. In this particular case, the bar could be supported at this nodal position without affecting the fundamental mode of vibration.

5.6   General Boundary Conditions for a Freely Vibrating Bar

Consider a freely vibrating bar with arbitrary loadings at each end. We shall establish the normal modes of vibration in terms of the mechanical impedances at both ends of the bar. Designating the mechanical impedance of the support at $x = 0$ by $\mathbf{Z}_{mi0}$ we express the force acting at this support due to the bar as

$$\mathbf{f}_0 = -\mathbf{Z}_{mi0}\,\mathbf{u}(0, t)$$

Here the minus sign is introduced to indicate that a positive (compressive) force generates an acceleration of the support in the $-x$-direction. But a positive compressive force at the end $x = L$ causes an acceleration of the other support to the right, in the positive x-direction; the force acting on this support is

$$\mathbf{f}_L = +\mathbf{Z}_{miL}\,\mathbf{u}(L, t)$$

where $\mathbf{Z}_{miL}$ represents the mechanical impedance of the support at $x = L$. The preceding two equations can be restated in terms of particle displacements by applying equation (5.3) to supplant the compressive forces and by expressing the particle velocity as $\mathbf{u} = \partial\xi/\partial t$:

$$\left(\frac{\partial\xi}{\partial x}\right)_{x=0} = \frac{\mathbf{Z}_{mi0}}{\rho_L c^2}\left(\frac{\partial\xi}{\partial t}\right)_{x=0} \tag{5.21}$$

$$\left(\frac{\partial\xi}{\partial x}\right)_{x=L} = \frac{\mathbf{Z}_{miL}}{\rho_L c^2}\left(\frac{\partial\xi}{\partial t}\right)_{x=L} \tag{5.22}$$

where $\rho_L = \rho\hat{A}$ is the linear density (kg/m) of the bar.

If the loads $\mathbf{Z}_{mi0}$ and $\mathbf{Z}_{miL}$ are purely reactive, there is no transient or spatial damping, and hence no loss of acoustical energy occurs. Equation (5.16) constitutes a proper solution. And because no loss of acoustical energy occurs a wave traveling in the $+x$-direction must equal the energy of a wave moving in the opposite direction. The absolute magnitudes of the complex wave amplitudes must therefore be equal (i.e., $|\mathbf{A}| = |\mathbf{B}|$). The boundary conditions (5.21) and (5.22) establish the phase angles of the complex amplitudes.

But if the mechanical impedances contain some measure of resistive components, a solution more general than that of equation (5.16) needs to be applied. As in the case of a freely vibrating string terminated by a resistive support, transient (or temporal) damping has to occur in the presence of resistance. The transient behavior of the bar is characterized by a complex angular frequency $\omega = \omega + i\beta$. The real portion of this frequency is the angular frequency ω; the imaginary part represents the transient absorption coefficient β. But no internal losses occur in the bar, so wave equation (5.7) still applies, and we infer the solution

$$\xi(x,t) = (\mathbf{A}e^{ikx} + \mathbf{B}e^{-ikx})e^{i\omega t} \tag{5.23}$$

where $\omega^2 = c^2 k^2$. If the losses are quite small we can use the approximation $\omega \approx ck$ to simplify the solution. Applying boundary conditions (5.21) for $x = 0$ and (5.22) for $x = L$ to (5.23) and making use of the approximation we obtain the following pair of equations

$$\mathbf{A} - \mathbf{B} = -\frac{\mathbf{Z}_{mi0}}{\rho_L c}(\mathbf{A} + \mathbf{B})$$

$$\mathbf{A}e^{-ikL} - \mathbf{B}e^{ikL} = \frac{\mathbf{Z}_{miL}}{\rho_L c}(\mathbf{A}e^{-ikL} + \mathbf{B}e^{ikL})$$

Solution of these preceding two equations by elimination of $\mathbf{A}$ and $\mathbf{B}$ results in the transcendental equation

$$\tan kL = i\,\frac{\dfrac{\mathbf{Z}_{mi0}}{\rho_L c} + \dfrac{\mathbf{Z}_{miL}}{\rho_L c}}{1 + \dfrac{\mathbf{Z}_{mi0}}{\rho_L c}\dfrac{\mathbf{Z}_{miL}}{\rho_L c}}$$

The characteristics of the vibration are determined from the complex impedances $\mathbf{Z}_{mi0}$ and $\mathbf{Z}_{mi}$. The solution of the preceding transcendental equation is rendered more difficult by the presence of any resistive component in either $\mathbf{Z}_{mi0}$ or $\mathbf{Z}_{miL}$, which produces a complex argument of the tangent.

5.7   Transverse Vibrations of a Bar

A bar not only vibrates longitudinally; it can also vibrate transversely, which is usually the case because the strains engendered by longitudinal motion give a virtually automatic rise to transverse strains as the result of the Poisson effect. A hammer blow aimed along the axis of a long thin bar supported at its center will usually result in principally transverse vibrations rather than the expected longitudinal effects, because it is not easy in the real world to avoid even the slightest eccentricity in the blow.

In our derivation of the transverse wave equation consider a straight bar of length L with a uniform bilaterally symmetric cross section $\hat{A}$. In Figure 5.6 a segment dx of the bar is shown bent (the bending is exaggerated to better illustrate the effect). The x-coordinate lies along the axis of the bar and the y-coordinate measures the transverse displacements of the bar from its unperturbed configuration. The bar behaves as a beam, that is, the upper part of the cross section stretches under tension and lower part becomes compressed. A *neutral axis NN'*, whose length remains unchanged, comprises the line of demarcation between compression and tension in the bar. If the cross-section of the bar is symmetrical about a horizontal plane the central axis of the bar will coincide with the neutral axis. The bending of the bar is gauged by the radius of curvature R of the neutral axis. Consider the length increment $\delta x = (\partial \xi / \partial x)\,dx$ due to the bending of a filament in the bar located at a distance r from the neutral axis. The longitudinal force df is found from

$$df = -E\,d\hat{A}\,\frac{\delta x}{dx} = -Ed\hat{A}\,\frac{\partial \xi}{\partial x} \qquad (5.24)$$

where $d\hat{A}$ is the cross-sectional area of the filament. Above the neutral axis NN' the value of δx is positive, so the force df becomes negative and thus is a tension. For filaments positioned below the neutral axis, δx is negative, resulting in a positive

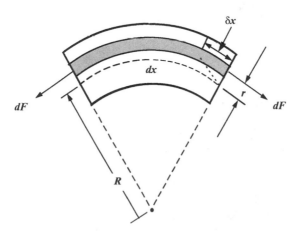

FIGURE 5.6. An element of a bar showing bending stresses and strains.

compressive force. From geometry we note that

$$\frac{dx + \delta x}{R + r} = \frac{dx}{R}$$

and this leads to $\delta x / dx = r/R$. Equation (5.2) now becomes

$$df = -\frac{E}{R} r \, d\hat{A}$$

The negative forces above the neutral axis cancel out the positive forces below the neutral axis, and hence the total longitudinal force $f = \int df$ equals zero. But a bending moment M occurs in the bar, that is,

$$M = \int r \, df = -\frac{E}{R} \int r^2 \, d\hat{A}$$

From classical mechanics we recognize the radius of gyration κ of the cross-sectional area $\hat{A}$, as defined by

$$\kappa^2 = \frac{\int r^2 d\hat{A}}{\hat{A}}$$

We now obtain

$$M = -\frac{\kappa^2 E \hat{A}}{R} \tag{5.25}$$

(For a bar with a rectangular cross-section, $\kappa = t/\sqrt{12}$ where t denotes the thickness of the bar. For a circular rod of radius a, the radius of gyration is given by $\kappa = a/2$.)

The radius of curvature R varies along the neutral axis, but the mathematics can be simplified by assuming the displacements y of the bar to be quite small, $\partial y / \partial x \ll 1$, which permits the use of the approximation

$$R = \frac{[1 + (\partial y / \partial x)^2]^{3/2}}{\partial^2 y / \partial x^2} \approx \frac{1}{\partial^2 y / \partial x^2}$$

Equation (5.25) now modifies to

$$M = -\kappa^2 E \hat{A} \frac{\partial^2 y}{\partial x^2} \tag{5.26}$$

The curvature shown in Figure 5.6 has a negative $\partial^2 y / \partial x^2$, and the bending moment is consequently positive. In order to get this type of curvature the torque must be applied to the left end of the segment in a counterclockwise (or positive angular) direction and the torque at the right end of the segment must be clockwise (or in the negative angular direction).

Shear forces as well as bending moments arise when a bar becomes distorted. In Figure 5.7 a shear force $F_y(x)$ acts upward (in the positive sense) on the left end of the element dx. An opposing shear force $-F_y(x + \Delta x)$ acts downward at the right end of the element. To sustain static equilibrium in the bent bar the shear forces and torsions acting on the element must counterbalance each other so that

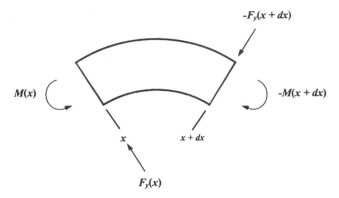

FIGURE 5.7. An element of the bar showing shear forces and bending moments.

there is no net turning momentum. Taking the left end of the element in Figure 5.7 as the reference pivot point, we obtain

$$M(x) - M(x + dx) = F_y(x + dx) \tag{5.27}$$

The terms $M(x + \Delta x)$ and $F_y(x + \Delta x)$ are now expanded in a Taylor's series about point x, with the result that equation (5.27) becomes:

$$F_y = -\frac{\partial M}{\partial x} = \kappa^2 E \hat{A} \frac{\partial y^3}{\partial x^3} \tag{5.28}$$

In equation (5.28) the second-order and higher terms in dx have been discarded.

In undergoing transverse vibrations the bar is in dynamic rather than static equilibrium. This requires that the right side of equation (5.27) must equal the rate of increase of the angular momentum of the segment. But as long as the displacement and the slope of the bar remain small the variations in angular momentum can be disregarded, and equation (5.28) should serve as a good approximation of the correlation between the displacement y and the acting force F_y. The net upward force dF_y in element dx is given by

$$dF_y = F_y(x) - F_y(x + \Delta x) = -\frac{\partial F_y}{\partial x} = -\kappa^2 E \hat{A} \frac{\partial^4 y}{\partial x^4} dx$$

The element undergoes an upward acceleration under the impetus of the force, and the equation of motion for the mass of the element, $\rho \hat{A} dx$, may now be written as

$$\rho \hat{A} dx \frac{\partial^2 y}{dt^2} = -\kappa^2 E \hat{A} \frac{\partial^4 y}{\partial x^4} dx$$

Setting $c = (E/\rho)^{1/2}$, as in the case of longitudinal waves, the last equation changes to:

$$\frac{\partial^2 y}{\partial t^2} = -\kappa^2 c^2 \frac{\partial^4 y}{\partial x^4} \tag{5.29}$$

Equation (5.29) for the transverse wave differs from the equation (5.7) for the longitudinal wave principally in the presence of the fourth partial derivative with respect to x. Solutions in the functional form of $f(ct - x)$ do *not* apply to transverse waves, as can be readily proved by direct substitution into equation (5.29). This means that transverse waves do not travel in the x-direction with constant speed c and unchanging shape.

Equation (5.28) can be solved by separation of variables by setting the complex transverse displacement **y** as

$$\mathbf{y} = \Psi(x)e^{i\omega t} \tag{5.30}$$

and inserting into equation (5.29). This yields a new total differential equation in which Ψ exists as a function of x only:

$$\frac{d^4\Psi}{dx^4} = \frac{\omega^2}{\kappa^2 c^2}\Psi$$

Setting

$$v = \sqrt{\kappa\omega c} \tag{5.31}$$

the fourth-order differential equation becomes

$$\frac{d^4\Psi}{dx^4} = \frac{\omega^4}{v^4}\Psi \tag{5.32}$$

The function Ψ may be assumed as an exponential of the form $\Psi(x) = Ae^{\gamma x}$ and substituted into equation (5.32). The result is

$$\gamma^4 = (\omega/v)^4$$

Four values of γ occur: $\pm(\omega/v)$ and $\pm(i\omega/v)$. The complete solution to (5.32) consists of the sum of the four solutions:

$$\Psi = \mathbf{A}e^{\omega x/v} + \mathbf{B}e^{-\omega x/v} + \mathbf{C}e^{i(\omega x/v)} + \mathbf{D}e^{-i(\omega x/v)}$$

wherein **A**, **B**, **C**, and **D** constitute complex amplitude constants. From equation (5.30) the solution for the displacements **y** can be written as

$$\mathbf{y} = e^{i\omega t}\left(\mathbf{A}e^{\omega x/v} + \mathbf{B}e^{-\omega x/v} + \mathbf{C}e^{i(\omega x/v)} + \mathbf{D}e^{-i(\omega x/v)}\right) \tag{5.33}$$

It should be noted that none of the terms in equation (5.33) contain a wave moving with a velocity c. The third term inside the parenthesis of equation (5.33) represents a wave disturbance moving to the left, and the fourth term represents a wave moving in the positive x-direction. The *phase speed* v is itself a function of the frequency, as attested by equation (5.31), so waves of differing frequencies will travel with different phase speeds. Higher-frequency waves will outpace the lower-frequency waves, and accordingly a complex wave containing a number of different frequencies will alter its shape along the x-axis. Each frequency component of a complex wave travels at its own speed v, which gives rise to a situation analogous to the transmission of light through glass, in which different component frequencies of the light beam travel with different speeds thereby causing dispersion. A vibrating bar thus acts as a *dispersive* medium for transverse waves.

The real part of equation (5.33) constitutes the actual solution of equation (5.29). We make use of the following hyperbolic and trigonometric identities

$$\sin y = (e^{iy} - e^{-iy})/2, \qquad \cos y = (e^{iy} + e^{-iy})/2$$
$$\sin(iy) = i \sinh y, \qquad \sinh(iy) = i \sin y$$
$$\cos(iy) = \cosh y, \qquad \cosh(iy) = \cos y$$

to recast equation (5.33) as

$$y = \cos(\omega t + \phi)\left(A \cosh \frac{\omega x}{v} + B \sinh \frac{\omega x}{v} + C \cos \frac{\omega x}{v} + D \sin \frac{\omega x}{v} \right) \quad (5.34)$$

Here A, B, C, and D are real constants that occur from the rearrangement of the original complex constants **A**, **B**, **C**, and **D**. The intricate relationships between the real set of constants and the set of complex constants are not really of much concern to us, because it is the application of the initial and boundary conditions that provides the evaluation of these constants. However, there are twice as many arbitrary constants in the transverse equation (5.34) as in the longitudinal wave equation (5.7), due to the fact that the former is of the fourth differential order rather than the second differential order. Therefore, twice as many boundary conditions are required, and this can be satisfied by specifying *pairs* of boundary conditions at the ends of the bars. The nature of the supports establishes the boundary conditions that generally fall into the categories of free and clamped ends.

5.8   Boundary Conditions for Transverse Vibrations

1. If the bar is rigidly clamped at one end, then both the displacement and the slope must be zero at that end at all times, and the boundary conditions are expressed as

$$y = 0, \qquad \partial y/\partial x = 0 \quad (5.35)$$

2. On the other hand, neither an externally applied moment nor a shear force may exist at a free end of a vibrating bar. But the displacement and the slope of the bar at a free end are not constrained, excepting for the mathematical stipulation they remain small. From equations (5.26) and (5.28) the boundary conditions become

$$\frac{\partial^2 y}{\partial x^2} = 0, \qquad \frac{\partial^3 y}{\partial x^3} = 0 \quad (5.36)$$

Case 1: Bar Clamped at One End

Consider a bar of length L that is rigidly clamped at $x = 0$, but is free at $x = L$. At $x = 0$, the two conditions of (5.35) apply, so $A = -C$ and $B = -D$. The general solution (5.35) reduces to

$$y = \cos(\omega t + \phi)\left[A\left(\cosh\frac{\omega x}{v} - \cos\frac{\omega x}{v}\right) + B\left(\sinh\frac{\omega x}{v} - \sin\frac{\omega x}{v}\right)\right]$$

Applying free-end conditions (5.36) at $x = L$ yields the following two sets of equations:

$$A\left(\cosh\frac{\omega L}{v} + \cos\frac{\omega L}{v}\right) = -B\left(\sinh\frac{\omega L}{v} + \sin\frac{\omega L}{v}\right)$$

$$A\left(\sinh\frac{\omega L}{v} - \sin\frac{\omega L}{v}\right) = -B\left(\cosh\frac{\omega L}{v} + \cos\frac{\omega L}{v}\right)$$

Both of the preceding two equations cannot hold true for all frequencies. In order to determine the permissible frequencies, one equation is divided into the other, thus canceling out the constants A and B. Ridding the resulting equation of fractional expressions by cross-multiplication and using the identities $\cos^2\theta + \sin^2\theta = 1$ and $\cosh^2\theta + 1 = \sinh^2\theta$, we obtain

$$\cosh\frac{\omega L}{v}\cos\frac{\omega L}{v} = -1$$

We can alter the last equation by using the identities

$$\tan\frac{\theta}{2} = \sqrt{\frac{1 - \cos\theta}{1 + \cos\theta}}, \qquad \tanh\frac{\theta}{2} = \sqrt{\frac{\cosh\theta - 1}{\cosh\theta + 1}}$$

and we now obtain:

$$\cot\frac{\omega L}{2v} = \pm\tanh\frac{\omega L}{2v} \tag{5.37}$$

The frequencies that correspond to the allowable modes of vibration can be found through the use of a microcomputer program that determines the intersections of the curves of $\cot\omega L/2v$ and $\pm\tanh\omega L/2v$, as shown in Figure 5.8. The frequencies of the permissible modes are given by

$$\frac{\omega L}{2v} = \zeta\frac{\pi}{4} \tag{5.38}$$

where $\zeta = 1.194, 2.988, 5, 7, \ldots$ with ζ approaching whole numbers for the higher allowed frequencies. Inserting $v = (\kappa\omega c)^{1/2}$ into (5.38), squaring both sides and solving for frequencies f we obtain

$$f = \zeta\frac{\pi\kappa c}{8L^2}$$

The constraint imposed by the boundary conditions leads to a set of discrete allowable frequencies, but the overtone frequencies are not harmonics of the fundamental. When a metal bar is struck in such a manner that the amplitudes of the vibration of some of the overtones are fairly strong, the sound produced has a metallic cast. But these overtones die out rapidly, and the initial sound soon evolves into a mellower pure tone whose frequency is the fundamental. This is

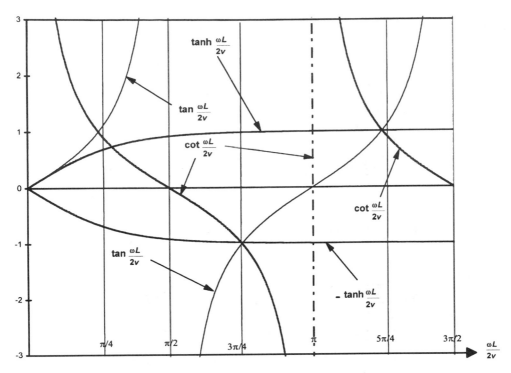

FIGURE 5.8. Trigonometric functions, used in equations (5.37) and (5.38), plotted as functions of $(\omega L/2v)$.

characteristic of the behavior of a tuning fork that emits a short metallic sound upon being struck before emitting a pure tone.

The distribution of the nodal points along the transversely vibrating bar is quite complex, with three distinct types of nodal points being identified mathematically. The clamping point of the bar constitutes one type, with conditions $y = 0$ and $\partial y/\partial x = 0$ at all times. Another group of points called *true nodes* is characterized by $y = 0$ and $\partial y/\partial x \approx 0$, and they are found near points of inflections on the bar. The spacing between these true nodes is very nearly (but not quite) $\lambda/2$. The third type of nodal point occurs at the node very near the free end of the bar, where $y = 0$, but the corresponding point of inflection where $\partial^2 y/\partial x^2 \approx 0$ does not coincide with that point but rather is moved out to the free end. The vibrational amplitudes do not equal each other at the various antinodes, but the greatest vibrational amplitude is that of the free end.

Case 2: Free–Free Bar

In the case of a bar that is free to move at both ends, the boundary conditions at $x = 0$ are satisfied by $A = C$ and $B = D$ as the result of applying (5.36). The same

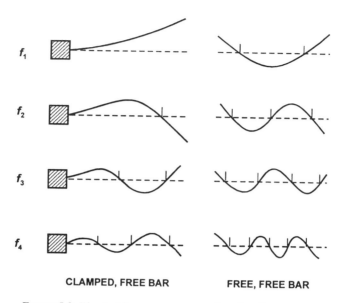

f_1

f_2

f_3

f_4

CLAMPED, FREE BAR   **FREE, FREE BAR**

FIGURE 5.9. The first four transverse modes of a vibrating bar.

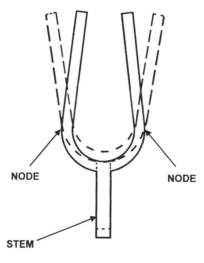

NODE   **NODE**

STEM

FIGURE 5.10. The tuning fork, essentially a U-shaped bar attached to a stem.

set of boundary conditions applied to the other end $x = L$ yields

$$\tan \frac{\omega L}{2v} = \pm\tanh \frac{\omega L}{2v}$$

which is turn gives a discrete set of allowable frequencies of the transverse vibration. The frequencies are given by

$$f = \xi \frac{\pi \kappa c}{8L^2}$$

where $\xi = 3.0112^2, 5^2, 7^2, 9^2, \ldots$. The overtones are *not* harmonics of the fundamental.

Figure 5.9 illustrates the transverse modes of a clamped-free bar and a free–free bar. In the free–free bar the modes corresponding to the fundamental f_1 and all additional odd-numbered frequencies. The odd-numbered frequencies f_3, f_5, and so on, are symmetric about the center of the bar. The slope $\partial y / \partial x$ is always zero at the center, which is a true antinode. But the even-numbered frequencies f_2, f_4, f_6, ... yield asymmetric modes of vibrations with respect to the center. In all modes the nodal points are found to be distributed symmetrically about the center. A bar may therefore be supported on a knife edge (or held by knife-edged clamps) at any nodal point without affecting the mode of vibration having a node at that point. A knife edge or a knife-edged set of clamps disallows displacement, but not a change in the slope that occurs at the node.

The xylophone consists of metal bars that are supported at the nodal locations of the fundamental. But the nodes of the associated overtones are unlikely to be located at the same points: the overtones quickly die out, leaving the pure tone of the fundamental. The concept of a free–free bar applies to a tuning fork that is essentially a U-shaped bar attached to a stem (Figure 5.10). The geometry of the fork and the mass-loading effect of the stem causes the nodes of the fundamental to be spaced closely near the stem. When the tuning fork is struck, overtones damp out in a very short time, leaving only the pure sinusoidal fundamental, much in the same manner as a xylophone. Because the stem shares the antinodal motion of the center of a free–free bar, the radiation efficiency of a tuning fork becomes greatly increased by touching the stem to a surface of a large area such as a counter top.

References

Fletcher, Neville H., and Thomas D Rossing. 1991. *The Physics of Musical Instruments*. New York: Springer-Verlag; pp. 53–60.

Kinsler, Lawrence E., Austin R Frey, Alan B. Coppens, and James V. Sanders. 1982. *Fundamentals of Acoustics*, 3rd ed. New York: John Wiley & Sons; Chapter 3.

Morse, Philip M., and K. Uno Ingard. 1968. *Theoretical Acoustics*. New York: McGraw-Hill; Chapter 5.

Reynolds, Douglas R. 1981. *Engineering Principles of Acoustics, Noise, and Vibration Control*. Boston: Allyn and Bacon; pp. 224–234.

Wood, Alexander. 1966. *Acoustics*. New York: Dover Publications; pp. 384–386.

Problems for Chapter 5

1. Show that a bar of length L that is rigidly fixed at $x = 0$ and totally free at $x = L$ will have only odd integral harmonic overtones.

2. Determine the fundamental frequency of the bar in Problem 1 if $L = 0.60$ m and the bar is made of steel. If a static force F is applied to the free end, so that the bar displaces a distance δ, and then is suddenly released, demonstrate that the amplitudes of the subsequent longitudinal vibrations are given by

$$A_n = 8\frac{\delta}{n^2\pi^2}\sin\frac{n\pi}{2}$$

Find these amplitudes for this steel bar with a cross-sectional area of 5.0 $(10)^{-5}$ m^2, under the effect of a force of 4500 N.

3. A steel bar is free to move at $x = 0$. It has 0.0002 m^2 cross-sectional area and 0.35 m length. A 0.20 kg load is placed at $x = 0.35$ m.
   (a) Determine the fundamental frequency of the longitudinal vibration of the mass-loaded bar.
   (b) Establish the position at which the bar may be clamped so as to minimize interference with the fundamental mode.
   (c) Find the ratio of the displacement amplitude of the free end to that of the mass-loaded end for the first overtone of the bar.

4. A steel bar having a mass of 0.05 kg and 0.25 m length is loaded at one end with 0.028 kg and 0.056 kg at the other end. Find the fundamental frequency of the system's longitudinal vibration and determine the location of the node in the bar. Also compute the ratio of the displacement amplitudes at the two ends of the bar.

5. Redo Problem 4 for an aluminum bar of the same length, but with a mass of 0.03 kg, subjected to the same end loadings,

6. Consider a thin bar of length L and mass M that is rigidly fixed at one end and free at the other. What mass m must be affixed to the free end in order to lower the fundamental frequency of longitudinal vibration by 30 percent from the fixed–free value?

7. A fixed–free bar of length L and mass m has a mechanical reactance, or stiffness, equal to $-is/\omega$ in the fixture. Develop an expression for the fundamental frequency of the longitudinal vibration.

8. A longitudinal force $F\cos\omega t$ drives a long thin bar at $x = 0$. The bar is free to move at $x = L$.
   (a) Obtain the equation for the amplitude of the standing waves occurring in the bar.
   (b) Obtain the expression giving the input mechanical impedance of the bar of length L.
   (c) Derive the expression for the input mechanical impedance of the same bar having an infinite length.
   (d) For the case of part (a) plot the amplitude of the driven end of the bar as a function of frequency over the range 100 Hz to 3 kHz, if the material of the

bar is aluminum, its length is 1.5 m, its cross-sectional area is 1.5 $(10)^{-4}$ m^2, and the amplitude of the driving force is 12 N.

9. Demonstrate that the dimensional units of $v = \sqrt{\kappa \omega c}$ are those of speed. At what frequency will the phase speed of transverse vibrations of a steel rod of 0.005 m diameter equal the phase speed of longitudinal vibrations in the rod?

10. Given an aluminum rod of 0.010 radius and length 0.4 m, what will be the fundamental frequency of free–free vibrations? Predict the displacement amplitude of the free ends if the displacement amplitude of the rod at its center is 2.5 cm.

6
Membrane and Plates

6.1 Introduction

In this chapter we are now applying two-dimensional wave equations to membranes and plates. In this group of physical applications, three dimensions are really involved in the theory which applies to two-dimensional surfaces such as drumheads and diaphragms of microphones or loudspeakers. Two spatial coordinates are required to locate a point on a vibrating surface, but the displacement generally occurs along the third spatial coordinate. However, the expansion of the general one-dimensional solution to the wave equation for a string or a bar can be extended easily to two dimensions. Again, boundary conditions determine the discreteness of the vibrational frequencies, but the peripheral geometry of a membrane constitutes a factor additional to the effects expected for different types of support. For certain membrane or plate geometries, the choice of an appropriate reference coordinate system to match the contour of the subject surface can greatly facilitate the solution of the wave equation.

In the following sections the wave equation for a membrane under tension is derived, and solutions are developed for various geometries and supports. The classic case of a kettledrum is included as a practical example that demonstrates the effect of damping; and forced vibrations are also taken into additional consideration.

The number of easily solvable two-dimensional problems is limited by the narrow choice of coordinate systems; and more complex cases can be treated by computerization, most notably through the application of finite-element methods.

6.2 Derivation of the Wave Equation for a Stretched Membrane

In order to develop a viably simple equation of motion for a vibrating membrane stretched under tension the assumption is made that the extremely thin membrane is uniform and the vibrational amplitudes remain small. The membrane is also deemed to be perfectly elastic, without any damping. Designate T as the tension

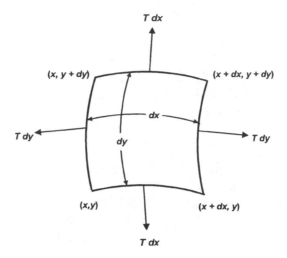

FIGURE 6.1. An element of the vibrating element.

of the membrane applied as a uniform force per unit length (e.g., in N/m) and ρ_S the surface density of the membrane as mass per unit area (kg/m^2). Thus, in Figure 6.1, an essentially two-dimensional membrane is stretched outward toward opposing sides of an element (each side of length dL) under the influence of a tension force $T\,dL$. A cartesian coordinate system is adopted, with the expanse of the quiescent membrane lying in the x-y plane and the transverse displacement of a point at (x, y) occurring in the z-direction as $z = z(x, y, t)$. In Figure 6.1 an element of area $dS = dx\,dy$ undergoes the effect of transverse forces acting in the x- and y-directions along the peripheral lengths of the element. The net force in the x-direction is given by

$$T\left[\left(\frac{\partial z}{\partial x}\right)_{x+dx} - \left(\frac{\partial z}{\partial x}\right)_{x}\right] dy = T\frac{\partial^2 z}{\partial x^2}\,dx\,dy$$

and similarly, the net force in the y-direction is $T(\partial^2 z/\partial y^2)\,dx\,dy$. These two terms add up to contribute to the acceleration $\partial^2 z/\partial t^2$ of the element's mass $\rho_S\,dx\,dy$, in accordance with the Newton's law of motion:

$$T\left(\frac{\partial^2 z}{\partial x^2} + \frac{\partial^2 z}{\partial y^2}\right) dy\,dy = \rho_S\,dx\,dy\frac{\partial^2 z}{\partial t^2}$$

and setting

$$c = \sqrt{\frac{T}{\rho_S}} \tag{6.1}$$

we obtain the classic two-dimensional wave equation

$$\frac{\partial^2 z}{\partial x^2} + \frac{\partial^2 z}{\partial y^2} = \frac{1}{c^2}\frac{\partial^2 z}{\partial t^2} \tag{6.2}$$

We can recast the wave equation (6.2) in the more general Laplacian format:

$$\nabla^2 z = \frac{1}{c^2}\frac{\partial^2 z}{\partial t^2} \tag{6.3}$$

Equation (6.2) is suitable for treatment of rectangular membranes, but equation (6.3) should be expressed in terms of polar coordinates to facilitate mathematical treatment of circular membranes:

$$\frac{\partial^2 z}{\partial r^2} + \frac{1}{r}\frac{\partial z}{\partial r} + \frac{1}{r^2}\frac{\partial^2 z}{\partial \theta^2} = \frac{1}{r^2}\frac{\partial^2 z}{\partial t^2} \tag{6.4}$$

For normal vibrational modes it is the standard mathematical procedure to assume that the solution to equation (6.3) consists of a spatially dependent function Ψ and a strictly time-dependent function $e^{i\omega t}$ (we dispense with the other function $e^{-i\omega t}$ as being superfluous for the current physical applications):

$$z = \Psi e^{i\omega t} \tag{6.5}$$

Inserting the above expression into equation (6.3) and setting $k = \omega/c$ yields the time-independent *Helmholtz equation*:

$$\nabla^2\Psi + k^2\Psi = 0 \tag{6.6}$$

of which solutions upon insertion into equation (6.5) yield normal modes of vibrations in a membrane of a given geometry and boundary conditions.

6.3   Rectangular Membrane with Fixed Edges

Consider a stretched rectangular membrane that is fixed at its four edges $x = 0$, $x = L_x$, $y = 0$, and $y = L_y$. The boundary conditions may be expressed as

$$z(0, y, t) = z(L_x, y, t) = z(x, 0, t) = z(x, L_y, t) = 0 \tag{6.7}$$

In the Cartesian format the solution $\mathbf{z}(x, y, t) = \Psi(x, y)e^{i\omega t}$ to equation (6.2) must derive from the Helmholtz equation given below for Cartesian coordinates:

$$\frac{\partial^2\Psi}{\partial x^2} + \frac{\partial^2\Psi}{\partial y^2} + k^2\Psi = 0 \tag{6.8}$$

But $\Psi(x, y)$ can be stated as the product of two singly dimensioned functions $X(x)$ and $Y(y)$ so that

$$\Psi(x, y) = X(x)\,Y(y)$$

and then Equation (6.8) transforms to

$$\frac{1}{X}\frac{\partial^2 X}{\partial x^2} + \frac{1}{Y}\frac{\partial^2 Y}{\partial y^2} + k^2 = 0 \tag{6.9}$$

Because the three terms of equation (6.9) cannot all sum to zero and the first and second terms are wholly independent of each other, we separate equation (6.9) into two separate differential equations, one wholly dependent on x and the other on y:

$$\frac{1}{X}\frac{\partial^2 X}{\partial x^2} + k_x^2 = 0$$

$$\frac{1}{Y}\frac{\partial^2 Y}{\partial y^2} + k_y^2 = 0 \qquad (6.10)$$

where k_x^2 and k_y^2 are constants related by

$$k_x^2 + k_y^2 = k^2$$

The solutions to the equation set (6.10) consists of sinusoids, with the result

$$\mathbf{z}(x, y, t) = \alpha \sin(k_x x + \phi_x)\sin(k_y y + \phi_y)e^{i\omega t} \qquad (6.11)$$

Here α represents the maximum displacement of the membrane in the transverse direction, and ϕ_x and ϕ_y are determined by boundary conditions. With the first and the third of the boundary condition set (6.7) we find that $\phi_x = \phi_y = 0$ and the remaining conditions necessitate that $\sin k_x L_x = 0$ and $\sin k_y L_y = 0$. Therefore, the normal modes occur from

$$\mathbf{z}(x, y, t) = \alpha \sin k_x x \sin k_y y e^{i\omega t} \qquad (6.12)$$

for which k_x and k_y turn out to be discrete values established by:

$$k_x = n\pi/L_x, \qquad n = 1, 2, 3, \ldots$$
$$k_y = m\pi/L_y, \qquad m = 1, 2, 3, \ldots$$

The frequencies of the allowed modes of vibrations are found from

$$f_{nm} = \frac{\omega_{nm}}{2\pi} = \frac{c}{2}\sqrt{\left(\frac{n}{L_x}\right)^2 + \left(\frac{m}{L_y}\right)^2} \qquad (6.13)$$

Equation (6.13) constitutes a fairly simple extension of the allowable frequencies of an idealized free-vibrating string to two-dimensional status.

The fundamental frequency is found by merely setting $n = m = 1$ in equation (6.12). The overtones corresponding to $m = n > 1$ will be harmonics of the fundamental frequency, but those in which $m \neq n$ (with either m or $n > 1$) may not necessarily be so. A number of possible modes in a rectangular membrane are illustrated in Figure 6.2. The shaded areas vibrate π radians out of time phase with the unshaded areas. Each normal mode is designated by an ordered pair (n, m), and the nodal lines are those with zero displacement at all times. In theory, rigid supports could be placed along these lines without affecting the nodal pattern for the associated specific frequency.

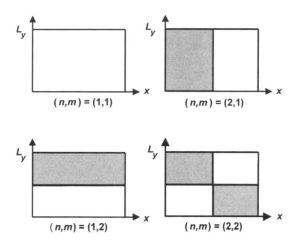

FIGURE 6.2. Four modes of a vibrating membrane. Lines located within the borders of the membranes constitute nodal loci where displacements are zero for these respective modes.

6.4   Freely Vibrating Circular Membrane with Fixed Rim

As mentioned in the foregoing it is preferable to adopt the polar coordinate version (6.4) of the wave equation (6.3) in order to treat the case of a circular membrane that is fixed at its rim. Accordingly, the zero displacement of the membrane's boundary at radius $r = a$ gives the boundary condition

$$z(a, \theta, t) = 0$$

The harmonic solution to equation (6.4) can be represented as the product of three terms, each of which is functions of only one variable:

$$z(r, \theta, t) = \mathbf{R}(r)\Theta(\theta)e^{i\omega t} \tag{6.14}$$

The boundary condition stipulated at the rim of the circular membrane now becomes

$$\mathbf{R}(a) = 0$$

and insertion into (6.4) yields the polar coordinate version of the Helmholtz equation,

$$\Theta\frac{d^2\mathbf{R}}{dr^2} + \frac{\Theta}{\mathbf{r}}\frac{d\mathbf{R}}{dr} + \frac{\mathbf{R}}{\mathbf{r}^2}\frac{d^2\Theta}{d\theta^2} + k^2\mathbf{R}\Theta = 0 \tag{6.15}$$

where $k = \omega/c$, as before. Equation (6.15) is then rearranged to effect the separation of two variables so we obtain

$$\frac{r^2}{\mathbf{R}}\left(\frac{d^2\mathbf{R}}{dr^2} + \frac{1}{r}\frac{d\mathbf{R}}{dr}\right) + (kr)^2 = -\frac{1}{\Theta}\frac{d^2\Theta}{d\theta^2} \tag{6.16}$$

The left-hand side of equation (6.16) is solely a function of r while the right depends only on θ. In order for the equality of equation (6.16) to prevail, both sides of the equation must be set equal to a constant m^2. Then we obtain from the right side of equation (6.16),

$$\frac{d^2\Theta}{d\theta^2} + m^2\Theta = 0$$

which in turn yields the harmonic solution

$$\Theta(\theta) = \cos(m\theta + \epsilon)$$

Here ϵ is the phase angle. The azimuthal coordinate is of periodic nature, repeating itself every 2π radians. In order that the displacement z be a single-valued function of position, $z(r, \theta, t)$ must then equal itself every 2π radians, that is, $z(r, \theta, t) = z(r, \theta + 2\pi, t)$, with the result that the constant m is constrained to integral values $m = 1, 2, 3, \ldots$. The left-hand portion of equation (6.16) therefore becomes the Bessel's differential equation:

$$\frac{d^2\mathbf{R}}{dr^2} + \frac{1}{r}\frac{d\mathbf{R}}{dr} + \left(k^2 - \frac{m^2}{r^2}\right)\mathbf{R} = 0 \qquad (6.17)$$

The solution of equation (6.17) is

$$\mathbf{R}(r) = \mathbf{A}J_m(kr) + \mathbf{B}Y_m(kr) \qquad (6.18)$$

where $J_m(kr)$ and $Y_m(kr)$ are, respectively, the transcendental *Bessel functions* of the *first* and the *second kind*, each of the order m. Bessel functions are oscillating functions with diminishing amplitudes for increasing kr. $Y_m(kr)$ approaches infinity as $kr \to 0$. The numerical values of Bessel functions and their properties are given in standard tables and advanced mathematical computer programs. An abbreviated set of Bessel formulas and tables are given in Appendix B. Because the circular membrane includes the origin $r = 0$ and the displacement of the membrane must remain finite at that point, it is necessary to set $\mathbf{B} = 0$,[1] reducing equation (6.18) to

$$\mathbf{R}(r) = \mathbf{A}J_m(k)$$

Applying the boundary condition $R(a) = 0$ requires that $J_m(ka) = 0$. Let us designate by q_{mn} those values of the argument ka at which the mth order Bessel function J_m equals zero. From $J_m(q_{mn}) = 0$ we can find those discrete values of k which are given by

$$k_{mn} = \frac{J_{mn}(q_{mn})}{a} = \frac{J_{mn}}{a}$$

---

[1] On the other hand, if an annular membrane stretched over region $a < r < b$ (thus excluding the origin $r = 0$) is considered, both Bessel functions must be retained in equation (6.18) to provide the two arbitrary constants needed for the boundary conditions at the inner and outer borders.

A number of values of q_{mn} that yield zeros of the Bessel functions are given in Appendix B. We can therefore write the normal modes of vibration as

$$z_{mn}(r, \theta, t) = A_{mn} J_n(k_{mn}r) \cos(m\theta + \varepsilon_{mn})e^{i\omega_{mn}t} \qquad (6.19)$$

with $k_{mn}a = q_{mn}$ and the natural frequencies found from

$$f_{mn} = \frac{1}{2\pi}\frac{q_{mn}c}{a} \qquad (6.20)$$

The real part of equation (6.19) describes the physical displacement in the normal mode (m, n) as follows,

$$z_{mn}(r, \theta, t) = A_{mn} J_m(k_{mn}r) \cos(m\theta + \epsilon_{mn}) \cos(\omega_{mn}t + \phi_{mn})$$

where $A_{mn} = \mathbf{A}_{mn} = e^{i\phi_{mn}}$. The arbitrary constant ϵ_{mn} is an azimuthal phase angle. For each normal mode, this constant of integration defines the directions along which the radial nodal lines of zero displacement occur, but the value of ϵ_{mn} depends on the value of the azimuthal angle at which the membrane is excited at $t = 0$. A number of the first few (and simpler) modes of vibration with $\epsilon_{mn} = 0$ are shown in Figure 6.3. Each mode is designated by the ordered pair of integers (m, n). Integer m governs the number of *radial nodal lines*, while integer n determines the number of *azimuthal nodal circles*. Because mode $(0, 0)$ would obviously be trivial,

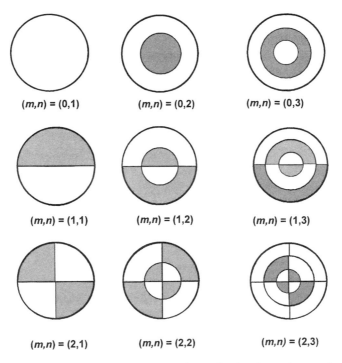

$(m,n) = (0,1)$ $(m,n) = (0,2)$ $(m,n) = (0,3)$

$(m,n) = (1,1)$ $(m,n) = (1,2)$ $(m,n) = (1,3)$

$(m,n) = (2,1)$ $(m,n) = (2,2)$ $(m,n) = (2,3)$

FIGURE 6.3. Vibration modes in a circular membrane fixed at its perimeter. A number of simpler modes are shown here.

TABLE 6.1. Relative Frequencies for Various Vibrational Modes.

$f_{01} = 1.0 f_{01}$	$f_{11} = 1.593 f_{01}$	$f_{21} = 2.135 f_{01}$
$f_{02} = 2.295 f_{01}$	$f_{12} = 2.917 f_{01}$	$f_{22} = 3.500 f_{01}$
$f_{03} = 3.598 f_{01}$	$f_{13} = 4.230 f_{01}$	$f_{23} = 4.832 f_{01}$

the case of $n = 1$ constitutes the least allowed value of n, and this corresponds to a mode of vibration where one azimuthal nodal circle occurs only at the rim of the membrane where $r = a$. Virtually the entire membrane vibrates in the z-direction in axisymmetric unison, with the maximum amplitude at $r = 0$ tapering off to zero at the boundary $r = a$.

For each nonzero value of m there exists a chain of allowed radial vibration modes of increasing frequency, as illustrated in Figure 6.3. At $m = 0$, $J_0(k_{0n}a)$ equals zero, which sets the conditions for the allowed frequencies. For $m = 1$, $J_1(k_{1n}a) = 0$ provides the allowed frequencies; and for $m = 2$, $J_2(k_{2n}a) = 0$ supplies the corresponding frequencies for that radial mode. Table 6.1 lists the frequencies f_{mn} of the circular membrane relative to the fundamental frequency f_{01} obtained from equation (6.20). From Table 6.1 it is seen that none of the overtones exists as a harmonic of the fundamental.

6.5 Case Study: Symmetric Vibrations of a Circular Membrane Fixed at Its Perimeter

The case of symmetric vibrations of a circular membrane fixed at its rim has many applications, so this situation holds great interest for us. Vibrational symmetry implies a solution to the wave equation (6.4) that is independent of θ, and we limit the solutions for this case to

$$z_{0n} = A_{0n} J_0(k_{0n}r)e^{i\omega_{0n}t} \tag{6.21}$$

and because only $m = 0$ applies, we discard this subscript and retain only the index n. The fixedness of the circular perimeter forecasts the boundary condition $z = 0$ at $r = a$. Therefore, $J_0(ka) = 0$, with the zeroes occurring as $k_n a = 2.405, 5.520, 8.654, 11/792, \ldots$. The fundamental frequency is derived from

$$f_1 = \frac{\omega_1}{2\pi} = \frac{k_1 c}{2\pi} = \frac{2.405}{2\pi a}\sqrt{\frac{T}{\rho_s}} \tag{6.22}$$

with the (nonharmonic) overtones to the fundamental frequencies obtained as

$$f_2 = \left(\frac{5.520}{2.405}\right)f_1 = 2.295 f_1,$$

$$f_3 = 3.58 f_1,$$

$$f_4 = 4.90 f_1, \ldots, \text{etc.}$$

For the real part of $z_1 = A_1 J_0(k_1 r)e^{i\omega t}$ the general expression for the membrane's displacement in the fundamental mode can be written as

$$z_1 = A_1 \cos(\omega_1 t + \phi_1) J_0 \left(2.405 \frac{r}{a} \right) \tag{6.23}$$

where A_1 is the maximum absolute value of the complex amplitude $\mathbf{A}_1$ at the center $r = 0$ of the membrane. To completely describe the symmetric vibrations the complete solution must be expressed as

$$z = \sum A_n \cos(\omega_n t + \phi_n) J_0(k_n t) \tag{6.24}$$

For the symmetric modes of vibration above the fundamental, nodal circles will exist at the inner radii at which $J_0(k_n r)$ vanishes. For example, the first overtone $J_0(k_2 r) = 0$ when $k_2 r = 5.520\,r/a = 2.405$, or $r = 0.436a$. In Figure 6.3, this mode of vibration is shown as an annulus surrounding the central circle. When the central circle of the membrane moves up the outer annulus moves down, and vice versa. Because portions of the membrane are moving upward at the same time the remaining areas move downward the efficiency of the sound output of a drum is low for the overtone frequencies.

A measure of the sound output of each mode is the average displacement amplitude of the surface when it is vibrating in that mode. For the nth symmetric mode, we can apply equation (6.24) to determine the average displacement amplitude $\langle \Psi_n \rangle$ over surface S as follows:

$$\begin{aligned} \langle \Psi_n \rangle &= \frac{1}{\pi a^2} \int_S A_n J_0(k_n r)\, dS \\ &= \frac{1}{\pi a^2} \int_0^a A_n J_0(k_n r) 2\pi r\, dr \\ &= \frac{2A_n}{k_n a} J_1(k_n a) \end{aligned} \tag{6.25}$$

For all nonsymmetric modes, we note that the angular dependence $\cos(m\theta + \epsilon)$ ensures that the average displacement is zero. Thus, using the prior double-subscript notation, we can state that $\langle \Psi_{mn} \rangle = 0$ for all $m \neq 0$.

It is of interest to find the value of average displacement amplitude $\langle \Psi_1 \rangle$ for the fundamental mode and to compare it with $\langle \Psi_2 \rangle$ for the first overtone. From equation (6.25):

$$\langle \Psi_1 \rangle = \frac{2A_1}{k_1 a} J_1(k_1 a) = \frac{2A_1}{2.405} = 0.432\, A_1$$

The motion of the membrane can be equated to the displacement of a rigid flat piston of radius a moving with an amplitude of $0.432\, A_1$. We can also apply equation (6.25) to find that $\langle \Psi_2 \rangle = -0.123\, A_2$. The negative sign denotes that the average displacement amplitude is opposite in direction to the displacement at the center. The fundamental node of vibration is thus seen to be more than three times as effective for displacing air as for the first overtone.

In real applications, such as loudspeakers, the amount of air displaced by the membranes, rather than the exact shape of the moving surface, determines the principle characteristics of generated sound waves. The radiating source can be depicted by an *equivalent simple piston* of area S_{eq}, and this piston moves through a displacement amplitude ζ_{eq} so as to sweep the volume displacement of the actual source. The volume displacement amplitude of the simple piston equivalent to the circular membrane vibrating in its fundamental mode is

$$S_{eq}\zeta_{eq} = 0.432\pi a^2 A_1$$

The nodal vibrations of actual membranes cannot be sustained with constant amplitudes because of damping forces caused by both internal friction and external forces associated with the radiation of acoustic energy. The amplitude of each mode tends to decay exponentially with time as $e^{-\beta_n t}$, where β_n represents the damping constant for mode n. This damping constant generally increases with frequency, with the result that higher frequencies damp out more quickly than does the fundamental.

6.6   Application of Membrane Theory to the Kettledrum

In addition to the damping forces mentioned above, other forces may act on a membrane and affect its vibration. The kettledrum is an example of the case of a membrane that covers a closed space in which changes of pressures occur as the entrapped volume of air changes in pressure incurred by the vibration of the drumhead. A similar situation occurs with the air atrapped behind the diaphragm of a condenser microphone.

The kettledrum consists of a membrane stretched taut over the open end of a hemispherical shell. As this membrane (i.e., the drumhead) vibrates, the air contained inside the shell undergoes alternative compressions and rarefactions. With the radial velocity of the transverse waves being considerably less than the speed of sound in air, the pressure arising from the alternative compression and decompression of the entrapped air is fairly uniform across the entire drumhead and depends only on the average displacement $\langle z \rangle$. With the radius of the drumhead designated by a, the incremental or displaced volume of the enclosed air is given by $dV = \pi a^2 \langle z \rangle$. Let us denote by V_0 the equilibrium or quiescent volume of the air enclosed in the kettledrum. The corresponding unperturbed pressure is P_0. The vibration of the air enclosed in the kettledrum is essentially an *adiabatic* process, with the result that the instantaneous pressure P and volume V are related to the quiescent values by:

$$PV^\gamma = P_0 V_0^\gamma = \text{const} \tag{6.26}$$

Here γ is the ratio of c_p, the specific heat of the contained air at constant pressure, to c_v, the specific heat at constant volume. Differentiating equation 6.26 yields the pressure deviation dP:

$$dP = -\frac{\gamma P_0}{V_0} dV = -\frac{\gamma P_0}{V_0} \pi a^2 \langle z \rangle \tag{6.27}$$

This gives rise to an incremental force $dP\, dx\, dy$ over incremental area $dx\, dy$ of the membrane. Modifying equation (6.3) to include this incremental force we obtain

$$\nabla^2 z - \frac{1}{c^2} \frac{\gamma P_0}{\rho_0 V_0} \pi a^2 \langle z \rangle = \frac{1}{c^2} \frac{\partial^2 z}{\partial t^2} \tag{6.28}$$

The term $\langle z \rangle$ is an integral function of all the permitted modes of vibration, which must also include the influences of their relative amplitudes and phases. But we can greatly simplify the solution of equation (6.28) by assuming only one mode of vibration and disregarding all of the other modes which constitute the general solution.

The average displacement is zero for all normal modes dependent upon θ; therefore, none of these modes are affected by the pressure fluctuation of the air inside the drum. We need only to consider the symmetric modes entailing the Bessel function J_0. The solution with only one frequency present is of the form depending only on the coordinate r,

$$z = \Psi e^{i\omega t}$$

Inserting the above into equation (6.28) yields

$$\frac{d^2 \Psi}{dr^2} + \frac{1}{r} \frac{d\Psi}{dr} + k^2 \Psi = \frac{\gamma P_0}{T V_0} \int_0^a 2\pi r \Psi \, dr \tag{6.29}$$

In order to establish the solution to the differential equation, we examine the salient features of equation (6.29). If the right-hand integral term were not present the solution would entail $J_0(kr)$. But the presence of this integral term involving the radius a suggests an additional term to the solution, one that is a function of a, namely, $J_0(ka)$. Moreover, the assumed solution should meet the boundary condition that $\Psi = 0$ at $r = a$, regardless of the value of k.

We now integrate the right side of equation (6.29) as follows

$$\frac{2\pi \gamma P_0}{T V_0} A \left[\frac{r J_1(kr)}{k} - \frac{r^2}{2} J_0(ka) \right]_0^a$$

$$= \frac{\gamma P_0}{T V_0} \pi a^2 A \left[\frac{2 J_1(ka)}{ka} - J_0(ka) \right] = \frac{\gamma P_0}{T V_0} \pi a^2 A J_2(ka) \tag{6.30}$$

Insertion of (6.30) into equation (6.29) provides the condition for the viability of the assumed solution

$$-k^2 J_0(ka) = \frac{\gamma P_0}{T V_0} \pi a^2 J_2(ka)$$

TABLE 6.2. Allowed Frequencies of a Kettledrum.

B	0	1	2	5	10
$k_1 a$	2.405	2.545	2.68	3.02	3.485
$k_2 a$	5.520	5.54	5.55	5.59	5.87
$k_3 a$	8.654	8.657	8.660	8.67	8.69

or

$$J_0(ka) = -\frac{B J_2(ka)}{(ka)^2}$$

where

$$B = \frac{\gamma P_0}{T V_0} \pi a^2$$

The parameter B is a dimensionless constant that compares the relative magnitudes of the restoring forces arising from the compression effects of the air trapped inside the drum with the tension applied to the drumhead. The value of B is small if either the volume or the tension in the membrane is quite large compared with the compressive pressure acting over the area of the membrane. In the limit where B approaches zero, the allowed frequencies become those corresponding to the freely vibrating circular membrane which was described earlier in this chapter. The allowed values of ka are listed in Table 6.2 for selected values of B ranging from 0 (which corresponds to an unimpeded vibrating circular membrane) to 10 (indicative of low drum volume or light drumhead tension). The effect of the additional term in equation (6.28), which is proportional to the displacement and therefore is indicative of membrane stiffness, is to elevate the allowed frequencies. The effect on the fundamental frequency is much more considerable than it is on the higher modes of vibration. This stems from the fact that the average displacement amplitude becomes smaller with increasingly higher modes of vibration with a consequently larger number of oppositely phased segments. It is also apparent that because pressure fluctuations inside the drum affect only the basic frequency modes z_{0n}, the tonal qualities of a kettledrum can be varied by parametric changes of the drum volume V_0 and the area πr^2 of the drumhead.

6.7   Forced Vibrations of a Membrane

Consider a circular membrane that is acted only on one side by a evenly distributed sinusoidal driving pressure $p = P \cos \omega t$. In complex notation the pressure is given by

$$\mathbf{p} = P e^{i\omega t}$$

and the equation of motion (6.3) becomes modified as follows:

$$\frac{\partial^2 z}{\partial t^2} = c^2 \nabla^2 z + \frac{P}{\rho_s} e^{i\omega t} \qquad (6.31)$$

Assume a steady-state solution

$$\mathbf{z} = \mathbf{\Psi} e^{i\omega t} \qquad (6.32)$$

which is then inserted into equation (6.31), resulting in

$$\nabla^2 \mathbf{\Psi} + k^2 \mathbf{\Psi} = \frac{P}{\rho_s c^2} = -\frac{P}{T} \qquad (6.33)$$

where $k = \omega/c$. In this situation of a driven membrane the angular frequency ω may have any value, and the wave number k is thus not limited to discrete sets of values which prevail in freely vibrating membranes.

The solution to equation (6.33) consists of two parts, one being a general solution of the homogeneous equation $\nabla^2 \mathbf{\Psi}_h + k^2 \mathbf{\Psi}_h = 0$ and the second being the particular solution $\mathbf{\Psi}_p = -P/(k^2 T)$. Then the complete solution can be written as

$$\mathbf{\Psi} = \mathbf{A} J_0(kr) - \frac{P}{k^2 T}$$

The immobility of the membrane at the rim $r = a$ provides the boundary condition $\mathbf{\Psi}(a) = 0$, and

$$\mathbf{A} = \frac{1}{J_0(ka)} \cdot \frac{P}{k^2 T}$$

The displacement of the membrane becomes

$$z(r, t) = \frac{P}{k^2 T} \left[\frac{J_0(kr)}{J_0(ka)} - 1 \right] e^{i\omega t} \qquad (6.34)$$

with the corresponding amplitude of the displacement at any position in the membrane given by

$$\mathbf{\Psi}(r) = \frac{P}{T} \left[\frac{J_0(kr) - J_0(ka)}{k^2 J_0(ka)} \right] \qquad (6.35)$$

From equation (6.35) it is seen that the amplitude of the displacement is directly proportional to the driving force P and inversely proportional to the tension T. The vibrational amplitude at any location on the membrane depends on the transcendental terms enclosed by the square bracket in equation (6.35). But if the driving frequency ω corresponds to any of the free oscillation frequencies of equation (6.22) and the overtones, the Bessel function $J_0(ka)$ assumes zero values, presaging infinite amplitudes. But damping forces occur in real cases, and these may be represented in equation (6.31) by a damping factor $-(R/\rho_s)(\partial z/\partial t)$ that limits the amplitudes to finite maximum values.

The average displacement $\langle z \rangle_s$ of the driven membrane is found by averaging over the surface area of the membrane:

$$\langle z \rangle_s = \frac{2\pi e^{i\omega t} \int_0^a \frac{P}{k^2 T} \left[\frac{J_0(kr)}{J_0(ka)} - 1 \right] r \, dr}{\pi a^2}$$

$$= \frac{P}{k^2 T} \frac{J_2(ka)}{J_0(ka)} e^{i\omega t} \qquad (6.36)$$

At low frequencies ka assumes a value less than unity, and the following approximations for Bessel functions hold true:

$$J_0(ka) \approx 1 - \frac{1}{4}(ka)^2$$

$$J_2(ka) = \frac{k^2 a^2}{8} \left(1 - \frac{k^2 a^2}{12} \right)$$

Introducing the above approximations into equation (6.36) yields the following expression for the average displacement at low frequencies:

$$\langle z \rangle_s \approx \frac{Pa^2}{8T} \left(1 + \frac{k^2 a^2}{6} \right) \qquad (6.37)$$

If we apply the situation as represented by equation (6.37) to the design of a condenser microphone, it is apparent that as long the driving frequency is sufficiently low, that is, $ka \ll 1$, the output of the microphone will be virtually independent of the frequency. No resonances should occur in that frequency range. The first resonance occurs at $ka = 2.405$. Because

$$k = \frac{2\pi f}{c} = 2\pi f \left(\frac{T}{\rho_s} \right)^{1/2}$$

and if we set the limiting frequency of the uniform microphone response to $ka < 1$, then

$$f < \frac{1}{2\pi a} \sqrt{\frac{T}{\rho_s}}$$

The upper frequency limit of the microphone can be elevated by either increasing the tension T or decreasing the radius a, all other factors being equal. But this also has the effect of lessening the amplitude of the average displacement $\langle z \rangle_s$ and, consequently, the voltage output of the microphone.

When a damping factor $-(R/\rho_s)(\partial z/\partial t)$ is included in equation (6.31) the resulting solution does not change except that k is replaced by $\mathbf{k}$, a complex expression represented by

$$\mathbf{k}^2 = \frac{\omega^2}{c^2} - \frac{i\omega R}{T}$$

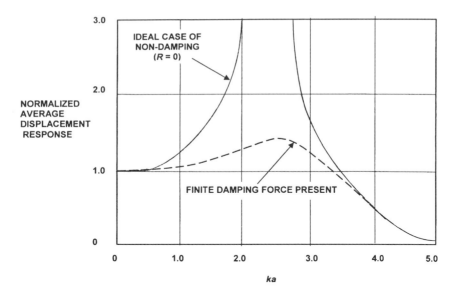

FIGURE 6.4. Plot showing the average (normalized) displacement response as a function of frequency. The effect of damping is also shown.

The presence of the imaginary component $-\omega R/T$ causes the average displacement to assume a finite value at resonance. Figure 6.4 displays the average displacement response $\langle \Psi \rangle_s$ of a freely vibrating dissipationless membrane, as computed through the use of equation (6.36). The amplitude assumes a value of infinity at $ka = 2.405$. Another curve that includes the effect of damping is also plotted, and the corresponding amplitude assumes a finite value at $ka = 2.405$. Both of these curves indicate zero responses at $ka = 5.136$, for which $J_2(ka) = 0$. If the frequency is increased beyond the first resonance value to approximately 1.60 times the first resonant frequency, a circular nodal line will appear near the rim of the membrane. As the frequency is increased, the nodular line moves inward as a circle of decreasing radius. The displacement of the membrane's center is out of phase with the driving force, while that of the membrane's outer portion remains in phase. As the driving frequency increases and the nodal circle shrinks, the average displacements of the two zones tend to cancel each other out. The cancellation becomes complete at $ka = 5.136$, and no displacement occurs across the entire surface of the membrane.

6.8 Vibrating Thin Plates

The principal difference between the vibration of a membrane and a thin plate is the restoring force in the former is due entirely to tension acting on the membrane, and in the latter there is no tension applied and the restoring force is attributed entirely to the inherent stiffness of the plate.

To keep matters simple, we consider only symmetrical vibrations of a uniform circular diaphragm. The appropriate equation, essentially equivalent to equation (6.36), but modified to include the effect of stiffness, is

$$\frac{\partial^2 z}{\partial t^2} = -\frac{\kappa^2 E}{\rho(1 - \mu^2)} \nabla^2(\nabla^2 z) \tag{6.38}$$

where ρ is the density of the material, μ is the Poisson's ratio, E is Young's modulus, and κ is the surface radius of gyration. For a circular plate of uniform thickness b the radius of gyration is given by

$$\kappa = \frac{b}{\sqrt{12}}$$

The elastic resistance to flexing provides the restoring force that acts on the circular plate. While there may be some temptation to consider the coefficient $-\kappa^2 E/[\rho(1 - \mu^2)]^2$ on the right-hand side of equation (6.38) as being analogous to $-\kappa^2 E/\rho$ in equation (5.28) for the transverse vibration of a bar, this is not strictly true because a sheet will curl up sideways as it is bend downward along its length. This is the *Poisson effect* in which the curling occurs from the lateral expansion as the longitudinal compression ensues from the bending of the plate. An increase in the effective stiffness is thereby produced. The Poisson ratio μ given in equation (6.38) is the negative ratio of the lateral strain $\partial \xi/\partial y$ to $\partial \zeta/\partial x$, that is,

$$\mu = -\frac{\partial \xi/\partial y}{\partial \varsigma/\partial x}$$

In order to keep the Poisson ratio a positive number, it is necessary to introduce the minus sign to counteract the effect that a positive longitudinal strain gives rise to a negative lateral strain of compression. The value of μ, which is a property of the material, may be obtained from standard tables and is generally of the value 0.3. In equation (6.38) the factor $(1 - \mu^2)^{-1}$ embodies the effective increase in the stiffness of the plate resulting from the curling.

In solving equation (6.38) it is assumed that

$$z = \Psi(r)^{i\omega t}$$

which is then substituted into the equation to give

$$\nabla^2(\nabla^2 \Psi) - K^4 \Psi = 0 \tag{6.39}$$

in which

$$K^4 = \frac{\omega^2 \rho(1 - \mu^2)}{\kappa^2 E}$$

The substitution of the Helmholtz equation $\nabla^2 \Psi = -K^2 \Psi$ into equation (6.39) indicates that if Ψ can satisfy the Helmholtz equation, it will also constitute a solution to equation (6.39). The Ψ in the relationship $\nabla^2 \Psi = K^2 \Psi$ will also satisfy equation (6.39), so the complete solution of this equation must be the sum of four independent solutions to

$$\nabla^2 \Psi \pm K^2 \Psi = 0 \tag{6.40}$$

Equation (6.40) with the positive sign is the Helmholtz equation with circular symmetry, which yields the solutions $J_0(Kr)$ and $Y_0(Kr)$. But the boundary condition that the displacement must be finite at $r = 0$ at the center of the plate requires that the latter solution must be scrapped. The solution of equation (6.40) with the negative sign yields $J_0(iKr)$ and $Y_0(iKr)$; the latter term is also discarded. The term $J_0(iKr)$ is a modified Bessel function of the first kind, generally written as $I_0(Kr)$.[2] The complete applicable solution of equation (6.39) is

$$\Psi = \mathbf{A} J_0(Kr) + \mathbf{B} I_0(Kr) \tag{6.41}$$

The function $J_0(Kr)$ is an oscillating function that damps out with increasing r while $I_0(Kr)$ increases continuously with r.

The manner in which the plate is supported determines the conditions which are used to evaluate the constants $\mathbf{A}$ and $\mathbf{B}$. A common type of support is one in which the circular plate is rigidly clamped at its periphery $r = a$. The boundary conditions therefore are

$$\Psi = 0 \quad \text{and} \quad \partial \Psi / \partial r = 0 \qquad \text{at } r = a$$

These yield

$$\mathbf{A} J_0(Ka) = \mathbf{B} I_0(Ka), \qquad -\mathbf{A} J_1(Ka) = \mathbf{B} I_1(Ka) \tag{6.42}$$

and through elimination of the constants $\mathbf{A}$ and $\mathbf{B}$ we obtain the transcendental equation which gives the permissible values of Ka:

$$\frac{J_0(Ka)}{J_1(Ka)} = -\frac{I_0(Ka)}{I_1(Ka)} \tag{6.43}$$

Both I_0 and I_1 remain positive for all values of Ka, so solutions occur only when J_0 and J_1 have opposite signs. The sequence of solutions satisfying equation (6.43) is

$$Ka = 3.20, 6.30, 9.44, 12.57, \ldots$$

The above can be approximated by $Ka = n\pi$, where $n = 1, 2, 3, ..$ This approximation improves with increasing values of n.

From the definition of K for equation (6.39) it is apparent that the frequency can be found from

$$f = \frac{\omega}{2\pi} = \frac{\kappa K^2}{2\pi} \sqrt{\frac{e}{\rho(1 - \mu^2)}}$$

---

[2] The modified Bessel functions $I_n(x)$ are solutions of the modified Bessel differential equation

$$\frac{d^2 y}{dx^2} + \frac{1}{x} \frac{dy}{dx} - \left(1 + \frac{n^2}{x^2}\right) y = 0$$

By setting $K = 3.20/a$, the fundamental frequency f_1 is found to be

$$f_1 = \frac{\omega_1}{2\pi} = \frac{3.2^2}{2\pi a^2} \frac{b}{\sqrt{12}} \sqrt{\frac{E}{\rho(1 - \mu^2)}} = 0.47 \frac{b}{a^2} \sqrt{\frac{E}{\rho(1 - \mu^2)}}$$

where b represents the thickness of the plate. The frequencies of the overtones are given by

$$f_2 = \left(\frac{6.3}{3.2}\right)^2 f_1 = 3.88 f_1, \qquad f_3 = 8.70 f_1, \quad \text{etc.}$$

These frequencies are spread out much further apart than those for the circular membrane.

For the fundamental mode of vibration, the displacement of a thin circular plate is given by

$$z_1 = \cos(\omega_1 t + \phi_1)\left[A_1 J_0\left(\frac{3.2}{a}r\right) + B_1 I_0\left(\frac{3.2}{a}r\right)\right]$$

From the boundary condition relationships of equation (6.42) the last expression becomes

$$z_1 = A_1 \cos(\omega_1 t + \phi_1)\left[J_0\left(\frac{3.2}{a}r\right) + 0.0555 I_0\left(\frac{3.2}{a}r\right)\right]$$

It is interesting to observe that the amplitude at the center $r = 0$ is $1.0555 A_1$, not A_1. If we compare the shape function represented by the bracketed terms on the right-hand side of the last equation with the corresponding shape function $J_0(2.405r/a)$ for the fundamental mode of a similarly sized vibrating circular membrane, it will be found that the relative displacement of the plate near its edge is considerably smaller than that of the membrane. Hence, the ratio of the average amplitude to the amplitude at the center is less than is the case for the membrane. The average displacement amplitude is given by

$$\langle \Psi_1 \rangle_s = 0.326 A_1 = 0.309 z_0$$

where $z_0 = 1.0555 A_1$ represents the amplitude at the center $r = 0$ of the plate. In the same manner we used to represent the membrane, the circular plate can be depicted by an equivalent flat piston so that

$$S_{eq} \zeta_{eq} = 0.309\pi a^2 z_0$$

Plates can also undergo loaded and forced vibrations. The mathematical treatments of these cases are analogous to those for membranes, and the response curves are similar to those shown in Figure 6.4. Large amplitudes will also occur at resonance frequencies unless there is appreciable damping.

The most apparent use of the vibrating thin plate is that of the telephone diaphragms (both receiver and microphone). While these diaphragms do not provide the flatter frequency responses or frequency range of membranes in condenser microphones, they do provide adequate intelligibility, are generally far more rugged in their construction and cheaper to manufacture. Sonar transducers used to generate underwater sounds less than 1 kHz constitute another class of vibrating plates; the signals are produced by the variations of an electromagnetic field in an electromagnet positioned closely to a thin circular steel plate.

References

Fletcher, Neville H., and Thomas D. Rossing. 1991. *The Physics of Musical Instruments*. New York: Springer-Verlag; Chapter 3.

Kinsler, Lawrence E., Austin R. Frey, Alan B. Coppens, and James V. Sanders. 1982. *Fundamentals of Acoustics*, 3rd ed., New York: John Wiley & Sons; Chapter 4.

Morse, Philip M., and Ingard, K. Uno. 1968. *Theoretical Acoustics*. New York: McGraw-Hill; Sections 5.2 and 5.3.

Reynolds, Douglas R. 1981. *Engineering Principles of Acoustics, Noise, and Vibration Control*. Boston: Allyn and Bacon; pp. 247–255.

Wood, Alexander. 1966. *Acoustics*. New York: Dover Publications; pp. 429–436.

Problems for Chapter 6

All membranes described below may be assumed to be fixed at their perimeters unless otherwise indicated.

1. Consider a square membrane, having dimensions $b \times b$, vibrating at its fundamental frequency with amplitude δ at its center. Develop an expression that gives the average displacement amplitude. Obtain a general expression for points having an amplitude of $\delta/2$. Plot at least five points from this general expression. Do these points fall in a circle?

2. A rectangular membrane has width b and length $3b$. Find the ratio of the first three overtone frequencies relative to the fundamental frequency.

3. Consider a circular membrane with a free rim. Develop the general expression for the normal modes and sketch the nodal patterns for the three normal modes with the lowest natural frequencies. Express the frequencies of these normal modes in terms of tension and surface density.

4. A circular aluminum membrane of 2.5 cm radius and 0.012 cm thickness is stretched with a tension of 15,000 N/m. Find the first three frequencies of free vibration, and for each of the frequencies determine any nodal circles.

5. Prove that the total energy of a circular membrane vibrating in its fundamental mode is equal to $0.135\pi\rho_s(a\omega A_f)^2$, where ρ_s is the area density, a is the radius of the membrane, ω is the angular frequency of the vibration, and A_f the fundamental amplitude at the center.

6. Steel has a tensile strength of 1.0 GPa ($= 10^9$ Pa) and aluminum, 0.2 GPa. Using these values as the maximum tensions, what will be the maximum fundamental frequency of a 2-cm-diameter circular membrane made up of each of these materials? Note: For thin membranes these fundamental frequencies are independent of the thicknesses.

7. A damping force is applied uniformly over the surface of a circular membrane. This damping force per unit area $= -\Im\, \partial y/\partial t$ should be introduced into the appropriate wave equation in a manner consistent with the dimensions of the terms of the equation. Solve the equation to demonstrate that the amplitudes of the free vibrations are damped exponentially as $e^{-1/2\,\Im t/P_s}$.

8. A kettledrum consists of a circular membrane of 50 cm diameter, with an area density of $1.0\,\mathrm{kg/m^2}$. The membrane is stretched under a tension of 10,000 N/m.
   (a) Determine the fundamental frequency of the membrane without a backing vessel.
   (b) Determine the fundamental frequency for the membrane with a backing vessel that is a hemispherical bowl of 25 cm radius. The vessel is filled with air at a pressure of 100 kPa, and γ (the ratio of specific heats) is 1.4.

9. An undamped membrane of 4 cm radius has an area density of $1.6\,\mathrm{kg/m^2}$ and is stretched to a tension of 1200 N/m. It is driven by a uniform pressure 7000 $\sin \omega t$ Pa applied over the entire surface.
   (a) Determine and plot the amplitude of the displacement at the center as a function of frequency ranging from 0 to 2 kHz.
   (b) Compute and plot the shape of the membrane when driven by the applied frequency of 600 Hz and the applied frequency of 1000 Hz.

10. A condenser microphone contains a circular aluminum diaphragm of 30 mm diameter and 0.02 mm thickness. Aluminum has a maximum tensile strength of 0.2 GPa.
    (a) What is the allowable maximum tension in newtons per meter in the diaphragm?
    (b) What will be the fundamental frequency under these conditions?
    (c) What will be the displacement of the diaphragm at its center under the impetus of a 500-Hz sound wave having a pressure amplitude of 1.5 Pa?
    (d) What will be the average displacement under the conditions of (c)?

11. If the volume of air trapped behind the diaphragm of the condenser microphone of the preceding problem is 2.5 10^{-7} m^3, by how much will the fundamental frequency be raised? Assume the normal air pressure to be equal to 100 kPa and $\gamma = 1.4$.

12. Use integration over the surface of a circular thin plate vibrating in its fundamental mode to show that the average displacement amplitude is $0.309A$, where A denotes the displacement in the center of the plate.

13. The diaphragm of a typical telephone receiver comes in the form of a circular sheet of steel, 4 cm in diameter and 0.18 mm thick.
    (a) What is the fundamental frequency if the diaphragm is rigidly clamped at its rim.
    (b) How will this fundamental frequency change if the diaphragm thickness is doubled?
    (c) What would happen to the fundamental frequency if the diameter of the diaphragm was increased by 50%?

14. Find the fundamental frequency of a vibrating circular steel plate that is clamped at its rim and is of 25 cm diameter and 0.55 mm thickness.

7
Pipes, Waveguides, and Resonators

7.1 Introduction

In dealing with strings, bars, and membranes in Chapters 4–6, we considered relatively simple geometric conditions. But when the sound waves are confined in a restricted amount of space, the situation becomes more complex. For example, as sound propagates inside a rigid-walled pipe with a wavelength that exceeds the radius of a rigid-wall pipe, the acoustic propagation inside the pipe becomes fairly planar. The resonance properties of the pipes driven at one end and closed off at other end constitute the basis for measuring acoustical impedances and absorption properties of materials. In our study of pipes we establish the models for physical analyses of wind musical instruments, organ pipes, and ventilation ducts. In larger spaces, where the dimensions may exceed wavelengths, two- and three-dimensional standing waves can occur. We shall treat the simple case of a waveguide with a uniform cross-section, and establish the concept of group speed and phase speed which occurs with a wave propagating inside a waveguide. The acoustic waveguide is very much analogous to the electromagnetic waveguides, and it finds applications in both surface-wave delay lines and in the propagation of sound in oceanic and atmospheric layers. We shall also consider the physics of a Helmholtz resonator.

7.2 The Simplest Enclosed System: Infinite Cylindrical Pipe

The simplest enclosed system inside which sound propagation occurs is an infinite cylindrical pipe with its axis parallel to the direction of the propagation of the plane wave in the enclosed medium. The pipe wall is assumed to be rigid, perfectly smooth, and adiabatic (i.e., no heat transfer occurs through the wall). The pipe thus has no effect on the wave propagation. A pressure wave generated by a piston moving in the x-direction can be expressed as

$$\mathbf{p}(\mathbf{x}, \mathbf{t}) = \mathbf{p}_0 e^{i(\omega t - kx)} \tag{7.1}$$

Here $\mathbf{p}_0$ is the maximum amplitude of the pressure. The volume flow is given by

$$U(x, t) = \frac{p_0 S}{\rho c} e^{i(\omega t - kx)}$$

where ω is the angular frequency, $k = 2\pi/\lambda = \omega/c$ is the wave number, S is the cross-sectional area of the pipe, ρ is the density of the fluid inside the pipe, and c is the propagation velocity.

7.3   Resonances in a Closed-Ended Pipe

In Figure 7.1 consider a pipe of length L and cross-sectional area S, filled with a fluid and sealed off at one end, $x = L$. Let the fluid inside the pipe be driven by a piston at $x = 0$. The pipe has a mechanical impedance $\mathbf{Z}_{nL}$. The piston vibrates harmonically at a sufficiently low frequency so that only plane waves are deemed to exist inside the pipe. The wave inside the pipe can be described by

$$\mathbf{p} = \mathbf{A}e^{i[\omega t + k(L-x)]} + \mathbf{B}e^{i[\omega t - k(L-x)]} \tag{7.2}$$

where $\mathbf{A}$ and $\mathbf{B}$ are established by the boundary conditions at $x = 0$ and $x = L$.

At $x = L$ the mechanical impedance of the wave must equal the mechanical impedance $\mathbf{Z}_{nL}$ at the termination so as to sustain the continuities of force and particles. The force of the fluid acting at the end of the pipe is $\mathbf{p}(L, t)S$, and the corresponding particle speed $u(L, t)$ derives from the integrated equation (2.22)

$$u = -\int 1/\rho \, \delta(\partial p/\partial x) \, dt.$$

Mechanical impedance $\mathbf{Z}_n$, expressed as

$$\mathbf{Z}_n = \frac{\mathbf{f}}{\mathbf{u}} \tag{7.3}$$

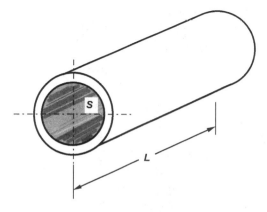

FIGURE 7.1. A pipe closed-ended at $x = L$.

represents a complex value that is the ratio of the complex driving force **f** to the complex speed **u** at the point where the force is applied. In the case of the finite pipe, the mechanical impedance at $x = L$ is given by:

$$\mathbf{Z}_{nL} = \rho_0 c S \frac{\mathbf{A} + \mathbf{B}}{\mathbf{A} - \mathbf{B}} \tag{7.4}$$

The value of the input mechanical impedance at $x = 0$ is expressed as

$$\mathbf{Z}_{n0} = \rho_0 c S \frac{\mathbf{A} e^{ikL} + \mathbf{B} e^{-ikL}}{1 + \frac{i \mathbf{Z}_{nL}}{\rho_0 c S} \tan kL} \tag{7.5}$$

Eliminating **A** and **B** by combining equations (7.4) and (7.5) yields

$$\frac{\mathbf{Z}_{n0}}{\rho_0 c S} = \frac{\frac{\mathbf{Z}_{nL}}{\rho_0 c S} + i \tan kL}{1 + \frac{\mathbf{Z}_{nL}}{\rho_0 c S} \tan kL} \tag{7.6}$$

The term $\rho_0 c S$ is the *characteristic mechanical impedance* of the fluid. The complex quantity $\mathbf{Z}_{n0}$ can be recast in terms of real and imaginary components, r and ψ, respectively:

$$\frac{\mathbf{Z}_{nL}}{\rho_0 c S} = r + i\psi \tag{7.7}$$

The ratio on the left-hand side of equation (7.7) constitutes a *normalized* impedance. Inserting equation (7.7) into equation (7.6) yields

$$\begin{aligned}
\frac{\mathbf{Z}_{n0}}{\rho_0 c S} &= \frac{(r + i\psi) + i \tan kL}{1 + i(r + i\psi) \tan kL} \\
&= \frac{r(\tan^2 kL + 1) - i[\psi \tan^2 kL + (r^2 + \psi^2 - 1)\tan kL - \psi]}{(\psi^2 + r^2)\tan^2 kL - 2\psi \tan kL + 1}
\end{aligned} \tag{7.8}$$

When $r = 0$, the input impedance $\mathbf{Z}_{n0}$ vanishes when the reactance vanishes, that is,

$$\frac{-i[\psi \tan^2 kL + (r^2 + \psi^2 - 1)\tan kL - \psi]}{(\psi^2 + r^2)\tan^2 kL - 2\psi \tan kL + 1} = 0 \tag{7.9}$$

and this results in

$$\psi \tan^2 kL + (\psi^2 - 1)\tan kL - \psi = 0 \tag{7.10}$$

that is,

$$\psi = -\tan kL \tag{7.11}$$

The impedance becomes infinite when

$$\psi \tan^2 kL + (\psi^2 - 1)\tan kL - \psi = 0$$

or

$$\psi = \cot kL$$

Let us briefly examine the situation for a constant driving force at $x = 0$. The vanishing of $\mathbf{Z}_{n0} = \mathbf{f}/\mathbf{u}_0$ indicates that the speed amplitude at the point of force application ($x = 0$) is infinite, the condition for mechanical resonance. Conversely, the input impedance reaching infinity means that the speed amplitude approaches zero, which describes the condition of *antiresonance*.

To obtain the condition of resonance in a pipe driven at $x = 0$ and sealed with a rigid cap at $x = L$, we let $|\mathbf{Z}_{nL}/\rho_0 c S|$ approach infinity in equation (7.4), giving

$$\frac{\mathbf{Z}_{n0}}{\rho_0 c S} = -i \cot kL$$

The reactance becomes zero when $\cot kL = 0$,

$$k_n L = (2n - 1)\pi/2, \qquad n = 1, 2, 3, \ldots$$

and so we obtain the set of resonant frequencies as we did for the forced-fixed string:

$$f_n = \frac{2n - 1}{4} \frac{c}{L}$$

With the odd harmonics of the fundamental constituting the resonance frequencies, the driven closed pipe contains a pressure antinode at $x = L$ and a pressure node at $x = 0$. This means that the driver must present a vanishing mechanical impedance to the tube.

7.4   The Open-Ended Pipe

Consider a pipe driven at $x = 0$ but open-ended at the other end $x = L$. The assumption that $\mathbf{Z}_{nL} = 0$ at $x = L$ (which would lead to resonances at $f_n = \frac{1}{2}nc/L$) is not valid, because the open end of the pipe radiates outward into the surrounding air. The appropriate condition must be $\mathbf{Z}_{nL} = \mathbf{Z}_r$, where $\mathbf{Z}_r$ is the radiation impedance at the open end of the tube. Also, the presence of a flange at the open end affects the exit impedance. Consider the case of a flange at the open end of a circular pipe of radius a. The flange is large with respect to the wavelength of the sound, which, in turn, is considerably larger than the tube radius ($\lambda \gg a$). This situation resembles a baffled piston in the low-frequency limit. From theory (Kinsler and Frey, 1962)

$$\frac{\mathbf{Z}_{nL}}{\rho_0 c S} = \frac{(ka)^2}{2} + i\frac{8}{3}\frac{ka}{\pi} \tag{7.12}$$

where the real component $r = (ka)^2/2$ and the imaginary component $\psi = 8ka/3\pi$ are both much less than unity and $r \ll \psi$. Under these conditions the solution to equation (7.9) yields $\tan kL = -\psi$ in order for resonance frequencies to occur. With the assumption $\psi \ll 1$, we obtain

$$\tan(n\pi - k_n L) = \frac{8}{3\pi} \approx \tan\frac{8ka}{3\pi} \tag{7.13}$$

where $n = 1, 2, 3, \ldots$. Hence

$$n\pi = k_n L + \frac{8k_n a}{3\pi} \tag{7.14}$$

The resonance frequencies therefore are

$$f_n = \frac{n}{2} \frac{c}{L + \frac{8}{3\pi}a} \tag{7.15}$$

All of the resonance frequencies are harmonics of the fundamentals. We also note that the denominator $L + 8a/3\pi$ constitutes the effective length of the pipe rather than the actual length L.

In the case of the *unflanged* pipe, the radiation impedance, indicated by both theory and experiment, is given approximately by

$$\frac{\mathbf{Z}_{nL}}{\rho_0 c S} = \frac{(ka)^2}{4} + i(0.6ka) \tag{7.16}$$

Here the end correction for the unflanged pipe equals $0.6a$, with the effective length being $L + 0.6a$. We also note that the end corrections do not depend on the frequency. Providing that $\lambda n \gg a$, the resonance frequencies of flanged and unflanged pipes constitute harmonics of the fundamental. Hence, the driving frequency of an open-ended organ pipe yield resonances that are harmonics of the driving frequency. The above exposition so far has dealt with pipes of constant cross sections. If a pipe is *flared* at the open end, as is the case with many wind instruments such as the clarinet and the oboe, the results are modified, and the resonances may not necessarily be harmonics of the fundamental. Variations in the flare design will emphasize or lessen certain harmonics present in the forcing function, thereby affecting the quality or timbre of the sound emanated by the pipe.

7.5   Radiation of Power from Open-Ended Pipes

Equation (7.4) may be revised to read:

$$\frac{\mathbf{B}}{\mathbf{A}} = \frac{\frac{\mathbf{Z}_{nL}}{\rho_0 c S} - 1}{\frac{\mathbf{Z}_{nL}}{\rho_0 c S} + 1} \tag{7.17}$$

When the termination impedance $\mathbf{Z}_{nL}$ is known the power transmission coefficient T_n can be established from

$$T_n = 1 - \left| \frac{\mathbf{B}}{\mathbf{A}} \right|^2 \tag{7.18}$$

Through the use of equation (7.16), equation (7.17) applied to the case of an open-ended pipe becomes

$$\frac{\mathbf{B}}{\mathbf{A}} = \frac{\left[1 - \frac{(ka)^2}{2}\right] - i\frac{8ka}{3\pi}}{\left[1 + \frac{(ka)^2}{2}\right] + i\frac{8ka}{3\pi}} \tag{7.19}$$

which is then inserted into equation (7.18), which now becomes

$$T_n = \frac{2(ka)^2}{\left[1 + \frac{(ka)^2}{2}\right]^2 + (ka)^2\left(\frac{8}{3\pi}\right)^2} \qquad (7.20)$$

Normally $ka \ll 1$, so the power transmission coefficient is quite small and it can be simplified to

$$T_n \approx 2(ka)^2 \qquad \text{(for flanged pipe)} \qquad (7.21)$$

From equation (7.18) it can be ascertained that $\mathbf{B}/\mathbf{A}$ is almost equal to -1. The pressure amplitude of the reflected wave is barely less than that of the incident wave. At $x = L$, its pressure differs by nearly $180°$ out of phase. On the other hand, the incident and reflected particle speeds remain nearly in phase at the opening of the pipe, so that location is (nearly) the antinode of the particle speed. Even though the amplitude of the particle speed at the opening is almost twice that of the incident wave, only a small percentage of the incident power transmits out of the flanged pipe. Thus, sources having dimensions small compared with the wavelength of the sound behave as inefficient radiators of sonic energy.

Inserting equation (7.16) for the unflanged pipe into equation (7.17), the transmission coefficient becomes

$$T_n = \frac{(ka)^2}{\left[1 + \frac{(ka)^2}{4}\right]^2 + (0.6ka)^2} \approx (ka)^2 \qquad (7.22)$$

By comparing equations (7.20) and (7.22) we perceive that adding a flange at the end of the pipe essentially doubles the radiation of sound at low frequencies. A gradual flare at the open terminal of the pipe will increase the low-frequency power transmission even more.

7.6   Standing Waves in Pipes

The existence of phase interference between transmitted and reflected waves inside a terminated pipe gives rise to a standing-wave pattern. The properties of the standing waves can be applied to gauge the load impedance. In equation (7.2) let us set

$$\mathbf{A} = A, \qquad \mathbf{B} = Be^{i\theta} \qquad (7.23)$$

where A and B are real, positive numbers, Combining equations (7.4) and (7.23) results in

$$\frac{\mathbf{Z}_{nL}}{\rho_0 cS} = \frac{1 + \frac{B}{A}e^{i\theta}}{1 - \frac{B}{A}e^{i\theta}} \qquad (7.24)$$

Inserting equation (7.23) into equation (7.2) we obtain for the pressure amplitude $p = |\mathbf{p}|$ of the wave

$$p = \sqrt{(A + B)^2 \cos^2[k(L - x) - \theta/2] + (A - B)^2 \sin^2[k(L - x) - \theta/2]}$$
(7.25)

At a pressure antinode, the pressure amplitude is $A + B$, while the amplitude pressure at the pressure node is $A - B$. The *standing-wave ratio* SWR occurs as the ratio of these two pressure amplitudes, respectively:

$$\text{SWR} = \frac{A + B}{A - B} \quad \text{or} \quad \frac{B}{A} = \frac{\text{SWR} - 1}{\text{SWR} - 1}$$
(7.26)

SWR is measured by probing the sound field along the pipe (also known as an *impedance tube*) with a tiny microphone to obtain the value of B/A. The phase angle θ can be found by measuring the distance of the first node from the end $x = L$. According to equation (7.25) the nodes are located at

$$k(L - x_n) - \frac{\theta}{2} = \left(n - \frac{1}{2}\right)\pi$$
(7.27)

For the first node ($n = 1$),

$$\theta = 2k(L - x_1) - \pi.$$
(7.28)

Example Problem 1

An impedance tube is found to have a SWR of 3 and the first node is 3/8 of the wavelength from the end. Find the normalized mechanical impedance at $x = L$.

Solution

$$L - x = 3\lambda/8$$

Hence from equation (7.28) $\theta = 2(2\pi/\lambda)(3\lambda/8) - \pi = \pi/2$. From equation (7.26)

$$\frac{B}{A} = \frac{3 - 1}{3 + 1} = \frac{1}{2}$$

Equation (7.24) now becomes:

$$\frac{Z_{nL}}{\rho_0 c S} = \frac{1 + \frac{1}{2}e^{i\pi/2}}{1 - \frac{1}{2}e^{i\pi/2}} = 0.60 + i0.80$$

Mechanical impedances at terminations occur as complicated functions of the frequencies, so it may be necessary to conduct measurements over the range of frequencies under consideration. Smith nomographs (Beranek, 1949) are useful tools to expedite computations for r and ψ, the real and imaginary components, respectively, of the impedances, from the measurements of the standing wave ratio and the position of the node most adjacent to the end.

The impedance tube is used to measure the reflective and absorptive properties of small sections of materials, such as acoustic tiles and sound control absorbers, mounted at the end of the tube.

7.7 The Rectangular Cavity

In Figure 7.2 a rectangular cavity is shown having dimensions L_x, L_y, L_z in the x-, y-, and z-directions, respectively. This parallelepiped can represent a simple auditorium or any other rectangular space that contains rigid walls, few windows and other openings. We assume the walls of the cavity to be perfectly rigid that $\hat{n} \cdot \bar{u} = 0$ at all of the boundaries (i.e., the walls will not move in the directions of their normals). This also means that $\hat{n} \cdot \nabla p = 0$, that is,

$$\left.\begin{array}{l} \left(\frac{\partial p}{\partial x}\right)_{x=0} = \left(\frac{\partial p}{\partial x}\right)_{x=L_x} = 0 \\ \left(\frac{\partial p}{\partial y}\right)_{y=0} = \left(\frac{\partial p}{\partial y}\right)_{y=L_y} = 0 \\ \left(\frac{\partial p}{\partial z}\right)_{z=0} = \left(\frac{\partial p}{\partial z}\right)_{z=L_z} = 0 \end{array}\right\} \tag{7.29}$$

Because acoustic energy cannot escape from a completely closed cavity with rigid walls, standing waves constitute the only appropriate solutions of the wave equation. Inserting

$$p(x, y, z, t) = X(x)Y(y)Z(z)e^{i\omega t} \tag{7.30}$$

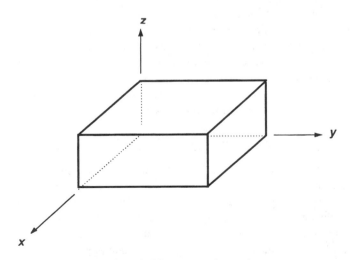

FIGURE 7.2. The rectangular cavity.

into the wave equation

$$\nabla^2 p = \frac{1}{c^2}\frac{\partial p}{\partial t} \tag{7.31}$$

and separating variables results in the following set of equations:

$$\left(\frac{d^2}{dx^2} + k_x^2\right)X = 0, \qquad \left(\frac{d^2}{dy^2} + k_y^2\right)Y = 0, \qquad \left(\frac{d^2}{dz^2} + k_z^2\right)Z = 0 \tag{7.32}$$

Here the separation constants are related as follows:

$$k^2 = \frac{\omega^2}{c^2} = k_x^2 + k_y^2 + k_z^2 \tag{7.33}$$

The boundary conditions of (7.29) stipulate cosine solutions, and equation (7.31) revises to

$$p_{lmn} = A_{lmn} \cos k_{xl} x \cos k_{ym} y \cos k_{zn} z e^{i\omega_{lmn}t} \tag{7.34}$$

where the components of k are

$$k_{xl} = \frac{l\pi}{L_x}, \quad l = 0, 1, 2, \ldots, \qquad k_{ym} = \frac{m\pi}{L_y}, \quad m = 0, 1, 2, \ldots,$$

$$k_{zn} = \frac{n\pi}{L_z}, \quad n = 0, 1, 2, \ldots \tag{7.35}$$

This leads to the quantitization of allowable frequencies of vibration:

$$\omega_{lmn} = \sqrt{\frac{l^2\pi^2}{L_x^2} + \frac{m^2\pi^2}{L_y^2} + \frac{n^2\pi^2}{L_z^2}} \tag{7.36}$$

The above gives rise to eigenfunctions of equation (7.34). Each eigenfunction is characterized by its own eigenfrequency (7.36) specified by the ordered integers (l, m, n). Equation (7.34), which is the solution to the wave equation (7.31), yields three-dimensional standing waves in the cavity with nodal planes parallel to the walls. The pressure varies sinusoidally between these nodal planes. In the same manner that a standing wave on a string could be resolved into a pair of traveling waves traveling in opposite directions, we can separate the eigenfunctions in the rectangular cavity into traveling plane waves. This is done by casting the solutions (7.34) into complex exponential form and expanding it as a sum of products:

$$\mathbf{p}_{lmn} = \frac{\mathbf{A}_{lmn}}{8} \sum e^{i\omega_{lmn}(\pm k_x x \pm k_y y \pm k_z z)} \tag{7.37}$$

where the summation is taken over all permutations of plus and minus signs. There are eight terms in all, each representing a plane wave traveling along the direction of its propagation vector $\hat{\mathbf{k}}$ which has projections $\pm k_{xl'} \pm k_{ym'} \pm k_{xn'}$ on the coordinate axes. The standing-wave solution results from the superposition of eight traveling waves (one into each quadrant) whose directions of travel are constrained by the boundary conditions. The methodology involving separation of variables can also be used to treat standing waves in other simple cavities, such as cylindrical and spherical cavities, with the eigenfunctions incorporating Bessel and Legendre functions.

7.8   Wave Guide with Constant Cross-Section

In Figure 7.3 a waveguide of rectangular cross section is assumed to have rigid side walls and a source of acoustic energy located at its boundary $z = 0$. There is no other boundary on the z-axis, which permits energy to propagate down the waveguide. This results in a situation where the wave pattern consists of standing waves in the transverse directions x and y and a traveling wave in the z-direction. The mathematical solution that contains applicable eigenfunctions is

$$p_{lmn} = A_{lmn} \cos k_{xl} x \cos k_{ym} y e^{i(\omega t - k_z z)} \tag{7.38}$$

Upon substituting equation (7.38) into the wave equation (7.31) we obtain the relationship

$$\frac{\omega^2}{c^2} = k^2 = k_{xl}^2 + k_{ym}^2 + k_z^2 \tag{7.39}$$

with permitted values of k_{xl} and k_{ym}, resulting from the boundary conditions of rigidity, these being

$$k_{xl} = \frac{l\pi}{L_x}, \quad l = 0, 1, 2, \ldots, \quad k_{ym} = \frac{m\pi}{L_y}, \quad m = 0, 1, 2, \ldots \tag{7.40}$$

We rearrange equation (7.39) to find k_z

$$k_z = \sqrt{\frac{\omega^2}{c^2} - k_{xl}^2 - k_{ym}^2} \tag{7.41}$$

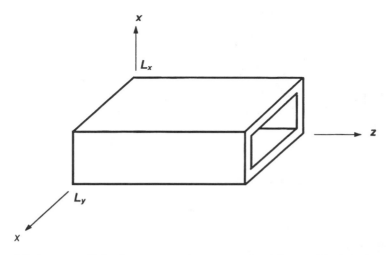

FIGURE 7.3. A wave guide having a rectangular cross-section. The travel is along the z-axis.

Because ω may have any value, equation (7.38) comprises a solution for all values of ω, in contrast to the totally enclosed cavity which allows for only quantitized frequencies. Setting

$$k_{lm} = \sqrt{k_{xl}^2 + k_{ym}^2} \qquad (7.42)$$

we can shorten equation (7.41) to

$$k_z = \sqrt{\frac{\omega^2}{c^2} - k_{lm}^2} \qquad (7.43)$$

The value k_z is real when $\omega/c > k_{lm}$. We then obtain a *propagating mode*, as the wave moves in the $+z$-direction. The cutoff frequency, which occurs when $\omega/c = k_{lm}$ so defining the limit for which k_z remains real, is given by

$$\omega_{lm} = ck_{lm} \qquad (7.44)$$

for the (l, m) mode. A frequency below the threshold value of ω_{lm} results in a purely imaginary value of k_z:

$$k_z \equiv \pm i \sqrt{k_{lm}^2 - \frac{\omega^2}{c^2}} \qquad (7.45)$$

We need to include the negative sign in equation (7.45) so that $\mathbf{p} \to 0$ as $z \to \infty$ and the eigenfunctions assume the form

$$p_{lm} = A_{lm} \cos k_{xl}x \cos k_{ym}y\, e^{-\left(\sqrt{k_{lm}^2 - \frac{\omega^2}{c^2}}\right)z} e^{i\omega t} \qquad (7.46)$$

Equation (7.46) represents a standing wave that decays exponentially with z. This form of eigenfunction is termed an *evanescent mode*—that is, no energy is propagated along the waveguide. If the frequency exciting the waveguide is just below the cutoff value of some particular mode, then this and higher modes are evanescent and are of little consequence at appreciable distances from the source. The lower modes may propagate energy and can be detected at large distances from the source.

Only plane waves can propagate in a rigid-walled waveguide if the frequency of the sound is sufficiently low. This frequency is less than $c/(2L)$, where L is the larger dimension of the rectangular cross section.

From equation (7.38), the phase speed for mode lm is not c but

$$c_p = \frac{\omega}{k_z} = \frac{c}{\sqrt{1 - \left(\frac{k_{lm}}{k}\right)^2}} = \frac{c}{\sqrt{1 - \left(\frac{\omega_{lm}}{\omega}\right)^2}} \qquad (7.47)$$

A physically meaningful solution can in written in the complex exponential form

$$\mathbf{p}_{lm} = \frac{1}{4} A_{lm} \sum_{\pm} e^{i(\omega t \pm k_{xl}x \pm k_{ym}y - k_z z)} \qquad (7.48)$$

The absence of the boundary at a location on the $+z$-axis necessitates a negative sign only for k_z, because there will be no reflected waves propagating in the $-z$

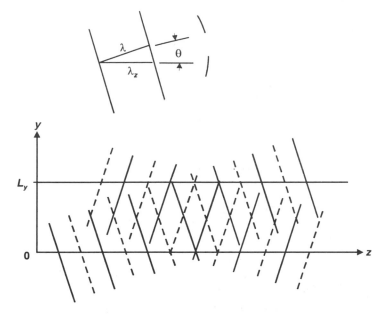

FIGURE 7.4. The surfaces of constant phase for two-component waves in the (0, 1) mode.

direction. The propagation vector **k** for each of the four traveling waves forms an angle θ with the z-axis, according to

$$\cos\theta = \frac{k_z}{k} = \sqrt{1 - \left(\frac{\omega_{lm}}{\omega}\right)^2} \qquad (7.49)$$

and the corresponding phase speed c_p is

$$c_p = \frac{c}{\cos\theta} \qquad (7.50)$$

The surfaces of constant phase for two component waves representing the (0, 1) mode are shown in Figure 7.4 for a rigid waveguide. The waves cancel each other precisely for $y = \frac{1}{2}L_y$, with the result that a nodal plane exists halfway between the walls. The waves remain in phase at the upper and the lower walls, so the pressure amplitude is maximized at these rigid boundaries. The apparent wavelength λ_z in the z-direction is given by

$$\lambda_z = \frac{\lambda}{\cos\theta} \qquad (7.51)$$

In the lowest mode (0, 0), $k_z = k$, and the four component waves form a single plane wave that travels down the axis of the waveguide with speed c. For all the other modes, the propagation vectors of the components waves generally form angles with the z-axis, with one aimed into each of the four forward quadrants. According to equation (7.47), at frequencies much greater than the cutoff of the $(l, m,)$ mode, that is, $\omega \gg \omega_{lm}$, the angle θ approaches zero and the waves are

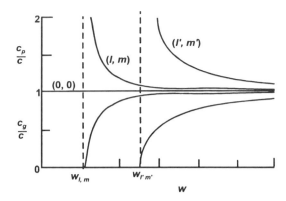

FIGURE 7.5. Group and phase speed as functions of radial frequency for three modes in a rigid-walled wave guide.

traveling fairly straight down the waveguide. As the input frequency approaches the cutoff value, θ increases with the result that the component waves move in increasingly transverse directions. In fact, when the frequency reaches the stage that $\omega = \omega_{lm}$, the component waves are traveling transversely to the axis of the waveguide. Each component wave carries energy along the waveguide through the process of continual reflection from the rigid walls (in the same manner a signal is transmitted through a fiberoptic line, bouncing from one wall back to the opposite wall down along the line). With the energy of the wave propagating at a speed c in the direction $\hat{\mathbf{k}}$, the corresponding speed c_g (the *group speed*) of the energy in the z-direction is given by the component of the plane wave velocity $\hat{c}$ along the waveguide axis:

$$c_g = c \cos \theta = c\sqrt{1 - \left(\frac{\omega_{lm}}{\omega}\right)^2} \qquad (7.52)$$

Given the driving frequency ω, each normal mode, in which $\omega_{lm} < \omega$, possesses its own set of values for c_p and c_g. Figure 7.5 illustrates the variation of the group and phase speeds as functions of frequency for three modes in a rigid-walled waveguide. The (0, 0) mode is simply the plane wave solution with its group and plane speeds equal to c for all driving frequencies.

7.9 Boundary Condition at the Driving End of the Waveguide

An impedance match must be made at the waveguide entry $z = 0$ so that the permissible mode solutions comply with the acoustic behavior of the active surface. If we know the pressure or velocity distribution of the driving source, these can be correlated with the behavior of $\mathbf{p}(x, y, 0, t)$ for the pressure or $\mathbf{z} \cdot \mathbf{u}(x, y, 0, t)$ for the velocity. If the pressure distribution of the source is given, say, then the boundary condition becomes

$$\mathbf{p}(x, y, z, t) = \mathbf{P}(x, y)e^{i\omega t} \tag{7.53}$$

However the left side of equation (7.53) can also be expressed as a superposition of the normal modes of the waveguide, that is,

$$\mathbf{p}(x, y, z, t) = \sum_{l,m} \mathbf{A}_{lm} \cos k_{xl}x \cos k_{ym}y e^{i(\omega t - k_z z)} \tag{7.54}$$

Setting $z = 0$, equation (7.53) becomes

$$\mathbf{P}(x, y) = \sum \mathbf{A}_{lm} \cos k_{xl}x \cos k_{ym}y \tag{7.55}$$

With equation (7.55) we can establish the values of $\mathbf{A}_{lm}$ by applying Fourier's theorem first for the x-direction and then for the y-direction.

7.10   Rigid-Walled Circular Waveguide

The treatment of the rigid-walled circular waveguide of radius $r = a$ is fairly straightforward, beginning with

$$\mathbf{P}_{ml} = \mathbf{A}_{ml} J_m(k_{ml}r) \cos(m\theta)e^{(i\omega t - k_z z)}$$

where (r, θ, z) are the customary cylindrical coordinates, J_m is the mth-order Bessel function, and

$$k_z = \sqrt{\frac{\omega^2}{c^2} - k_{lm}^2}$$

where k_{ml} is found from the boundary conditions. Because $r \cdot \nabla p = 0$ at the wall $r = a$,

$$k_{ml} = \frac{j'_{ml}}{a}$$

where j'_{lm} are the zeros of $dJ_m(z)/dz$. When the values of k_{ml} are determined, the applicable results developed for the rectangular waveguide may be applied by substituting the values of k_{ml} for the circular waveguide. As with the rectangular unit, the (0, 0) mode of the circular waveguide is a plane wave that propagates with phase velocity $c_p = c$ for all $\omega > 0$. For frequencies below $f_{1,1}$, only plane waves can propagate in a rigid-walled circular waveguide.

7.11   The Helmholtz Resonator

Many analyses of acoustic devices become simplified with the assumption that the wavelength in the propagation fluid is much greater than the dimensions of the devices. If the wave length is indeed much larger in *all* dimensions, the acoustic variable may be time varying, but it is virtually independent of the distance within

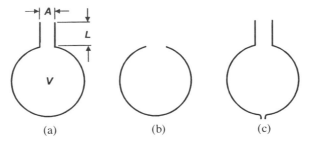

FIGURE 7.6. Three simple Helmholtz resonators.

the confines of the device. In such a case, the device can be viewed as a harmonic oscillator with one degree of freedom; and in the long-wavelength limit, such a device can be considered a *lumped acoustic element*. An example of this type of device is the *Helmholtz resonator*, illustrated in Figure 7.6 (three types are shown). The resonator is a rigid-wall cavity of volume V, with a neck of area A and length L.

According to theory, if $\lambda \gg L$, the fluid in the neck behaves somewhat as a solid plug that vibrates back and forth. As the fluid (usually air) moves back and forth, acoustic energy becomes converted to heat as a result of friction along the neck. These losses can be increased by placing a light porous material across the neck or by placing the material inside the volume. Maximum sound absorption occurs at the resonant frequency of the mass of air in the neck and the "spring" supplied by the air resistance inside the enclosed volume. Very little sound is absorbed at other frequencies. However, for necks greater than 1 cm in diameter, the viscous losses are considerably less than those associated with radiation. For the purpose of this analysis we can ignore the effects of viscosity in analyzing the Helmholtz resonator viewed as an analogous spring–mass system. The air in the neck has a total effective mass

$$m = \rho_0 A L' \tag{7.56}$$

where the effective neck length L' is longer than the physical length L because of its radiation mass loading. We have seen earlier in this chapter that at low frequencies a circular opening of radius a is loaded with a radiation mass equal to that of the fluid occupying area πa^2; and effective length $8a/(3\pi)$ equals $0.85a$ when the neck terminates in a wide flange or $0.6a$ for an unflanged terminal. We assume the mass loading at the inner end of the neck is equivalent to a flanged termination. Then

$$L' = L + 2(0.85a) = L + 1.7a \quad \text{(flanged outer end)}$$
$$L' = L + (0.85 + 0.6)a = L + 1.5a \quad \text{(unflanged outer end)} \tag{7.57}$$

Now consider the neck of the resonator to be fitted with an airtight piston. When the piston travels a distance δ, the volume of the cavity changes by $\Delta V = -\delta S$.

Then

$$\frac{\Delta\rho}{\rho} = -\frac{\Delta V}{V} = \frac{A\delta}{V}$$

and from the thermodynamic relation

$$p = \rho_0 c^2 \frac{\Delta p}{p}$$

Combining the last two equations gives

$$p = \rho_0 c^2 \frac{A\delta}{V} \tag{7.58}$$

The force $f = pA$ necessary to execute the displacement is $\rho_0 c^2(A^2/V)\delta$. The effective stiffness S (from the spring formula $f = S\delta$) is

$$S = \rho_0 c^2 \frac{A^2}{V} \tag{7.59}$$

The fluid moving in the neck radiates sound into the surrounding medium in the same manner as an open-ended pipe. For wavelengths much larger than the radius of the neck, the radiation resistance [cf. equations (7.12) and (7.16)] is

$$R_r = \rho_0 c \frac{kA^2}{2\pi} \qquad \text{(flanged)} \tag{7.60a}$$

or

$$R_r = \rho_0 c \frac{kA^2}{4\pi} \qquad \text{(unflanged)} \tag{7.60b}$$

The sound wave impinging on the neck opening is represented as an instantaneous driving force with a pressure amplitude P:

$$\mathbf{f} = APe^{i\omega t} \tag{7.61}$$

We then can write the differential equation for the inward displacement δ of the fluid "plug" in the neck as

$$m\frac{d^2\delta}{dt^2} + R_r\frac{d\delta}{dt} + S\delta = APe^{i\omega t} \tag{7.62}$$

This last equation is analogous to that of a sinusoidally driven oscillator which has analogous solutions. Solution of equation (7.62) gives a complex displacement δ. The real part of the driving force represents the actual driving force $AP\cos\omega t$, and the real part of the complex displacement represents the actual displacement. Because $\mathbf{f}$ is periodic with angular frequency ω, $\delta = Be^{i\omega t}$ must be the solution where B exists as a complex constant. Then inserting the solution into equation (7.62) results in

$$(-B\omega^2 m + iB\omega R_r + BS)e^{i\omega t} = APe^{i\omega t}$$

Solving for **B** provides the complex displacement

$$\delta = \frac{APe^{i\omega t}}{i\omega\left[R_r + i\left(m\omega - \frac{S}{\omega}\right)\right]} \tag{7.63}$$

Differentiating equation (7.63) yields the complex speed of the fluid in the resonator's neck:

$$\vec{\mathbf{u}} = \frac{APe^{i\omega t}}{\mathbf{R}_r + \mathbf{i}\left(\mathbf{m}\omega - \frac{S}{\omega}\right)} \tag{7.64}$$

The impedance found from the relation $\mathbf{Z}_m = \mathbf{f}/\mathbf{u}$, and making use of equations (7.61) and (7.64) yield

$$Z_m = R_r + i\left(m\omega - \frac{S}{\omega}\right) \tag{7.65}$$

which means that the mechanical reactance is

$$X_m = m\omega - \frac{S}{\omega}$$

Resonance occurs when the reactance becomes zero, that is,

$$\omega_0 = \sqrt{\frac{S}{m}} = \sqrt{\frac{\rho_0 c^2 A^2 / V}{\rho_0 L' A}} = c\sqrt{\frac{A}{L'V}} \tag{7.66}$$

We note that the resonance frequency depends on the volume of the cavity, not the shape. Experimentation with differently shaped resonators having the same $S/L'V$ ratios indicate identical resonant frequencies. This holds true as long as all dimensions of the cavity are appreciably less than a single wavelength and the opening is quite small. There are additional frequencies in Helmholtz resonators higher than that given by equation (7.66), which arise from standing waves in the cavity rather than from the oscillatory motion of the mass of air in the neck. These overtone frequencies are not harmonics of the driving fundamental, and the first overtone may be several times the frequency of the fundamental.

The sharpness (or narrowness of spread) of the resonance of a Helmholtz resonator is define by the *quality factor Q*:

$$Q = \frac{m\omega_0}{R_r} = 2\pi\sqrt{V\left(\frac{L'}{A}\right)^3} \tag{7.67}$$

Assumptions have been made that there are no losses other than those arising from acoustic radiation and the termination of the neck is flanged.

In conducting his experiments of complex musical tones, Helmholtz used a series of resonators of the type illustrated in Figure 7.6(c), with small nipples opposite the necks of the resonators. He used a graduated series of resonators, with differing volumes and neck sizes to achieve a wide range of resonant frequencies. When an incident sound wave contains a frequency component that corresponds to the resonant frequency of a Helmholtz device, a greatly amplified signal will be generated within the cavity of the resonator, and it can be detected aurally by connecting a short flexible tube from the nipple to the ear. These phenomena lead to the definition of pressure amplification, which is the ratio of the acoustic pressure amplitude P_c inside the cavity to the external driving pressure amplitude P. Equation (7.58) provides the pressure amplitude P_c. At resonance [cf. equation (7.63)]

$$|\delta| = \frac{PA}{\omega_0 R_r}$$

Then applying equations (7.60) and (7.66), we find for the resonator with a flanged neck that

$$\frac{P_c}{P} = 2\pi \sqrt{V\left(\frac{L'}{A}\right)^3} = Q$$

The gain, therefore, is the same as the value of the quality factor defined in equation (7.67), and we see the Helmholtz resonator behaves as an amplifier of gain Q at resonance, a fact that explains why a loudspeaker mounted in a closed chamber can be regarded as a Helmholtz resonator in which the air's reactance and the speaker cone mass constitute the effective mass of the system. The air resistance within the box and the cone's stiffness combine to provide the effective stiffness, while the effective resistance results from the sum of that attributable to the radiation of acoustic energy and that arising from the mechanical resistance of the speaker cone.

Example Problem 2

A Helmholtz resonator is a sphere of 8 cm internal diameter. If it is to resonate at 450 Hz in air, what is the hole diameter that should be drilled into the sphere?

Solution

Applying equation (7.66) we have

$$\omega_0 = 450 \times 2\pi \text{ rad} - s^{-1} = c\sqrt{\frac{A}{L'V}} = 344 \text{ m} - s^{-1}\sqrt{\frac{\pi a^2}{1.5a \times \frac{4}{3}\pi(.04\,\text{m})^3}}$$

which results in $a = 0.0087$ m $= 0.865$ cm, or 1.73 cm diameter. Here we assumed an unflanged "pipe" with an effective value of $L' = 1.5a$.

References

Beranek, Leo L. 1986. *Acoustics*. New York: American Institute of Physics; Chapters 5 and 6.

Beranek, Leo L. 1949. *Acoustic Measurements*. New York: John Wiley & Sons; pp. 317–321.

Fletcher, Neville H. and Thomas D. Rossing. 1993. *The Physics of Musical Instruments*. New York: Spring-Verlag; pp. 150–155.

Kinsler, Lawrence E., and Austin R. Frey. 1962. *Fundamentals of Acoustics*, 2nd ed. New York: John Wiley & Sons; Chapter 8.

Kinsler, Lawrence E., Frey, Austin R., Coppens, Alan B., and Sanders, James V. 1982. *Fundamentals of Acoustics*, 3rd ed. New York: John Wiley & Sons; Chapters 9 and 10.

Olson, Harry F. 1967. *Music, Physics, and Engineering*, 2nd ed. New York: Dover Publications; Chapter 4.

Wood, Alexander. *Acoustics*. 1960. New York: Dover Publications; pp. 103–110.

Problems for Chapter 7

1. Determine the minimum length of a pipe in which the input mechanical reactance is equal to the input mechanical resistance for the frequency of 700 Hz. Assume the loading impedance is four times the pipe's characteristic mechanical impedance $p_0 c A$.

2. A condenser microphone is constructed by stretching a diaphragm across one end of a hollow cylinder with a diameter of 2 cm and length of 0.75 cm. The other end of the cylinder is open. Determine the ratio of the pressure at the diaphragm (considered to be a rigid plane) to the pressure at the open end as a function of frequency from 64 Hz to 2 kHz.

3. A pipe is 1.2 m long and has a radius of 6 cm. It is being driven at one end by a piston (negligible mass). The other, open end of the pipe terminates in a flange.
   (a) Find the fundamental resonance of the system
   (b) If the piston has a displacement of 1 cm when being driven at 500 Hz, what is the amount of acoustic power being transmitted by plane waves traveling to the open end of the pipe?
   (c) What is the acoustic power (in watts) being radiated out at the open end of the pipe?

4. Consider a pipe that is 1.2 m long with a radius of 6 cm. It is being driven at one end by a piston of 0.02 kg mass and the same radius as the pipe. The other end of the pipe terminates in an infinite baffle.
   (a) For 200-Hz frequency, find the mechanical impedance of the piston of the pipe, inclusive of the loading effect of the air inside the pipe.
   (b) For the above frequency, determine the amplitude of the force necessary to drive the piston with a displacement amplitude of 0.50 cm.
   (c) What will be the amount of acoustic wattage that emanates out through the open end of the pipe?

5. A 70 dB (re 20 μPa) 1 kHz signal is incident normally to a boundary between the air and another medium having a characteristic impedance of 780 Pa s/m.
   (a) Determine the effective root mean square pressure amplitude of the reflected waves.
   (b) Determine the effective root mean square pressure of the transmitted waves.

(c) How far from the boundary is the location where pressure amplitude in the pattern of standing waves equals the pressure amplitude of the incident wave?

6. The speed of sound in water is 1480 m/s. Consider a series of 2960-Hz plane waves in water normally incident to a concrete wall. It can be assumed that all of the sound energy is completely absorbed by the wall. The pattern of standing waves results in a peak pressure amplitude of 30 Pa at the wall and a pressure amplitude of 10 Pa at the nearest pressure nose at a distance of 50 cm from the wall.

    (a) What is the ratio of the intensity of the reflected wave to that of the incident wave?

    (b) Find the specific acoustic impedance of the wall.

7. An acoustic signal consisting of 400-Hz plane wave is normally incident to an acoustical tile surface having a complex impedance of $1500 - i\,3000$.

    (a) Find the standing-wave ratio in the resultant pattern of standing waves.

    (b) Determine the locations of the first four nodes.

8. A loudspeaker is fitted to one end of an air-filled pipe of 12 cm radius to generate plane waves inside the pipe. The far end of the pipe is closed off by a rigid cap. A 5-kHz signal is fed into the loudspeaker. The measured standing wave ratio of pressure at one location in the pipe is 7. At another location inside the pipe 50 cm downstream of the pipe, the measured standing wave ratio is 9. Derive an equation that includes these ratios and the distance between them which can be used to establish the absorption constant for the signal being propagated inside the pipe. To simplify matters, assume that absorption coefficient to be $\alpha \ll 1$.

9. Prove analytically (assuming $ka \ll 1$) that the frequencies of the resonance and those near antiresonance of the forced-open tube with damping correspond to the frequencies of maximum and minimum power dissipation, respectively.

10. Determine the lowest normal mode frequency of a fluid-filled cubic cavity (of length L to a side) that consists of five rigid walls and one pressure release side. Plot the pressure distribution associated with that mode.

11. Given a rigid-walled rectangular room with dimensions $6.50\,\text{m} \times 5.65\,\text{m} \times 3\,\text{m}$, calculate the first ten eigenfunctions. Assume the sound propagation speed $c = 344$ m/s.

12. A concrete water sluice measures measures 25 m wide and 8 m deep. It is completely filled with water. Find the cutoff frequency of the lower mode of propagation, assuming the concrete to be totally rigid.

8
Acoustic Analogues, Ducts, and Filters

8.1 Introduction

In this chapter we deal with lumped and distributed acoustic elements, applying electrical and mechanical analogues to acoustic behavior in order to treat differing duct geometries, acoustic filters, and networks. The reflection and transmission of sound waves at piping interfaces, where the acoustic impedance changes, are analogous to the behavior of current waves in a transmission line at locations where the electrical impedance undergoes change.

A simple mechanical system can often be converted into analogous electrical systems and solved in analogous terms. The motion of a fluid compares to the behavior of current in an electrical circuit, with pressure gradient between two points playing the role of voltage across the corresponding parts of the circuit. In terms of electricity the impedance is voltage divided by the current, which corresponds to the effect of lumped elements of inductance, capacitance, and resistance. In acoustics the *acoustic impedance* $\mathbf{Z}$ of a fluid acting with acoustic pressure $\mathbf{p}$ on a surface of area A is given by

$$\mathbf{Z} = \frac{\mathbf{p}}{\mathbf{U}} \tag{8.1}$$

where $\mathbf{U}$ represents the *volume velocity* of the fluid in the acoustic element of interest. $\mathbf{U}$ is not truly a vector; it is a speed representative of a scalar quantity, unlike velocity which is a magnitude defined with an indicated direction. The *acoustic impedance* $\mathbf{Z}$ defined by equation (8.1) is a complex quantity.

The *specific acoustic impedance* $\mathbf{z}$ is given by

$$\mathbf{z} = \frac{\mathbf{p}}{\mathbf{u}} \tag{8.2}$$

where $\mathbf{u}$ is the *particle* velocity, not the volume velocity. The specific acoustic impedance, which is useful for treating the transmission of acoustic waves from one medium to another, is a characteristic of the propagation medium and the type of propagating wave. The acoustic impedance, defined by equation (8.1) as the ratio of pressure to volume velocity, is applied to treat acoustic radiation from vibrating surfaces and the transmission of this radiation across lumped acoustic elements and through ducts and horns. These two impedances are related to each other by

$$\mathbf{Z} = \frac{\mathbf{z}}{A} \tag{8.3}$$

If a vibrating surface is driven with a velocity $\mathbf{u}$, with a force $\mathbf{f}$ acting on the fluid, the *radiation impedance* $\mathbf{Z}_r$ is given by the ratio of the force to the speed:

$$\mathbf{Z}_r = \frac{\mathbf{f}}{\mathbf{u}} \tag{8.4}$$

This type of impedance constitutes a part of the mechanical impedance $\mathbf{Z}_m$ of the vibrating system. Radiation impedance relates to specific impedance at a surface as follows:

$$\mathbf{Z}_r = \mathbf{z}A \tag{8.5}$$

It is useful for dealing with coupling between acoustic waves and a driving surface or a driven load.

8.2   Lumped Acoustic Impedance

In the use of lumped parameters, advantage is taken of the assumption that the signal wavelength is larger than all principle dimensions which allows for further simplification. Each acoustic parameter may be time varying, but it becomes virtually independent of distance over the spatial extent of the device. When we consider *lumped* or *concentrated impedances* rather than distributed impedances, we define that impedance in a segment of the acoustic system as the (complex) ratio of the pressure difference $\mathbf{p}$ (which propels that segment) to the resultant volume velocity $\mathbf{U}$. The unit of acoustic impedance is Pa-s/m^3, and it is often referred to an *acoustic ohm*.

Example Problem 1

Consider a Helmholtz resonator described by differential equation (7.62) which is repeated below:

$$m\frac{d^2\delta}{dt^2} + R_r\frac{d\delta}{dt} + S\delta = APe^{i\omega t} \tag{8.6}$$

and recast the system as a lumped acoustic impedance with an electric analog.

Solution

Divide equation (8.6) by surface area A and apply the fact that $\mathbf{U} = (d\delta/dt)A$, which will result in

$$\mathbf{Z} = \frac{\mathbf{p}}{\mathbf{U}} = R + i\left(M\omega - \frac{1}{\omega C}\right) \tag{8.7}$$

where

$$R = \frac{R_r}{A^2} = \frac{\rho_0 c k^2}{2\pi}, \qquad M \equiv \frac{m}{A^2} = \frac{\rho_0 L'}{A}, \qquad C \equiv \frac{A^2}{S} = \frac{V}{\rho_0 c^2}$$

and the stiffness is given by

$$S = \rho_0 c^2 \frac{A^2}{V},$$

according to equation (7.59). We also assume here a flanged resonator. In electrical terms this constitutes a RLC series circuitry, where the inductance L is the electrical analog to M. This electrical analogue to the Helmholtz resonator is illustrated in Figure 8.1. As a further aid to analytical treatment of lumped parameters, Figure 8.2 summarizes the fundamental analogous elements of acoustic, mechanical, and electrical systems. The inertance M in the acoustic system is represented by a "plug" of fluid that is sufficiently short so that all particles in the fluid can be depicted as moving in phase under the impetus of sound pressure. The compliance C of the acoustic system is represented by an enclosed volume incorporating a stiffness. A number of different situations can contribute to resistance; we can represent acoustical resistance in the conventional manner by a narrow slits inside a pipe segment.

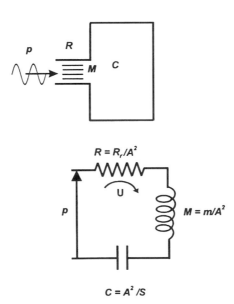

FIGURE 8.1. The electrical analogue to the Helmholtz resonator.

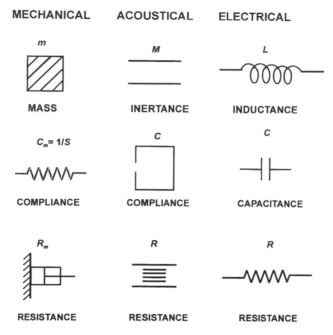

FIGURE 8.2. Fundamental mechanical, acoustical, and electrical analogues.

8.3 Distributed Acoustic Impedance

What if one or more of the principal dimensions of an acoustic system is of the same order of magnitude as a wavelength? In this case it may not be possible to treat the system as one possessing lumped parameters. The alternative is to analyze it as a system having distributed physical constants. Consider the very simple case of plane waves propagating through a pipe in the positive x-direction. The characteristic impedance of the pipe is given by the ratio of acoustic pressure to particle speed; and the acoustic impedance at any cross section A of the pipe is

$$Z = \frac{\mathbf{p}}{\mathbf{U}} = \frac{1}{A}\frac{\mathbf{p}}{\mathbf{u}} = \frac{\rho_0 c}{A} \qquad (8.8)$$

This case of propagation inside a pipe is electrically equivalent to high frequency currents traveling along a transmission line which has an inductance *per unit length* L_s and a capacitance *per unit length* C_s. The corresponding input electrical impedance is $\sqrt{L_s/C_s}$. In compliance with the electrical analogy, we may consider the fluid in the pipe to have a distributed inertance M_s per unit length and a distributed compliance C_s per unit length. It also follows that the distribution of mass per unit length of the pipe can be represented by $m_s = \rho_0 A$. The acoustic inductance per unit length becomes $M_s = m_s/A^2 = \rho_0/A$.

We shall now find the mechanical stiffness per unit length. When the fluid is compressed adiabatically by a small linear displacement $\delta \ell$, then $p = \rho_0 c^2 (\delta \ell / \ell)$

and the force providing this impetus is pA. Hence the stiffness becomes $S = pA/\delta\ell$; and the stiffness per unit length is $S_s = \rho_0 c^2 A$. The mechanical compliance C_m relates to acoustic compliance C by $C = A^2 C_m$, and on a per unit length basis it follows that $C_s = A/(\rho_0 c^2)$. By analogy the acoustic impedance of the pipe is given by

$$Z = \sqrt{\frac{M_s}{C_s}} = \frac{\rho_0 c}{A}$$

which corresponds to equation (8.8).

8.4   Waves in Pipes: Junctions and Branches

Consider a sound wave traveling in the positive x-direction, represented by

$$\mathbf{p}_i = \mathbf{a} e^{i(\omega t - kx)} \tag{8.9}$$

is incident upon a point $x = 0$, where the acoustic impedance changes from $\rho_0 c/A$ to some complex value $\mathbf{Z}_0$. At this point a reflected wave

$$\mathbf{p}_r = \mathbf{b} e^{i(\omega t + kx)} \tag{8.10}$$

will be produced and travel in the negative x-direction. It is our task to find the power reflection and transmission coefficients for this point. The acoustic impedance at any point in the pipe is given by

$$\mathbf{Z} = \frac{\mathbf{p}_i + \mathbf{p}_r}{\mathbf{U}_i + \mathbf{U}_r} = \frac{\rho_0 c}{A} \frac{\mathbf{a} e^{-ikx} + \mathbf{b} e^{+ikx}}{\mathbf{a} e^{-ikx} - \mathbf{b} e^{+ikx}} \tag{8.11}$$

which, at $x = 0$, reduces to

$$\mathbf{Z}_0 = \frac{\rho_0 c}{A} \frac{\mathbf{a} + \mathbf{b}}{\mathbf{a} - \mathbf{b}} \tag{8.12}$$

Equation (8.12) can be rearranged to:

$$\frac{\mathbf{b}}{\mathbf{a}} = \frac{\mathbf{Z}_0 - \frac{\rho_0 c}{A}}{\mathbf{Z}_0 + \frac{\rho_0 c}{A}} \tag{8.13}$$

The sound power reflection coefficient R_p, which yields the fraction of the incident power that is reflected, is given by

$$R_p = \left| \frac{\mathbf{b}}{\mathbf{a}} \right|^2 = \frac{\left(R_0 - \frac{\rho_0 c}{A} \right)^2 + X_0^2}{\left(R_0 + \frac{\rho_0 c}{A} \right)^2 + X_0^2} \tag{8.14}$$

In equation (8.14) we have set $\mathbf{Z}_0 = R_0 + iX_0$. The sound power transmission $T_p = 1 - R_p$ represents the portion of the incident sound power that travels past $x = 0$. We obtain

$$T_p = \frac{4 R_0 \rho_0 c / A}{\left(R_0 + \frac{\rho_0 c}{A} \right)^2 + X_0^2} \tag{8.15}$$

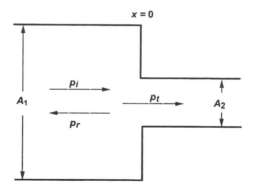

FIGURE 8.3. Transmission and reflection of a plane wave at the junction $x = 0$ between two pipes of different cross-sections.

Example Problem 2

Apply the above equations to plane waves in a pipe of cross section area A_1 that is mated to a pipe of cross-sectional area A_2, as shown in Figure 8.3. The second pipe is of infinite extent or the proper length so that no reflected wave is returned from its far terminus. Assume that the wavelength is considerably larger than the diameter of either pipe so that the region of complicated flow at the junction, where the wave adjusts from cross-sectional area to another, is considerably smaller than the wavelength itself.

Solution

Under the conditions stipulated above, the acoustic impedance seen by the incident wave at the junction is $Z_0 = \rho_0 c / A_2$. Inserting this value of Z_0 into equations (8.14) and (8.15), we have

$$R_p = \left(\frac{A_1 - A_2}{A_1 + A_2}\right)^2 \quad \text{and} \quad T_p = \frac{4 A_1 A_2}{(A_1 + A_2)^2} \tag{8.16}$$

Note that if the above pipe is closed at $x = 0$, S_2 equals 0 and then Z_0 becomes infinity, which results in a reflection coefficient of unity. On the other hand, if the pipe is open at $x = 0$, the impedance at the junction is not zero but corresponds to the impedance given by equation (7.12) for an unflanged pipe.

In Figure 8.4 we have the more complex case of a pipe that branches into two pipes, each with its own input impedance. Let the junction be located at the origin. The pressures produced by the waves in the three pipes at $x = 0$ are represented by

$$\mathbf{p}_i = \mathbf{a} e^{i\omega t}, \qquad \mathbf{p}_r = \mathbf{b} e^{i\omega t}$$

$$\mathbf{p}_1 = \mathbf{Z}_1 \mathbf{U}_1 \, e^{i\omega t}, \qquad \mathbf{p}_2 = \mathbf{Z}_2 \mathbf{U}_2 \, e^{i\omega t} \tag{8.17}$$

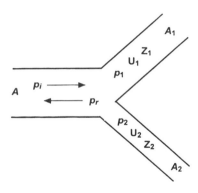

FIGURE 8.4. A three-way junction.

Here **a** and **b** denote the amplitudes of the incident and reflected waves, respectively; and $\mathbf{Z}_1$, $\mathbf{Z}_2$ and $\mathbf{U}_1$, $\mathbf{U}_2$ the input impedances and volume velocity complex amplitudes in branches 1 and 2. Again under the assumption of a large wavelength so that the impact of branching remains confined to a small region at the juncture, we apply the condition of continuity of pressure so that

$$\mathbf{p}_i + \mathbf{p}_r = \mathbf{p}_1 = \mathbf{p}_2 \tag{8.18}$$

Likewise, the continuity of the volume velocity demands that

$$\mathbf{U}_i + \mathbf{U}_r = \mathbf{U}_1 + \mathbf{U}_2 \tag{8.19}$$

which is analogous to Kirchhoff's law of electric currents. Dividing equation (8.19) by the equation (8.18) yields the impedance relationship

$$\frac{1}{\mathbf{Z}_0} = \frac{1}{\mathbf{Z}_1} + \frac{1}{\mathbf{Z}_2} \tag{8.20}$$

The reciprocal of an impedance, $\mathbf{Z}^{-1}$, is called the *admittance*. Equation (8.20) shows that the combined admittance $1/\mathbf{Z}_0$ is equal to the sum of the admittances of the two branches 1 and 2.

Example Problem 3

An infinitely long pipe of cross sectional area A has a branch at $x = 0$ that presents an given impedance of $\mathbf{Z}_g$. Find the appropriate power reflection and transmission coefficients.

Solution

In this case we consider this pipe to have two branches, one corresponding to the given impedance and the other to an infinite pipe that does not provide any reflections but does present an impedance of $\rho_0 c / A$. Applying equation (8.20) we obtain

$$\frac{\mathbf{b}}{\mathbf{a}} = \frac{\frac{\rho_0 c}{2A}}{\frac{\rho_0 c}{2A} + \mathbf{Z}_g} \tag{8.21}$$

The ratio of pressure amplitude $\mathbf{a}_t$ of the wave transmitted beyond $x = 0$ along the infinite pipe to the pressure amplitude of the incident wave is found by inserting (8.21) into equation (8.18), yielding

$$\frac{\mathbf{a}_t}{\mathbf{a}} = \frac{\mathbf{Z}_g}{\mathbf{Z}_g + \frac{\rho_0 c}{2A}} \tag{8.22}$$

We resolve the acoustic impedance $\mathbf{Z}_g$ of the branch into real and imaginary components, that is, $\mathbf{Z}_g = R_g + i X_g$. The reflection and transmission coefficients become

$$R_p = \left|\frac{\mathbf{b}}{\mathbf{a}}\right|^2 \frac{\left(\frac{\rho_0 c}{2A}\right)^2}{\left(\frac{\rho_0 c}{2A} + R_g\right)^2 + X_g^2} \tag{8.23}$$

$$T_p = \left|\frac{\mathbf{a}_t}{\mathbf{a}}\right|^2 = \frac{R_g^2 + X_g^2}{\left(\frac{\rho_0 c}{2A} + R_g\right)^2 + X_g^2} \tag{8.24}$$

The portion T_{pg} of the power that is transmitted into the branch is $T_{pg} = 1 - R_p - T_p$, or

$$T_{pg} = \frac{\frac{R_g \rho_0 c}{A}}{\left(\frac{\rho_0 c}{2A} + R_g\right)^2 + X_g^2} \tag{8.25}$$

If R_g has a positive finite value, some acoustic energy is dissipated at the branch and some is transmitted beyond the junction, no matter what the value of X_g is. When either R_g or X_g greatly exceeds $\rho_0 c/A$, nearly all of the incident power is transmitted past the branch. At the other extreme, when $R_g = X_g = \infty$, which corresponds to no branch, the power transmission is unity.

8.5   Acoustic Filters

Advantage can be taken of the fact that a side branch can attenuate sound energy transmitted in a pipe. The input impedance of the side branch determines whether the system can behave as a low-pass, high-pass or band-pass filter. We shall now consider each of these filters.

1. Low-Pass Filters. Figure 8.5 illustrates the construction of a simple low-pass filter, essentially consisting of an enlarged segment of a pipe of cross-sectional area A_1 and length L in a pipe of cross-section A. At sufficiently low frequencies ($kL \ll 1$), this filter may be viewed as a side branch with acoustic compliance $C = V/(\rho_0 c^2)$, where $V = A_1 L$ represents the volume of the expansion chamber.

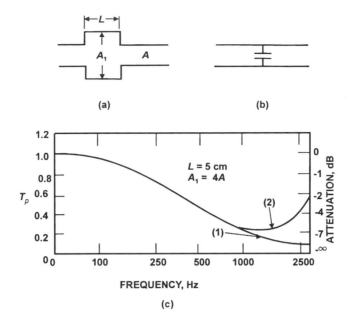

FIGURE 8.5. (a) A simple low-pass acoustic filter, (b) its analogous electrical filter, and (c) the corresponding power transmission curves for the acoustic filter of (a).

The acoustic impedance of this type of branch is pure reactance, hence

$$R_g = 0, \qquad X_g = -\frac{1}{C\omega} = \frac{\rho_0 c^2}{A_1 L \omega}$$

Inserting the above into the expression for the transmission coefficient, equation (8.24) yields

$$T_p = \frac{1}{\left(\frac{A_1}{2A}kL\right)^2 + 1} \tag{8.26}$$

Equation (8.26) plotted in Figure 8.5 indicates that as the frequency approaches zero, the transmission coefficient approaches unity (100% transmission), but as the frequency becomes higher, this coefficient tends toward zero. Curve 1 is for a expansion chamber 5 cm in length and a cross-sectional ratio of $A_1/A = 4$. However, equation (8.25) does not apply to $kL > 1$.

In order to treat the case of the above acoustic filter for $kL > 1$, the incident, reflected, and transmitted waves in the three regions of the pipe must be related to one another by the fact that continuity in pressure and volume velocity must occur at the two junctions of the pipe. This results in a power transmission coefficient expressed as follows:

$$T_p = \frac{4}{4\cos^2 kL + \left(\frac{A_1}{A} + \frac{A}{A_1}\right)^2 \sin^2 kL} \tag{8.27}$$

In Figure 8.5, curve 2 constitutes the plot of equation (8.27) for the same filter system used to obtain curve 1. At low values of frequencies, that is, $kL \ll 1$, the two curves basically coincide. Equation (8.27), which is physically more valid, exhibits a minimum transmission coefficient of

$$T_p(\text{minimum}) = \left(\frac{2A_1 A}{S_1^2 + S^2} \right)^2 \qquad (kL = \pi/2) \tag{8.28}$$

in the case where the length of the filter segment equals a quarter-wavelength. Beyond this saddle point, T_p gradually rises with increasing frequency until it reaches 1.0 (100%) at $kL = \pi$. At even higher frequencies the transmission coefficient vacillates through a series of maxima and minima until ka (a is the radius of the through pipe) becomes somewhat larger than unity. From this point on, the transmission coefficient remains at unity. This characteristic of the transmission coefficient leveling out at unity is also shared by high-pass and band-pass filters.

Equation (8.28) may also be used to treat the constriction-type low-pass filter illustrated in Figure 8.6, because it does not matter if A_1 is larger or smaller than A. The decrease in the area can be viewed as introducing an inertance in series with the pipe, but the validity of this analog also extends over a limited range of frequencies, as with the case of the expanded-area low-pass filter of Figure 8.5.

In the real world of filter design (of mufflers, sound-absorption plenum chambers for ventilating systems, etc.) the filter cross-section cannot be radically different in value from the cross-sectional area of the pipe. As was demonstrated in the curves of Figure 8.5, a limited range of frequencies exists for the practical operation of the filters.

2. High-Pass Filters. A high-pass filter can be constructed by attaching a short length of pipe as a branch to a main pipe, in effect creating an orificed Helmholtz resonator. Both the radius a and length L of this appendage are small compared to a wavelength. Equations (7.60b) and (7.57) for the unflanged resonator apply to the branch impedance of this orifice given by

$$\mathbf{Z}_g = \frac{k\rho_0 c^2}{4\pi} + i \frac{\rho_0 \omega L'}{\pi a^2} \tag{8.29}$$

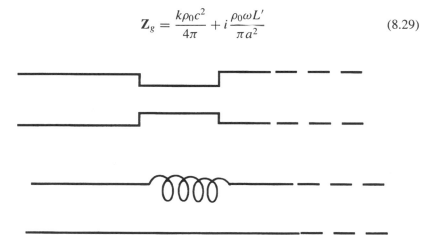

FIGURE 8.6. A pipe with a constriction and its electrical analogue.

where $L' = L + 1.5a$. The first term in the right-hand side of equation (8.29) stems from the radiation of sound through the orifice into the external medium, and the second is attributable to the inertance of the gas in the orifice. The ratio of the branch acoustic resistance to its acoustic reactance is $R_g/X_g = 1/4\,ka^2/L'$. Because it has been assumed that $ka \ll 1$, we can neglect the acoustic resistance in comparison with the acoustic reactance in the use of equation (8.24) to find the power transmission coefficient $T_{p,}$, with the result:

$$T_p = \frac{1}{\left(\frac{\pi a^2}{2AkL'}\right)^2 + 1} \tag{8.30}$$

We observe that this transmission coefficient is virtually zero for low frequencies and it rises to nearly unity at higher frequencies, as shown in Figure 8.7. The halfway point at which the transmission coefficient is 50% is reached when

$$k = \frac{\pi a^3}{2AL'}$$

The presence of a single orifice turns a pipe into a high-pass filter. If the radius of the orifice is increased, the attenuation of the low-frequency components also increases. If a pipe contains several orifices separated by only a fraction of a wavelength, these orifices can be treated as a group acting with their equivalent parallel impedance. But if the distances between the orifices constitute an appreciable portion of the wavelength, the system becomes analogous to an electrical network of filters or to a transmission line which has a number of impedances shunted across

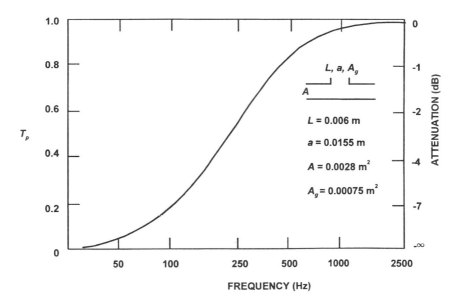

FIGURE 8.7. Attenuation produced by an orifice-type branch, resulting in high-pass transmission.

it, spaced apart at wide intervals. Waves reflected from these different orifices then become out of phase with respect to each other, and equation (8.30) no longer remains valid. Electrical filter theory must then be utilized to compute the transmission coefficient. As a rule, a number of orifices strategically placed apart can attenuate at low frequencies more effectively than a single orifice of equal total area.

The sound power transmission coefficient T_{pg} into a single orifice is approximated by

$$T_{pg} = \frac{2k^2 A}{\pi\left[\left(\frac{2AkL'}{\pi a^2}\right)^2 + 1\right]} \tag{8.31}$$

The filtering action of an orifice is principally that of the reflection of energy back toward the source, not so much the loss of acoustic energy out of the pipe through the orifice into the ambient medium.

A common example of the application of orifices is the control of the behavior of a wind instrument such as a flute or a saxophone. When such an instrument is played in its fundamental register, all or nearly all of the orifices some distance beyond the mouthpiece are kept open by the player. The diameters of these orifices nearly equal the bore of the tube, essentially shortening the effective length of the instrument. The acoustic energy reflected from the first open orifice generates a pattern of standing waves between the first open orifice and the mouthpiece. The flute behaves like an open pipe, with the wavelength approximately equal to twice the distance between the first orifice and the opening of the mouthpiece. A clarinet or a saxophone contains a vibrating reed at the mouthpiece, which approximates the conditions of the closed end of a tube. In this case the wavelength will equal nearly four times the distance from the reed to the first open orifice.

Both the reed-type (clarinet, saxophone, coronet, etc.) and tubular instruments (flute, recorder, piccolo, etc.) contain a number of harmonics with those of the reed instruments being primarily odd harmonics (characteristic of closed pipes). When higher notes are played on either type of instruments, the fingering of these holes become more complicated, with some orifices beyond the first orifice closed and some others opened. The fingering of these orifices controls the standing-wave patterns that correspond to specific notes.

3. Band Pass Filters. A side branch in the form of a long pipe rigidly capped at its far end or a fully enclosed Helmholtz resonator (shown in Figure 8.8) contains both inertance and compliance, so it will behave as a band-pass filter. Apart from almost negligible viscosity losses, no net dissipation of acoustic energy occurs from the pipe into the resonator. All energy absorbed by the resonator during some phase of the acoustic cycle is returned to pipe during other phases of the cycle so $Rg = 0$. Denoting the opening area by $A_g = \pi a^2$, the neck length by L and the volume of the resonator by V, the branch reactance X_g is expressed as

$$X_g = \rho_0\left(\frac{\omega L'}{A_g} - \frac{c^2}{\omega V}\right)$$

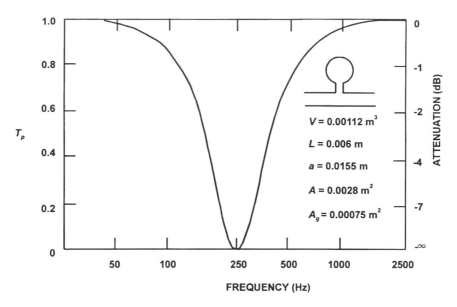

FIGURE 8.8. The effect of a Helmholtz resonator branch, resulting in band-pass transmission.

which is then inserted into equation (8.24) to yield the following transmission coefficient

$$T_p = \left[1 + \frac{c^2}{4A^2 \left(\frac{\omega L'}{A_g} - \frac{c^2}{\omega V} \right)} \right]^{-1} \tag{8.32}$$

The resonant frequency occurs when the transmission coefficient become zero, that is,

$$\omega_0 = c \sqrt{\frac{A_g}{L'V}}$$

which corresponds to the resonant frequency of a Helmholtz resonator. When this frequency occurs, large-volume velocities prevail in the neck of the resonator, and all acoustic energy that transmits into the resonator returns to the main pipe in such a manner as to be reflected back to the source. The plot of the power transmission coefficient in Figure 8.8 is fairly typical for a bandpass resonator.

Equation (8.32) serves well for a resonator that has a relatively large neck radius. Narrower and longer constrictions will cause the transmission coefficient to deviate from the prediction of equation (8.32), unless consideration is taken of the viscous dissipation that manifest itself more with such geometries.

4. *Filter Networks*. The design procedure for acoustic networks, which incorporate resonators, orifices, and divergence and convergence of pipe areas, is rendered easier by analogy with the deployment of electronic filters. The sharpness of cutoff

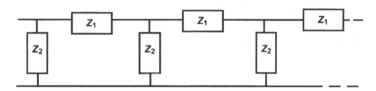

FIGURE 8.9. A ladder-type network used as a filter.

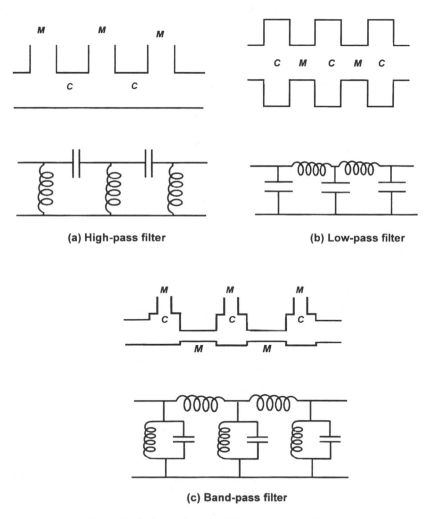

(a) High-pass filter   (b) Low-pass filter

(c) Band-pass filter

FIGURE 8.10. Examples of ladder-type acoustic filters.

of an electrical filter system, for example, can be enhanced by using the ladder-type network of Figure 8.9. This network is constructed by using the reactances of one type of impedance $\mathbf{Z}_1$ in series with the line and the reactance of another type of impedance $\mathbf{Z}_2$ shunted across the line. The standard theory of wave filters states that a nondissipative repeating structure such as that illustrated in Figure 8.9 causes a marked attenuation of all frequencies except those for which the ratio $\mathbf{Z}_1/\mathbf{Z}_2$ meets the condition

$$0 > \frac{\mathbf{Z}_1}{\mathbf{Z}_2} > -4. \tag{8.33}$$

Several examples of acoustic ladder-type filters are displayed with the electrical analogues in Figure 8.10. The condition of (8.33) provides the following cutoff frequencies:

$$f = \frac{1}{4\pi\sqrt{MC}}$$

for the high-pass filter of Figure 8.10(a), and

$$f = \frac{1}{\pi\sqrt{MC}}$$

for the low-pass filter of Figure 8.10(b). The behavior projected by the filters of Figure 8.10 applies only to wavelengths considered to be large compared to the dimensions of the filter. At higher frequencies, deviation of behavior predicted by electrical analogues becomes more prominent, because the filter begins to manifest the properties of distributed parameters rather than those of lumped parameters.

References

Beranek, Leo J. 1986. *Acoustics*. New York: The American Institute of Physics; Chapter. 3.
Kinsler, Lawrence E., Austin R. Frey, Alan B. Coppens, and James V. Saunders. 1982. *Fundamentals of Acoustics*, 3rd ed. New York: John Wiley & Sons; Chapters 9 and 10.

Problems for Chapter 8

In the following problems, consider the fluid medium to be air at the standard conditions of 1 atm and 20°C, unless otherwise stated.

1. It is desired to make a Helmholtz resonator out of a sphere with a diameter 12 cm for 350 Hz.
   (a) What should be the diameter of the hole in the sphere?
   (b) What should be the pressure amplitude of the incident acoustic plane wave at 350 Hz if it is to produce an excess internal pressure of 25 μbar?
   (c) If the hole is doubled in area, what will be the resonant frequency?
   (d) What will be the resonance frequency if two independent separate holes, each of the diameter found in part (a), are drilled in the sphere?

2. Consider a loudspeaker system that is rigid-walled and back-enclosed. Its inside dimensions are 40 cm × 55 cm × 44 cm. The front panel of cabinet is 4 cm thick and it has a 22 cm diameter hole cut out to accommodate a loudspeaker.

   (a) What is the fundamental frequency of this cabinet considered as a Helmholtz resonator?

   (b) A direct-radiating loudspeaker having a cone of 22 cm diameter and 0.008 kg mass and a suspension system of 1100 N/m stiffness is mounted in the cabinet. Find the resonance frequency of the cone. It may be assumed that the effective mass of the system is that of the cone and that of the air moving in the opening of the cabinet. The effective stiffness is the sum of the stiffness of the cone and that of the cabinet.

   (c) What would be the resonant frequency of the loudspeaker if it is not mounted in the cabinet and it has no air loading?

   (d) Find the acoustic power emitted if the cone is driven with an amplitude of 2.5 mm at the frequency established in part (b).

   (e) Under the conditions same as that of (d) what is the amplitude of the force acting on one of the 55 cm × 44 cm panels?

3. A rectangular room has internal dimensions of 3.0 m × 5.0 m × 2.6 m and walls of 12 cm thickness. A door that opens into the room has dimensions of 2.2 m × 0.8 m. Assume that the inertance of the door opening is equivalent of a circular opening of equal area.

   (a) Find the resonance frequency of the room considered an a Helmholtz resonator.

   (b) Find the acoustic compliance of the room.

   (c) Find the inertance of the door opening.

   (d) What is the acoustic impedance presented to the sound source at 30 Hz inside the room? Consider only the compliance of the room and the inertance of the door opening.

4. Pipe 1 having cross section A_1 connects to pipe 2 with cross section A_2.

   (a) Obtain an expression for the ratio of intensity of the waves transmitted into the second pipe to that of the incident waves.

   (b) Under what conditions will the transmitted energy exceed the incident energy?

   (c) Develop a general expression for the standing-wave ratio (SWR) produced in pipe 1 in terms of relative areas A_1 and A_2.

5. Consider two pipes that are connected to each other, but separated by a thin rubber diaphragm. Pipe 1 is filled with a fluid with characteristic impedance $\rho_1 c_1$ and pipe 2 contains a fluid with characteristic impedance $\rho_2 c_2$. A plane wave travels in pipe 1 in the positive x direction toward pipe 2.

   (a) Obtain an expression for the power transmission ratio from pipe 1 into pipe 2.

   (b) Under what conditions does 100 percent power transmission occur?

6. An infinitely long pipe of cross-sectional area A has a side branch, also infinitely long, of cross-sectional area A_b. The main pipe is transmitting plane waves with

frequencies such that their wave lengths are much larger than the diameter of either pipe.

(a) Develop an equation for the transmission coefficient in the main pipe.

(b) Do the same thing for the branch pipe.

(c) Let the area of the main pipe be twice as much as that of the branch pipe. Obtain numerical values for the transmission coefficient into each pipe. Do the sum of the two coefficients equal unity? If not, where did the remaining power go?

7. A ventilating duct in a basement has a square cross-sectional area, 35 cm to the side. In order to quiet the duct in part, a Helmholtz resonator–type filter is constructed about the duct by drilling a hole of 9 cm radius in one wall of the duct, leading into a surrounding closed chamber of volume V.

(a) What is the volume V necessary to most effectively filter sounds at a frequency of 30 Hz?

(b) What will be the sound power transmission coefficient of the filter at 45 Hz? 60 Hz?

8. Demonstrate that the radius r of the hole drilled into a pipe of radius r_0 that result in a 50% sound power transmission coefficient at a frequency f is given by:

$$r = \frac{64}{3} \frac{r_0^2 f}{c}$$

9. A 400-Hz plane wave of 0.2 W power propagates down an infinitely long pipe of 5 cm diameter. Find the power reflected, the power transmitted along the pipe, and the power transmitted out through a simple orifice of 1.2 cm diameter.

10. A 5-cm pipe carries water. It is planned to filter out plane waves traveling in the water by using sections of pipes 10 cm in diameter to serve as expansion-chamber type of filters.

(a) Find the minimum length of the filter section that will most effectively filter out a sound of 1 kHz.

(b) If it is desired to have a filtering action lessen the level of intensity by 35 dB, how many sections must be used? Disregard the effects of any interaction between the individual filter sections.

9
Sound-Measuring Instrumentation

9.1 Introduction

Acoustic measurements constitute an essential step in order to establish the status of an acoustic environment and develop a systematic approach toward modifying the environment, and to set up criteria for improvements. Other more recently developed methodologies of acoustical measurements entail the studies of material properties and medical diagnoses. A wide variety of instrumentation exists, ranging from a simple sound-pressure-level meter to real-time spectrum analyzers interfaced with sophisticated computer systems. Instruments may be portable for field use; and recorded field data can be later evaluated in more elaborate systems. New instruments are being developed continually—and with new advances in digital technology arriving on the scene on a daily basis, it is not inconceivable that more versatile and "user-friendly" devices will become available at even lower prices.

This chapter deals principally with instruments intended for the audio range of frequencies. Much of the salient aspects of instrumentation in this chapter, such as the principal performance requirements and methodologies for evaluating data, also apply to the specialized underwater instruments and ultrasonic sensing devices which are described in Chapters 15 and 16, respectively.

9.2 Principal Characteristics of Acoustical Instruments

The most important performance characteristics of acoustical instruments are the frequency response, dynamic range, crest factor capability, and response time. It is also desirable that a measuring device or system have a negligible (or at least predictable) effect or influence on the variable being measured.

Frequency response refers to the range of frequencies that an instrument is capable of correctly measuring the relative amplitudes of the subject variable within acceptable limits of accuracy. Measurement accuracy depends on the instrumentation type and the quality of design and manufacture. A typical limit for the flatness of response for microphones may be ±2 dB or better; in contrast, the response curve of a high-quality loudspeaker may deviate ±5 dB over its rated frequency range.

Dynamic range defines the range of signal amplitudes an instrument is capable of handling in the process of responding and measuring accurately. A sound-level meter, for example, that can measure a minimum of 10 dB and a maximum of 150 dB, covers a dynamic range of 140 dB.

Crest factor capability denotes an instrument's capacity to measure and distinguish instantaneous peaks. The crest factor itself is the ratio of the instantaneous peak sound pressure to the root-mean-square sound pressure.

Response time refers to the rapidity with which a measuring instrument responds to abrupt changes in signals. An oscilloscope display of a square wave response will result in a trapezoidal display if the relayed signals came from a loudspeaker that requires a longer reaction time to respond to a square-wave pulse.

9.3   Microphones

Microphones serve as transducers by receiving and sensing pressure fluctuations and converting them into electrical signals that are relayed to other electronic components. The quality of a microphone determines the accuracy of a measurement system. A top-caliber measuring (or a sound reproduction) system can be undermined by the use of a microphone that is of lesser caliber. Four principal types of microphones are used in measurement procedures, namely, the dynamic, ceramic, electret, and condenser microphones.

Dynamic microphones produce an electric signal through the motion of a coil connected to a diaphragm in a magnetic field. They are in effect loudspeakers working in reverse, receiving an acoustic signal and converting it into an electric pulsation rather than the other way around. Because of their low impedance, they can be used in applications that necessitate the use of long cables for connection to auxiliary instrumentation. But they cannot be used in the vicinity of devices that emit magnetic fields (e.g., transformers and motors). Also, they generally have longer response times and more limited frequency response, but they can be constructed to withstand rough handling and high humidity.

Ceramic microphones consist of a piezoelectric (ceramic) element attached to the rear of a diaphragm. Sound pressure causes the diaphragm to vibrate, exerting a varying force on the ceramic element. The piezoelectric crystal generates an electric signal from the oscillating strains imposed by the action of a moving diaphragm. These microphones are rugged, relatively inexpensive, reliable, have high capacitance and good dynamic range, and do not require a polarizing voltage that both electret and condenser microphones need. But the high-frequency response may be lacking, and operating temperature range may be limited.

The *electret* (or *electret-condenser*) microphone, illustrated in Figure 9.1, consists of a self-polarized metal-coated plastic diaphragm. Sound pressure causes the diaphragm to move relative to a back plate, varying the capacitance and producing a signal. While relatively impervious to high humidity, this type of microphone does not equal the frequency response of a condenser microphone and is not likely to withstand temperature extremes very well. An alternative form of the electret microphone is to deposit the electret film onto the stationary plate and use the thin metal foil as a moving diaphragm.

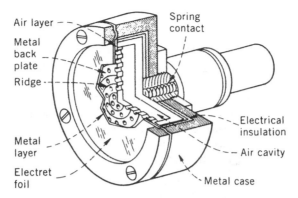

Air layer

Metal back plate

Ridge

Metal layer

Electret foil

Spring contact

Electrical insulation

Air cavity

Metal case

FIGURE 9.1. A cutaway view of an electret microphone showing the principal components. A thin electret polymer foil is suspended over a perforated backplate.

Condenser microphones are capacitance-varying devices, each one consisting of two electrically charged plates with an air gap between them. The thin diaphragm serves as the plate that deflects under the influence of changes in sound pressure, causing the gap to vary. The resulting change in the capacitance is converted into an electric signal. Figure 9.2 illustrates the principal components of a condenser microphone. A capillary tube behind the plates provides the air bleed to provide air pressure equalization with the ambient. These devices have relatively low capacitance, require a polarizing voltage supplied by a preamplifier which also provides the appropriate impedance for connecting the microphone to a measuring system. Because of their long-term stability, superb high-frequency response, and insensitivity to vibration excitation, condenser microphones are generally preferred for precision measurements. They function well in extreme temperature and pressure environments, but they are affected adversely by high humidity (which causes electrical leakage), and their diaphragms are fragile.

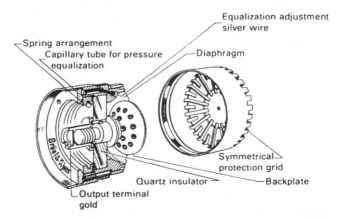

Equalization adjustment silver wire

Spring arrangement
Capillary tube for pressure equalization

Diaphragm

Symmetrical protection grid

Quartz insulator

Backplate

Output terminal gold

FIGURE 9.2. Principal elements of a condenser microphone cartridge. (Courtesy of Brüel & Kjær Division of Spectris Technologies, Inc.)

9.4   Sensitivity

In addition to frequency response, sensitivity constitutes one of the principal characteristics of a transducer. In general there is a trade-off between sensitivity and frequency response. Small microphones tend to have lower sensitivity, but operate at both low and high frequencies, while large microphones possess high sensitivity but are useful mainly at lower frequencies. Microphone diameters are typically $1/4$, $1/2$, and 1 in. (1, 2, and 4 cm) in diameter. The frequency response of a precision 1-in. condenser microphone is virtually flat to 20 kHz. while that of a $1/4$-in. microphone is fairly flat to approximately 100 kHz. We define microphone sensitivity S by

$$S = \frac{\text{electrical output}}{\text{mechanical input}}$$

The microphone sensitivity level L_s, also called simply "sensitivity," is defined as

$$L_s = 10 \log \left(\frac{\frac{E_{\text{out}}}{p}}{E_{\text{re}}} \right)^2 \text{ dBV/mbar} = 20 \log \left(\frac{E_{\text{out}}}{p} \right) \text{ dBV/mbar} \qquad (9.1)$$

where

E_{out} = output voltage into instrument

E_{re} = reference voltage (1 V for an incident sound pressure of 1.0 μbar)

p = rms pressure on the microphone

One microbar (μbar) is equal to 0.1 Pa. Equation (9.1) can be rearranged to obtain the output voltage:

$$E_{\text{out}} = p 10^{L_s/20} \qquad (9.2)$$

We recall the definition of equation (3.22) for sound pressure level (SPL), L_p, and recast (9.2) as

$$E_{\text{out}} = 0.0002 \times \left(10^{L_p/20} \right) \left(10^{L_s/20} \right) = 0.0002 \times 10^{(L_p + L_s)/20} \text{ V} \qquad (9.3)$$

Example Problem 1

A microphone has sensitivity rating of $L_s = -50$ dBV/μbar. Find the output voltage for a sound pressure level of 85 dB.

Solution

Equation (9.3) is applied to yield

$$E_{\text{out}} = 2 \times 10^{-4} \times 10^{(85-50)/20} \text{ V} = 0.0112 \text{ V} = 11.2 \text{ mV}$$

Microphone sensitivities typically range between 0.5 μV/μbar (-125 dBV/μbar) and 3 mV/μbar (-50 dBV/μbar). A microphone used for general sound-level-pressure measurements in the frequency range of 10 Hz to 20 kHz might have a sensitivity of 3 mV/μbar (-50 dBV/μbar). But if it is desired to measure in the frequency range extending beyond 100 kHz, a special microphone designed for this purpose would have a sensitivity of only 0.5 μV/μbar (-125 dBV/μbar).

9.5 Selection and Positioning of Microphones

Figure 9.3 illustrates a flow chart for the selection of microphones and their orientation during use on the basis of the nature of the sound field. Two standards apply here, one by the American National Standards Institute (ANSI) which calls for microphones with random-incidence response and other by the International Electrotechnical Commission (IEC) which specifies free-field microphones.

The *free field* occurs as a region that is not subjected to reflected waves, as is the case an open field or in an anechoic chamber. The presence of a microphone in the sound field disturbs the field. A microphone designed to compensate for this disturbance is called a *free-field microphone*. In order to obtain maximum accuracy in measurements, the free-field microphone should be pointed toward the noise source [0° incidence, as shown in Figure 9.4(a)]. Microphone sensitivities are also stated in terms of mV/Pa.

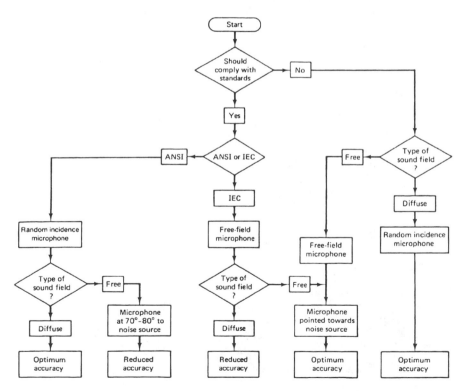

FIGURE 9.3. Flow chart for the selection of microphones and their orientiation. (Courtesy of Brüel & Kjær Division of Spectris Technologies, Inc.)

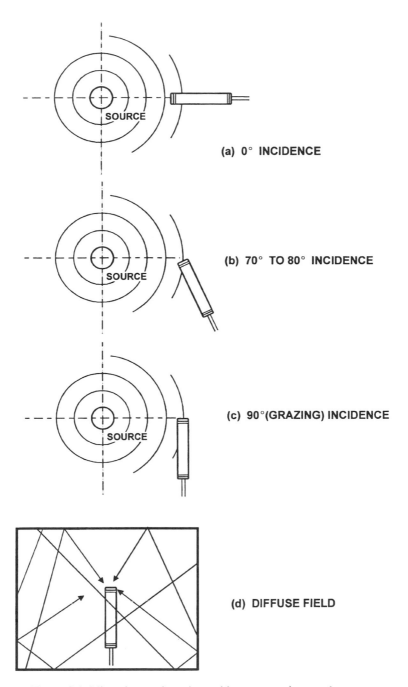

FIGURE 9.4. Microphone orientations with respect to the sound source.

The disturbance of the sound field by the presence of the microphone depends on the sound frequency, the direction of propagation, and the size and shape of the microphone. At higher frequencies, where the wavelength of the sound is small compared with the principal dimensions of the microphone, reflections from the microphone cause the pressure acting on the microphone diaphragm to differ from the actual free-field sound pressure that is supposed to be measured. Because a wave length of 1 in. corresponds to 13,540 Hz, a 1-in. microphone will not provide accurate free-field measurements of noise in the frequency range in the neighborhood of 13 kHz and above. Even at 6 kHz, the error for a 1 in. microphone can exceed 2 dB.

The converse of the free field is the *diffuse field* which occurs as the result of multiple reflections. A *random-incidence microphone* is utilized in measuring sound in diffuse fields; it is *omnidirectional* in that it responds uniformly to sound arriving from all angles simultaneously [cf. Figure 9.4(d)].

Pressure microphones are designed to yield a uniform frequency response to the sound field including the disturbance produced by the microphone's presence. This type of microphone may also be used in diffuse fields. Using a free-field microphone in a diffuse field will result in lessened accuracy unless special circuitry in the measurement system provides compensating corrections. As shown in Figure 9.4(b), when a random-incidence microphone is used to measure sound in a free field, the unit should be place at an incidence angle of 70°–80° to the source. A pressure microphone should be positioned at an incidence angle of 90° (often referred to as the *grazing incidence*) in a free field [Figure 9.4(c)]. Microphone placement become more critical as the increasing sound frequencies approach the upper limits of accuracy.

For a windy environment, special precautions should be taken to obtain the proper data. The wind rushing past a microphone produces turbulence, generating pressure fluctuations resulting in low-frequency noise. In the use of A-weighted measurements, no precautions are necessary for winds below 5 mph (8 km/h), because the A-weighting network attenuates greatly at low frequencies. But for C-scale or linear sound level measurements, a wind screen should be employed for any sort of wind. The device should also be used for wind speeds above 5 mph in A-weighted measurements. Wind screens are typically open-celled polyurethane foam spheres that are placed over the microphones. However, these may not be too effective if the wind speed exceeds 20 mph (30 km/h).

Changes in atmospheric conditions, namely, temperature and pressure, may necessitate corrections. Figure 9.5 illustrates a chart for determining corrections due to deviations from 1 standard atmosphere (30 in Hg or 762 mm Hg) and 20 °C, based on the following expression:

$$C_{T,P} = 10 \log \left(\sqrt{\frac{F + 460}{528}} \left(\frac{30}{B} \right) \right)$$

$$= 10 \log \left(\sqrt{\frac{C + 273.3}{293.3}} \left(\frac{760}{B'} \right) \right) \tag{9.4}$$

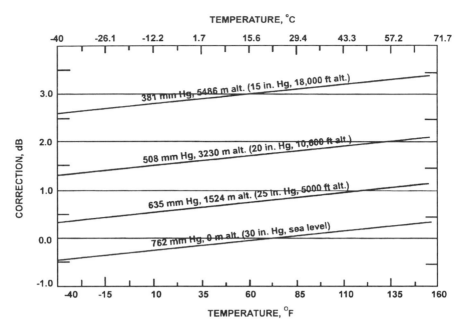

FIGURE 9.5. Sound pressure level corrections for conditions which deviate from the standard atmospheric pressure and temperature of 760 mm Hg and 20°C.

where

$C_{T,P}$ = correction factor to be added to the sound pressure level, dB

F, C = temperature in °F or °C, respectively

B, B' = barometric in inches Hg or mm Hg, respectively

Example Problem 2

For a sound-pressure-level reading of 92 dB is taken at the standard conditions of 1 atm. and 68 °F, predict the readings at 10 °F and at 110 °F for the same atmospheric pressure.

Solution

Applying equation (9.4)

$$C_{T,P} = 10\log\left(\sqrt{\frac{10+460}{528}}\left(\frac{30}{30}\right)\right) = -0.25 \text{ dB} \qquad \text{(for 10 °F)}$$

$$C_{T,P} = 10\log\left(\sqrt{\frac{110+460}{528}}\left(\frac{30}{30}\right)\right) = +0.17 \text{ dB} \qquad \text{(for 110 °F)}$$

At the lower temperature the reading would be $(92 - 0.25) = 91.8$ dB and at the higher temperature, $(92 + 0.17) = 92.2$ dB. Because the deviations are within the instrument error, the application of such corrections is not very meaningful.

The presence of reflecting surfaces affects measurements with the use of microphones. For example, the presence of a person near the microphone will disturb the sound field, and it would be advisable to place the microphone on a tripod and monitor the instrumentation from a distance, say at least several wavelengths of the lowest frequency present. Also when a long cable is used to connect the microphone to other instruments, care must be taken to avoid noise generated by the motion of different segments of the cable with respect to each other. The cable should be constrained from moving and isolated from vibration to the greatest degree possible. The cable should also be well-shielded from stray electromagnetic fields; but it is preferable that the tendency to noise generation be cut down by incorporating the preamplifier and the microphone into a single unit, thereby yielding a considerably greater signal-to-noise ratio (S/N ratio). As a rule, stronger signals are more impervious to outside influences than would be the case if the signals were not pre-amplified.

9.6   Vector Sound Intensity Probes

A sound intensity probe that is used to measure vector sound intensity is illustrated in Figure 9.6(a), and a block diagram in Figure 9.6(b) shows the components of this vector sound-measuring system. The device contains two microphones mounted face to face. Other modes of mounting are possible, including side-by-side and back-to-back. The probe measures sound-pressure levels at two different points simultaneously. With the microphone spacing constituting a given factor, the sound pressure gradient can be determined and the particle velocity in a given direction can be calculated. The intensity vector component in that direction may then be established. The microphone spacing should be considerably smaller than the wavelength of the sound being measured in order to yield valid results from two-point measurements. For example, in order to sustain an accuracy of ± 1 dB, the upper frequency limit for two $1/4$-in. microphones set apart 12 mm from each other is approximately 5 kHz. If the microphones are spaced apart by only 6 mm, the corresponding upper frequency limit is 10 kHz.

9.7   The Sound Level Meter

The sound level meter (SLM), a most valuable means of assessing noise environments, amplifies the signal from a sensing microphone and processes the information for visual display or information storage. It is generally portable and battery operated. The quality ratings of sound level meters are specified by the

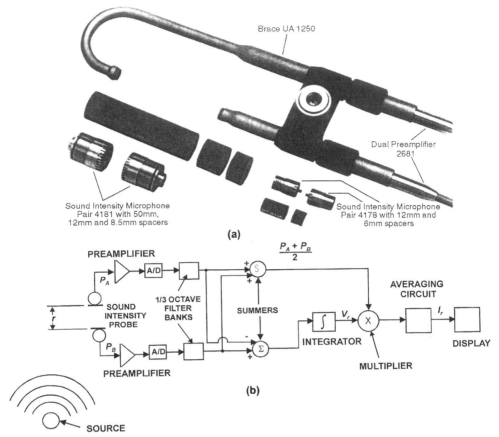

FIGURE 9.6. Sound intensity probe. (a) View of a sound intensity probe and its components. (b) Block diagram of the vector intensity measurement device. (Courtesy of Brüel & Kjær Division of Spectris Technologies, Inc.)

American National Standards Institute (ANSI) and the International Electrotechnical Commission (IEC) according to the precision of these instruments. Types 1, 2, and 3 are respectively termed "precision," "general purpose," and "survey." In addition, type 0 has been specified by IEC for laboratory reference standard. ANSI includes the suffix S in its standard to designate special-purpose meters, e.g., meters equipped with only A-weighting. Measurement precision depends on a number of factors, including meter calibration, method of surveying, and frequency content of the noise being measured. Type 1 meters generally should measure with 1 dB accuracy and are employed to obtain accurate data for noise control purposes. The corresponding error for type 2 meters, which are used for quick surveys may not exceed 2 dB.

Figure 9.7 shows a view of a digital-readout SLM. The contour of the meter case slopes away from the microphone in order to minimize reflections from its surfaces. Both types 1 and 2 generally incorporate A-, B-, and C-weighting networks.

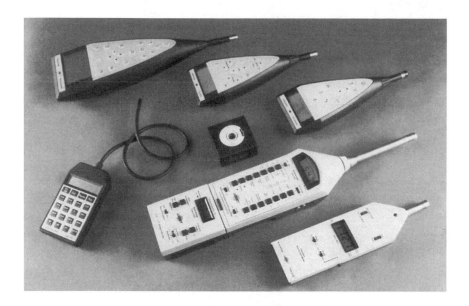

(a)

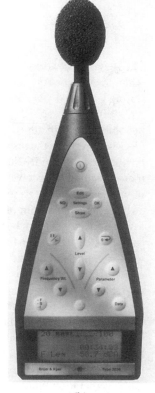

FIGURE 9.7. (a) Sound-level meters. Some models come equipped with built-in octave or 1/3-octave filters for band measurements or with provisions for attaching separate filter units to perform such measurements (the largest model in the group is shown here equipped with a detachable filter unit). A sound level calibrator used to check the operational integrity of the meters is shown near the center of the sound-level meter grouping. The unit in the lower right hand corner of this group is a noise dosimeter, (b) The key features of a basic digital readout sound-level meter. (Courtesy of Brüel & Kjær Division of Spectris Technologies, Inc.)

(b)

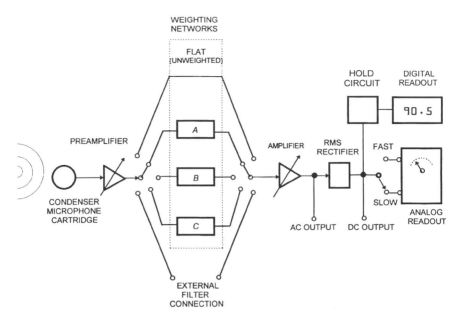

FIGURE 9.8. Block diagram of the principal elements of a sound-level meter.

The A and C networks are far more commonly used than B-weighting, particularly when low-frequency acoustic energy is present. Also a "fast" or "slow" response setting is generally available. The former setting, corresponding to a time constant $RC \approx 0.125$ s, responds more quickly to changes in noise levels, but the readings become more difficult to ascertain with very rapid fluctuations. The slow setting ($RC \approx 1$) reduces the response speed, and a better grasp of typical sound levels can be obtained for rapidly fluctuating sounds. More elaborate versions of sound level meters include 1/3-octave and octave filters to enable sound pressure measurements in various frequency bands.

In the block diagram of Figure 9.8, a typical SLM is shown to contain the following components: a 1-in. or ^{1}/2-in. microphone feeding a preamplifier (which functions as a cathode follower) which in turns relays to one of the weighting networks that is selected by a switch. The weighted (or unweighted, if none of the weighting curves has been selected) signal is then amplified and then passes through a root-mean-square rectifier, becomes converted to logarithmic (i.e., decibel) form, and is fed to either a digital or analog readout device. Some SLM models contain output jacks so that AC and DC signals from the meter can serve as inputs to other instruments, such as Fast Fourier Transform (FFT) analyzers, printers, or graphic plotters. By taking advantage of the PCMCIA modular technology, a meter can be designed to provide a variety of other functions such measuring L_{eq}, L_n, reverberation times, and so on.

9.8   Proper Procedures for Using the Sound-Level Meter

To ensure that a sound-level meter is in proper working order, an acoustic calibrator should be employed just prior to beginning a series of measurements and after completing the series. The calibrator is a single-tone battery-driven device that fits over the microphone and produces a precise reference sound pressure level for calibrating the meter. The calibrator, illustrated in Figure 9.9, employs a zener-stabilized oscillator to provide impetus to a piezoelectric driver element that causes a diaphragm to vibrate at 1 kHz $\pm 1.5\%$. The diaphragm produces a sound pressure level of 1 Pa (corresponding to rms SPL of 94 $\pm$ 0.3 dB) in the coupler volume. Calibrations operate at 1 kHz, the international reference frequency for weighting networks. Therefore, no correction is required for calibrating instruments for weighted and unweighted measurements.

Periodically, every six months to a year depending on frequency and rigor of usage, both the SLM and the calibrator should be rechecked with calibration instruments which, in turn, have been calibrated periodically in a manner traceable to direct comparison with the standards set up by the U.S. National Institute of Science and Technology (NIST) or the appropriate standards bureau in another nation. Updated certifications of calibration are necessary for both SLM and its calibrator as proof of the accuracy and reliability of the measurement devices. While executing measurements and entering data, one should also record sufficient information to identify all measuring instruments at a later date, should the need arise to prove the accuracy of the instruments at the time of its questioned usage.

The simplest case of noise measurement is that of steady noise, say a motor running at steady speed in a factory. In this situation it would be necessary to take only a few sound-level readings at the worker's ear and to check the measured

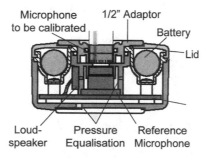

FIGURE 9.9. Cross-section of an oscillator-driven sound-level calibrator. This model produces a 1-kHz tone at a SPL of 94 dB. (Courtesy of Brüel & Kjær Division of Spectris Technologies, Inc.)

values with established permissible levels. In the more common case of fluctuating levels, moving noise sources or receivers, and so on, it may be necessary to take more readings over longer time intervals. Other types of meters or measurement procedures may be more suitable for sustained noise exposures, such as the use of the integrating sound-level meter and the dosimeter, both of which are described in the following sections.

9.9   The Integrating Sound-Level Meter

In Chapter 3, the equivalent sound level was defined as

$$L_{eq} = 10 \log \frac{1}{T} \int_0^T 10^{L/10} \, dt$$

or in terms of N sound-level measurements taken during N equal intervals:

$$L_{eq} = 10 \log \left(\frac{1}{N} \sum_{i=1}^{N} 10^{L_i/10} \right) \tag{9.5}$$

Integrating sound-level meters are based on the application of equation (9.5) in measurements of fluctuating noise over a considerable interval. For example, a meter may be preset to measure in intervals of 1 s over a total time period of 15 minutes, thus calling for 900 individual sets of measurements. A measurement over 12 hours may be programmed for 720 intervals of one minute each. More elaborate meters designed for 24-hour surveillance can compute the day-night equivalent sound level in which a 10 dB penalty is automatically added to noise levels occurring between the hours of 10 P.M. and 7 A.M. The integrating sound-level meter is also capable of measuring the sound exposure level (SEL), which characterizes a single event on the basis of both the pressure level and the duration. This parameter is defined by

$$\text{SEL} = 10 \log \left(\int_0^T \frac{p_{rms}^2}{p_{ref}^2} \, dt \right) = 10 \log \left(\int_0^T 10^{L/10} \, dt \right) \tag{9.6}$$

where T is measured in seconds. If we consider a 2-s burst of sound at the rms pressure of 1 Pa, use of equation (9.6) will indicate an SEL of 97 dB. Comparison of equations (9.5) and (9.6) leads to the following relationship between SEL and L_{eq}:

$$\text{SEL} = L_{eq} + 10 \log T \tag{9.7}$$

Example Problem 3

What is the SEL for an equivalent sound-level pressure of 98 dB(A) lasting 8 seconds?

Solution

From equation (9.7)

$$\text{SEL} = 98 + 10 \log 8 = 107 \text{ dB}(A)$$

Equation (9.6) provides the means of computing SEL when the integrating sound level meter does not offer an automatic SEL mode.

9.10   Dosimeters

It is not generally a practical matter to continuously trail an industrial worker performing his or her duties in order to gauge the amount of noise exposure, particularly if that individual moves from place to place and is exposed to varying degrees of noise levels in the course of the day. A more convenient method of measuring the total exposure is to have this person wear a either a *personal sound exposure meter* or a *noise dose meter* (also called a *noise dosemeter* or *noise dosimeter*). The personal sound exposure meter, which is schematically described in Figure 9.10, consists of a small microphone attached to a small extension cord so that it may be mounted close to the ear, a tiny amplifier that incorporates an *A*-weighting network, and a circuit that squares and integrates the electrical signal with respect to time. The unit normally includes a sound exposure indicator which can come separate from the wearable unit. This device must be able to respond to a wide range of frequencies and sound levels without the presence of a manual control. A *latching overload indicator* is also incorporated to provide a warning that excessive sound pressure levels are occurring within the frequency range of the instrument. The unit must be small, battery-powered, tamperproof, and constructed to withstand harsh environments.

   The noise dosimeter is a device designed to measure the percentage of the maximum daily noise dose permitted by regulations. Its functional elements are given in Figure 9.11. Both the noise dosimeter and the personal sound exposure meter share the same requirements for operation under severe working conditions; and

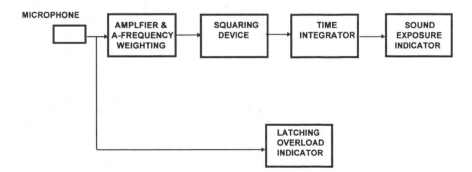

FIGURE 9.10. Schematic of a personal noise exposure meter.

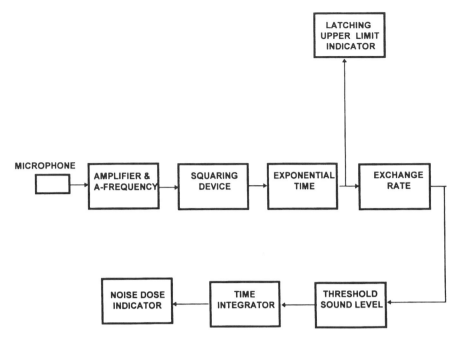

FIGURE 9.11. Schematic of a noise dosimeter.

both types of instruments include a microphone, an amplifier, A-weighting network, a squaring device, a time integrator, and an indicator. The personal sound exposure meter contains a latching overload but the noise dosimeter may include a latching upper-limit indicator. Additionally, the dosimeter must include exponential time-weighting usually incorporating the *slow response* characteristic and a manufacturer-specified threshold sound level,[1] neither of which is found in the personal sound exposure meter.

American National Standard ANSI SI.25 gives the specifications for a noise dosimeter. Slow-time weighted, A-weighted sound pressure is integrated with a 5-dB exchange rate[2] in accordance with the regulations of U.S. Occupational Safety and Health Administration (OSHA) and Mine Safety and Health Administration (MSHA). The exchange rate may also be 3 or 4 dB depending on application. ANSI SI.25 also specifies limits for the effects of changes in air pressure, temperature,

---

[1] *Threshold sound level*, stated in decibels, is the A-weighted sound level specified by the manufacturer of the noise dosimeter below which the instrument provides no significant indication. The threshold sound level should be at least 5 dB less than the criterion sound level.

[2] The terminology *exchange rate*, expressed in decibels, refer to the change in decibels required to double (or halve) the exposure time in order to maintain the same amount of noise exposure. For example, with the 5 dB exchange rate, exposure to 90 dB for 8 hours is equivalent to exposure to 95 dB for 4 hours or to 100 dB for 2 hours.

vibration, and magnetic field on the noise dosimeter. The British standard BS 6402 calls for a personal sound-level meter to measure sound exposure, that is, the timed-integrated A-weighted sound pressure level with a 3-dB exchange rate and without exponential time weighting. A latching overload indicator is also mandated by the British standard to provide warning that the sound level received at the microphone has exceed the measuring range of the instrument (up to 132 dB peak).

An IEC international standard for personal sound exposure meters defines a wider operating range for a type 2 integrating averaging SLM. Under this specification, measurements are made of the exposure produced by impulsive, fluctuating, and intermittent sounds over a range of A-weighted sound levels from 80 dB to 130 dB. Excessive input sound levels should trigger the mandatory overload indicator.

9.11   Noise Measurement in Selected Frequency Bands, Band Pass Filters

A valuable means of analyzing noise is the evaluation of sound in each frequency band. If a machine emits a noise that indicates it is malfunctioning, analysis of the sound output according to frequency can provide vital clues as to which component of the machine is defective. This situation calls for the use of a spectrum analyzer, which is a device that analyzes a noise signal in the frequency domain by electronically separating the signal into various frequency bands. This separation is executed through the use of a set of filters. A filter is a two-port electrical network with a pair of terminals are each port. A filter can be constructed with as few as two passive electrical components (e.g., a resistor and a capacitor), or it can be more complex, involving a large number of passive components or a combination of passive elements operating in conjunction with active components (e.g., op amps). Analog filters embody electronic circuitry tuned to pass certain frequencies, whereas digital filters make use of active electronic elements.

Figure 9.12 illustrates the effects of ideal and real filters. An ideal bandpass filter is a circuit that transmits only that part of the input signal within its bandpass $(f_1 < f < f_2)$ and completely attenuate all of the components at all frequencies outside of the bandpass $(f < f_1$ and $f > f_2)$. The ideal low-pass filter passes all signals up to frequency f_{lp} and rejects all signals having frequencies above f_{lp}. The ideal high-pass filter passes all signals above frequency f_{hp} and rejects all of the contents of the input signal below f_{hp}. In the real world, filters alter the shapes of the input signals to some degree. The amplitude and phase characteristics of the filter can be ascertained by computing the *transfer function* (or *filter response*), which is the ratio of the filter output to the filter input for all possible values of frequency. Both ideal and actual frequency responses are shown in Figure 9.12. The actual filters have characteristics of the form shown on the right-hand side of Figure 9.12. In the case of the low-pass filter of Figure 9.13(a), the amplitude

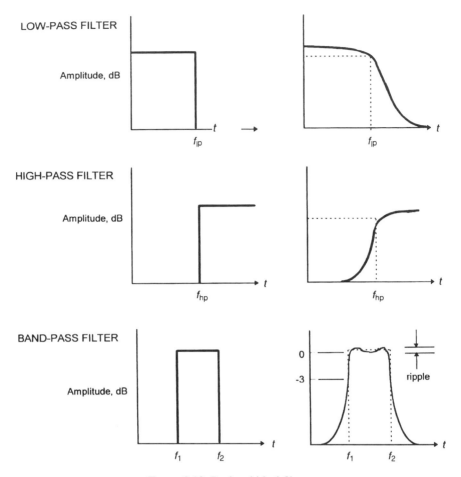

FIGURE 9.12. Real and ideal filters.

transfer function for the circuit is expressed as

$$|H(\omega)| = \left|\frac{E_o}{E_i}\right| = \frac{1}{\sqrt{1 + (\omega RC)^2}} \tag{9.8}$$

where the subscripts i and o to voltage E denote the input and output values, respectively; R is the resistance in ohms and C the capacitance in farads. Equation (9.8) can be written to yield in decibels the characteristic of the form shown plotted on the semi-logarithmic Figure 9.13(b) as follows:

$$|H'(\omega)| = 20\log\left|\frac{E_o}{E_i}\right| = 20\log\left[\frac{1}{\sqrt{1 + (\omega RC)^2}}\right] \text{ dB}$$

$$= -20\log\sqrt{1 + (2\pi f RC)^2} \text{ dB} \tag{9.9}$$

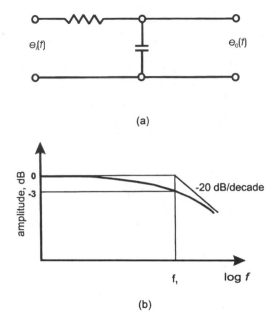

FIGURE 9.13. (a) Circuit for a single low-pass filter and (b) frequency response of the low-pass filter.

Equation (9.9) becomes

$$|H'(\omega)| \approx -20 \log 1 = 0 \qquad \text{for } (\omega RC)^2 \ll 1$$

that is, for relatively small frequency values the amplitude transfer function is 0 dB. In the case of large values of frequency, that is, $(\omega RC)^2 \gg 1$, equation (9.9) becomes

$$|H'(\omega)| \approx -20 \log(2\pi f RC)$$

The amplitude characteristic therefore exists as a slope of 20 dB per decade. The intersection f_1 of the two asymptotes is the *cutoff frequency* given by

$$f_1 = \frac{1}{2\pi RC}$$

According to equation (9.9) the amplitude decreases by $1/\sqrt{2}$ at the cutoff frequency, or 3 dB. Because power is proportional to the square of the pressure amplitude, the cutoff frequency is also the *half-power point*. The amplitude responses of complex filters can be obtained in a similar manner.

Filters may be grouped in bands so that *serial analysis* is rendered possible by switching manually or automatically from one band to the next. The *octave band analyzer* constitutes an example of a commonly used serial analyzer. This device is so-called because it is capable of resolving the noise signal spectrum into frequency bands that are one octave in width. We have already seen from Chapter 3

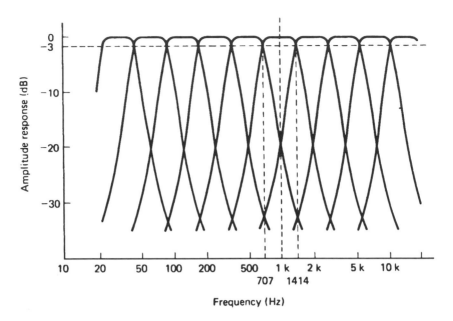

FIGURE 9.14. Filter response typical of an octave-band filter set.

that the center frequency of an octave band is $\sqrt{2}$ times the lower cutoff frequency, and the upper cutoff frequency is twice the lower cutoff frequency. The octave bands in the audio range are designated by their center frequencies, and are standardized internationally at 31.5, 63, 125, 250, 500, 1000, 2000, 4000, 8000, 16,000 Hz. Figure 9.14 illustrates the frequency response typical of an octave-band filter.

In order to obtain more details of the noise spectrum, filters with bandwidths narrower than one octave must be used. *Narrow band analyzers* constitute excellent tools for diagnosing noises in industrial environments. Typical relative bandwidth analyzers use 1/3-octave and 1/12-octave bandwidths, with the former being more common. These analyzers resolve the noise spectrum into third octaves and twelfth octaves. Denoting f_1 as the lower cutoff frequency, f_2 as the upper cutoff frequency, and f_0 the center frequency, the relationship between the upper and lower cutoff frequencies for a $1/n$th octave filter is

$$f_2 = 2^{1/n} f_1 \qquad (9.10)$$

The center frequency is the geometric mean of the product of the upper and lower cutoff frequencies,

$$f_0 = \sqrt{f_1 f_2} \qquad (9.11)$$

and the bandwidth **bw** is expressed as follows:

$$\mathbf{bw} = f_2 - f_1 = \left(2^{1/2n} - 2^{-1/2n}\right) f_0 \qquad (9.12)$$

The smaller the bandwidth, the more detailed the analysis, and it obviously follows that the equipment becomes costlier. The advantages of narrower-band analyzers become more apparent in the detection of prominent components of extremely narrow-band noise.

9.12   Real-Time Analysis

Real-time analysis entails the evaluation of a signal over a specified number of frequency bands simultaneously. This type of parallel process is schematically illustrated in Figure 9.15. Most real-time analyzers rely on digital filtering of a sampled time series. A fluctuating noise signal is converted by a microphone assembly into an electric signal that is entered simultaneously into a parallel set of filters, detectors, and storage components. The scanned output is displayed as a bar graph of sound pressure level versus frequency on a cathode ray tube (CRT) monitor or another type of display such as liquid crystal display (LCD) or plasma screen. The display may be refreshed several times per second as the level and frequency content of the input signal change. The scanning interval may be adjusted to meet different conditions, becoming longer to cover the cycling of a specific machine operation or shortened to evaluate impact sounds.

A considerable variety of features are available on current real-time analyzers. Some features may include alphanumeric displays, choice of weighting and linear (unweighted) sound levels, linear and exponential averaging, time constants, spectrum storage for recall and comparison with other data, and integrated or external software packages for further analysis of analyzer output. A number of real-time analyzers operate on batteries, and models are available that provide a choice of battery or AC operation.

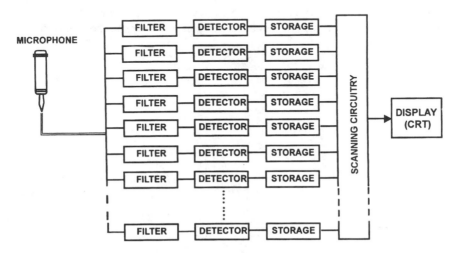

FIGURE 9.15. Schematic of the processing of a signal through a parallel bank of filters in a real-time analyzer.

9.13   Fast Fourier Transform Analysis

The Fast Fourier Transform (FFT) technique, rendered feasible by the advent of microcomputers, employs both digital sampling and digital filtering. Instead of relying on band-pass filters to measure the analog amplitudes that formulate a signal's spectrum, the FFT analyzer executes an efficient transformation of the signal from the time domain to the frequency domain. As it is capable of executing nearly any analysis function on the signal at high speed, the FFT technique is an extremely powerful analytical method. The FFT analyzer captures a block of sampled data of finite length (generally 1024 or 2048 samples) for a processing interval. In transforming from the time domain to the frequency domain, the Fourier transform relates a function of time $g(t)$ to a function of frequency $F(\omega)$ in the following manner:

$$F(\omega) = \int_{-\infty}^{+\infty} g(t)e^{-i\omega t}\,dt \tag{9.13}$$

In measuring noise, a microphone assembly generates a voltage proportional to sound pressure. A *time series* is formed when the voltage is sampled at equal intervals, as shown in Figure 9.16. In order to transform this series into the frequency domain, equation (9.13) must be reformulated into the *discrete Fourier transform* (DFT), given by

$$F(k) = \frac{1}{N} \sum_{n=0}^{n=N-1} g(n)e^{-2\pi i k n/N} \tag{9.14}$$

The matrix format of equation (9.14) (Randall, 1977) is

$$\mathbf{F} = \frac{1}{\mathbf{N}}[\mathbf{A}]\mathbf{g} \tag{9.15}$$

where $\mathbf{F}$ is a column array of N complex frequency components, $[\mathbf{A}]$ is a square matrix of unit vectors, and $\mathbf{g}$ is a column array of N samples in the time series.

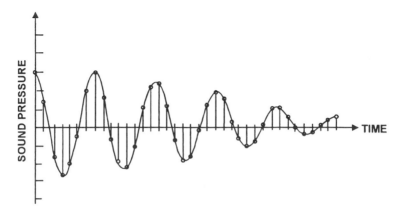

FIGURE 9.16. Sound pressure sampling at discrete intervals.

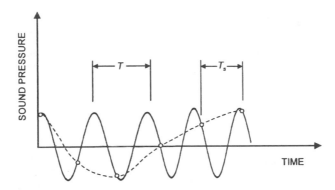

FIGURE 9.17. The effect of aliasing in generating a "false" wave.

Direct evaluation of the discrete Fourier transform obviously requires a large amount of number crunching, and the DFT methodology was made possible only in recent years by the development of efficient data processors. The *FFT algorithm*, developed by Cooley and Tukey originally for implementation on mainframe computers, but now utilized even in portable FFT analyzers, cuts down on the number of calculations required to find a discrete Fourier transform.. The Cooley–Tukey algorithm rearranges the [A] matrix of equation (9.15) by interchanging rows and factoring, with the result that reduces memory requirements and saves computing time. Computing time can be cut even more so by tabulating sine and cosine values.

In the FFT process, the signal is modified in three ways, giving rise to the three potential traps of the FFT, these being the *aliasing*, *leakage*, and the so-called *picket fence effect*. Aliasing is the apparent measurement of a false or incorrect frequency. Higher frequencies (after sampling) appear as lower ones, as in Figure 9.17 in which a solid line represent a sound wave pressure with a period T. The wave is sampled at intervals of T_s corresponding to the sampling frequency of $f_s = 1/T_s$. The frequency of the dotted line that results from the sampling rate can then be incorrectly identified as the frequency of the solid line. This ambiguity can be avoided by having a sampling rate at least twice the frequency of the highest frequency present in the signal, that is,

$$f_s > 2f_{max}$$

This minimum sampling frequency $2 \times f_{max}$ is called the *Nyquist frequency*. Use of very steep antialiasing filters (typically 120 dB/octave) prevents frequencies that cannot be adequately sampled from being analyzed and renders it possible to utilize a major portion of the computed spectrum (e.g., 400 lines out of 512 calculated). In most FFT units this is done automatically when the frequency range is selected. (This step also sets the appropriate sampling frequency.)

The effect of leakage, or *window error*, becomes apparent when the power in a single-frequency component appears to leak into other frequency bands. Two possible causes exist for window error: the signal is not fully contained within the observation window or it is not periodic within the window. The simplest case is that of a monofrequency sinusoidal wave, which should yield only one

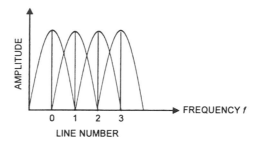

FIGURE 9.18. Illustration of the picket fence effect. If the frequency coincides with a line, it is indicated at its full level. Otherwise, it becomes represented at a lower level if the frequency falls between the lines.

frequency component in the FFT analysis if there are an integral number of periods of the sine wave in the finite record length. But a nonintegral number of periods will generally occur, and the cyclic repetition will yield a signal whose spectrum covers a range of frequencies. Leakage may be mitigated by forcing the signal in the data window to correspond to an integral number of period of all significant frequency components through a process called *order* or *tracking analysis*, where the sampling rate is related directly to basic frequency of the noise-generating process (such as the shaft speed of a machine) and in *modal analysis* measurements where the analyzer cycle synchronizes with the periodic excitation signals.

The picket fence effect (Figure 9.18), which is not unique to FFT analysis, occurs in any set of discrete fixed filters. The magnitude of the amplitude error occurring from this effect depends on the degree of overlap of adjacent filter characteristics, a consideration that influences the selection of a data window.

9.14   Data Windows and Selection of Weighting Functions

If a steady pure-tone of an unknown frequency is to be analyzed, it would usually be sampled over a short time interval that is termed *window duration*. A *rectangular window* will pass a portion of the input signal without adjustment. That short segment obtained is presumably representative of the original signal, and this would also hold true if that segment embodies an integer number of periods of the original signal. The effect of window duration arising from the scanning of (usually) noninteger number of periods is depicted in Figure 9.19. Such a signal would be analyzed as if the segment iterated itself as in the figure. The discontinuities at the ends of the segments will engender a frequency spectrum differing from the true frequency content of the original signal. Aperiodic signals may also be misinterpreted on account of the finite length of sampling.

The weighting or window function modifies the shape of the observation window by tapering off both the leading and trailing edges of the data. Because the weighting function is used to obviate the effects of duration limiting, it increases

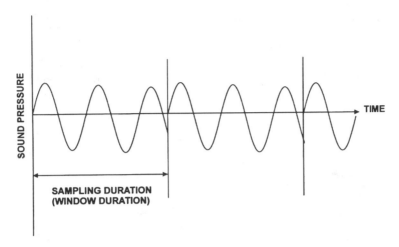

FIGURE 9.19. Effect of window deviations in FFT scanning of noninteger number of periods.

the effective bandwidth. This increase is stated in terms of the *noise bandwidth factor* (NBF) defined as

$$\text{NBF} = \frac{\text{effective bandwidth with window function}}{\text{effective bandwidth without window function}} \quad (9.16)$$

The ideal value of NBF is unity. Another important consideration in the use of weighting functions is the *side-lobe ratio* (SLR) or *highest side-lobe level*, expressed as

$$\text{SLR} = \frac{\text{most sensitive out-of-band response}}{\text{centre of bandwidth response}} \text{ dB} \quad (9.17)$$

A greater negative value of SLR (in decibels) is more desirable. But the selection of the two parameters given by equations (9.16) and (9.17) requires a compromise, that is, a tradeoff between the steepness of the filter characteristic on one hand and the effective bandwidth on the other. Table 9.1 lists the popular types of

TABLE 9.1. Properties of Data Windows.

Window Type	Maximum Sidelobe, dB	Sidelobe Falloff, dB/decade	Noise Bandwidth (relative to line spacing)	Maximum Amplitude Error, dB
Rectangular	−13.4	−20	1.00	3.9
Hanning	−32	−60	1.50	1.4
Hamming	−43	−20	1.36	1.8
Kaiser–Bessel	−69	−20	1.80	1.0
Truncated Gaussian	−69	−20	1.90	0.9
Flattop	−93	0	3.70	0.1

FUNCTION	SHAPE	NBF(Hz)	SLR(dB)

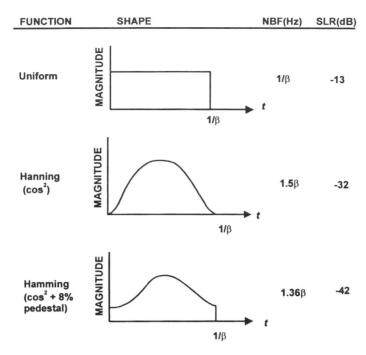

FIGURE 9.20. Typical weighting functions for FFT analysis, with corresponding noise band factor (NBF) and side lobe ratio (SLR).

data windows and their parametric values. The *Hanning window* (one period of a cosine squared function) constitutes a good choice for stationary signals, since that function has a zero value and slope at each end and thus renders a gradual transition over the discontinuity of data. Figure 9.20 illustrates the rectangular, Hanning, and Hamming weighting functions.

9.15   Resolution

The smallest increment in a parameter that can be displayed by a measurement system is the *resolution* for that system. In FFT analysis, the resolution is expressed by

$$\beta = \frac{f_s}{N} = \frac{2f_R}{N} \tag{9.18}$$

where

β = the resolution that defines the frequency increment between lines in a spectrum (in Hz)

f_s = sampling frequency, the reciprocal of sampling period T_s

N = number of samples in the original time series

f_R = frequency range, normally from 0 to Nyquist frequency

Some versions of FFT analyzers incorporate a "zoom" capability that displays selected portions of a spectrum with finer reduction. Normally, in order to increase the resolution (which means that β must decrease) the number of samples must be larger. This is not always feasible because either the processing time will increase or the block size is limited by machine memory capacity. The zoom capability overcomes this limitation by permitting the spectrum analyzer to concentrate its entire resolution, whether it be 200 or 800 or 1000 lines, on a small frequency interval selected by the user. While arbitrarily fine resolution is achievable, a compromise must be effected between resolution and the required sampling time, as attested by equation (9.18). Note that according to this equation, if the frequency resolution for a signal is 10 Hz, the sampling time is 0.1 s, but if the frequency resolution is changed to 0.1 Hz the corresponding sampling time will be 10 s.

A number of FFT analyzers incorporate large digital processing capacities that can present measurement results in the format of three-dimensional plots (or "waterfall" or "cascade" plots) with the vertical coordinate representing the amplitudes of the spectra as functions of the other two coordinates, one representing the frequency f and the other time t. The scan presented on a display can show a "running" plot that moves in the direction of increasing t. This type of display is most useful in observing the behavioral characteristics of transient sounds.

9.16   Measurement Error

Noise is usually random in nature, that is, its sound pressure level cannot be predicted for any instant. But statistical means can be used to describe random noise. If the noise is relatively constant in level and frequency content, then it may be deemed a *stationary random process*, one in which statistical parameters are invariant with respect to time. A machine operating in a constant cyclic manner may emit different levels of sound, with corresponding changes in frequency content, for each successive instant of the cycle, but a measurement interval over a group of cycles will yield a consistent spectral distribution over time. Analytic data based on a very short interval that is less than the length of a single cycle is certain to yield misleading results.

Let us consider noise with an idealized Gaussian or normal probability distribution. The standard deviation ε, which is the uncertainty in the rms signal divided by the long-term average rms signal, relates to the ideal filter bandwidth **bw** and the averaging time T, according to

$$\varepsilon = \frac{1}{2\sqrt{\mathbf{bw} \times T}} \tag{9.19}$$

The *measurement* uncertainty is *twice* that given by equation (9.19).

Example Problem 4

A normal distributed noise is to be analyzed in 20 Hz bands, with the measurement uncertainty not to exceed 4%. Find the necessary averaging time T.

Solution

Using equation (9.19) rearranged to give the averaging time, we obtain:

$$T = \frac{1}{\text{bw} \times \varepsilon^2} = \frac{1}{20 \times 0.04^2} = 31.3\,\text{s}$$

9.17 Sound Power

Sound power denotes the rate per unit time at which sound energy is radiated. This rate is expressed in watts. The sound power L_w level defined by

$$L_w = 10\log\left(\frac{W}{W_0}\right)$$

is given in decibels. W is the power of the sound energy source in watts and $W_0 = 1$ pW is the standard reference power in watts.

The principal advantage of using sound power level rather than the sound pressure level given by equation (3.22) to describe noise output of stationary equipment is that the sound power output radiated by a piece of equipment is independent of its environment. The sound energy output of the machine will not change if that unit is moved from place to place, provided it operates in the same manner. Sound power is used primarily to describe stationary equipment, but it is not generally used to rate mobile equipment because the operational situations may be too highly variable, as is the case with construction equipment.

Sound power may be measured directly using sound intensity instrumentation or indirectly, either determined from the rms sound pressures at a number of microphone locations spatially averaged over an appropriate surface enclosing the source in a free field over a reflecting plane (as exemplified by the use of a semianechoic chamber) or in a totally free field (a full anechoic chamber), or averaged over the volume of a reverberation chamber in which the measurements are conducted.

Some sources are omnidirectional, that is, they radiate sound uniformly in all directions. Most sources are highly directional, radiating more sound energy in some directions than in others. Hence, the *directivity*, or *directional characteristic*, constitute an important descriptor of a sound source. In a free field or anechoic chamber, the directivity is readily apparent, owing to the absence of reflections. But in a highly reflective environment, such as that of an echo chamber, the multiple reflections that occur render the directivity less important and the sound field becomes more uniform.

9.18   Measurement of Sound in a Free Field Over a Reflecting Plane (Semianechoic Chamber)

We can summarize the measurement procedure in a free field over a reflecting plane as follows:

1. The source is surrounded with an imaginary surface of area S, either a hemisphere of radius r or a rectangular parallelepiped.
2. The area of the hypothetical surface is calculated. $S = 2\pi r^2$ in the case of a hemispherical surface, and $S = ab + 2(ac + bc)$ in the case of the parallelepiped having length a, width b, and height c.
3. The sound pressure is measured at designated points of the imaginary surface.
4. The average sound pressure level $\bar{L}_p$ is computed from the measured results of the previous step. This is found from

$$\bar{L}_p = 10 \log \left[\frac{1}{N} \sum_{i=1}^{N} 10^{L_i/10} \right] \tag{9.20}$$

where N is the total number of measurements and L_i denotes the measured value of the SPL at the designated point i.
5. The sound power level is then calculated from

$$L_w = \bar{L}_p + 10 \log \left(\frac{S}{S_0} \right) \tag{9.21}$$

where S_0 is the reference area of 1 m^2.

The above procedure applies only if the source is not too large, that is, the radius r of the hypothetical hemisphere should be at least 1 m and at least twice the largest dimension of the source (or the perpendicular distance between the source inside the imaginary parallelepiped and a measurement surface is at least 1 m), and the background noise level is more than 6 dB below that of the source. The rectangular parallelepiped setup is preferred for large rectangular sources.

Figure 9.21 shows the designated points on the hemispherical surface where the microphones are located, The corresponding points for the rectangular parallelepiped are given in Figure 9.22. These designated points are associated with equal areas on the surface of the hemisphere or the rectangular parallelepiped. The SPL is usually measured at the designated points with A-weighting or in octave and partial octave bands, with the meter set in the slow-response mode.

The applicable international standards for acceptable sound power measurement techniques under semianechoic conditions are given in ISO 3744 and ISO 3745. Adjustments in the values of the measured SPL should be made for the presence of background noise. Equations (9.20) and (9.21) are used to convert the measurements into the desired values of averaged SPL, $\bar{L}_p$, and sound power level L_w.

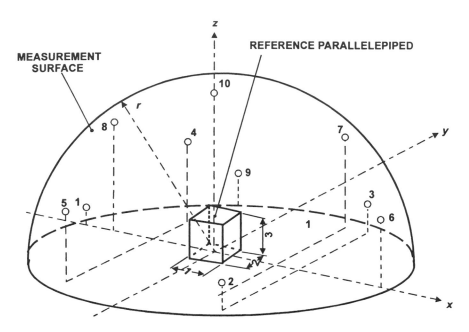

FIGURE 9.21. Locations of the microphones on the surface of an imaginary hemisphere surrounding a source whose sound power level is to be measured according to ISO 3744.

The directivity index **DI** of the source may be computed from measurements in a semianechoic chamber from the following equation:

$$\mathbf{DI} = L_{pi} - \bar{L}_p + 3 \, \text{dB} \tag{9.22}$$

where L_{pi} is the sound pressure level measured at point i, located on the measuring surface and defining the direction along which the DI is desired at a distance from the source.

9.19   Measurement of Sound Power Level in a Free Field (Full Anechoic Chambers)

The procedure for measuring sound power in a free field is basically the same as that for a free field with a reflecting surface (semianechoic condition), with some modifications. In this case, the source is centered in a hypothetical sphere of radius r and surface area $S = 4\pi r^2$. The sound pressure levels are measured at specific points on the spherical surfaces; these points are stipulated by ISO 3745 and shown in Figure 9.23, and defined in terms of Cartesian coordinates in Table 9.2. Equations (9.20) and (9.21) also apply to yield the average sound pressure level $\bar{L}_p$ and the sound power L_w. The surface area ratio S/S_0 in equation (9.21) now refers to two spheres, the hypothetical one (of radius r) used for placement of measuring

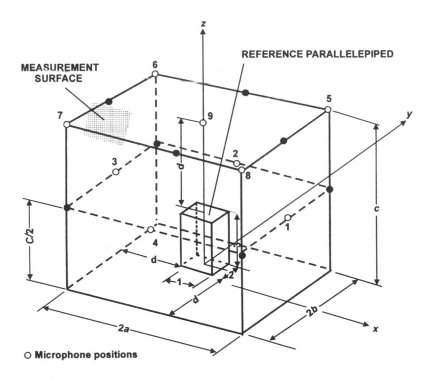

○ **Microphone positions**

● **Additional microphone positions**

FIGURE 9.22. Locations of the microphones on the surface of an imaginary parallelepiped surrounding a source whose sound power level is to be measured according to ISO 3744.

sensors and the other a reference sphere with an area of unity, with the result that this ratio is equal to simply $4\pi r^2$. Equation (9.21) becomes simply:

$$L_w = \bar{L}_p + 20\log r + 10.99 \tag{9.23}$$

where all dimensions are stated in SI units. The rectangular measurement surface is not generally used for measurements in a free field. As specified by ISO 3745, the radius of the test sphere should be at least twice as large as the major source dimension, but never less than 1 meter. A large source will necessitate a very large anechoic chamber for measurement purposes.

9.20 Sound Power Measurement in a Diffuse Field (Reverberation Chamber)

Reverberation chambers are rooms with extremely reflective walls, ceilings, and floors. If the surfaces are made nonparallel to each other, standing waves can

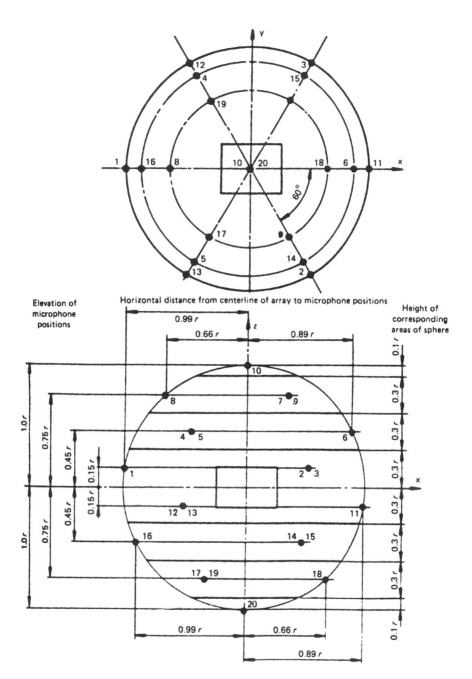

FIGURE 9.23. The locations of microphones on an imaginary spherical surface surrounding the source in a free field, according to ISO 3755.

TABLE 9.2. Microphones Array Positions[a] in a Free Field According to ISO 3745.

No.	x/r	y/r	z/r
1	−0.99	0	0.15
2	0.50	−0.86	0.15
3	0.50	0.86	0.15
4	−0.45	0.77	0.45
5	−0.45	−0.77	0.45
6	0.89	0	0.45
7	0.33	0.57	0.75
8	−0.66	0	0.75
9	0.33	−0.57	0.75
10	0	0	1.0
11	0.99	0	−0.15
12	−0.50	0.86	−0.15
13	−0.50	−0.86	−0.15
14	0.45	−0.77	−0.45
15	0.45	0.77	−0.45
16	−0.89	0	−0.45
17	−0.33	−0.57	−0.75
18	0.66	0	−0.75
19	−0.33	0.57	−0.75
20	0	0	−1.0

[a] This table lists the Cartesian coordinates (x, y, z) with the origin located at the center of the source. The vertical axis is perpendicular to the horizontal plane $z = 0$.

be avoided. When a steady noise source is operating, the sound field is diffuse everywhere in the room except in the immediate vicinity of the source, as the result of sound waves reflecting back and forth between room surfaces until they die out. A steady noise source generates sound and builds up the sound pressure level from ambient level in the room, until an equilibrium sound pressure level is reached. This occurs when new acoustic energy emanating from the source offsets the dissipation of reflected sound energy in the slightly elastic reflecting surfaces (however hard they may be) and in the slightly viscous air of the chamber and through energy leakage from the room. The applicable international standards for measuring the sound power levels in reverberation chambers are ISO 3741 and ISO 3742, the former giving details for broadband noise and the latter for discrete-frequency and narrow-band sources. The relationship between the sound power of a source and the average reverberant sound level can be determined and allows calculation of sound power.

The procedure for measuring the sound power level in an reverberation chamber may be summarized as follows:

1. The reverberation time T_{60} is measured by using a standard technique prescribed by ISO 354.[3]

---

[3] Reverberation time T_{60} for an enclosure is the time interval required for a sound pressure level to drop 60 dB after the sound source has been stopped. More detail treatment of T_{60} is rendered in Chapter 11.

2. The room volume V and the total surface area A of the test chamber are calculated from its internal dimensions. The barometric pressure B is measured but this constitutes only a small influence on the sound power level of the source.

3. The average sound pressure level $\bar{L}_p$ in the room is obtained by sweeping a microphone at steady speed over a path at least 3 m in length while its output signal is measured and averaged on a mean-square-pressure basis. The sweep time should be at least 10 s for the 200 Hz band and higher and 30 s for lower frequencies. Measurements in the near field (close to the source) and very near the chamber surfaces should be avoided. The microphone should not be located closer to a surface than one-half the wavelength of the lowest pertinent frequency. Alternatively, the average value may be obtained by averaging the output of an array of three fixed microphones spaced a distance of $\lambda/2$ apart (wavelength λ corresponds to the lowest frequency of interest). $\bar{L}_p$ constitutes the average band pressure level corrected for the background noise (i.e., the sound level that exists in the measurement chamber when the source is not operating).

4. The contribution of each frequency band to the sound power level L_w of the source, is calculated by using the following equation:

$$L_w = \bar{L}_p - 10 \log \left(\frac{T_{60}}{1\,\text{s}} \right) + \log \left(\frac{V}{1\,\text{m}^3} \right) + \log \left(1 + \frac{S\lambda}{8V} \right)$$

$$- 10 \log \left(\frac{B}{1000\,\text{mbar}} \right) - 14 \ \text{dB} \qquad (9.24)$$

where wavelength T corresponds to the center frequency of the frequency band of interest. The A-weighted sound power level L_{wA} may be computed from octave-band or one-third-octave band levels, according to ISO 3741, Annex C.

9.21   Substitution (or Comparison) Method for Measuring Sound Power Level

A simpler method than the direct method described above can be used to determine the sound power level L_w of an unknown source by comparing two measurements, without the necessity for knowing the reverberation time of the test chamber. In fact, this method does not even require the use of a special laboratory chamber, and it can be applied in situations where it would be impracticable to move a large piece of machinery to a laboratory. Commercially available reference sources provide known values of sound power level L_{wr} for each octave band and one-third octave band. Reference sound sources are classified into three types: aerodynamic sources, electrodynamics sources, and mechanical sources. Aerodynamic sources, the most prevalent type, consist of a specially designed fan or blower wheel driven by a motor. Technical requirements are listed in international standard ISO 6926.2. The comparison method, defined by ISO 3741, is as follows:

1. The procedures for measuring the average of the sound pressure levels $\bar{L}_p$ are conducted in the test room in the same manner described above for the direct

method. $\bar{L}_p$ is measured for the unknown source, and then the unknown sound source is turned off.

2. The reference source is turned on, and $\bar{L}_{pr}$ is measured for the reference source. The reference sound sources should be mounted on the floor at least 1.5 m away from another sound-reflecting surface, such as a wall or the unknown source being evaluated.
3. The sound power level for the source undergoing measurement can be simply computed from

$$L_w = \bar{L}_p + (L_{wr} - \bar{L}_{pr}) \text{ dB} \tag{9.25}$$

9.22 Alternation Method for Measuring Sound Power Level

A reference source with adjustable sound power—for example, a wideband source that generates pink noise (i.e., the same level in each band) or an octave band filtered noise—is used in the alternation method for measuring sound power. A readout meter indicates the reference source power output. The procedure is as follows:

1. The noise source being tested is operated in a diffuse field. The spatial average sound power level $\bar{L}_p$ is measured in each octave band.
2. The reference source replaces the noise source. The reference source is adjusted until it produces the same sound level as the tested source did in the first octave band. The sound power level L_w indicated on the reference source meter is recorded. This procedure is repeated for each of the other octave bands.
3. The sound power levels noted in the reference source meter are the sound power levels for the unknown source for each octave band. The total sound power can be computed from

$$\bar{L}_w = 10 \log \sum_{i=1}^{N} 10^{\bar{L}_{wi}/10} \tag{9.26}$$

where $\bar{L}_{wi}$ is the spatial average sound power level for the ith octave.

9.23 The Addition Method for Measuring Sound Power Level

The addition method is useful for situations where the machine under test cannot be conveniently shut down, for example, a power station generator. As with the alternation method, an adjustable reference source is used. The procedure is as follows:

1. The spatial average sound power level $\bar{L}_p$ generated by the subject machine is measured in each octave band.
2. The reference source is placed near the machine and both are operated simultaneously. The reference source is adjusted until it produces an additional 3 dB in

the sound pressure level in each octave band than when the machine was operating by itself. The reference source is then producing a sound power level in that band equal to that generated by the machine under test. The sound power level indicated on the meter of the reference power source is recorded. The procedure is repeated for each of all the other octave bands.
3. The total sound power level can now be calculated from the use of equation (9.26).

9.24   Magnetic Tape Recorders and Data Acquisition Systems

Magnetic recorders provide a permanent record of noise data taken in situ for subsequent analyses on instruments located elsewhere and for archival purposes. Currently, digital recorders are being used for acoustical data acquisition. Very few analog recorders (either AM or FM) are currently sold.

Computer technology has advanced to the stage where measured data and its analytical results can be acquired and digitally stored in computer memory or on floppy disks, recordable CD-ROMs, and removable cartridges or disks. A $3^1/_2$-in. removable disk can hold 100 MB or more; and newer versions of recording units are moving into the gigabyte range. It is conceivable that magnetic tape recording will be in time be totally supplanted by computerized acquisition devices and digital data storage units. Transient sounds, analyzed on a cascade-type FFT analyzer, are more conveniently archived in a nonvolatile memory medium such as a random-access removable cartridge than they would be on a magnetic tape (which obviously does *not* provide random access) for the purpose of later retrieval.

9.25   Integration of Measurement Functions in Computers

Advances in computer technology and software development tools make it possible to integrate measurement functions into a personal computer equipped with the appropriate acquisition boards, high-quality sound cards, and the applicable sensors. A specially equipped personal computer can perform DSP-based (digital signal processing) signal generation, filtering, and spectrum analysis. Thus, a single personal computer can replace a whole rack of dedicated analog hardware units linked together by BNC cables. In testing the performance of a loudspeaker through traditional analog means, a sine wave generator provides a signal to the loudspeaker and a calibrated microphone picks up the loudspeaker acoustical output and feeds it to a spectrum analyzer, which then relays the data to a display or a recorder. The dedicated functions of the traditional hardware—the signal generation, filtering, analysis, and data handling—can now be performed by software in such a manner that the personal computer constitutes the test platform. The user interface can even be made to emulate familiar analog instrumentation controls using standardized Windows controls.

Measurement procedures with a PC require a high-quality D/A (digital-to-analog) converter for transforming the digital representation of a sine sweep created by the program into analog signal, and an equally good A/D converter for transforming the measured analog signals back into the digital domain for analytical purposes. Most sound cards provide at least 16-bit resolution, and some sound cards can even measure frequency response from DC to 20 kHz with an accuracy of ± 0.25 dB and distortion as low as 0.003%.

The SoundCheck™ PC-based electroacoustical measurement system, essentially a software package, requires only a computer, sound card, amplifier, microphone, and microphone power supply. The SoundWare™ software features a family of "virtual instruments" that perform the functions of a signal generator, voltmeter, oscilloscope, spectrum analyzer, and real-time analyzer. The PC is rendered capable of measuring frequency, time, phase response, total harmonic distortion (THD), impedance, as well as performing other electrical tests. Such a system can be applied to evaluating loudspeakers, microphones, telephones, hearing aids, headsets, and other communication devices. Programming of test sequences permits the PC to be used not only for research and development work, but also in high-speed production testing and inspection procedures.

References

American National Standard Institute. 1966. *ANSI Standard Specification for Octave, Half-Octave and Third-Octave Band Filer Sets*, S1.11-1966.

American National Standard Institute. 1978. *ANSI Standard Specification for Personal Noise Dosimeters*, ANSI S1.25-1978.

British Standards Institution, 1983. *Personal Sound Exposure Meters*, BS6402.

Cooley, J. W., and J. W. Tukey. 1965. An algorithm for the machine calculation of complex Fourier series. *Math. Comp.* 19, 90: 297–301.

Crocker, Malcolm J. (ed.). 1997. *Encyclopedia of Acoustics*, vol. 4. New York: John Wiley & Sons; Part XVII, pp. 1837–1879, and Part XVIII, pp. 1933–1944.

Crocker, Malcolm J. (ed.). 1998. *Handbook of Acoustics*. New York: John Wiley & Sons; Part XV, pp. 1308–1353, and Part XVI, pp. 1411–1422. (The *Handbook* is essentially Crocker's *Encyclopedia* minus a number of sections. The sections cited here are identical to those given in the preceding reference.)

Fraden, Jacob. 1993. *AIP Handbook of Modern Sensors: Physics, Designs and Applications*. New York: American Institute of Physics; Chapters 3, 4, 9, and 11. (While this useful handbook covers devices for sensing temperature, light, humidity and chemical processes, the chapters cited here refer to topics applicable to acoustic measurements.)

Harris, Cyril M. (ed.). 1991. *Handbook of Acoustical Measurements and Noise Control*, 3rd ed. New York: McGraw-Hill; Chapters 5, 8, 9, 11–15.

Hixson, E. L., and I. Busch-Vishniac. 1997. Transducer principles, *Encyclopedia of Acoustics* (Crocker, Malcolm J., ed.), vol. 4, New York: John Wiley and Sons; Part XVIII, Chapter 159.

International Electrotechnical Commission. 1985. *IEC 804: Integrating-Averaging Sound Level Meters*. CH-1211 Geneva 20, Switzerland: International Electrotechnical Commission.

International Organization for Standardization. 1981. *ISO 3744: Engineering Methods for Free-Field Conditions over a Reflecting Plane.* CH-1211 Geneva 20, Switzerland: International Organization for Standardization.

International Organization for Standardization. 1977. *ISO 3745: Precision Methods for Anechoic and Semi-Anechoic Rooms.* CH-1211 Geneva 20, Switzerland: International Organization for Standardization.

International Organization for Standardization. 1975. *ISO 3741: Precision Methods for Broad-Band Sources in Reverberation Rooms.* CH-1211 Geneva 20, Switzerland: International Organization for Standardization.

International Organization for Standardization. 1975. *ISO 3742: Precision Methods for Discrete-Frequency and Narrow-Band Sources in Reverberation Rooms.* CH-1211 Geneva 20, Switzerland: International Organization for Standardization.

International Organization for Standardization. 1985. *ISO 354: Measurement of Sound Absorption in a Reverberation Room.* CH-1211 Geneva 20, Switzerland: International Organization for Standardization.

International Organization for Standardization. 1990. *ISO 6926.2: Requirements on the Performance and Calibration of Reference Sound Sources.* CH-1211 Geneva 20, Switzerland: International Organization for Standardization.

International Organization for Standardization. 1979. *ISO 3746: Survey Method.* CH-1211 Geneva 20, Switzerland: International Organization for Standardization.

Randall, Robert B. 1977. *Application of B & K Equipment to Frequency Analysis*, 2nd ed. Nærum, Denmark: Brüel and Kjær.

Sessler, G. M., and J. E. West. 1973. *Journal of the Acoustical Society of America.* pp. 1589–1600.

Temme, S. F. 1998. Virtual Instruments for Audio Testing. Presented at the 105th Convention of the Audio Engineering Society. *Audio Engineering Society Preprint 4894 (J-8).*

U.S. Department of Labor, Occupational Safety, and Health Administration. 28 June 1983. Occupational Noise Health Standard. *Code of Federal Regulations*, title 29, part 1910, sec. 1910.95 (29 CFR 1910.95), *Federal Register* 48: 29687–29698.

U.S. Mine Safety and Health Administration. 29 June 1982. *Code of Federal Regulations*, title 30, part 70, subpart F, and part 71, subpart I, *Federal Register* 47: 28095–28098.

Wilson, Charles E. 1989. *Noise Control*, New York: Harper & Row; Chapter 3 (probably the most outstanding chapter in this text).

Wong, George S. K., and Tony F. W. Embleton. 1995. *AIP Handbook of Condenser Microphones: Theory, Calibration, and Measurements*, New York: American Institute of Physics. (A great reference for those interested in the history of condenser microphones and in learning further details on their uses. Chapter 2 describes the classic Western Electric 640AA, the de facto standard for all microphones.)

Problems for Chapter 9

1. If a microphone has a sensitivity of -60 dBV/μbar and an output voltage of 21.6 mV is measured, what is the sound pressure level responsible for that output voltage?

2. Determine the value of the output voltage of a microphone with a sensitivity level of -100 dBV/μbar if it is exposed to a sound pressure level of 85 dB.

3. A sound pressure level reading of 95 dB is taken at the standard conditions of 1 atm and 68°F. What will be the readings at 20°F and at 95°?

4. Some laboratory notes indicated that a noise level reading resulted in L_{eq} = 95 dB(A) and an SEL of 107 dB(A). How long did this noise level reading last?

5. Assuming a sixth octave analyzer could be made available, determine the first three center frequencies after 16 Hz. Also determine the bandwidth.

6. Calculate the lower and upper cutoff frequencies and the bandwidth for the octave band where the center frequencies are respectively f_0 = 63 Hz and f_0 = 500 Hz.

7. If it is desired to analyze noise 10-Hz bandwidths with an uncertainty of 3% what must be the necessary averaging time?

8. If we are sampling through the FFT process a signal that ranges from 15 Hz to 30 kHz, what must be the minimum sampling frequency?

9. Why is a spectrum analyzer necessary and possibly more useful in dealing with noise machinery?

10. In a FFT analysis, it is desired to have a resolution of 1 Hz. If there are 1000 samples in the time series, what must be the minimum frequency sampling rate?

10
Physiology of Hearing and Psychoacoustics

10.1 Human Hearing

The mechanism of the human ear has been a source of much wonder for physiologists, who are progressing well beyond the fragmentary knowledge of the past by continually uncovering new marvels about how the human ear really functions. Our hearing mechanism is a complex system that consists of many subsystems. The mechanism of the brain's processing of auditory stimuli is probably beginning to be understood, with the relatively recent discovery that a part of the hearing system is metabolic in nature. Much of the pioneering work was performed by Georg von Békésy (1899–1972) who received the 1961 Nobel Prize in Medicine or Physiology for his investigations, particularly those entailing the frequency selectivity of the inner ear. His work indicated that the frequency selectivity ranked far poorer than the ear actually exhibits, but William Rhode found much greater selectivity in his work with live animals (von Békésy worked with dead animals, which most likely accounted for the difference). It is now realized that frequency selectivity of the inner ear fades within minutes after the metabolism ceases.

A young, healthy human is capable of hearing sounds over the frequency range of 20 Hz to 20 kHz, with a frequency resolution as small as 0.2%, Thus, we can discern the difference between a tone of 1000 Hz and one of 1002 Hz. With normal hearing, a sound at 1 kHz that displaces the eardrum less than 1 Å (angstrom) can be detected, in fact, less than the diameter of a hydrogen atom! The intensity range of the ear spans extremes from the threshold at which softest sounds can be detected to the roar of a fighter jet taking off, thus covering an intensity range of approximately 100,000,000 to 1. The ear acts as a microphone in the process of collecting acoustic signals and relaying them through the nervous system into the brain. The ear (cf. Figure 10.1) subdivides into three principal areas: the *outer*, *middle*, and *inner* ear.

The outer ear consists of a *pinna* which serves as a sound-collecting horn and the auditory canal that leads to the inner ear. The collected sound enters the ear through the opening (the *meatus*) into the auditory canal that forms a tube approximately $3/4$ cm in diameter and 2.5 cm in length. The canal terminates at the *tympanic membrane* (eardrum). Under the impetus of sound the eardrum vibrates, causing three bones linked in a ossicular chain—namely, the *malleus* (or hammer), the *incus* (anvil), and the *stapes* (stirrup)—to oscillate sympathetically. At the lowest resonance (3 kHz) of the auditory canal, the sound pressure level at the eardrum is about 10 dB greater than it is at the entry into the canal. Because a resonance curve tends to be broad, human hearing tends to be more sensitive to sound in the range of approximately 2 to 6 kHz, as the consequence of the resonance being centered at 3 kHz. The diffraction of sound waves inside the head has the effect of causing the sound pressure level at the eardrum to exceed the free-field sound pressure level by as much as 20 dB for some specific frequencies.

The eardrum itself is a thin, semitransparent diaphragm that completely seals off the canal, marking the inner boundary of the outer ear and the outer boundary of the middle ear. This membrane is quite flexible at its center and is attached at its perimeter at the terminus of the auditory canal, thus demarcating the entrance to the middle ear. The middle ear, lined with a mucous membrane, constitutes an air-filled cavity of about 2 cm^3 in volume, which contains the three ossicles (bones), namely, the malleus, incus, and the stapes, forming a bony bridge from the external ear to the middle ear. These bones are supported by muscles and ligaments. The malleus is attached to the eardrum; the incus connects the malleus and the stapes. The last bone in the chain, the stapes, covers the *oval window*. The *Eustachian tube*, which is normally closed, opens in the process of swallowing or yawning to equalize the air pressure on each side of the eardrum; this is a tube approximately 37 mm in length that connects the middle-ear cavity with the pharynx at the rear of the nasal cavity.

Just below the oval window lies another connection between the middle and inner ears, the membrane-covered *round window*. Between the oval and round windows is a rounded osseous projection, formed by the basal turn of the cochlea, called the *promontory*. A canal encasing the facial nerve is situated just above the oval window.

The structures to the right of the oval and round windows in Figure 10.1 are collectively called the *inner ear* (also called *labyrinth*), which is comprised of a number of canals hollowed out of the petrous portion of the temporal bone. These interconnecting canals contain fluids, membranes, sensory cells, and nerve elements. Three principal parts exist in the inner ear: the *vestibule* (an entrance chamber), the *semicircular canals*, and the *cochlea*. The vestibule connects with the middle ear through the oval and the round windows. Both of these windows are effectively sealed, by the action of the stapes and its support on the oval window and the presence of a thin membrane in the round window, thus preventing the loss of the liquid filling the inner ear. The semicircular canals play no role in the process of hearing but they do provide us with a sense of balance. The *cochlea*, shown in enlarged detail in Figure 10.2, is the sensory system that converts the vibratory energy of sound into electrical signals to the brain for the detection and interpretation of that sound. The cochlea can be described as a 3.5-cm long tube of roughly circular cross-section, wound about 2½ times in a snail-like coil. This tube's cross-sectional area decreases in a somewhat uneven manner from its base to its apex. Its total volume is about $5 \times (10)^{-2}$ cm^3.

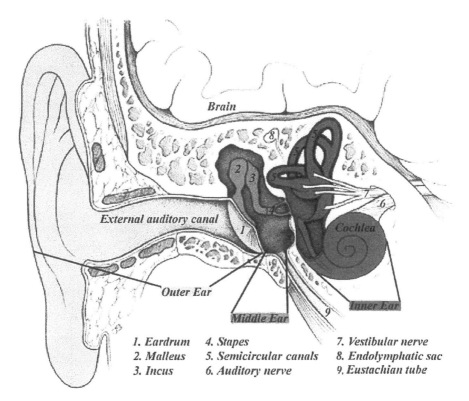

FIGURE 10.1. Coronal section of the right ear. (From Internet site of the Center for Sensory and Communication Disorders, Northwestern University, funding by U.S. National Institute of Health.)

The coils of the cochlea surrounds an area called the *modiolus*; and the membranous labyrinth of the cochlear sector of the inner ear divides into three ducts or galleries (*scalae*). The *cochlear duct* (*ductus cochlearis*) runs the length of the spiraling cochlea, and because it occupies the central portion of the cochlea's interior, it has been termed the *scala medi* (i.e., *middle gallery*), whose walls effectively partition the cochlea into two longitudinal channels, the *scala vestibuli* (or *upper gallery*) and the *scala tympani* (*lower gallery*). The only communication between the two galleries occurs through the *helicotrema*, a small opening at the apex of the cochlea. The other ends of the upper and lower galleries terminate in the oval and round windows, respectively.

Figure 10.3 shows an enlarged view of the cochlear duct. This duct is bounded by Reissner's membrane, the basilar membrane, and the *stria vascularis*. The basilar membrane extends from the bony *spiral lamina*, a ledge extending from the central core of the cochlea, to the *spiral ligament*. The length of the basilar membrane is about 32 mm long, from the base to the apex of the cochlea; the width varies from about 0.05 mm at the base to about 0.5 mm at the apex; and

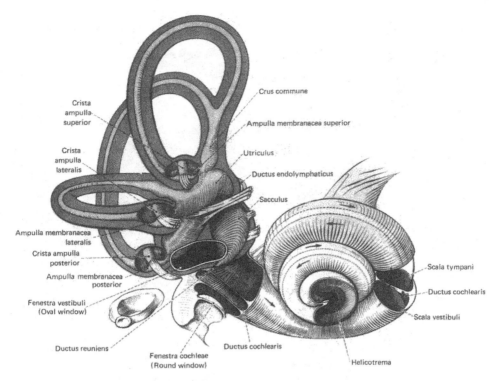

FIGURE 10.2. The membranous semicircular canals showing the cristae within the ampullae. (From "The Internal Ear," *What's New*, 1957, Abbott Laboratories. Reproduced with permission of the publisher.)

the membrane gradually becomes thinner at it nears the apex. Positioned on the basilar membrane is the *organ of Corti*. This organ, shown in detail in Figure 10.4, consists of some structural cells (e.g., *Dieter's cells* and *Hensen's cells*), the rods and tunnels of Corti, and two types of hair cells on top of which lies the *tectorial membrane*. The tunnel of Corti, isolated from the endolymph, contains the fluid *cortilymph*.

The hair cells constitute the sensory cells for hearing. The *inner hair cells* are arranged in a single row on the modiolar side of the tunnel of Corti, and the *outer hair cells* exist in three parallel rows on the strial side of the tunnel of Corti. The inner hair cells are round and squat; their upper surfaces contain about 50–70 hairs called stereocilia. The outer hair cells, which look more reedlike than do the inner hairs, contain about 40–150 stereocilia arranged in a W-shaped pattern. There are 3000–3500 ciliated cells in the single row of inner hair cells and a total of 9,000–12,000 ciliated cells in the three rows of the outer hair cells. The hair cells connect to some 24,000 transverse nerve fibers in a complex network leading into the central core of the cochlea. The nuclei of these nerve fibers form the *spiral ganglion*, which unite to form the *cochlear branch of the VIIIth nerve*.

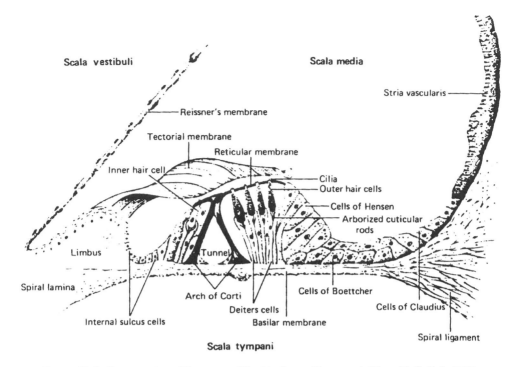

FIGURE 10.3. Cross-section of the organ of Corti in the cochlear canal. (From H. Gulick. 1971. *Hearing: Physiology and Psychophysics.* New York: Oxford University Press. Reproduced with permission of the publisher.)

The cochlear branch joins with the vestibular branch to form the *VIIIth cranial nerve*, also called the *auditory* or *vestibulocochlear nerve*. The VIIIth cranial nerve along with the VIIth (facial) cranial nerve proceeds in a helical fashion through the *internal auditory meatus* to nuclei in the brain stem. From the brain stem the auditory pathway extends through various nuclei to the cerebral cortex in the temporal lobes of the brain.

The VIIIth cranial nerve functions primarily as sensory nerve, that is, it conveys sensory information from the cochlea and the vestibular system to the brain.

10.2   The Mechanism of Hearing

Sound waves are directed by the pinna into the auditory canal. The longitudinal changes in air pressure of the sound wave propagate to the ear drum, causing it to vibrate. Because the handle of the malleus is imbedded in the ear drum, the ossicular chain is set into vibration. These tiny bones vibrate as a unit, elevating the energy from the eardrum to the oval window by a factor of 1.31:1. Sound energy is further enhanced by the difference in area between the eardrum and the stapes footplate by a factor of approximately 14. Multiplying this effective areal difference of 14

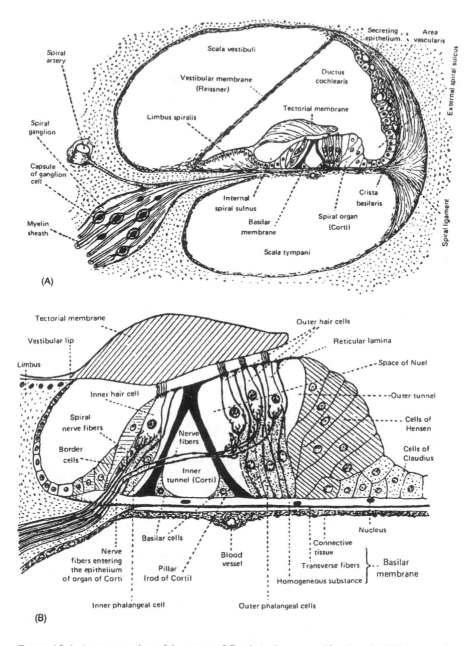

FIGURE 10.4. A cross-section of the organ of Corti. (a) Low magnification. (b) higher magnification. (From A. T. Rasmussen. 1947. *Outlines of Neuro-Anatomy*, 3rd edition. Dubuque, IA: Wm. C. Brown Company.)

by the lever action of the ossicular chain (1.3) yields an energy increase of 18.3:1, which translates into an amplification factor of 25.25 dB on the SPL scale. The middle ear acts as a transformer, by changing the energy collected by the eardrum into greater force and less excursion, thereby matching the impedance of the air to the impedance of the inner ear's fluid.

Because the fluid of the inner ear is virtually incompressible, provision for relief of the pressure produced by the movement of the footplate of the stapes is provided by the interaction of the oval window and the round window, with the fluid motion from the oval to the round window being transmitted through the cochlear duct. When the footplate of the stapes pushes into the perilymph of the scala vestibuli, the vestibular membrane (or membrane of Reissner) bulges into the cochlear duct, causing movement of the endolymph within the cochlear duct and displacement of the basilar membrane. Von Békésy's experiments with cochlear models led him to formulate a theory that the displacement of the basilar membrane is in the form of a traveling wave that proceeds from the base to the apex of the cochlea. The maximum amplitude of the wave occurs at a point along the basilar membrane corresponding to the frequency of the stimulus, that is, each point of the basilar membrane corresponds to a specific value of the simulating frequency.

The cilia of the hair cells are embedded in the gelatinous tectorial membrane, so that when the basilar membrane is displaced, it generates a "shearing" force on the cilia. The sidewise motion of the cilia creates an alternating electrical current, also referred to as the *cochlear microphonic* (CM), *cochlear potential* or the *Wever–Bray effect*[1]. The deflection of the hair cells triggers responses in the neurons connected to the hair cells. Impulses are borne along the nerve fibers to the main trunk of the cochlear portion of the VIIIth nerve and onward to the brain. This is how the cerebral cortex eventually "hears" the vibration that strike the eardrum.

Research over the past decade indicates that the cochlea does not act passively. Active processes occur that indicate that energy is being added to the cochlea through mechanisms that are not yet fully understood. More energy is contained in the cochlea than that from the sound going into it. One phenomenon that has been identified is that of *otoacoustic emission*, which is a sound in the external ear canal believed to have originated from vibrations within the cochlea and propagated back through the middle ear. Otoacoustic emissions can be measured by placing a miniature microphone in the ear canal. A spontaneous otoacoustic emission is identified as a constant low-level sound that occurs spontaneously in half of normal ears. When a high-level click is introduced to the ear, an *evoked otoacoustic emission* occurs some 5 msec later as a low-level sound. Also, when two different tones are presented at high levels to a normal ear, the otoacoustic emission occurs in the form of new tones generated at frequencies other than the two original frequencies. These new tones are termed *distortion products*.

---

[1] So named after the two investigators who discovered in 1930 that speech delivered to a cat's ear could be understood when the CM signal was picked up from the cochlear nerve and amplified.

When the sound striking the eardrum is sufficiently loud, the middle ear muscles contract reflexively. This *acoustic reflex* occurs as a contraction in the stapedius muscle, which results in a pull against the ossicular chain and a reduction in the energy transmitted through the oval window into the perilymph in the vestibule. The largest amount of reduction in sound due to acoustic reflex—ranging approximately from 20 to 30 dB—occurs for low frequencies. Above 2 kHz this acoustic reflex is fairly negligible.

10.3   Hearing Loss

Nearly a quarter of the population between the 15 and 75 years of age suffer some type of hearing impairment. Impaired hearing, which is often caused by infectious diseases or overexposure to loud noise or simply the process of aging, is common enough to be on a par with the onset of poor vision. When hearing loss occurs in early childhood, its consequences become more obvious than when it occurs in adulthood. A child's progress in learning and developing social relationships may be hindered and the child may even be deemed "not too bright" if professional help and guidance are not forthcoming. The principal problem of hearing loss, regardless of the age of the affected individual, occurs in the diminution of a person's ability to understand speech.

Even milder forms of hearing loss early in life can generate great difficulty, particularly for children who developed within normal limits but are not doing well in school, due to their being inattentive. Such moderate hearing losses are not uncommon and may even be on the increase due to heightened exposure to "rock" music. When a mild hearing loss is corrected, the child often becomes "like a different person." Fortunately, many of the hearing impaired can be helped through the use of hearing aids.

The gradual waning of hearing capabilities affect adults in a more underhanded manner. Most people with age-induced or noise-induced hearing impairment first lose hearing acuity at high frequencies, making it difficult for them to distinguish consonants, especially s versus f, and t versus z. Such persons must strain harder to understand conversations. Going to the movies, listening to lectures, conversing with friends, and other pleasures become stressful chores. This can result in an individual's becoming withdrawn from his friends and relatives. Some of these patients can be helped through counseling and rehabilitation, but no cure exists for most cases of sensorineural deafness.

Hearing loss falls into two principal categories: *conductive* and *sensorineural*. Conductive hearing loss occurs from any condition that impedes the transmission of sound through the external or the middle ear. Sound waves are not transmitted effectively to the inner ear because of some blockage in the auditory canal, interference in the eardrum, the ossicular chain, the middle ear cavity, the oval window, the round window, or the Eustachian tube. For example, damage to the middle ear, which carries the task of transmitting sound energy effectively, or the Eustachian tube, which sustains equal air pressure between the midear cavity and the external canal, may result in mechanical deficiency of sound transmission. In pure conductive hearing loss, no damage exists in the inner ear or the neural system. Conductive hearing losses are generally treatable.

Sensorineural deafness, which is a far more accurate term than the ambiguous terms "nerve deafness" and "perceptive deafness," describes the effect of damage that lies medial to the stapedial footplate—in the inner ear, the auditory nerve, or both. In the majority of cases sensorineural deafness is not curable. The term "sensory hearing loss" is applied when the damage is localized in the inner ear. Applicable synonyms are "cochlear" or "inner-ear" hearing loss. "Neural" hearing loss is the proper terminology to describe the result of damage in the auditory nerve proper, anywhere between its fibers at the base of the hair cells and the auditory nuclei. This category also encompasses the bipolar ganglion of the eighth cranial nerve.

Mixed hearing loss results from conductive hearing loss accompanied by a sensory or a neural (or a sensorineural) loss in the same ear. Otologic surgery may help in cases of mixed hearing loss in which the loss is primarily conductive accompanied by some sensorineural damage to a lesser degree.

Functional hearing loss, which occurs far less frequently than the hearing loss types described above and presents a greater diagnostic challenge in clinics, constitutes the condition in which the patient does not seem to hear or to respond, yet the handicap cannot be attributable to any organic pathology in the peripheral or the central auditory pathways. The basis for this type of hearing difficulty may be caused by entirely emotional or psychological etiology. Psychiatric or psychological therapy may be called for, rather than otological treatment.

Central hearing loss, or *central dysacusis*, remains mystifying to otologists, although information about this type of hearing defect is accumulating. Patients suffering this type of condition cannot interpret or understand what is being said, and the cause of the difficulty does not lie in the peripheral mechanism, but somewhere in the central nervous system. In central hearing loss, the problem is not a lowered pure-tone threshold, but the patient's ability to interpret what he or she hears. It is obviously a more complex task to interpret speech than to respond to pure-tone signals; consequently, the tests necessary to diagnose central hearing impairment must stress measuring the patient's ability to process complex auditory information. It requires an extremely skilled, intuitive otologist to make an accurate diagnosis.

10.4   Characteristics of Hearing

If the sound is audible, the amplitude of the sound is said to be *above threshold*; and if the sound is inaudible, the amplitude is considered to be *below threshold*. The amplitude of the sound at the transition point between audibility and inaudibility is defined as the *threshold of hearing*. When sound amplitude exceeds threshold, the sound is processed and perceived as having certain qualities including loudness, pitch, and a variety of other perceptive traits such as information. The study of auditory perception in relation to the physical characteristics of sound defines the field of *psychoacoustics*.

Sensitivity

The ear is not equally sensitive to all frequencies. The absolute sensitivity of the ear, defined by its threshold, depends on a variety of factors, the most important of which is the sound pressure level and the frequency of the sound. The resonance of the ear canal, the level effect of the ossicles, and the difference between the surface area of the eardrum and of the stapes footplate all affect the intensity of the sound that actually penetrates the cochlea.

An *audiometer* that generates signals of varying frequency and intensity is used to measure an individual's hearing sensitivity. The signals produced by the audiometer can be directed either to earphones or to a loudspeaker in an anechoic chamber. As it is far more difficult to ascertain the intensity of the sound at the level of the cochlea, and such a determination would not accurately represent how an individual hears under normal circumstances, we generally specify hearing sensitivity in terms of thresholds for sounds of various frequencies of which sound-pressure levels were determined in a sound field without the listener present. Figure 10.5 maps the hearing sensitivity of the normal young human ear over a range of frequencies. The solid curve, referred to as the *minimum audible field*, or *MAF*, describes the minimum intensities that can be detected when the listener

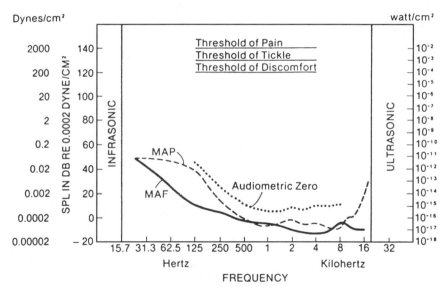

FIGURE 10.5. The area of human audibility. The two lower curves represent the lowest (best) thresholds of hearing of young adults. The solid curve is the minimum audible field (MAF) and the dashed curve is minimum audible pressure (MAP). (From *American National Standard Specification for Audiometers*, ANSI S3.6-1969, revised 1989, American National Standards Institute. New York: Acoustical Society of America.) The dotted curve represents the current standard for the audiometric zero. The upper three lines represent averages for sensations of discomfort, tickle, and pain. The ordinates define intensity in terms of pressure in dynes/cm², sound pressure level in dB, and power flow in watt/cm².

is situated before a loudspeaker at a prescribed distance. Both of the listener's ears are stimulated simultaneously by the sound source (i.e., the loudspeaker).

However, most clinical work in audiology entails measurements in reference to a single ear rather than to both ears. This is usually performed by directing the test signals to the appropriate earphone of a headset rather than to a loudspeaker. The use of a headset as opposed to exposure to a loudspeaker considerably modifies the listening situation. For example, the resonant frequency of the ear canal is shifted because both ends of the canal are sealed in contrast to the situation when the canal is open to the sound field. Moreover, the placement of the earphones may give rise to unwanted physiological noise that can interfere with the detection of low-frequency sounds. Also, the method of calibrating sound from a loudspeaker differs from that for calibrating sound from an earphone. Because of these differentiating and other factors, the measurement of thresholds through the use of headphones is called *minimum audible pressure*, or *MAP*. The MAP measurements are contrasted with MAF measurements in Figure 10.5. The threshold curve for the MAP condition appears to be several dB higher (i.e., showing lower sensitivity) than the MAF curve, a situation referred to as the "missing 6 dB." This can be attributed to the fact that using both ears in a sound field enhances sensitivity, in contrast to listening with only one ear under an earphone; and other factors occur such as the diffraction of sound around the head in a sound field, the different resonances of the external ear canal, and so on.

The two curves of Figure 10.5 represent thresholds that are two standard deviations below the mean, that is, the curves represent the thresholds of approximately 2.5% of young adults (16–25 years of age) determined by examination to be otologically normal. These curves are based on data given in two separate studies performed four years apart conducted by the National Physical Laboratory in Great Britain. Tables 10.1 and 10.2 list the mean and standard deviations reported in these two studies and the data points (two standard deviations below the means) on which the MAP and MAF curves are based.

The intensities defining an audiometric zero at each of the standard frequencies on a pure-tone audiometer are represented by the dotted curve of Figure 10.5. These intensity values were established by international agreement among scientists as being representative of the average minimum audible sound-pressure levels for young adult ears and have been incorporated in the standards for audiometric calibration throughout most of the world. This standard, as the result of being adopted in 1969 by the American National Standards Institute, is referred to as *ANSI-1969 standard* and it is listed with the MAP values in Table 10.1.

Loudness

While the sensation of loudness correlates to the amplitude of the sound above threshold, loudness is not perceived by the human ear in equal measures as the amplitude increases over different frequencies above the threshold. Individual judgment constitutes the deciding factor in ascertaining the degree of loudness. This had led to the development of equal loudness contours which are curves

TABLE 10.1. The Means and Standard Deviations and the Data Points (Two Standard Deviations Below the Means) on Which MAP Curves of Figure 10.5 Were Based.

Frequency, Hz or kHz	Mean SPL, dB	σ, dB	SPL at 2σ Below Mean	ANSI-1969
80 Hz	61.0	8.0	45.0	
125	45.5	6.8	31.9	45.5
250	28.0	7.3	13.4	24.5
500	12.5	6.5	−0.5	11.0
1 kHz	5.5	5.7	−5.9	6.5
1.5	8.5	6.1	−3.7	6.5
2	10.5	6.1	−1.7	8.5
3	7.0	5.9	−4.8	7.5
4	9.5	6.9	−4.3	9.0
6	10.5	9.1	−7.7	8.0
8	9.0	8.7	−8.4	9.5
10	17.0	9.0	−1.0	
12	20.5	9.6	1.3	
15	39.0	10.7	17.6	
18	74.0	21.9[a]	30.2	

[a] Calculated from reported standard of error of the mean. Data on mean sound pressure levels and standard deviations (σ) for 80 Hz through 15 kHz taken from Dadson and King. The data for 18 kHz were taken from Harris and Myers.

TABLE 10.2. The Means and Standard Deviations and the Data Points (Two Standard Deviations Below the Means) on Which the MAF Curves of Figure 10.5 Were Based.[a]

Frequency, Hz or kHz	Mean SPL, dB	σ, in dB	SPL at 2σ Below Mean
25 Hz	63.5	8.0	47.5
50	43.0	6.5	30.0
100	25.0	5.0	15.0
200	15.0	4.5	6.0
500	5.5	4.5	−3.5
1 kHz	4.5	4.5	−4.5
2	0.5	5.0	−9.5
3	−1.5	6.0	−13.5
4	−5.0	8.0	−21.0
6	4.5	8.5	−12.5
8	13.5	8.5	−3.5
10	16.5	11.5	−6.5
12	13.0	11.5	−10.0
15	24.5	17.0	−9.5

[a] Data on mean sound pressure levels and standard deviations (σ) taken from Robinson and Dadson.

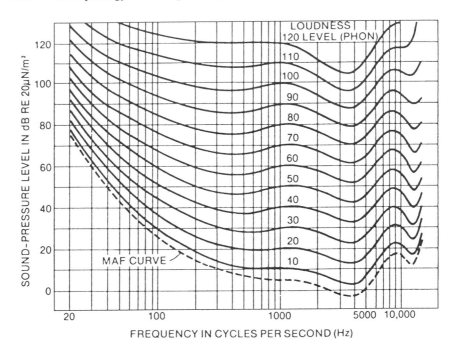

FIGURE 10.6. Equal loudness contours. (From Peterson, A. P. G., and E. E. Gross. *Handbook of Noise Measurement*. 1980. Concord, MA: General Radio.)

connecting SPL points of equal loudness for a number of frequencies, as judged by tested listeners. These curves, also called *phon*[2] *curves*, are constructed by asking subjects to judge when tones of various frequencies are considered equal in loudness to a 1-kHz of a given SPL. The official definition (ANSI, 1973) of the phon specifies binaural (two-ear) listening to the stimuli in a sound field. Equal-loudness contour curves are given in steps of 10 phons in Figure 10.6, with the dashed MAF curve from Table 10.2 included in the plot as a threshold reference.

As an example of how humans perceive sound, consider a 30-Hz tone at 95 dB SPL. It would be judged by a typical listener as being as equally loud as a 1000-Hz tone at 70 dB SPL or a 5000-Hz tone at 65 dB SPL. As sound is steadily increased in intensity above the threshold, it will eventually cause the listener to experience physiological discomfort. A further increase in the intensity produces a tickling sensation in the ear; and an additional increase in intensity causes the listener to experience pain. These three levels constitute, respectively, the thresholds of *discomfort*, *tickle*, and *pain*, which are represented by the upper three lines of Figure 10.5. While these threshold values of 120, 130, and 140 dB SPL represent statistical averages for young adult ears, different individuals have different

---

[2] A phon is a unit of loudness that, at the reference frequency of 1 kHz, is equated to the decibel scale.

tolerance thresholds; but these values do not differ markedly from the statistical averages. Thus, in Figure 10.5, the region between the discomfort and the audiometric zero constitute the usable dynamic range of hearing for humans.

Pitch

The sensation of *pitch* is obviously related to the frequency of the tone. The actual pitch of a sound is affected by other factors, including the sound pressure level and the presence of component frequencies. Pitch perception is a complex process, one that is not yet fully understood. Pitch elicited by some sounds may evoke the same aural response whether or not the fundamental frequency is present. The average adult male voice carries a fundamental between 120 and 150 Hz and that of a typical adult female lies between 210 and 240 Hz. Yet we generally find it easy to distinguish between male and female voices even though the telephones does not transmit frequencies much lower than 300 Hz. Somehow we are able to compensate aurally for the fundamental frequency missing from the signal passing through the telephone receiver.

The spectrum of a sound generates a psychological sensation of quality. This permits us to distinguish the difference, say, between a trumpet and an English horn playing the same note. This is because of the differences in their respective sound spectra (i.e., frequency content or the presence of overtones). which are, in turn, are functions of the complex vibrations and resonance modes inherent in these instruments' respective structures. We are also able to discern different speech sounds because of the differences in the sound spectra. Even over the telephone, individual voices are recognizable owing to differences in their frequency content.

Masking

Masking is said to have occurred when the audibility of a sound is interfered with by the presence of noise or other background sound. The "cocktail party" effect, which makes it difficult to carry on a private conversation against a backdrop of other people's chatter, is a prime example of masking. Speech becomes unintelligible by the presence of excessive background noise. Although masking is almost always undesirable, broadband noise may be purposely introduced into an office environment to make conversation unintelligible to potential eavesdroppers in an adjacent office. Because most of the intelligence in speech is generally contained in the frequency range between 200 Hz and 6 kHz, noise in that frequency range is most obtrusive in terms of speech masking. But excessively loud noise in any frequency band can adversely affect speech intelligibility by causing such an overload of the auditory system that one cannot effectively discriminate speech from the prevailing total signal. Consonants essential to conveying verbal information tend to be pronounced softly, so they become readily indiscernible in the presence of noise.

A person speaking normally produces an unweighted sound level of 55 to 70 dB at 1 meter. It is more taxing for that person to speak more loudly for a sustained time. A typical maximum voice effort, resulting in a shout, produces about 90 dB at

1 meter. Speech intelligibility generally improves when the speaker and the listener are near each other and if the speaker increases the signal-to-noise ratio by talking louder. Maximum intelligibility usually can be obtained if the unweighted level of the speech is between 50 and 75 dB at 1 m from the speaker. Speaking more loudly does not always guarantee greater intelligibility, even though the signal-to-noise ratio (defined as the intensity of the signal divided by the intensity of the noise) may be increased, because the formation of speech sound above 75 dB may degrade sufficiently that there is little or no improvement in intelligibility. If a listener is familiar with the words and the dialect used, intelligibility will be greater. It is for this reason that critical communications, particularly those of air controllers, are based on a limited vocabulary. In ordinary face-to-face conversation, the listener has the additional luxury of making out the context of the words by observing the speaker's facial expressions and gestures.

10.5   Prediction of Speech Intelligibility: The Articulation Index

In order to assess the effect of noise on speech communication, it is necessary to conduct speech-intelligibility tests with actual speakers and listeners in the presence of interfering noise. The test materials may be sentences, digits, disyllabic words, monosyllabic words, or nonsense syllables. The listeners are scored according to the percentage of the speech materials heard correctly. The background interfering noise is generally recorded and played back in the testing laboratory.

From such experiments came the realization that speech intelligibility is a function of the intensity and the frequency characteristics of the interfering noise. Regarding the S/N ratio, Licklider and Miller stated that the S/N should exceed 6 dB for satisfactory communication, although the presence of speech may be detected for S/N as low as −18 dB. If the intensity of the signal (speech) exceeds the noise, the sign of the S/N value is plus; conversely, a negative value of S/N indicates that the noise is more intense than the signal.

Figure 10.7 maps the effect of white noise on the thresholds of detection and intelligibility of running speech. According to this figure, which was developed by Hawkins and Stevens based on extensive tests making use of running speech and white noise, the threshold of intelligibility occurs when the level of the speech exceeds the noise level by about 6 dB (S/N of 6 dB). As the sound pressure level of the noise is increased above this value, the threshold of intelligibility is proportionally increased so that the S/N value of 6 dB remains fairly constant over a wide range of intensities. For other kinds of speech materials and different masking noises, the relationship between the threshold of intelligibility and the level of the interfering noise may not necessarily remain the same. At a S/N of −18 dB running speech can be detected but not understandable.

Other but simpler methods have been developed for measuring the effect of interfering noise on the intelligibility of speech. A principal method of predicting speech intelligibility is the *articulation index*, or *AI*, which is a value that ranges

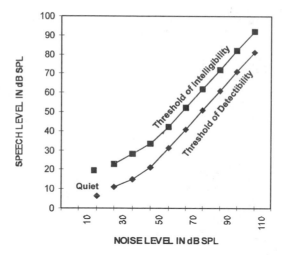

FIGURE 10.7. The effect of white noise on the thresholds of detectability and intelligibility of running speech. (From Hawkins, J. E., Jr., and S. S. Stevens. 1950. The masking of pure tones and of speech by white noise. *Journal of the Acoustical Society of America.*)

from 0.0 to 1.0 and represents the proportion of the speech spectrum that occurs above the noise. French and Steinberg of Bell Laboratories developed the concept of articulation index on the basis of the assumption that most of the intelligence in speech is contained in the frequency bands between 200 and 6100 Hz. The articulation index can be calculated from the levels of the masking signal and the speech level in the frequency bands. The contribution of each frequency band to speech intelligibility is defined as 12 dB plus the sound level of the speech less the masking level. Each frequency band's contribution is limited to the range between 0 and 30 dB. The sound level of the speech signal is based on a long-term energy average in each frequency band; and each frequency band contribution is multiplied by a weighting factor. The sum of the weighted contributions divided by 10,000 yields the articulation index.

Table 10.3, based on the division of the speech spectrum into one-third octaves, provides the data necessary to calculate articulation index. The first column lists the center frequency of the one-third octave band, the second column gives the typical male voice long-term average speech level plus 12 dB at 1 m distance. The weighting factor for each one-third octave band is listed in the third column. If the masking noise has been measured only in full octave bands, Table 10.4 may be used instead of Table 10.3.

Example Problem 1: Calculation of AI

Let us compute the articulation index of a male voice speaking at a normal level 1 m from the listener in the presence of pink noise that contributes 47 dB in each octave band.

TABLE 10.3. Weighting Factors as a Function of Center Frequency for One-Third Octave Band-Based Calculation of Articulation Indexes. Speech Levels Given in the Second Column are Those for a Typical Male Voice at 1 Meter.

Center Frequency (Hz)	Speech Level (+12 dB)	Weighting Factor
200	67	4
250	68	10
315	69	10
400	70	14
500	68	14
630	66	20
800	65	20
1000	64	24
1250	62	30
1600	60	37
2000	59	37
2500	57	34
3150	55	34
4000	53	24
5000	51	20

TABLE 10.4. Weighting Factors for One Octave Band-Based Calculation of Articulation Indexes. This Is for the Typical Male Voice Level at 1 Meter.

Center Frequency (Hz)	Speech Level (+12 dB)	Weighting Factor
250	72	18
500	73	50
1000	78	75
2000	63	107
4000	58	83

Solution

Table 10.4 is used for this calculation. The 47-dB octave-band noise level is subtracted from the values in the second column, and the difference (up to a maximum of 30 dB) is multiplied by the weighting factors of the third column. The resulting weighted contributions are added and divided by 10,000, yielding the articulation index. Table 10.5 gives details of the calculations, with the values given in the fifth column produced by subtracting 47 dB from the speech level

TABLE 10.5. Calculations for the Articulation Index in the Sample Problem.

Center Frequency (Hz)	Speech Level (SL) (+12 dB)	Weighting Factor (WF)	Noise (N)	SL − N = DIFF	WF × DIFF
250	72	18	47	25	450
500	73	50	47	26	1300
1000	78	75	47	30[a]	2250
2000	63	107	47	16	1712
4000	58	83	47	11	913

[a] The value of (speech level + 12 dB-noise level) must fall between 0 and 30, so the actual difference of 31 dB is limited to 30 dB.

of the second column; and each of the values in the fifth column is multiplied by the weighting factor of the third column to yield the figures listed in the sixth column.

$$\text{articulation index} = \frac{450 + 1300 + 2250 + 1712 + 913}{10,000} = 0.6625$$

The articulation index is 0.6625, or 66.25%.

10.6   Speech Interference Level (SIL)

Measurements to obtain data for articulation indexes require special laboratory equipment for determination of S/N in a number of frequency bands. A simpler procedure of estimating the effect of noise on verbal communication makes use of octave-band levels as measured in a typical noise survey. The parameter that is called the speech *interference level*, abbreviated *SIL*, can be obtained by computing the arithmetic average of octave band levels in the three octave bands of 600–1200, 1200–2400, and 2400–4800 Hz. However, current practice uses the arithmetic level in the "preferred" octave bands with center frequencies at 500, 1000, and 2000 Hz. The speech interference level defined thusly is referred to as *PSIL*.

The speech interference level PSIL = 68 dB has been identified as the level at which reliable speech communication is barely possible in a normal male voice at a distance of 0.3 m (or 1 ft) outdoors. For a male speaker talking in a raised voice, a very loud voice, or in a shout, the speech interference levels have been identified respectively as PSIL = 74, 80, and 86 dB. A female speaker, on the average, has PSIL levels 5 dB less than the corresponding values for a male. Table 10.6 lists the PSIL (in dB) at which effective speech communication is barely possible. The table is based on minimally reliable communication, in which about 60% of the communication of uttered numbers and words out of context can be discerned. In order to roughly approximate PSIL in terms of dBA, 7 dB can be added to the values of PSIL.

TABLE 10.6. PSIL (in dB) at Which Effective Speech Communication Is Barely Possible.

Distance (m)	Normal		Raised		Very Loud		Shouting	
	M	F	M	F	M	F	M	F
0.3	68	63	74	69	80	75	86	81
1	58	53	64	59	70	65	76	71
2	52	47	58	53	64	59	70	65
3	48	43	54	49	60	55	66	61
4	46	41	52	47	58	53	64	59

Example Problem 2: SIL

Background noise levels for an industrial plant were measured to be 62, 65, and 74 dB, respectively in the 500-, 1000-, and 2000-Hz center frequency bands. What are the implications for speech interference at a distance between a speaker and a listener standing 1 m apart?

Solution

To solve this problem, the arithmetic average of the noise level in three bands are first determined. This will be $(62 + 65 + 74)/3 = 67$ dB. From Table 10.6 we establish that reliable speech is barely possible for a male speaking in a raised voice or a female speaking in a very loud voice.

References

American National Standards Institute. 1969. *Methods for Calculation of the Articulation Index* (ANSI S3.5-1969, revised 1989). New York: Acoustical Society of America.

American National Standards Institute. 1989. *ANSI 1969: American National Standard Specification for Audiometers* (ANSI S3.6-1969, revised 1989). New York: Acoustical Society of America.

American National Standards Institute. 1986. *ANSI 1973: American National Psychoacoustic Terminology* (ANSI S3.20-1973, revised 1986). New York: Acoustical Society of America.

Crocker, Malcolm J. (ed.). 1997. *Encyclopedia of Acoustics*, vol. 2. Parts XI and XII.

Dadson, R. S., and J. H. King. 1952. A determination of the normal threshold of hearing and its relation to the standardization of audiometers. *Journal of Laryngology and Otology* 66: 366–78. Reproduced in "Forty Germinal Papers in Human Hearing." J. Donald Harris (ed.). Groton, CT: *The Journal of Auditory Research*, 1969; pp. 600–601.

French N. R., and J. C. Steinher. 1947. Factors governing the intelligibility of speech sounds. *Journal of the Acoustical Society of America* 19: 90–119.

Harris, J. Donald, and C. K. Myers. February 1971. Tentative audiometric hearing threshold level standards from 8 through 19 Kilohertz. *Journal of the Acoustical Society of America* 49: 600–601.

Hawkins, J. E., Jr., and S. S. Stevens. 1950 . The masking of pure tones and speech by white noise, *Journal of the Acoustical Society of America* 22: 12.

Kinsler, Lawrence E., Austin R. Frey, Alan B. Coppens, and James V. Sanders. 1982. *Fundamentals of Acoustics*, 3rd ed.. New York: John Wiley & Sons; Chapter 11.

Kleine, Ulrich, and Manfred Mauthe. Spring 1994. Designing circuits for hearing instruments, *Siemens Review R&D Special.*

Kryter, Karl. D. 1970. *The Effects of Noise on Man.* New York: Academic Press.

Lipscomb, David M. 1974. *Noise: The Unwanted Sound.* Chicago: Nelson-Hall.

Neby Hayes A., and Gerald R. Popelka. 1992. *Audiology*, 6th ed.. New York: Prentice-Hall.

Robinson, D. W., and R. S. Dadson. 1956. A re-determination of the equal loudness relations for pure tones. *British Journal of Applied Physics* 7: 166–81. Reproduced in "Forty Germinal Papers in Human Hearing." J. Donald Harris (ed.). Groton, CT: *The Journal of Auditory Research*, 1969; pp. 600–601.

Sandlin, R. E. 1988. *Handbook of Hearing Aid Amplification, Vol. I: Theoretical and Technical Considerations.* Boston: College Hill Press.

Sandlin, R. E. 1990. *Handbook of Hearing Aid Amplification, Vol. II: Clinical Considerations and Fitting Practices.* Boston: College Hill Press.

Sataloff, Robert T., and Joseph Sataloff. 1993. *Hearing Loss*, 3rd ed., New York: Marcel Dekker.

Schow, Ronald L., and Michael A. Nerbonne. 1989. *Introduction to Aural Rehabilitation*, 2nd ed. Boston: Allyn and Bacon.

Silman, Shlomo, and Carol A. Silverman. 1991. *Auditory Diagnosis: Principles and Applications.* New York: Academic Press.

Von Békésy, Georg. 1978. *Experiments in Hearing.* Huntington, NY: Kreiger.

Problems for Chapter 10

1. Find the articulation index for a male speaker at a normal level 1 m from a listener if we have the following background noise spectrum:

Center frequency, Hz	Noise level, dB
200	42
250	39
315	44
400	46
500	48
630	38
800	30
1000	26
1250	20
1600	23
2000	18
2500	15
3150	14
4000	10
5000	10

2. Determine the suitability of an industrial plant for carrying out conversations between two males, one female and one male, and two females under the following background noise conditions:

Center frequency	Background noise level, dB
500	55
1000	63
2000	72

3. Find the PSIL from the following octave-band noise spectrum:

Center Frequency, Hz	63	125	250	500	1k	2k	4k	8k
Band pressure level, dB	52	56	57	62	55	51	50	45

4. Determine the voice level necessary to effectively communicate over a distance of 4 ft with a background of
   (a) 63 dB
   (b) 72 dB
   (c) 85 dB

5. Why does the U.S. Occupational Safety and Health Administration prohibit impact noises of 130 dB or more even if the overall sound pressure level is less than 90 dB during the course of an 80-hr day?

6. Express 35 dB in terms of pressure in pascals and in terms of intensity (W/cm^2). At 62.5 Hz, is this considered audible for normal hearing? If so, how much above the MAP is the sound pressure level?

7. How loudly must one speak, in terms of decibels, with a white noise background of 50 dB in order to be understood? How loud must the speech be in order to be detected, if not necessarily understood?

8. At 500 Hz, a 40-dB tone sounds as equally loud as a 5-kHz tone. What is the dB level of that 5 kHz tone?

11
Acoustics of Enclosed Spaces: Architectural Acoustics

11.1 Introduction

Although people have gathered in large auditoriums and places of worship since the advent of civilization, architectural acoustics did not exist on a scientific basis until a young professor of physics accepted an assignment from Harvard University's Board of Overseers in 1895 to correct the abominable acoustics of the newly constructed Fogg Lecture Hall. Through careful (but by present-day standards, rather crude) measurements with the use of a Gemshorn organ pipe of 512 Hz, a stopwatch, and the aid of a few able-bodied assistants who lugged absorbent materials in and out of the lecture hall, Wallace Sabine established that the reverberation characteristics of a room determined the acoustical nature of that room and that a relationship exists between quality of the acoustics, the size of the chamber, and the amount of absorption surfaces present. He defined a reverberation time T as the number of seconds required for the intensity of the sound to drop from a level of audibility 60 dB above the threshold of hearing to that threshold of inaudibility. To this day reverberation time still constitutes the most important parameter for gauging the acoustical quality of a room. The original Sabine equation

$$T = \frac{0.049V}{\sum_i S_i \alpha_i}$$

is deceptively simple, as effects such as interference or diffraction and behavior of sound waves as affected by the shape of the room, presence of standing waves, and normal modes of vibration, are not embodied in the equation. Here V is the room volume in cubic feet, S_i the component surface area, and α_i the corresponding absorption coefficient. On the basis of his measurements Sabine was able to cut down the reverberation time of the lecture hall from 5.6 seconds through the strategic deployment of absorbing materials throughout the room. This accomplishment firmly established Sabine's reputation, and he became the acoustical consultant for Boston Symphony Hall, the first auditorium to be designed on the basis of quantitative acoustics.

In this chapter we shall examine the behavior of sound in enclosed spaces, and develop the fundamental equations that are used in optimizing the acoustics of auditoriums, music halls, lecture rooms. We shall also study the means of improving room acoustics through installation of appropriate materials. This chapter concludes with descriptions of a number of outstanding acoustical facilities.

11.2   Sound Fields

The distribution of acoustic energy, whether originating from a single or multiple sound source in an enclosure, depends on the room size and geometry and on the combined effects of reflection, diffraction, and absorption. With the appreciable diffusion of sound waves due to all of these effects it is no longer germane to consider individual wave fronts, but to refer to a *sound field*, which is simply the region surrounding the source. A *free field* is a region surrounding the source, where the sound pattern emulates that of an open space. From a point source the sound waves will be spherical, and the intensity will approximate the inverse square law. Neither reflection nor diffraction interferes with the waves emanating from the source. Because of the interaction of sound with the room boundaries and with objects within the room, the free field will be of very limited extent.

If one is close to a sound source in a large room having considerably absorbent surfaces, the sound energy will be detected predominantly from the sound source and not from the multiple reflections from surroundings. A free field can be simulated throughout an entire enclosure if all of the surrounding surfaces are lined with almost totally absorbent materials. An example of such an effort to simulate a free field is the extremely large anechoic (echoless) chamber at Lucent Technologies Bell Laboratories in Murray Hill, New Jersey, shown in the photograph of Figure 11.1. Such a chamber is typically lined with long wedges of absorbent foam or fiberglass, and the "floor" consists of either wire mesh or grating suspended over wedges, suspended above the "real" floor underneath. Precisely controlled experiments on sound sources and directivity patterns of sound propagation are rendered possible in this sort of chamber.

A *diffuse field* is said to occur when a large number of reflected or diffracted waves combine to render the sound energy uniform throughout the region under consideration. Figure 11.2 illustrates how diffusion results from multiple reflections. The degree of diffusivity will be increased if the room surfaces are not parallel so there is no preferred direction for sound propagation. Concave surfaces with radii of curvature comparable to sound wavelengths tend to cause focusing, but convex surfaces will augment diffusion. Multiple speakers in amplifying auditoriums are used to achieved better diffusion, and special baffles may be hung from ceilings to deflect sound in the appropriate directions.

Sound reflected from walls generates a *reverberant field* that is time dependent. When the source suddenly ceases, a sound field persists for a finite interval as the result of multiple reflections and the low velocity of sound propagation. This

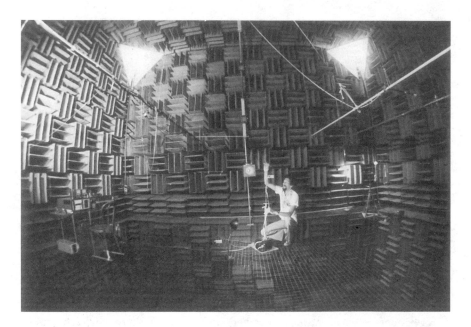

FIGURE 11.1. Photograph of the large anechoic chamber at the Lucent Bell Laboratory in Murray Hill, NJ. Dr. James E. West, a former president of the Acoustical Society of America, is shown setting up test equipment. (Courtesy of Lucent Technologies.)

residual acoustic energy constitutes the reverberant field. The sound that reaches a listener in a fairly typical auditorium can be classified into two broad categories: the direct (free field) sound and indirect (reverberant) sound. As shown in Figure 11.3, the listener receives the primary or direct sound waves and indirect or reverberant sound. The amount of acoustic energy reaching the listener's ear by any single reflected path will be less than that of the direct sound because the reflected path is longer than the direct source–listener distance, which results in greater divergence; and all reflected sound undergoes an energy decrease due to the absorption of even the most ideal reflectors. But indirect sound that a listener hears comes from a great number of reflection paths. Consequently, the contribution of reflected sound to

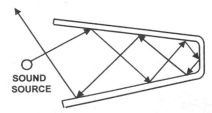

FIGURE 11.2. Sound diffusion resulting from multiple reflections.

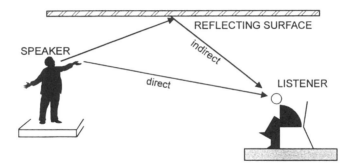

FIGURE 11.3. Reception of direct and indirect sound.

the total intensity at the listener's ear can exceed the contribution of direct sound particularly if the room surfaces are highly reflective.

The phases and the amplitudes of the reflected waves are distributed randomly to the degree that cancellation from destructive interference is fairly negligible. If a sound source is operated continuously the acoustic intensity builds up in time until a maximum is reached. If the room is totally absorbent so that there are no reflections, the room operates as an anechoic chamber, which simulates a free field condition. With partial reflection, however, the source continues to add acoustic energy to the room, which is partially absorbed by the enclosing surfaces (i.e., the walls, ceiling, floor and furnishings) and deflected back into the room. For a source operating in a reverberation chamber the gain in intensity can be considerable—as much as ten times the initial level. The gain in intensity is approximately proportional to the reverberation time; thus it can be desirable to have a long reverberation time to render a weak sound more audible.

11.3   Reverberation Effects

Consider a sound source that operates continuously until the maximum acoustic intensity in the enclosed space is reached. The source suddenly shuts off. The reception of sound from the direct ray path ceases after a time interval r/c, where r represents the distance between the source and the reception point and c the sound propagation velocity. But owing to the longer distances traveled, reflected waves continue to be heard as a reverberation that exists as a succession of randomly scattered waves of gradually decreasing intensity.

The presence of reverberation tends to mask the immediate perception of newly arrived direct sound unless the reverberation drops 5–10 dB below its initial level in a sufficiently short time. Reverberation time T, the time in seconds required for intensity to drop 60 dB, offers a direct measure of the persistence of the reverberation. A short reverberation time is obviously necessary to minimize the masking effects of echoes so that speech can be readily understood. However, an extremely short reverberation time tends to make music sound harsher—or less "musical"—

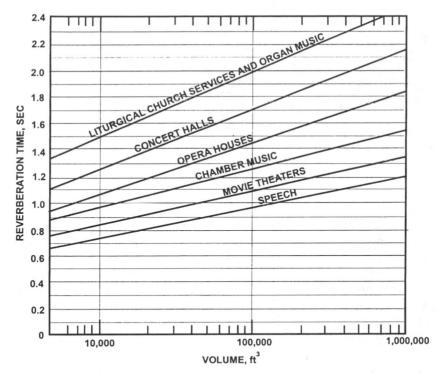

FIGURE 11.4. Typical reverberation times for various auditoriums and functions.

while excessive values of reverberation time T can blur the distinction between individual notes. The choice of T, which also depends on the room volume, therefore represents an optimization between two extremes.

Figure 11.4 represents the accumulation of optimal reverberation time data as functions of intended use and enclosure volume. Lower values of T occur from increased absorption of sound in the surfaces of the enclosures. Hard surfaces such as ceramic tile floors and mirrors tend to lengthen the reverberation time. In addition to reverberation time, the ability of a chamber or enclosure to screen out external sound minimizes annoyance or masking effects. The acoustic transmission of walls, treated in Chapter 12 constitutes a major factor in enclosure design. A short reverberation time with its attendant high absorption tends to lessen the ambient noise level generated by external sound that penetrate the walls of the enclosure.

11.4   Sound Intensity Growth in a Live Room

We now apply the classic ray theory to deal with a sound source operating continuously in an enclosure, which will yield results in fairly good agreement with experimental measurements. The process of absorption in the medium or the enclosing surfaces prevents the intensity from becoming infinitely large. Absorption in the medium is fairly negligible in medium- and small-sized enclosures, so the ultimate intensity depends on the absorption power of the boundary surfaces. If the enclosure's boundary surfaces have high absorption the intensity will quickly achieve the maximum which exceeds only slightly the intensity of the direct ray. If the enclosure has highly reflective surfaces, that is, low absorption, a "live" room ensues; the growth of the intensity will be slow and appreciable time will have elapsed for the intensity to reach its maximum.

After a sound source is started in a live room, reflections from the wall become more uniform in time as the sound intensity increases. With the exception of close proximity to the source, the energy distribution can be considered uniform and random in direction. In reality a signal source having a single frequency will result in standing-wave patterns, with resultant large fluctuations from point to point in the room. But if the sound consists of a uniform band of frequencies or a pure tone warbling over at least a half octave, the interference effects of standing waves are obliterated.

Referring to Figure 11.5, we establish the relationship between *intensity* (which represents the energy flow) and energy density of randomly distributed *acoustic energy*. In the figure dS represents an element of the wall surface and dV the volume element in the medium at a distance r from dS. The distance r makes an angle θ with the normal NN' to dS. Let the average acoustic energy density E (in W/m^3) be assumed uniform throughout the region under consideration. The acoustic energy in incremental volume dV is $E\,dV$. The surface area of the sphere

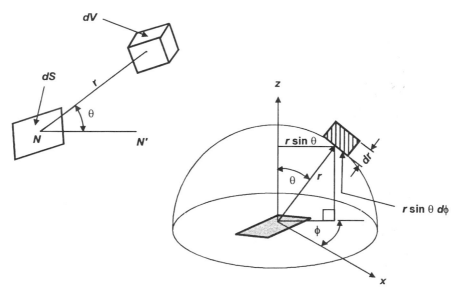

FIGURE 11.5. Geometric configuration for setting up the relationship between energy density and intensity of sound.

of radius r encompassing dV is $4\pi r^2$. The projected area of dS on the sphere is $\cos\theta\, dS$. The portion of the total energy contained in dV is given by the ratio $dS\cos\theta/4\pi r^2$. The energy from dV that strikes dS directly becomes

$$dE = \frac{E\, dV\, dS\cos\theta}{4\pi r^2} \tag{11.1}$$

Now consider the volume element dV as being part of a hemisphere shell of radius r and thickness dr. The acoustic energy rendered to S by the complete shell is found by assuming a circular zone of radius $r\sin\theta$ (with θ treated as a constant) in Figure 11.5 and integrating over the entire surface of the shell. The volume of the resultant element is $2\pi r\sin\theta\, r\, dr\, d\theta$ from $\theta = 0$ to $\theta = \pi/2$, and equation (11.1) yields

$$\Delta E = \frac{E\, dS\, dV}{2}\int_0^{\pi/2}\sin\theta\cos\theta\, d\theta = \frac{E\, dS\, dr}{4}$$

This energy arrives during time interval $t = dr/c$. Hence, the rate of acoustic energy impinging dS from all directions is

$$\frac{\Delta E}{t} = \frac{Ec\, dS}{4}$$

or $Ec/4$ per unit area, which is therefore the intensity I of the diffused sound at the walls. This is also equal to one-fourth of a plane wave of energy intensity I incident at a normal angle onto a plane. The intensity I of the diffuse sound at the wall becomes

$$I = \frac{Ec}{4} \tag{11.2}$$

11.5   Sound Absorption Coefficients

All materials constituting the boundaries of an enclosure will absorb and reflect sound. A fraction α of the incident energy is absorbed and the balance $(1 - \alpha)$ is reflected. Reflection is indicated by the reflection coefficient r defined as

$$r = \frac{\text{amplitude of reflected wave}}{\text{amplitude of incident wave}}$$

Because the energy in a sound wave is proportional to the square of the amplitude, the sound absorption coefficient α and the reflection coefficient are related by

$$\alpha = 1 - r^2$$

The value of the sound absorption coefficient α will vary with the frequency of the incident ray and the angle of incidence. Materials comprising room surfaces are subject to sound waves that impinge upon them from many different angles as the result of multiple reflections. Hence, published data for absorption coefficients are for "random" incidence as distinguished from "normal" or "perpendicular" incidence.

The angle-absorption correlation appears to be of somewhat erratic nature, but at high frequencies the absorption coefficients in some materials is roughly constant at all angles. At low frequencies the random-incidence absorption tends to be greater than for normal incidence. However, as Table 11.1 shows, α varies considerably with frequency for many materials, and the absorption coefficients are

TABLE 11.1. Absorption Coefficients.

	Octave-Band Center Frequency (Hz)					
	125	250	500	1000	2000	4000
Brick, unglazed	0.03	0.03	0.03	0.04	0.05	0.07
Brick, unglazed, painted	0.01	0.01	0.02	0.02	0.02	0.03
Carpet on foam rubber	0.08	0.24	0.57	0.69	0.71	0.73
Carpet on concrete	0.02	0.06	0.14	0.37	0.60	0.65
Concrete block, coarse	0.36	0.44	0.31	0.29	0.39	0.25
Concrete block, painted	0.10	0.05	0.06	0.07	0.09	0.08
Floors, concrete or terrazzo	0.01	0.01	0.015	0.02	0.02	0.02
Floors, resilient flooring on concrete	0.02	0.03	0.03	0.03	0.03	0.02
Floors, hardwood	0.15	0.11	0.10	0.07	0.06	0.07
Glass, heavy plate	0.18	0.06	0.04	0.03	0.02	0.02
Glass, standard window	0.35	0.25	0.18	0.12	0.07	0.04
Gypsum, board, $\frac{1}{2}$ in.	0.29	0.10	0.05	0.04	0.07	0.09
Panels, fiberglass, $1\frac{1}{2}$ in. thick	0.86	0.91	0.80	0.89	0.62	0.47
Panels, perforated metal, 4 in. thick	0.70	0.99	0.99	0.99	0.94	0.83
Panels, perforated metal with fiberglass insulation, 2 in. thick	0.21	0.87	1.52	1.37	1.34	1.22
Panels perforated metal with mineral fiber insulation, 4 in. thick	0.89	1.20	1.16	1.09	1.01	1.03
Panels, plywood, $\frac{3}{8}$ in.	0.28	0.22	0.17	0.09	0.10	0.11
Plaster, gypsum or lime, rough finish on lath	0.02	0.03	0.04	0.05	0.04	0.03
Plaster, gypsum or lime, smooth finish on lath	0.02	0.02	0.03	0.04	0.04	0.03
Polyurethane foam, 1 in. thick	0.16	0.25	0.45	0.84	0.97	0.87
Tile, ceiling, mineral fiber	0.18	0.45	0.81	0.97	0.93	0.82
Tile, marble or glazed	0.01	0.01	0.01	0.01	0.02	0.02
Wood, solid, 2 in. thick	0.01	0.05	0.05	0.04	0.04	0.04
Water surface	nil	nil	nil	0.003	0.007	0.02
One person	0.18	0.4	0.46	0.46	0.51	0.46
Air	nil	nil	nil	0.003	0.007	0.03

Note: The coefficient of absorption for one person is that for a seated person (m^2 basis). Air absorption is on a per cubic meter basis.

generally measured at six standard frequencies, 125, 250, 500, 1000, 2000, and 4000 Hz. Absorption occurs as the result of incident sound penetrating and becoming entrapped in the absorbing material, thereby losing its vibrational energy which converts into heat through friction. Ordinarily the values of α should fall between zero for a perfect reflector and unity for a perfect absorber. Measurements of $\alpha > 1.0$ have been reported, owing possibly to diffraction at low frequencies and other testing condition irregularities.

Let $\alpha_1, \alpha_2, \alpha_3, \ldots, \alpha_i$ denote the absorption coefficient of different materials of corresponding areas $S_1, S_2, S_3, \ldots, S_i$ forming the interior boundary planes (viz. the walls, ceiling, and floor) of the room as well as any other absorbing surfaces (e.g., furniture, draperies, people, etc.). The average absorption coefficient $\hat{\alpha}$ for an enclosure is defined by:

$$\bar{\alpha} = \frac{\alpha_1 S_1 + \alpha_2 S_2 + \alpha_3 S_3 + \cdots + \alpha_i S_i}{S_1 + S_2 + S_3 + \cdots + S_i} = \frac{A}{S} \tag{11.3}$$

where A represents the total absorptive area $\sum \alpha_i S_i$, and S the total spatial area.

11.6   Growth of Sound with Absorbent Effects

The rate W of sound energy being produced equals the rate of sound energy absorption at the boundary surfaces of the room plus the rate at which the energy increases in the medium throughout the room. This may be expressed as a differential equation governing the growth of acoustic energy in a live room:

$$V \frac{dE}{dt} + \frac{AcE}{4} \equiv W \tag{11.4}$$

The solution for E in equation (11.4) is

$$E = \frac{4W}{Ac} \left(1 - e^{-(Ac/4V)t}\right) \tag{11.5}$$

with the initial condition that the sound source begins operating at $t = 0$. From the relationship of equation (11.2) the intensity becomes

$$I = \frac{W}{A} \left(1 - e^{-(Ac/4V)t}\right) \tag{11.6}$$

and from equation (3.58) the energy density is

$$E = \frac{p^2}{2\rho_0 c^2} \tag{11.7}$$

The mean square acoustic pressure becomes

$$p^2 = \frac{4W\rho_0 c}{A} \left(1 - e^{-(Ac/4V)t}\right) \tag{11.8}$$

Equation (11.8) is analogous to the one describing the growth of direct current in an electric circuit containing an inductance and a resistance. The time constant of the acoustic process is $4V/Ac$. If the total absorption is small and the time constant is large, a longer time will be necessary for the intensity to approach its ultimate value of $I_\infty = W/A$. The ultimate values of the energy density and mean square acoustic pressure are given by

$$E_\infty = \frac{4W}{Ac}, \qquad p_\infty^2 = \frac{4W\rho_0 c}{A}$$

A number of caveats pertain to the use of equation (11.8). In order that the assumption of an even distribution of acoustic energy be cogent, a sufficient time t must have elapsed for the initial rays to undergo several reflections at the boundaries. This means approximately 1/20 of a second should have elapsed in a small chamber; and the time must approach nearly a full second for a large auditorium. The final energy density, being independent of the size and shape of the room, should be the same at all points of the room and dependent only upon the total absorption A. But equation (11.6) does not hold for spherical or curved rooms which can focus sounds; neither is equation (11.8) applicable to rooms having deep recesses nor to oddly shaped rooms or rooms coupled together by an opening, and nor to rooms with some surfaces of extraordinarily high absorption coefficients α (these cause localized lesser values of energy densities).

11.7   Decay of Sound

We can now develop the differential equation describing the decay of uniformly diffuse sound in a live room. The sound source is shut off at time $t = 0$, meaning $W = 0$ at that instant. E_0 denotes the uniformly distributed energy density at that instant. From equation (11.4)

$$\frac{AcE}{4V}dt = dE \tag{11.9}$$

and the solution to equation (11.9) becomes

$$E = E_0 e^{-(Ac/4V)t} \tag{11.10}$$

The intensity I at any time t after the cessation of the sound source is related to the initial intensity I_0 by

$$\frac{I}{I_0} = e^{-(Ac/4V)} \tag{11.11}$$

Applying the operator $10\log$ to both sides of equation (11.11) results in

$$\Delta IL = 10\log e^{-(Ac/4V)t} = \frac{10}{2.3}\ln e^{-(Ac/4V)}t = -\frac{1.087\,Ac}{V} \tag{11.12}$$

where ΔIL denotes the intensity level change in decibels. The intensity level in a live room decreases with elapsed time at a constant decay rate D (in dB/s),

$$D = \frac{1.087\,Ac}{V}$$

Following Sabine's definition, we define the reverberation time T as the time required for the sound level in the room to decay by 60 dB

$$T = \frac{60}{D} = \frac{55.2\,V}{Ac} \tag{11.13}$$

Expressing volume V in m^3 and area S used to compute A in m^2 and setting sound propagation speed $c = 343$ m/s, equation (11.13) becomes

$$T = \frac{0.161\,V}{A} \tag{11.14}$$

Equation (11.14) becomes for English units

$$T = \frac{0.049\,V}{A} \tag{11.15}$$

where volume V is rendered in ft^3 and A in ft^2 (or *sabins*, with 1 sabin equal to 1 ft^2 of absorption area αS). (One *metric sabin* is equal to 1 m^2 of absorption area.) It becomes apparent here that the reverberation time for a room can be controlled by selecting materials with the appropriate acoustic absorption coefficients. The absorption coefficient of a material can be measured by the introduction of a definite area of the absorbent material in a specially constructed live room or *reverberation* (or *echo*) *chamber*. A photograph of such a chamber is given in Figure 11.6.

Example Problem 1: Reverberation Prediction

A room 8 m in length, 4 m in width, 2.8 m high contains four walls faced with gypsum boards. The only exceptions to the wall area are a glass window 1 m $\times$ 0.5 m and a plywood-paneled door 2.2 m $\times$ 0.6 m. In addition the door has a gap underneath 1.5 cm high. The ceiling is mineral fiber tile and the floor is hardwood. In order to estimate the reverberation time of the room at 500 Hz we make use of the data in Table 11.1.
Predict the reverberation time T.

Solution

The absorption area (in m^2) is found as follows:

$$
\begin{aligned}
A = \sum S_i \alpha_i &= [2(8 \times 2.8) + 2(4 \times 2.8) - 2.2(0.6) - (0.015)(0.6) - 1.0(0.5)] \\
&\quad \times 0.05 + (2.2)(0.6)(0.17) + (0.015)(0.6)(1) + 1.0(0.5)(0.18) + (4)(8)(0.81) \\
&\quad + (4)(8)(0.10) = 32.70 \text{ m}^2
\end{aligned}
$$

Applying equation (11.14):

$$T = \frac{0.161 \times 8 \times 4 \times 2.8}{32.70} = 0.44 \text{ s}$$

The gap at the bottom of the door is treated as a complete sound absorber with a coefficient of unity. From the above estimated value of the reverberation time of 0.44 s and a chamber volume of 89.6 m^3, the room may be suitable for use as a classroom, according to Figure 11.4.

FIGURE 11.6. Photograph of a reverberation chamber. (Courtesy of Eckel Industries, Inc.)

11.8   Decay of Sound in Dead Rooms

The derivation of equation (11.14) was based on the assumption that a sufficient number of reflections occur during the growth or decay of sound and also that the energy of the direct sound and the energy of the fractional amount of sound reflected were both sufficient to ensure a uniform energy distribution. In the case of anechoic chambers, where the absorption coefficient of the materials constituting the boundaries is very close to unity, it is apparent that the derivations of the preceding equations for growth and decay of sound are not applicable. The only energy present is the direct wave emanating from the sound source. The reverberation time must be zero, whereas application of equation (11.14) would yield a finite reverberation time of $0.161 V/S$, where S is simply the total area of the interior surfaces of the chamber. Thus, it is apparent that equation (11.14) would be increasingly in error as the average sound absorption coefficient increases. If the average value of the absorption coefficient exceeds 0.2, equation (11.14) will be in error by approximately 10 percent.

A different approach to ascertaining the decay of sound in a dead room, which was developed by Eyring, is to consider the multiplicity of reflections as a set of image sources, all of which are considered to exist as soon as the real source begins. Let $\bar{\alpha}$, found from the relationship $\bar{\alpha} = (\sum \alpha_i S_i)/\sum S_i$, denote the average sound absorption coefficient of the room's boundary materials. The growth of acoustic energy at any point in the room results from the accumulation of successive increments from the sound source, from the first-order (single reflection) images with strengths $W(1 - \bar{\alpha})$, from the second-order (secondary reflection) images with strengths $W(1 - \bar{\alpha})^2$, and so on, until all the image sources of appreciable strengths have rendered their contributions. When the true sound source is stopped, the decay of the sound occurs with all the image sources stopped simultaneously along with the source. The energy decay in the room occurs from successive losses of acoustic radiation from the source, then from the first-order images, the second-order images, and so on.

Eyring derived the following equation for the growth in acoustic energy density:

$$E = \frac{4W}{-cS \ln(1 - \bar{\alpha})}\left[1 - e^{\frac{cS \ln(1-\bar{\alpha})t}{4V}}\right] \tag{11.16}$$

The above equation is very similar to equation (11.5) excepting that the total room absorption is given by

$$a = -S \ln(1 - \bar{\alpha}) \tag{11.17}$$

Here S is the total area of the boundary surfaces of the room. In a like fashion the analogy to equation (11.6) for the decay of sound energy is given by

$$E = E_0 e^{\frac{cS \ln(1-\bar{\alpha})t}{4V}}$$

and the decay rate in dB/sec is expressed as:

$$D = -\frac{1.08cS \ln(1 - \bar{\alpha})}{V}$$

with the reverberation time expressed by:

$$T = \frac{0.161V}{-S \ln(1 - \bar{\alpha})}$$

For small values of absorption ($\alpha \ll 1$) the term $\ln(1 - \alpha)$ may be replaced by α, the first term in an infinite series. This results in recovering the Sabine formula for live rooms. It should be also noted that the coefficient 0.161 for the Sabine and the Eyring formulas, which is based on the speed of sound at 24°C, will vary according to air temperature. The coefficient becomes somewhat higher at lower air temperatures and vice versa.

Another formula for determining the reverberation time of a room lined with materials of widely ranging absorption coefficients was developed by Millington and Sette. The Millington–Sette theory indicates that the total room absorption is given by

$$A = \sum - S_i \ln(1 - \alpha_i)$$

which yields the reverberation time

$$T = \frac{0.161V}{\sum - S_i \ln(1 - \alpha_i)}$$

11.9   Reverberation as Affected by Sound Absorption and Humidity in Air

We have not previously considered the effect of absorption of sound and humidity in air on reverberation times. The volume of air contained in very large auditoriums or a place of worship can absorb an amount of acoustic energy that cannot be neglected as in the case with smaller rooms. If a room is small, the number of reflections from the boundaries is large and the amount of time the sound wave spends in the room is correspondingly small. In this situation acoustic energy absorption in the air is generally not important. In very large room volumes the time a wave spends in the air between reflections becomes greater to the extent that absorption of energy in air no longer becomes negligible. The reverberation equations must now include the effect of air absorption, particularly at higher frequencies (>1 kHz).

Sound waves lose some energy through viscous effects during the course of their propagation through a fluid medium. The intensity of a plane wave lessens with distance according to the equation

$$I = I_0 e^{-2\beta x} = I_0 e^{-mx}$$

Here $m = 2\beta$ represents the attenuation coefficient of the medium. Some texts use α rather than β to denote the attenuation constant of the medium; we eschew its use in order to avoid confusion with α used in this chapter to denote the absorption coefficient of a surface. During time interval t, a sound wave travels a distance $x = ct$, and the preceding equation may be revised to read

$$I = I_0 e^{-(\beta/4V + m)ct}$$

The expression for the reverberation time becomes

$$T = \frac{0.161V}{A + 4mV} \tag{11.18}$$

where the constant m is expressed in units of m^{-1}. The total surface absorption A is given either by equation (11.3) or by equation (11.17) depending on whether that room fits into the category of being an acoustically live or dead chamber. As the room volume V becomes larger, the second term in the denominator of equation (11.18) increases in magnitude, as air absorption becomes more significant, due to increasing path lengths between the walls. Because m also increases with frequency, air absorption also becomes more manifest at higher frequencies (above 1 kHz) than at lower frequencies. The values of m are rendered in

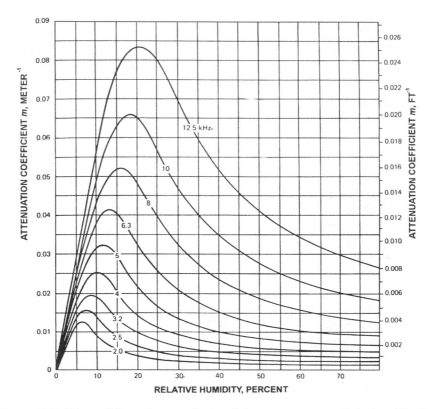

FIGURE 11.7. Values of the total attenuation coefficient m vs. percent relative humidity for air at 20°C and normal atmospheric pressure for frequencies between 2 kHz and 12.5 kHz. Values are rendered in both SI units and U.S. Customary System units. (After Cyril M. Harris. 1966 and 1967.)

Figure 11.7[1] as a function of humidity for various frequencies at a normal room temperature of 20°C. More details, also given in tabular form, for a range of air temperatures and humidities are given in the NASA report (1967), prepared by Cyril M. Harris, listed at the end of this chapter. It is seen from Figure 11.7 that the effect of humidity reaches a maximum in each of the given frequencies in the 5%–25% relative humidity range and then trails off at higher humidities.

Example Problem 2

Find the air absorption at a frequency of 6300 Hz and 25% relative humidity for a room volume of 20,000 m³.

---

[1] The plot of Figure 11.7 applies to *indoor* sound propagation, not to outdoor propagation which includes meterological effects not present indoors.

Solution and Brief Discussion

From Figure 11.7 the value of m is equal to 0.026 per meter. The air absorption $A_{air} = 4mV$ in equation (11.18) is equal to $4 \times (20,000 \text{ m}^3) \times 0.026 \text{ m}^{-1} = 2080 \text{ m}^2$. If we consider absorption at 500 Hz the effect of air absorption would be negligible in comparison.

11.10   Early Decay time (EDT10)

A modification of the reverberation time T_{60} is the *early decay time*, or EDT10, which represents the time interval required for the first 10 dB of decay to occur, multiplied by six to produce an extrapolation to 60 dB decay Originally proposed by V. Jordan (1974), EDT10 is based on early psychoacoustical research, and according to Cremer and Muller (1982), "the latter part of a reverberant decay excited by a specific impulse in running speech or music is already masked by subsequent signals once it has dropped by about 60 dB."

11.11   Acoustic Energy Density and Directivity

In order to account for uneven distribution of sound in some sources, we express the sound intensity $I(\text{W/m}^2)$ due to a point source of power $W(\text{W})$ in the direct field (i.e., reflections are not considered) as

$$I = \frac{WQ(\theta, \phi)}{4\pi r^2} \tag{11.19}$$

where r is the distance (m) from the source and $Q(\theta, \phi)$ is the *directivity factor*. The directivity factor $Q(\theta, \phi)$ equals unity for an ideal point source which emits sound evenly in full space. For an ideal point source above an acoustically reflective surface, in an otherwise free half-space, $Q(\theta, \phi)$ equals 2. The sound or acoustic energy density is the sound energy contained per m^3 at any instant. In the direct field in full space, the direct sound energy density $D_D(\text{W} \cdot \text{s/m}^3)$ is given by

$$D_D = \frac{I}{c} = \frac{WQ(\theta, \phi)}{4\pi r^2 c} \tag{11.20}$$

where c is the speed of sound in m/s.

11.12   Sound Absorption in Reverberant Field:
The Room Constant

The product IS gives the rate of acoustic energy striking a surface area S; and $IS \cos \theta$ gives that rate for the incidence angle θ. In an ideal reverberant field, with equal probability for all angles of incidence, the average rate of acoustic energy striking one side of the surface is given by $IS/4$. The power absorbed by the surface having an absorption coefficient α is

$$\text{power absorbed} = \frac{\alpha IS}{4} = \frac{\alpha c D_R S}{4}$$

where D_R denotes the reverberant sound field density. In a fairly steady state condition the power absorbed is balanced by the power supplied by the source to the reverberant field. This is the portion of the input power W which remains after one reflection:

$$\text{power supplied} = W(1 - \alpha)$$

The steady-state condition results in:

$$\frac{c D_R S \alpha}{4} = W(1 - \alpha)$$

which we rearrange to obtain the energy density in the reverberant field,

$$D_R = 4W \frac{1 - \alpha}{\alpha c S} = \frac{4W}{cR} \tag{11.21}$$

where the room constant R is by definition

$$R = \frac{\alpha S}{1 - \alpha}$$

In most cases, the boundaries of the actual enclosure and other objects inside the enclosure are constructed of different materials with differing absorption coefficients. The room constant R of the enclosure is then described in terms of mean properties by

$$R = \frac{S_T \bar{\alpha}}{1 - \bar{\alpha}}$$

where

$$R = \text{the room constant (m}^2)$$
$$S_T = \text{total surface area of the room (m}^2)$$
$$\bar{\alpha} = \text{mean sound absorption coefficient} = \sum \alpha_i S_i / S_T$$

11.13   Sound Levels Due to Direct and Reverberant Fields

Near a point of nondirectional sound source, the sound intensity is greater than from afar. If the source is sufficiently small and the room not too reverberant, the acoustic field very near the source is independent of the properties of the room. In other words, if a listener's ear is only a few centimeters away from a speaker's mouth, the room surrounding the two persons has negligible effect on what the listener hears directly from the speaker's mouth. At greater distances from the source, however, the direct sound decreases in intensity, and, eventually the reverberant sound predominates.

If we are more than one-third wavelength from the center of a point source the energy density of a point r is given by equation (11.20) for the direct sound field. Combining the equations for the direct and the reverberant sound intensities equations (11.20) and (11.21) we get the total sound intensity I given by

$$I = W \left[\frac{Q(\theta, \phi)}{4\pi r^2} + \frac{4}{R} \right] \tag{11.22}$$

It is assumed that reverberant sound comes from nearly all directions in a fairly even distribution. The modes generated by standing waves must be rather insignificant; otherwise the assumption of uncorrelated sound is not valid and equation (11.22) will not truly constitute the proper model for the actual sound field.

The sound pressure level within the room can now be found from

$$L_p = 10\log \left(\frac{I}{I_{\text{ref}}} \right) = 10\log \left[W \left(\frac{1}{4\pi r^2} + \frac{4}{R} \right) \right] + 120$$

$$= L_W + 10\log \left(\frac{1}{4\pi r^2} + \frac{4}{R} \right) \tag{11.23}$$

for an ideal point source emanating equally in all directions. The reference sound intensity I_{ref} is equal to 10^{-12} W/m^2, and the sound power level L_W of the source is defined as

$$L_W = 10\log \left(\frac{W}{10^{-12}} \right)$$

which is given in dB re 1 pW. For an ideal point source over an acoustically reflective surface

$$L_P = 10\log \left[W \left(\frac{1}{2\pi r^2} + \frac{4}{R} \right) \right] + 120 = L_W + 10\log \left(\frac{1}{2\pi r^2} + \frac{4}{R} \right) \tag{11.24}$$

Equations (11.23) and (11.24) are based on the fact that the absorption coefficients do not vary radically from point to point in the room, and the source is not close to reflective surfaces. If the sound power and room absorption characteristics are assigned for each frequency band, the sound pressure level L_P can be determined for each frequency band in dB/octave, dB/one-third octave, and so on. If the sound power level is A-weighted, and if the room constant is based on frequencies in the same range as the frequency content of the source, the sound power level will be expressed in dB(A).

Example Problem 3

Predict the reading of a sound-pressure-level meter 12 m from a source having a sound power level of 108 dB(A) re 1 pW in a room with a room constant $R = 725$ m^2 (at the source frequencies). The source is mounted directly on an acoustically hard floor.

Solution

We apply equation (11.24) as follows:

$$L_p = L_W + 10 \log \left(\frac{1}{2\pi r^2} + \frac{4}{R} \right)$$

$$= 108 + 10 \log[(2\pi \times 12^2 \text{m}^2)^{-1} + 4/725m^2]$$

$$= 86.2 \text{ dB(A)}$$

If the meter would be placed at $R = 3$ m from the source, the SPL meter reading for L_p will increase to 91.7 dB(A).

11.14 Design of Concert Halls and Auditoriums

Ideally, the main objective of auditorium design is to get as many members of the audience as close as possible to the source of the sound, because sound levels decrease with increasing distances from the sound source. A good visual line of sight usually results in good acoustics, so stepped seating becomes desirable for larger rooms seating more than 100 people. Reverberation should be controlled in order to provide optimum reinforcement and equalization of sound. For speech the room design should provide more in the way of direct sound augmented by reflections, while the articulation of successive syllables must be sustained. Rooms for music have typically longer reverberation times because the requirements for articulation are not as stringent, and more enhancement of the sound is desirable.

The aim of the design of a listening type of facility is to avoid the following acoustic defects (Siebein, 1994):

- *Echoes*, particularly those from the rear walls of the facility. Echoes can be lessened or eliminated by placing absorbent panels or materials on the reflecting walls or introducing surface irregularities to promote diffusion of the sound.
- *Excessive loudness* can occur from prolonged reverberation. Again, the proper deployment of absorbent materials should alleviate this problem.
- *Flutter echo* results from the continued reflection of sound waves between two opposite parallel surfaces. This effect can be especially pronounced in small rooms; and this can be countervened by splaying the walls slightly (so as to avoid parallel surfaces) or using absorbent material on one wall.
- *Creep* is the travel of sound around the perimeter of domes and other curved surfaces. This phenomenon is also responsible for whispering galleries in older structures with large domed roofs.
- *Sound focusing* arises when reflections from concave surfaces tend to concentrate the sound energy at a focal point.
- *Excessive* or *selective absorption* occurs when a material that has a narrow range of acoustical absorption is used in the facility. The frequency that is absorbed is lost, resulting in an appreciable change in the quality of the sound.

• *Dead spots* occur because of sound focusing or poorly chosen reflecting panels. Inadequate sound levels in specific areas of the listening facility can result.

11.15   Concert Halls and Opera Houses

Three basic shapes exist in the design of large music auditoriums, namely, (1) rectangular, (2) fan-shaped, and (3) horseshoe, all of which are illustrated in floor plans of Figure 11.8. A fourth category is the "modified arena," nearly elliptical in shape. The Royal Albert Hall (constructed in 1871) in London, the Concertgebouw (1887) in Amsterdam, the Sidney Opera House, and the Colorado Symphony Hall in Denver are examples of this type of facility.

The *rectangular hall* is quite traditional, and it has been built to accommodate both small and large audiences. But these halls will always generate cross reflections (flutter echoes) between parallel walls. Sound can also be reflected from the rear walls back to the stage, depending on balcony layout and the degree of sound

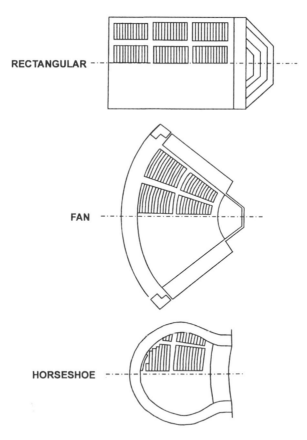

FIGURE 11.8. Three basic hall configurations: rectangular, fan-shaped, and horseshoe-shaped.

FIGURE 11.9. A view of the Boston Symphony Hall from the stage. Wallace Clement Sabine was the principal acoustic consultant for this facility. (Courtesy of the Boston Symphony Orchestra.)

absorption. These reflections can help in the buildup of sound and provides a reasonable degree of diffusion in halls of modest interior dimensions. A considerably larger hall can result in standing-wave resonances and excessive flutter echoes.

It is interesting to note that the first music hall to be designed from a scientific viewpoint, by none other than Wallace C. Sabine, was the Boston Symphony Hall (1900), views of which are given in Figures 11.9 and 11.10. The structure contains a high, textured ceiling and two balconies extending along three walls. The volume is 602,000 ft^3; seating capacity, 2631; and the reverberation time in the 500–1000 Hz range is 1.8 s (occupied). Another example of a great rectangular hall is the venerable Grösser Musikvereinssaal (1870) in Vienna which has a reverberation time of 2.05 (occupied) in a volume of 530,000 ft^3. Its superior acoustics can be attributed to its relatively small size, a high ceiling, irregular interior surfaces, and the plaster interior (Beranek, 1996).

A *fan-shaped hall* accommodates, through its spread, a larger audience within closer range from the sound source (stage). It features nonparallel walls that eliminate flutter echoes and standing waves, and most audience members can obtain a pleasing balance between direct and reflected sounds. A disadvantage in terms of early time delay gap is the distance from the side walls. Often it is necessary to add a series of inner reflectors or canopies hanging from ceilings over the proscenium area to maintain articulation and other acoustical characteristics. Many architects in the United States have resorted to the fan-shaped hall design in order to accommodate larger audiences while retaining an appreciable degree of both visual and

FIGURE 11.10. The stage area of the Boston Symphony Hall. (Photograph courtesy of the Boston Symphony Orchestra.)

aural coupling to the stage area. Relatively modern examples of this design are the Dorothy Chandler Pavilion in Los Angeles; the Orchestra Hall, Chicago; the Eastman Theater, Rochester, New York; and the Kleinhaus Music Hall in Buffalo, New York.

Over a number of centuries *horseshoe-shaped structures* have been used as the preferred design for opera houses and concert halls of modest seating capacity. This design provides for a greater sense of intimacy, and the textures of convex surfaces promote adequate diffusion of sound. Multiple balconies allow for excellent line of sight and short paths for direct sound. The La Scala Opera House (Figure 11.11) in Milan is probably the most notable example of the horseshoe design. Other celebrated examples of the horseshoe design are the Carnegie Hall in New York City and the Academy of Music in Philadelphia.

Nearly all concert halls have balconies, which were designed to accommodate additional seating capacity within a smaller auditorium volume, so that listeners can sustain an intimate relationship with the stage. The depths of the balconies generally do not exceed more than twice their vertical "window" (opening) to the stage. In fact a smaller ratio is desirable to minimize undue sound attenuation at the rear wall. A rule of thumb in contemporary acoustical design: the depth of the balcony should not exceed 1.4 times its outlook to the stage at the front of the balcony.

FIGURE 11.11. A view of the La Scala Opera House in Milan, Italy.

In all types of auditorium design, ceilings constitute design opportunities for transporting sound energy from the stage to distant listeners. In Figure 11.12, it is shown how a ceiling can convey sound to the listeners without imposing a great time difference between direct and ceiling-reflected sound. Floor profile is also important in establishing the proper ratio of direct to indirect sound. Splays on the side walls have proven effective in promoting diffusion and uniformity of loudness. Rear walls generally should be absorbent to minimize echoes being sent back to the stage.

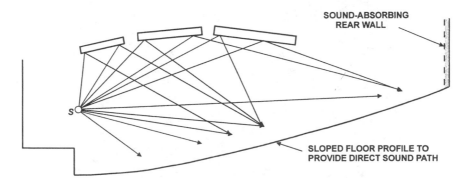

FIGURE 11.12. Transmission of sound to all areas of an auditorium through the ceiling and floor profiling.

Many concert halls have been built throughout the world, some with outstanding acoustics and others resulting in dismal sound. Among the features common to all of the aurally superior halls are a limited audience capacity (generally 2800 seats or less), extreme clarity of sound so that the audience can clearly distinguish the individual instruments of the orchestra without loss of fullness or blending of tones associated with reverberation. A good hall also allows the orchestra to hear itself.

Table 11.2 contains a summary of the characteristics of a number of prominent musical facilities all over the world. Figure 11.13 shows the stage view of the Orchestra Hall in Minneapolis (1974), which is patterned after the classical rectangular design. The hall contains a slanted floor with 1590 seats, and an additional 983 seats are located in the three stepped balcony tiers, making for a total audience capacity of 2573. A random pattern of plaster cubes cover the ceiling, providing effective diffusion of sound throughout the hall. This overlay of cubes also continues down the back wall, behind the stage, as shown in Figure 11.13.

FIGURE 11.13. The stage area of the Minnesota Orchestra Hall in Minneapolis. (Courtesy of the Minnesota Orchestral Association. Photograph by Tom W. McElin.)

TABLE 11.2. Reverberation Times of Leading Concert Halls and Auditoriums.

	Volume (ft^3)	Seating Capacity	Reverberation Time[a]	
			Occupied	Unoccupied
United States				
Baltimore, Lyric Theatre	744,000	2616	1.47	2.02
Boston Symphony Hall	662,000	2631	1.8	2.77
Buffalo, Kleinhans Music Hall	644,000	2839	1.32	1.65
Cambridge, Kresge Auditorium	354,000	1238	1.47	1.7
Chicago, Aric Crown Theatre	1,291,000	5081	1.7	2.45
Cleveland, Severance Hall	554,000	1890	1.7	1.9
Detroit, Ford Auditorium	676,000	2926	1.55	1.95
New York, Carnegie Hall	857,000	2760	1.7	2.15
Philadelphia Academy of Music	555,000	2984	1.4	1.55
Purdue University Hall of Music	1,320,000	6107	1.45	1.6
Rochester, New York, Eastman Theatre	900,000	3347	1.65	1.82
Austria				
Vienna, Grosser Musikvereinssaal	530,000	1680	2.05	3.6
Belgium				
Brussels, Palais des Beaux-Arts	442,000	2150	1.42	1.95
Canada				
Edmonton and Calgary, Alberta Jubilee Halls	759,000	2731	1.42	1.8
Vancouver, Queen Elizabeth Theatre	592,000	2800	1.5	1.9
Denmark				
Tivoli Koncertsal	450,000	1789	1.3	2.25
Finland				
Turku, Konserttisali	340,000	1002	1.6	1.95
Germany				
Berlin, Musikhochschule Konzertsaal	340,000	1340	1.65	1.95
Bonn, Beethovenhalle	555,340	1407	1.7	1.95
Great Britain				
Edinburgh, Usher Hall	565,000	2760	1.65	2.52
Liverpool Philharmonic Hall	479,000	1955	1.5	1.65
London, Royal Albert Hall	3,060,000	5080	2.5	3.7
London, Royal Festival Hall	755,000	3000	1.47	1.77
Israel				
Tel Aviv, Frederic R. Mann Auditorium	750,000	2715	1.55	1.97
Italy				
Milan, Teatro Alla Scala	397,300	2289	1.2	1.35
Netherlands				
Amsterdam, Concertgebouw	663,000	2206	2.0	2.4
Sweden				
Gothenburg, Konserthus	420,000	1371	1.7	2.0
Switzerland				
Zurich, Grosser Tonhallesaal	402,500	1546	1.6	3.85
Venezuela				
Caracas, Aula Magna	880,000	2660	1.35	1.8

[a] At 500–1000 Hz.

Wood paneling partially covers the walls, and both the flooring of the stage and audience areas are wood. This concert hall is notable for its clarity, dynamic range, and balance.

Another acoustic success among the contemporary musical facilities is the Kennedy Center for the Performing Arts which opened in Washington, D.C. in 1971. The Center consists of a single structure that contains a 2759-seat concert hall, a 2319-seat opera house, and a 1142-seat theater. The location was environmentally challenging, for the Center is situated on a site near the Potomac River, in close proximity to the Washington Ronald Reagan National Airport on the other side of the river. Both commercial and private aircraft fly as low as a few hundred feet directly over the roof, and occasionally helicopters pass by along the Potomac River at rooftop levels. In addition, vehicular traffic runs across the river and directly beneath the plaza of the Center.

To deal with these external noise sources, the Center was constructed as a box-within-a-box, so that each of the auditoriums are totally enclosed within an outer shell. The columns within each auditorium are constructed to isolate interior ceilings, walls, and floors from both airborne and mechanical vibrations. The double-wall construction generally consists of 6-in. solid high-density blocks separated by a 2-in. air gap. The huge windows in the Grand Foyer facing the river consist of $^1/_2$-in. and $^1/_4$-in. thick glass sheets separated by a 4-in. air gap. Resilient mounts are used to isolate interior noise sources (e.g., transformers, air-conditioning units, etc.). Ductworks are acoustically lined, flexible connectors are used, and special doors are installed at all auditorium entrances, together with "sound locks" between foyer and the auditoriums.

The rectangular concert hall (Figure 11.14) encompasses a volume of 682,000 ft^3 and accommodates an audience of 2759. Large contoured wall surfaces and a coffered ceiling abet the diffusion of sound at low and high frequencies. The eleven massive crystal chandeliers, each weighing 1.3 metric tons, donated by the Norwegian government, also contribute to the diffusion. The balconies are purposely shallow to prevent a reduction in sound below the balcony overhang. Unoccupied, the concert hall has a reverberation time of 2.2 s at 500 Hz and 2.0 s at 1 kHz; the corresponding values are 2.0 s and 1.8 s for the fully occupied hall.

Located in downtown Seattle, Washington, the newly opened (in September 1998) Benaroya Hall contains two spaces for musical performances: a 2500-seat main auditorium (Figure 11.15) and a more intimate 540-seat recital hall. The main auditorium, the S. Mark Taper Foundation Auditorium, is a classic rectangular configuration with the stage enclosed in a permanent acoustic "shell." LMN (Loschky, Marquardt & Nesholm) Architects and the acoustical consultant, Cyril M. Harris, combined the shoebox design with state-of-the-art materials to achieve maximum warmth and balance.

The location of Benaroya Hall in a busy sector of Seattle posed special challenges to the designers. They had to contend with a railroad tunnel running diagonally beneath the auditorium and a nearby underground bus tunnel. A slab of concrete more than 6 ft (2 m) thick, 80 ft wide, and 430 ft long was poured under the hall to swallow the sound from the tunnels. In order to combat other exterior noises,

FIGURE 11.14. The interior of the John F. Kennedy Opera House in Washington, DC. (Courtesy of the John F. Kennedy Center.)

FIGURE 11.15. The interior of the Benaroya Hall in Seattle, WA. (Courtesy of the Seattle Symphony. Artist: Bill Hook)

the designers essentially encased a building within a building. The auditorium, weighing 27 million pounds, rests on 310 rubber pads, which absorb vibration from the tunnels. The pads are 15 in. square and are composed of four layers of natural rubber sandwiched with $1/8$-in. steel plates.

All electrical, plumbing and other noise-generating equipment are located outside the auditorium box. Any penetration of the box is made with flexible connections. Water is known to transmit sound very well, so the fire sprinkler system is left dry and it will flood with water only when a fire is detected. The ventilation system is connected to the outside by a sound trap, which channels air through narrow openings between perforated aluminum boxes of sound insulation. Very large vents collect air below the floor and move it slowly behind the auditorium to another sound trap and then to fans. The basic idea of the ventilation system is to move a high volume $(85,000 \text{ ft}^3/\text{min})$ of air at low speed, eliminating noise created by fast-moving air in conventional systems.

Instead of frame construction, the walls are built of precast concrete panels. The heavy mass helps to cut down building vibration and provides a stiff, hard surface to reflect concert sound. Side walls, back walls, and ceiling are covered with paneling shaped like truncated pyramids to reflect sound at various angles (diffusion). Randomly spaced wood blocking behind the angled paneling creates framed sound boxes that reflect both high and low frequency sounds so that no tone is eliminated from the music. Side walls are covered with particle boards veneered with a dense, fine-grained hardwood from a single Makore tree. The ceiling is suspended from the roof by hundreds of metal strips. The ceiling is coated with $1\text{-}1/2$ inches of plaster in irregularly shaped panels to diffuse sound. The plaster is sufficiently dense to prevent the ceiling from vibrating. House lighting is imbedded in the ceiling to minimize sound leakage. Access to the light bulbs is achieved above the ceiling through heavy, removable plaster caps.

11.16   Band Shells and Outdoor Auditoriums

Over the past several decades there has been an increasing trend toward outdoor concerts, either at bandshells or in semi-open structures. These types of structures are more economical to construct than full-fledged indoor concert halls, and they also meet the criteria of providing an informal setting for audiences seeking entertainment in a usually rural environment, away from the metropolitan areas.

It is generally not possible for a large orchestra to play effectively in open air. The use of a bandshell becomes necessary, as this permits the members of a musical group to hear one another and directs the music to the audience area. The bandshell site should also be carefully selected. Ideally, the region should be isolated from the noise of passing traffic and overhead aircraft. The topology of the land also ranks important in providing the proper acoustics. If the land can be contoured properly, there can be appreciably less attenuation of the sound than would be the case if the bandshell were located on flat ground.

FIGURE 11.16. The Hollywood Bowl in Los Angeles, CA. (Courtesy of the Los Angeles Symphony Orchestra.)

The Hollywood Bowl (Figure 11.16) is an example of an orchestra shell strategically located in a natural hollow. The only reflected sound is that reflected from the shell, but the stage distances are sufficient short so that the sound is heard without any discernable time delay gap. However, shells can never equal the dynamic range of sound power and sonority that are achieved in an enclosed reverberant concert hall. The use of high quality amplification systems is often necessary at many outdoor concerts.

Another type of "outdoor" structure is the "music shed," a semi-enclosed structure designed specifically for musical performances. The Tanglewood Music Shed in the Berkshires region of Massachusetts (Figure 11.17) is the summer home of the Boston Symphony Orchestra, and the quality of its acoustics exceeds that of any bandshell for an audience of 6000 persons. An additional 6000 people on the lawn adjacent to the shed can listen to the music through the open segments of the pavilion. The canopy of the interior projects and diffuses sound through a volume of 1,500,000 ft^3. The ceiling is constructed of 2-in. thick wooden planks; the side and rear walls are $3/4$-in. fiberboard; and the floor is simply packed earth. When occupied, the shed embodies a reverberation time of 2 s in the frequency range of 500–1000 Hz, which is quite excellent considering the rustic nature of its construction. The Tanglewood shed is the precursor to similar structures at Wolf Trap in Vienna, Virginia, the Performing Arts Center at Saratoga Springs, New York (the summer home of the Philadelphia Symphony Orchestra), and the Blossom Music Center near Cleveland, Ohio.

FIGURE 11.17. A view looking toward the interior of the Tanglewood Music Shed in Lenox, MA. The ceiling canopy reflects sound, adding dynamic range and brilliance. (Photograph by William Mercer. Courtesy of the Boston Symphony Orchestra.)

11.17   Subjective Preferences in Sound Fields of Listening Spaces

Beyond the Sabine realm of architectural acoustics which is based on the relatively simple but effective formula $T_{60} = (0.161V)/\sum A_i\alpha_i$., other considerations come into play in determining optimal configurations for listening spaces (Ando, 1998). This involves combining the elements of psychoacoustics, modeling of the auditory-brain system, and mapping of subjective preferences. The physical properties of source signals and sound fields in a room are considered, in particular the autocorrelation function (ACF) which contains the same information as power density spectrum but it is adjusted to account for hearing sensitivity. Effective duration of the normalized ACF is defined by the delay τ_e at which the envelope of the normalized ACF becomes one-tenth of its maximal value. The response of the ear includes the effects of time delay due not only to the room's acoustical characteristics, but also the spatially incurred differences in the signals reaching the right ear and the left ear. The difference in the sounds arriving at the ear is measured by the "interaural cross-correlation function," or IACF, which is defined by

$$\text{IACF}(\tau) = \frac{\int_{t_1}^{t_2} p_L(t)p_R(t+\tau)\,dt}{\sqrt{\int_{t_1}^{t_2} p_L^2\,dt \int_{t_1}^{t_2} p_R^2\,dt}} \tag{11.25}$$

where L and R denote entry to the left and right ears, respectively. The maximum possible value of IACF is unity. The time $t = 0$ is the time of the arrival of the *direct* sound from the impulse radiated by a source. Integration from 0 to t_2 msec includes the energy of the direct sound and any early reflections and reverberant sounds occurring during the t_2 interval. Because there is a time lapse of about 1 msec for a sound wave to impinge the other side of the ear after impinging one side, it is customary to vary τ over the range from -1 msec to $+1$ msec. In order to obtain a single number that measures the maximum similarity of all waves arriving at the two ears with the time integration limits and the range of τ, it is customary to choose the maximum value of IACF, which is then called the interaural cross-correlation coefficient (IACC).

Different integration periods are used for IACC. The standard ones include $IACC_A$ ($t_1 = 0$ to $t_2 = 1000$ msec), and $IACC_{E(arly)}$ (0–80 msec), $IACC_{L(ate)}$ (80 to 1000 msec). The early IACC is a measure of the apparent source width ASW and the late IACC is a measure of the listener envelopement LEV. IACC is generally measured by recording on a digital tape recorder the outputs of two tiny microphones located at the entrances to the ear canals of a person or a dummy head, and quantifying the two ear differences with a computer program that performs the operation of equation (11.25). $IACC_A$ is determined with a frequency band width of 100 Hz to 8 kHz and for a time period of 0 to about 1 sec.

Subjective attributes for a sound field in a room have been developed experimentally with actual listeners. The simplest sound field is considered first, a situation that consists of the direct sound and a single reflection acting in lieu of a set of reflections. The data obtained are based on tests in anechoic chambers (which allowed for simulation of different concerts halls) with normal hearing subjects listening to different musical motifs. From these subjective tests, the optimum design objectives are established, namely, the listening level, preferred delay time, preferred subsequent reverberation time (after the early reflections), and dissimilarity of signals reaching both ears (involving IACC).

These factors are each assigned scalar values and then combined to yield a subjective preference that can vary from seat to seat in a concert hall. Some rather interesting results of investigations include the fact that the right hemisphere of the brain is dominant for "the continuous speech," while the left hemisphere is dominant when variation occurs in the delay time of acoustic reflection. The left hemisphere is usually associated with speech and time-sequential identifications, while the right hemisphere is allied with nonverbal and spatial identifications. A proposed model for the auditory-brain system has been developed (Ando, 1998) that incorporates a subjective analyzer for spatial and temporal criteria and entails the participation of the left and the right hemispheres of the brain. The power density spectra in the neural processes in the right and left auditory pathways yield sufficient information to establish autocorrelation functions.

It is obvious that different individuals are likely to have different subjective preferences with respect to the same musical program, so seating requirements can differ, with respect to the preferred listening level and to the initial time delay, and even lighting. For example, evaluations were conducted for a performance of

Handel's *Water Music* with 106 listeners providing the input on their preferences with respect to listening level, reverberation time, and IACC. The information thus obtained can provide insight into how the acoustic design of a concert hall and a multipurpose auditorium can be accomplished. Procedures for designing sound fields include consideration of temporal factors, spatial factors, the effect of sound field on musicians, the conductor, stage performers, listener and the archetypal problem of fusing acoustical design with architecture. Multi-purpose auditoriums present bigger challenges, some of which have been met very well and many which have not. In the design procedure, a number of factors other than acoustical include measurable quantities such as temperature, lighting levels, and so on, and other less tangible determinants that can be aesthetically evocative.

References

American Society for Testing and Materials. 1981. "Test Method for Sound Absorption and Sound Absorption Coefficients by the Reverberation Room Method." ASTM C423-81a.

Ando, Yoichi. 1985. *Concert Hall Acoustics*. Berlin: Springer-Verlag.

Ando, Yoichi. 1998. *Architectural Acoustics: Blending Sound Sources, Sound Fields, and Listeners*. New York: Springer-Verlag, 1998. (A followup to Ando's previous text. The author, arguably Japan's most prominent architectural acoustician, introduces a theory of subjective preferences, based on a model of the auditory cognitive functioning of the brain.)

Bies, D. A., and C. H. Hansen. 1996. *Engineering Noise Control: Theory and Practice*, 2nd ed. London: E. and F. N. Spon.

Beranek, Leo L. 1962. *Music, Acoustics, and Architecture*. New York: John Wiley & Sons.

Beranek, Leo L. July 1992. Concert hall acoustics—1992. *Journal of the Acoustical Society of America* 92, 1: 1–39.

Beranek, Leo. *Concert and Opera Halls: How They Sound*. 1996. Woodbury, NY: Acoustical Society of America, 1996. (A tremendous compendium of the acoustical characteristics of musical auditoria and how they were achieved. This volume is a must-have text for architectural acousticians and aficionados of architectural acoustics.)

Cremer, L., and H. A Muller. 1982. *Principles and Applications of Room Acoustics*, Vol. 1 and 2. English translation with additions by T. J. Schultz. New York: Applied Science Publishers.

Crocker, Malcolm J. (ed.). 1997. *Encyclopedia of Acoustics*, vol. 3. New York: John Wiley & Sons; Chapters 90–93, pp. 1095–1128.

Eyring, C. F. 1930. Reverberation time in "dead rooms." *Journal of the Acoustical Society of America* 1: 217–241.

Harris, Cyril M. 1966. Absorption of sound in air vs. humidity and temperature. *Journal of the Acoustical Society of America* 40: 148–49.

Harris, Cyril M. January 1967. Absorption of sound in air vs. humidity and temperature, NASA contractor report NASA CR-647. Springfield, VA: Clearinghouse for Federal Scientific and Technical Information.

Harris, Cyril M. (ed.) 1991. *Handbook of Acoustical Measurements and Noise Control*, 3rd ed. New York: McGraw-Hill.

International Standardization Organization. 1963. Measurement of Sound Absorption in a Reverberation Room. ISO R354-1963.

Jordan, V. 1974. Presented at the 47th Audio Engineering Society convention, Copenhagen.

Kinzey, B. Y., Jr., and S. Howard, 1963. *Environmental Technologies in Architecture*. Englewood Cliffs, NJ: Prentice-Hall.

Knudsen, Vern O., and Cyril M. Harris. 1978 (reissue). *Acoustical Designing in Architecture*. Woodbury, NY: Acoustical Society of America. (A classic text by two of the giants in field of architectural acoustics.)

Kutruff, H. 1979. *Room Acoustics*, 2nd ed. London: Applied Science.

Kuttruff, H. 1991. *Room Acoustics*, 3rd ed. London: Elsevier Applied Science.

Millington, G. 1932. A modified formula for reverberation. *Journal of the Acoustical Society of America* 4: 69–82.

Morse, Philip M., and Richard H. Bolt. 1944. Sound waves in rooms. *Review of Modern Physics* 16: 65–150.

Sabine, Wallace Clement. *Collected Papers on Acoustics*. Reissued 1993. Los Altos, CA: Peninsula Publishing. (Included are the original papers that laid the foundation of modern architectural acoustics.)

Sette, W. J. 1933. A new reverberation time formula. *Journal of the Acoustical Society of America* 4: 193–210.

Seattle Times. June 21, 1998. "Benaroy Hall: orchestrating the sound of music"; p. A9.

Siebein, Gary W. June 1994. "Acoustics in Buildings: A Tutorial on Architectural Acoustics," presented at the 127th Annual Meeting of the Acoustical Society of America, Cambridge, MA.

Problems for Chapter 11

1. The amplitude of the reflected wave is one-half that of the incident wave for a certain angle and frequency. What is the reflection coefficient? What is the corresponding sound absorption coefficient?

2. Find the average sound coefficient at 125 Hz and 2 kHz for a wall constructed of different materials as follows:

Area, m^2	Material
85	Painted brick
45	Gypsum board on studs
35	Plywood paneling
4	Glass window

3. Find the time constant of a room that is 8 m × 10 m × 2.5 m and an average sound absorption coefficient of 0.34.

4. A room has dimensions 3.5 m high × 30 m length × 10.0 m width. Two of the longer walls consists of plywood paneling; the rear wall is painted brick and the front has gypsum boards mounted on studs. The ceiling is of acoustic paneling and the floor is carpeted. For 250 Hz and 125 people in the audience:

   (a) Determine the reverberation time from the Sabine equation.

   (b) Compute the room constant R.

   (c) Comment on the suitability of the room for use as an auditorium.

5. An auditorium has dimensions 6.0 m high × 22 m length × 15.5 m width. The floor is carpeted, and one of the longer walls has gypsum boards mounted on studs along the entire length, while the other three walls and the ceiling are constructed of plaster. For 500 Hz:
   (a) determine the reverberation time from the Sabine equation.
   (b) compute the room constant R.
6. In Problem 5, there is a set of two swinging doors 2.5 m high and 2.5 m total width. However there is open cracks along the bottom and between the doors 1 cm wide. In addition there is an open window on the gypsum boarded wall that is 0.8 m × 1 m. Find the effect of these openings on the reverberation time of the auditorium.
7. A room has dimensions 4.0 m high × 18 m length × 9.5 m width. The floor is carpeted, and one of the longer walls has gypsum boards mounted on studs along the entire length, while the other three walls are constructed of painted brick. The ceiling is plaster. For 500 Hz:
   (a) determine the reverberation time from the Sabine equation.
   (b) compare the reverberation time obtained from the Eyring formula with that obtained in part (a) above.
   (c) compute the room constant R.
8. An isotropic source emitting sound power level of 106 dBA re 1 pW is operating in the room of Problem 4. What sound pressure level (on the A-weighted scale) will be registered by a meter 5.2 m from the source? If 125 people are *not* seated in the room, how will it affect the reading? Assume that no one blocks the path of direct propagation between the meter and the signal source.
9. Why does an anechoic chamber provides what we call a free field?

12
Walls, Enclosures and Barriers

12.1 Introduction

Acoustics constitutes an important factor in building design and in layouts of residences, plants, and offices. A building not only protects against inclement weather; it must also provide adequate insulation against outside noises from transportation and other sources. Interior walls and partitions need to be designed to prevent the intrusion of sound from one room into another. Exposure of workers to excessive occupational noises can be decreased by construction of appropriate barriers or enclosures around noisy machinery. A shell constructed of the densest materials may be the most effective barrier against noise transmission, but such an enclosure can lose a major portion of its effectiveness if there are weak links that tend to promote sound transmission rather than hinder it. For example, a large proportion of sound energy enters a building through its windows, even closed ones; and many cracks and crevices inevitable in real structures permit sound to enter the structure's interior.

12.2 Transmission Loss and Transmission Coefficients

Sound absorption materials tend to be light and porous. Sound isolation materials, however, generally are massive and airtight, thereby forming effective sound insulation structures between the noise source and the receiver. When airborne sound impinges on a wall, some of the sound energy is reflected, some energy is absorbed within the wall structure, and some energy is transmitted through the wall. Sound pressure against one side of the wall may cause the wall to vibrate and transmit sound to the other side. The amount of incident energy transmitted to the wall depends on the impedance of the wall relative to the air. The amount of the sound transmitted through the wall that is finally transmitted to the air on the receiver side also depends on the impedance of the wall relative to air. A double-wall construction, which incorporates air space between panels, repeats the process between the two panels, which results in even more insulation against sound.

Most construction materials transmit only a small amount of acoustic energy, with the major portion of the energy undergoing reflection or conversion into heat due to impedance mismatch, absorption within the material, and damping. Heavy walls, usually of masonry, allow very little of the sound to pass through, owing to their high mass-per-unit-area. A wall of gypsum boards mounted on both sides of a stud frame provides effective insulation against sound penetration due to energy losses resulting during the passage of sound from air to solid to air between the two panels to solid and thereon to air on the other side of the wall.

A measure of sound insulation provided by a wall or some other structural barrier is given by the *transmission loss* TL given in units of dB by

$$TL = 10 \log \left(\frac{W_1}{W_2} \right) \tag{12.1}$$

where W_1 represents the sound power incident on the wall, and W_2 is the sound power transmitted through the wall. Because the frequency loss varies with the frequency of the sound, it is usually listed for each octave band or one-third octave band. The fraction of sound power transmitted through a wall or barrier constitutes the *sound transmission coefficient*, τ, written as

$$\tau = \frac{W_2}{W_1} \tag{12.2}$$

Combining equations (12.1) and (12.2) we obtain

$$TL = 10 \log \left(\frac{1}{\tau} \right) \tag{12.3}$$

or

$$\tau = 10^{-TL/10} \tag{12.4}$$

In Figure 12.1 we consider the case of a plane wave approaching a panel in the yz-plane (i.e., the plane is normal to the x-axis) at an angle of incidence θ, Subscripts I, R, and T are used to identify, respectively, the incident wave, the reflected wave, and the transmitted wave. Arrows indicate the directions of the wave propagation. The wave equation for a plane wave

$$\frac{\partial^2 p}{\partial t^2} = c^2 \nabla^2 p$$

bears the solutions

$$\left. \begin{aligned} p_I &= \mathbf{P}_I e^{ik(ct - x \cos \theta - y \sin \theta)} \\ p_R &= \mathbf{P}_R e^{ik(ct + x \cos \theta - y \sin \theta)} \\ p_T &= \mathbf{P}_T e^{ik(ct - x \cos \theta - y \sin \theta)} \end{aligned} \right\} \tag{12.5}$$

Here the real part of p denotes the sound pressure, $\mathbf{P}$ is the (complex) pressure amplitude, and k is the wave number.

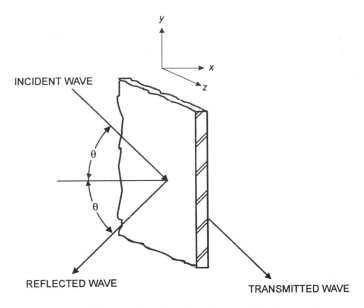

FIGURE 12.1. Transmission loss through a panel in the y-z plane.

12.3 Mass Control Case

Consider the panel as being quite thin, that is, its thickness is considerably smaller than one wavelength of the sound in air; and let us also neglect the material stiffness and damping in the panel. We can stipulate the conditions of (a) the continuity of velocities normal to the panel and (b) a force balance that occurs inclusive of the inertial force. From application of the first condition we express the particle velocity u at the panel as

$$u_{\text{panel}} = u_T \cos \theta = u_I \cos \theta - u_R \cos \theta \qquad (12.6)$$

When the sound pressure and the particle velocity are in-phase they are related by

$$Z = \frac{p}{u} = \rho c \qquad (12.7)$$

where Z denotes the characteristic resistance, which represents a special case of specific acoustic resistance.

Setting $x = 0$ at the panel, we apply equation (12.7) and equation set (12.5) to equation (12.6) to obtain

$$\mathbf{P}_T = \mathbf{P}_I - \mathbf{P}_R \qquad (12.8)$$

The second condition applied over a unit surface area of the panel gives

$$p_I + p_R - p_T - m a_{\text{panel}} = 0 \qquad (12.9)$$

where m represents the mass density per unit area of the panel and a_{panel} the acceleration normal to the surface. Through the use of equation (12.7) we can correlate the panel velocity with the transmitted particle velocity

$$u_{panel} = P_T \frac{\cos\theta}{\rho c}$$

Using the last of the equation set (12.5) in the preceding expression and differentiating with respect to time, we now obtain

$$a_{panel} = ikc\mathbf{P}_T e^{\frac{ik(ct - x\cos\theta - y\sin\theta)\cos\theta}{\rho c}}$$

$$= \frac{i\omega\mathbf{P}_T \cos\theta}{\rho c}$$

Inserting the above expression into equation (12.9) and applying equations (12.5) at the panel, where $x = 0$, the amplitudes of the pressures are related to each other by

$$\mathbf{P}_I + \mathbf{P}_R - \mathbf{P}_T = \frac{im\omega\mathbf{P}_T}{\rho c}\cos\theta \qquad (12.10)$$

The pressure term $\mathbf{P}_R$ in equation (12.10) may now be eliminated through the use of equation (12.8). The ratio of transmitted pressure amplitude to incident pressure amplitude may now be obtained as

$$\frac{P_T}{P_I} = \frac{1}{1 + \frac{im\omega\cos\theta}{2\rho c}}$$

We now express the sound transmission coefficient as:

$$\tau = \frac{p_{rms(T)}^2}{p_{rms(I)}^2} = \frac{|P_T|^2}{|P_I|^2} = \frac{1}{1 + \left(\frac{im\omega\cos\theta}{2\rho c}\right)^2}$$

From the definition of equation (12.3), we obtain the mass law transmission loss equation

$$TL = 10\log\left(\frac{1}{r}\right) = 10\log\left[1 + \left(\frac{m\omega\cos\theta}{2\rho c}\right)^2\right] \qquad (12.11)$$

Setting the angle of incidence equal to zero, equation (12.11) reduces to:

$$TL_0 = 10\log\left[1 + \left(\frac{m\omega}{2\rho c}\right)^2\right] \qquad (12.12)$$

which constitutes the statement for the normal incidence mass law for approximating transmission loss of panels with sound at 0° angle of incidence.

12.4   Field Incidence Mass Law

In the situation of transmission loss between two adjoining rooms, the sound source in one room may produce a reverberant field. The incident sound emanating from the source may strike the wall at all angles between $0°$ and $90°$. For field incidence it is customary to assume that the angle of incidence lies between $0°$ and $72°$, which results in a field incidence transmission loss of approximately

$$TL = TL_0 - 5\,dB$$

We can modify the mass law equation by converting the angular frequency of sound (rad/s) into cyclic frequency (Hz), that is, $\omega/2\pi = f$, by setting the acoustic impedance for air $\rho c = 407$ rayls, and assuming $m\omega/(2\rho c) \gg 1$. This gives the field incidence mass law equation

$$TL = 20\log(fm) - 47\,dB \tag{12.13}$$

where m denotes the mass per unit area (kg/m^2).

Equation (12.13) indicates for the mass-controlled frequency region that transmission loss of a panel increases by 6 dB per octave, Doubling either the panel thickness or the density of the panel will also engender an additional 6 dB loss at a given frequency. While equation (12.13) is useful for the prediction of the material acoustical behavior, laboratory or field testing should be conducted to measure the transmission loss of actual structures under real environmental conditions.

Example Problem 1

A wall is considered to have its sound transmission mass-controlled. Plot the transmission loss as a function of the product of frequency and mass per unit area.

Solution

Using equation (12.13), which expresses the field incidence mass law for typical conditions we can write

$$TL - 20\log(fm) - 47\,dB$$

For the normal incidence mass law

$$TL = 20\log(fm) - 42\,dB$$

Figure 12.2 shows the semilog plot of both relationships as two straight lines. It should be noted here that these values tend to be considerably higher than those of the actual transmission losses.

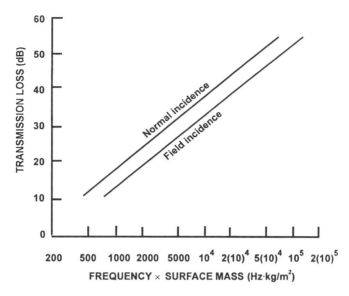

FIGURE 12.2. Semi-log plot of transmission loss versus product of frequency and surface mass.

12.5   Effect of Frequencies on Sound Transmission Through Panels

Panel bending stiffness constitutes the governing factor in low-frequency sound transmission. The panel resonances play the principal role in determining the nature of transmission of higher-frequency sounds. The panel may be considered mass-controlled in the frequency range from twice the lowest resonant frequency to below the critical frequency (discussed below).

Ver and Holmer (1971) develop the sound transmission coefficient equation for panels manifesting significant bending stiffness and damping, which is given as follows:

$$\tau = \frac{1}{\left[1 + \eta\left(\frac{m\omega\cos\theta}{2\rho c}\right)\left(\frac{B\omega^2\sin^4\theta}{mc^4}\right)\right]^2 + \left[\left(\frac{m\omega\cos\theta}{2\rho c}\right)\left(1 - \frac{B\omega^2\sin^4\theta}{mc^4}\right)\right]^2} \quad (12.14)$$

The panel thickness is assumed to be small compared with the wavelength of the wavelength of the incident sound, B = panel bending stiffness (N · m), η = composite loss factor (dimensionless), and m = panel surface density (kg/m^2).

12.6   Coincidence Effect and Critical Frequency

In propagating though panels and other structural elements, sound can occur as longitudinal, transverse, and bending waves. Bending waves give rise to the coincidence effect. In Figure 12.3 a panel is shown with an airborne sound wave of

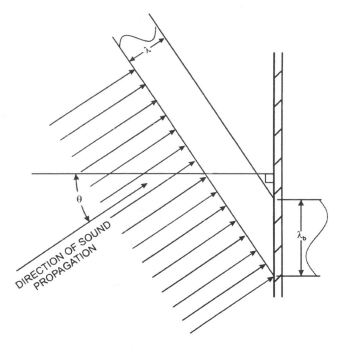

FIGURE 12.3. The coincidence effect effected by bending waves.

wavelength λ incident at angle θ. A bending wave of wavelength λ_b is excited in the panel. The propagation velocity of bending waves depends on the frequency, with higher velocities occurring at higher frequencies. The coincidence effect occurs when

$$\theta = \theta^* = \arcsin\left(\frac{\lambda}{\lambda_b}\right) \tag{12.15}$$

or

$$\sin\theta^* = \frac{\lambda}{\lambda_b}$$

With the asterisk indicating the occurrence of coincidence, θ^* denotes the coincidence angle. Under such circumstances, the sound pressure on the surface of the panel falls into phase with bending displacement. This results in a highly efficient transfer of acoustic energy from the airborne sound waves in the source side of the panel to bending waves in the panel, and thence to airborne sound waves in the receiving room on the other side of the panel. This is a highly undesirable situation, if the panel is meant to prevent the transmission of sound from one room to another as a noise control measure.

Figure 12.4 shows an idealized plot of transmission loss for a panel as a function of frequency, with stiffness-controlled, resonance-controlled, mass-controlled, and

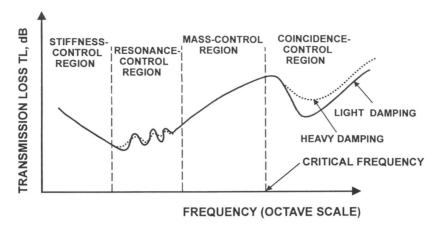

FIGURE 12.4. Idealized plot of transmission loss versus frequency.

coincidence-controlled regions. The transmission loss curve drops considerably in the region beyond the critical frequency owing to the coincidence effect.

From equation (12.15) it can be deduced that the coincidence effect cannot occur if the wavelength λ of the airborne sound is greater than the bending mode wavelength λ_b in the panel. The minimum coincidence frequency, or the *critical frequency*, exists at the critical airborne sound wavelength

$$\lambda = \lambda^* = \lambda_b$$

where the critical frequency f^* is given by

$$f^* = \frac{c}{\lambda^*}$$

which corresponds to the grazing incidence $\theta = 90°$. Also, there is a critical angle θ^*, at which coincidence will occur, for any frequency above the critical frequency.

Figure 12.5 contains a plot of critical frequencies against thickness for a number of common construction materials. It must be realized that the critical frequency often falls in the range of speech frequencies, rendering some partitions nearly useless for providing privacy and preventing speech interference.

From equation (12.14), it is seen that the coincidence effect depends on characteristics of the plate or panel and of the airborne sound wave. At coincidence

$$\frac{B\omega^2 \sin^4 \theta^*}{mc^4} = 1$$

and inserting the above condition into equation (12.14) yields the sound transmission coefficient for the coincidence condition

$$\tau = \tau^* = \frac{1}{\left(1 + \frac{m\omega\eta \cos\theta^*}{2\rho c}\right)^2}$$

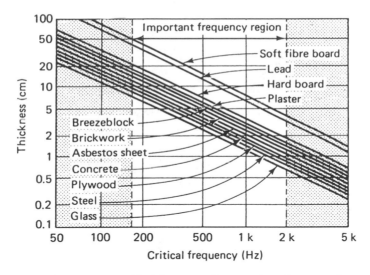

FIGURE 12.5. Thickness versus critical frequency for a number of construction materials. (*Source:* Brüel & Kjær. 1980. *Measurements in Building Acoustics.* Nærum, Denmark: Brüel & Kjær; p. 19.

The corresponding transmission loss for the coincidence condition is

$$\text{TL} = \text{TL}^* = 10\log\left(\frac{1}{\tau^*}\right) = 20\log\left(1 + \frac{fm\eta\cos\theta^*}{\rho c}\right)$$

From the above equations, it would appear that $\tau^* = 1$ and $\text{TL}^* = 0$ for undamped panels (in which the loss factor $\eta = 0$). The above transmission loss equation is premised on the theoretical behavior of an infinite plate; and the finite boundaries of actual structures such as windows and walls and the presence of damping in real construction materials will produce a different response to sound waves.

12.7   The Double-Panel Partition

A single-panel wall can exhibit resonant frequencies that fall below the range of speech frequencies. Most walls are constructed of two panels with an airspace between them, and they may yield low-frequency resonances in the speech range. A typical interior partition consists of two gypsum board panels (ranging in thickness from $\frac{1}{2}$ to $\frac{3}{4}$ in.) separated 3-$\frac{1}{2}$ in. by 2 × 4[1] wooden or metals studs. In Figure 12.6 a double-panel configuration is shown; the two panels of mass per unit area m_1 and m_2, respectively, are separated by an airspace h. Because the air entrapped between the two panels behaves as a spring, a spring-mass analogy, shown in Figure 12.7,

---

[1] Lumber sizes are given by figures that are almost always nominal rather than representative of actual values. A 2 × 4 stud generally measures 1-$\frac{3}{4}$ × 3-$\frac{1}{2}$ inches.

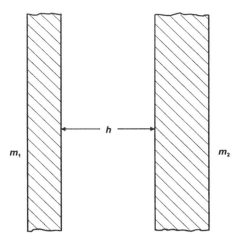

FIGURE 12.6. A double-panel partition.

can be applied, with k representing the spring constant between two masses. The wall response mode can be depicted by two masses vibrating at the same frequency. A node (i.e., a "motionless" point on the spring exists, thus effectively resolving the spring into two springs with spring constants (or spring rates) k_1 and k_2, The natural frequency of the dual-mass system is given by

$$f_n = \frac{\sqrt{\frac{k_1}{m_1}}}{2\pi} = \frac{\sqrt{\frac{k_2}{m_2}}}{2\pi} \qquad (12.16)$$

The individual spring constants are related to the spring constants of the composite spring by

$$\frac{1}{k} = \frac{1}{k_1} + \frac{1}{k_2} \qquad (12.17)$$

Eliminating k_2 between equations (12.16) and (12.17) results in

$$k_1 = k\frac{m_1 + m_2}{m_2}$$

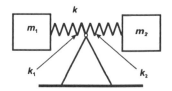

FIGURE 12.7. Spring-mass analogy of the double-panel partition of Figure 12.6.

Inserting the last expression into the first portion of equation (12.16) yields the natural frequency f_n of the system:

$$f_n = \frac{\sqrt{k\frac{m_1+m_2}{m_1 m_2}}}{2\pi} \tag{12.18}$$

In order to establish the effective spring constant of the air between the panels, it will be assumed that an isentropic process constitutes the action of sound waves, because the pressure changes occur too rapidly for an isothermal process to occur. From thermodynamics, for an isentropic process the pressure p of air (considered an ideal gas) varies in the following manner

$$pv^\gamma = \text{const} \tag{12.19}$$

where v represents the specific volume (equal to the reciprocal of density) and γ, the ratio of specific heats. Differentiating equation (12.19) gives

$$\frac{dp}{dv} = -\gamma\frac{p}{v} \tag{12.20}$$

Sound waves cause the ambient pressure and specific volume of air to vary only slightly from the quiescent values of p_0 and v_0, and thus the instantaneous values p and v differ very little from p_0 and v_0 in equation (12.20). Because the mass of air entrapped between the panels remains constant, the ratio of panel displacement (arising from the sound pressure pushing on the panels) dh to the original (quiescent) spacing h_0 should equal the ratio of the change of specific volume to the quiescent value of the specific volume:

$$\frac{dh}{h_0} = \frac{dv}{v_0} \tag{12.21}$$

But the spring constant represents the force per unit displacement, or for a unit area of the wall

$$k = -\frac{dp}{dh} \tag{12.22}$$

Combining equations (12.21) and (12.22) and dropping the subscripts (because $h_0 \approx h$, $p_0 \approx p$), we now have

$$k = -\frac{\gamma p}{h}$$

which, upon inserting into equation (12.19), gives

$$f_n = \frac{\sqrt{\gamma p\frac{m_1+m_2}{hm_1 m_2}}}{2\pi}$$

Setting $\gamma = 1.4$ and $p = 101.325$ kPa (the standard atmosphere), the low-frequency resonance of the double-panel wall may now be found from

$$f_n = 60\sqrt{\frac{m_1 + m_2}{hm_1m_2}} \tag{12.23}$$

The surface masses m_1 and m_2 are given in kg/m² and the panel spacing h in meters.

Example Problem 2

Predict the transmission loss for a 8-in. wall of poured concrete for 800-Hz sound. The concrete has a density of approximately 150 lb/ft³ (2406 kg/m³).

Solution

Surface density m = 8 in. × 1 m/(39.37 in.) × 2406 kg/m³ = 488.9 kg/m²

Using field incidence law equation (12.13)

$$TL - 20 \ \log(fm) - 47 = 20 \ \log(800 \times 488.9) - 47 = 65 \ dB$$

Laboratory tests indicate a TL closer to 50 dB. Actual installations may yield even lower values because of flanking noise transmission paths.

Example Problem 3

Find the resonant frequency of a double-panel wall constructed of a 15-kg/m² and a 20-kg/m² gypsum board with 9-cm airspace.

Solution

Applying equation (12.23),

$$f_n = 60\sqrt{\frac{m_1 + m_2}{hm_1m_2}} = 60\sqrt{\frac{15 + 20}{.09(15)(20)}} = 68 \ Hz$$

12.8   Measuring Transmission Loss

Recasting equation (12.1) for transmission loss TL in terms of vector sound intensity we write

$$TL = 10 \log\left(\frac{\mathbf{I}_I}{\mathbf{I}_T}\right)$$

where $\mathbf{I}_I$ = incident vector sound intensity and $\mathbf{I}_T$ = transmitted vector sound intensity. For sound pressure level and particle velocity in phase, sound intensity is given by

$$I = \frac{p_{rms}^2}{\rho c}$$

with the result that transmission loss could be restated as

$$TL = 10 \log \left(\frac{p_{rms I}^2}{p_{rms T}^2} \right)$$

But the above equation is not usable for measurement purposes, because the distinction cannot be made between rms pressures attributable to incident sound, transmitted sound, and reflected sound.

In order to determine the transmission loss, it is required to set up a source chamber and a receiving chamber, with a panel of the specimen material installed in a window between the two chambers. Such a setup is shown in Figure 12.8. The chambers must be designed so as to minimize the sound transmission paths other than that through the evaluation specimen. Sound level is measured in the source chamber and the receiving chamber with a microphone system, a measuring

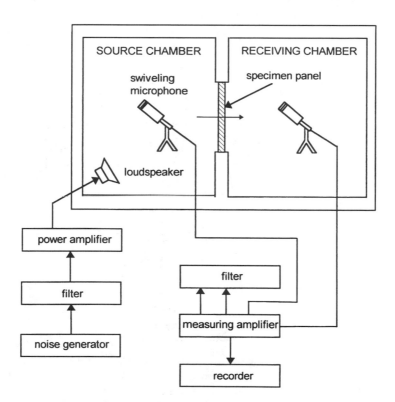

FIGURE 12.8. Setup for a transmission loss measurement.

amplifier, and filter. The measured sound is filtered to mitigate the effect of background noise on the measurement procedures, with the filter settings being identical for the generated sound and the measuring sound filters. The sound level may be averaged through rotating the microphone about its base on the mounting tripod. The transmission loss is found by applying the measured results in the following equation:

$$\text{TL} = L_S - L_R + 10 \log \left(\frac{A_m}{A_R} \right)$$

where

L_s = average sound level in the source chamber

L_R = average sound level in the receiving chamber

A_m = area of the material under investigation

A_R = equivalent absorption area of the receiving room

Another measurement setup is given in Figure 12.9. A pink noise generator, filters, audio power amplifier, and speaker constitute the sound source system. A computer-controlled analyzer incorporates a random noise generator and provisions for automatic remote control of the microphone boom for spatial and temporal averaging. Such an analyzer can measure in one-third octave bands the following parameters: source chamber level, receiving chamber background SPL, receiving chamber SPL, and receiving chamber reverberation time.

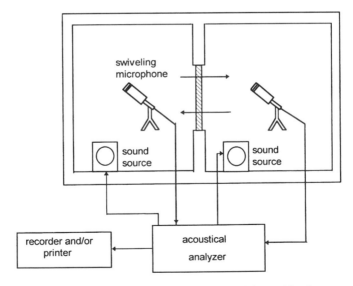

FIGURE 12.9. An alternative to the setup of Figure 12.8, which provides for automatic evaluation of transmission loss. The sound source can consist of a pink noise generator, filters, power amplifier, and speaker. The analyzer, which can generate random noise, can automatically position a microphone boom for spatial and temporal averaging.

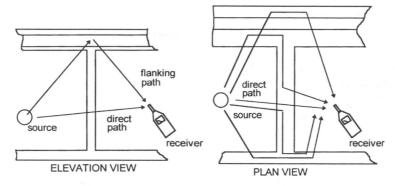

FIGURE 12.10. Examples of flanking.

12.9   Flanking

Flanking occurs when noise from a sound source is transmitted through paths other than direct transmission through a wall between the source chamber and the receiver on the opposite side of the wall. Among the causes of flanking are acoustical leaks through cracks around doors, windows and electrical outlets, passage of the sound through suspended ceilings with resulting reflection into adjacent rooms, HVAC (heating/ventilation/air-conditioning) ducts, floor and ceiling joists. The insulation effect of a partitioning wall is effectively decreased by flanking and acoustic leaks, examples of which are shown in Figure 12.10.

12.10   Combined Sound Transmission Coefficient

A wall may contain a number of elements such as windows, doors, openings, and cracks. The effective or combined sound transmission coefficient depends on the values of the sound transmission coefficients of the individual elements and their respective areas. Assuming the incident sound to be fairly uniform over the individual elements, we can compute the combined sound transmission coefficient from

$$\tau_{\text{combined}} = \frac{\sum_{i=1}^{n} A_i \tau_i}{\sum_{i=1}^{n} A_i} \tag{12.24}$$

where i designates the ith element of the subdivided wall area A and the corresponding value of τ. For cracks and open area, $\tau = 1$, which is to say the sound transmission through such crevices may be considered to be virtually unimpeded. The corresponding transmission loss of the composite wall then becomes

$$\text{TL}_{\text{combined}} = 10 \log \left(\frac{1}{\tau_{\text{combined}}} \right) = 10 \log \left(\frac{\sum_{i=1}^{n} A_i}{\sum_{i=1}^{n} A_i \, 10^{-\text{TL}_i/10}} \right) \tag{12.25}$$

Example Problem 4

A 6.0 m × 2.5 m wall is for the most part constructed of 8-cm thick dense poured concrete with a transmission loss of 52 dB at 500 Hz. An opening in the wall is provided for a 2.2 m × 1.0 m wood door with a transmission loss of 25 dB at 500 Hz. There is a crack across the width of the door that is 1.0 cm high. Estimate the effective transmission loss of the composite wall at 500 Hz.

Solution

The total area A of the wall is $6.0 \times 2.5 = 15.0$ m^2, but the concrete portion of the wall is $15.0 - 2.2 \times 1.0 - .01 \times 2.2 = 12.778$ m^2. Applying equation (12.25)

$$TL_{combined} = 10 \log \frac{15.0}{12.778 \cdot 10^{-55/10} + 2.2 \cdot 10^{-25/10} + .022 \cdot 10^{-0/10}}$$
$$= 27.1 \text{ dB}$$

Suppose that a door with TL = 45 dB was installed instead. What would be the new value of the composite TL for the wall?

Solution

$$TL_{combined} = 10 \log \frac{15.0}{12.778 \cdot 10^{-55/10} + 2.2 \cdot 10^{-45/10} + .022 \cdot 10^{-0/10}}$$
$$= 28.3 \text{ dB}$$

This example shows that negligible improvement will occur because the crack constitutes the principal means of negating the combined effectiveness of the acoustic insulation of the concrete wall and the door. The use of an acoustical door that effectively seals the area will result in an appreciably improved value of the transmission loss.

12.11   Noise Insulation Ratings

Two principal methods of rating sound insulation are discussed in this section, namely, *Sound Transmission Class (STC)* and *Shell Isolation Rating (SIR)*. The former designation constitutes a single-number description of noise insulation effectiveness of a structural element and it is widely used to describe the characteristics of interior partitions or walls with respect to noise falling in the range of speech and music frequency. The SIR methodology was developed by Pallett et al. at the U.S. National Bureau of Standards (now National Institute of Science and Technology) in order to establish a simple system to predict the capacity of building shells to attenuate transportation noise.

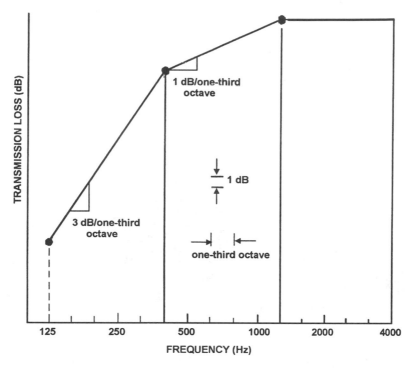

FIGURE 12.11. The standard STC contour.

Sound Transmission Class (STC)

The values of STC is computed from transmission loss values measured in one-third octave bands in the 125-Hz to 4-kHz range. Figure 12.11 illustrates the standard STC contour. This contour begins at 125 Hz, sloping upward at 3 dB per one-third octave (i.e., 9 dB per octave) until it reaches 400 Hz. The slope of the contour then changes at 400 Hz to 1 dB per one-third octave (3 dB per octave), and it remains constant to 1.25 kHz. In the range from 1.25 to 4 kHz the slope is zero. The STC single-number classification of a specific wall is designated by the value of TL at 500 Hz. If the values of TL for the wall at the sixteen $\frac{1}{3}$-octave points are known for the region from 125 Hz to 4 kHz and plotted, the STC for the wall is established by superimposing the contour of Figure 12.11 upon the TL curve so that (a) there is not more than 8-dB deficiency between the TL and the STC contour at any $\frac{1}{3}$-octave frequency and (b) the total deficiency between the STC contour (i.e., the value of the STC contour less the value of the TL curve summed at all $\frac{1}{3}$-octave frequencies from 125 Hz to 4 kHz) must be less than or equal to 32 dB. When the curve is adjusted to meet these two criteria, the STC value of the wall it taken to be equal to the value of the TL of that contour at 500 Hz.

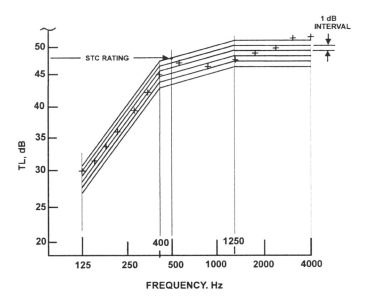

FIGURE 12.12. STC overlay on a TL vs. frequency plot.

Example Problem 5

Find the STC value of the TL curve of Figure 12.12.

Solution

The standard STC contour is overlaid on the TL plot and positioned to satisfy the two criteria described above. From the value of the contour at 500 Hz, it is seen that the STC rating of the wall is approximately 48 dB.

Table 12.1 lists the STC ratings of a number of structural elements.

Shell Isolation Rating (SIR)

The SIR method stems from intensive studies conducted by investigators at the National Institute of Standards and Technology in the effort to develop a simple system to predict the attenuation of A-weighted transportation noise by building shells. This method is based on statistical studies, and +3 dB per octave was chosen as the standard SIR contour. The STC contour method is generally preferred for describing the noise attenuation effectiveness of building structures in the presence of nontransportation sounds, viz. speech, television, and radio. The SIR system is favored for the evaluation of noise reduction provided by building shells against transportation noise.

An element-SIR (or member-SIR) is an estimate of the difference in the A-weighted sound levels when a structural element or member is placed between the transportation noise source and the receiver. It is assumed that all noise transmission

TABLE 12.1. Sound Transmission Class (STC) Ratings of Common Construction Materials.

Description	STC
4-in. (100-mm) thick brick wall with airtight joints	40
4-in. (100-mm) thick brick wall with gypsum board on one side	45–50
8-in. (200-mm) thick brick wall with gypsum board on one side	50–60
4-in. (100-mm) thick hollow concrete block with airtight joints	36–41
4-in. (100-mm) thick hollow concrete block with gypsum board on one side	42–48
8-in. (200-mm) thick hollow concrete block with airtight joints	46–50
8-in. (200-mm) thick hollow concrete block with gypsum board on both sides	50–55
2 × 4-in. (nominal) wood or metal stud wall with gypsum board on both sides, spackled at joints, floor, and ceiling	33–43
Single-glazed window	20–30
Double-glazed window	26–44
Hollow core wood or steel door	20–27
Specially mounted acoustical door	38–55
Double-walled soundproof room, 12-in. (300-mm) thick walls including airspace	70
Quilted fiberglass mounted on vinyl or lead (limp Mass) barrier septrum	21–29

occurs through the subject element. A *room-SIR* (or *composite SIR*) refers to the estimate of the A-weighted sound level difference caused by the presence of all members that act as noise transmission paths between the source and the receiver.

An estimate of the member-SIR may be obtained from a set of transmission loss measurements in one-third octave or octave bands in the following manner:

1. The transmission loss is plotted against frequency.
2. The +3 db per octave SIR reference contour is plotted on an overlay sheet, which is then superimposed on the TL vs. f plot in the highest position so that the sum of the deficiencies is less the twice the number of test frequencies. A deficiency is defined as the number of decibels by which the SIR reference value exceeds the measured TL of the member.
3. The value of TL at 500 Hz on the SIR reference curve constitutes the value of the SIR of the member.

The statistical quality of SIR prediction is improved by using a larger number of TL laboratory measurements, and the $\frac{1}{3}$-octave band measurements will generally yield more reliable results than those based on octave bands. Manufacturers' catalogues frequently report the $\frac{1}{3}$-octave TL measurements used to evaluate STC. An example of an SIR determination is given in Figure 12.13 for the TL plot of Figure 12.12. Much of the tedium of SIR contour fitting can be eased through the use of computer plotting.

The field incidence mass law, expressed by equation (12.13), can be utilized to extrapolate data. If the transmission loss is known for all applicable frequencies for a structural element 1, and it is desired to find the TL of a similar element with a different surface mass, then

$$\mathrm{TL}_2 = \mathrm{TL}_1 + 20 \log \left(\frac{m_2}{m_1} \right)$$

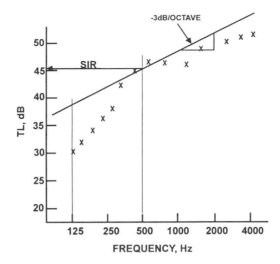

FIGURE 12.13. SIR Determination for TL vs. f plot of Figure 12.12.

If the relationship holds in the applicable frequency range, the values of SIR are similarly related.

Example Problem 6

A shell constructed of 5-in. thick concrete with 55 lb/ft^2 surface mass carries a rating of SIR 43. Estimate the SIR rating for an 8-in. thick concrete with a surface mass of 92 lb/ft^2.

Solution

$$\text{SIR}_{8\text{-inch}} = \text{SIR}_{5\text{-inch}} + 20\log\left(\frac{m_{8\text{-inch}}}{m_{4\text{-inch}}}\right)$$

$$= 45 + 20\log\left(\frac{92}{55}\right) = 49.5 \text{ db}$$

Table 12.2 gives the SIR values of a number of structural elements.

12.12   Noise Reduction of a Wall

Let us consider two rooms that are separated acoustically by a partition of area S_w in Figure 12.14. L_{p1} and L_{p2} denote the spatially averaged values of the sound pressure levels in room 1 and room 2, respectively. We assume the noise source in room 1 produces a purely reverberant field near the partition. This occurs if the

TABLE 12.2. Shell Isolation Rating (SIR).

Description	Weight (lb/ft^2)	SIR Rating
Dense poured concrete or solid block walls		
4 in. thick	50	41
6 in. thick	73	43
8 in. thick	95	46
12 in. thick	145	49
16 in. thick	190	51
Hollow concrete block walls		
6 in. thick	21	41
8 in. thick	30	43
Brick veneered frame walls	—	48–53
Stuccoed frame walls	—	34–52
Frame walls with wood siding	—	33–40
Metal walls, curtain walls	—	25–39
Shingled wood roof with attic	10	40
Steel roofs	—	36–51
Fixed, single-glazed windows (the higher SIR values apply to windows with special mountings and laminated glass)	—	22–39
Fixed double-glazed windows (the higher SIR values apply to windows with large spaces between the glass and special mountings)	—	22–48
Double-hung windows		
Single-glazed	—	20–24
Double-glazed	—	20–29
Casement windows, single-glazed	—	19–29
Horizontal sliding windows		16–24
Glass doors	2.3–3	24
Wood doors (the higher SIR values apply to weatherstripped doors)		
Hollow core	1.2–5	14–26
Solid core	4–5	16–30
Steel doors (the higher SIR values apply to acoustical doors)	4–23	21–50

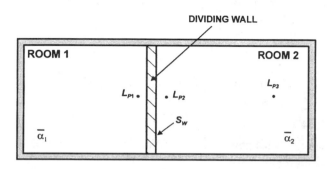

FIGURE 12.14. Two rooms separated by a dividing wall of area S_w.

sound pressure level can be described by equation (11.21)

$$L_{p1} = L_{w1} + 10 \log \left(\frac{4}{R_1} \right)$$

where $4/R_1 \gg Q/(4\pi r^2)$, R_1 is the room constant of room 1, and r is the distance from the source in room 1 to the position of measured L_{p1}. This assumption permits us to simplify the analysis by assuming a constant sound pressure over the entire area of the dividing wall.

The partition absorbs a certain amount of power W_α from the reverberant field in room 1, namely,

$$W_\alpha = W_r \left(\frac{S_w \alpha_w}{S_1 \bar{\alpha}_1} \right) \tag{12.26}$$

where

$$W_\alpha = \text{power absorbed by the dividing wall}$$
$$W_r = \text{power in the reverberant field}$$
$$S_w = \text{area of the wall}$$
$$\alpha_w = \text{absorption coefficient of the wall}$$
$$S_1 = \text{total surface area of room 1}$$
$$\bar{\alpha}_1 = \text{average absorption coefficient}$$

Let us assume that all the power incident upon the wall will be absorbed, that is, $\alpha_w = 1$. The portion of the power W_1 of the source in room 1 that becomes the power in the reverberant field is given by

$$W_r = (1 - \bar{\alpha})W_1$$

Substituting the above into equation (12.26) gives

$$W_\alpha = (1 - \bar{\alpha})W_1 \frac{S_w}{S_1} \tag{12.27}$$

The power W_α absorbed by the wall can be expressed in terms of the room constant R_1, so equation (12.27) becomes

$$W_\alpha = W_1 S_w \frac{1 - \bar{\alpha}_1}{S \bar{\alpha}_1} = \frac{W_1 S_w}{R_1}$$

which can now be combined with equation (12.2) to yield

$$W_2 = \frac{W_1 S_w \tau}{R_1} \tag{12.28}$$

where W_2 represents the power transmitted into room 2 and τ is the transmission coefficient of the wall.

We can consider the direct field in the region near the partition in room 2 to be a uniform plane wave progressing outward from the radiating wall. The energy in the direct field equals the product of the power transmitted into the room multiplied by the time required for the plane wave to transverse the room. This time is given by $t = L/c$, where L denotes the length of room 2. The energy density δ_{d2} of the direct field is given by the directed field energy divided by the room volume $V = S_w L$:

$$\delta_{d2} = \frac{W_2}{V} \frac{L}{c} \tag{12.29}$$

From equation (11.21), the reverberant energy is given by

$$\delta_{r2} = \frac{4W_2}{c R_2} \tag{12.30}$$

Combining the last two equations gives the total energy δ_2 density near the wall:

$$\delta_2 = \left(\frac{1}{S_w} + \frac{4}{R_2} \right) \frac{W_2}{c} \tag{12.31}$$

Equation (12.28) can now be used to eliminate W_2 in equation (12.31) to yield

$$\delta_2 = \frac{4}{R_1} \frac{W_1}{c} \left(\frac{1}{4} + \frac{S_w}{R_2} \right) \tau \tag{12.32}$$

The mean square pressure in room 2 is given in terms of energy density by

$$p_2^2 = \rho_0 c^2 \delta_2$$

which is then inserted into equation (12.32) to yield

$$p_2^2 = \frac{4W_1}{R_1} \rho_0 c \tau \left(\frac{1}{4} + \frac{S_w}{R_2} \right) \tag{12.33}$$

From the use of the definitions

$$L_{p2} = 10 \log \left(\frac{p_2}{20\,\mu\mathrm{Pa}} \right)^2 \quad \text{and} \quad L_{W1} = 10 \log \left(\frac{W_1}{10\,pW} \right)$$

equation (12.33) becomes

$$L_{p2} = L_{w1} + 10 \log \left(\frac{4}{R_1} \right) - 10 \log \left(\frac{1}{\tau} \right) + 10 \log \left(\frac{1}{4} + \frac{S_w}{R_2} \right) \tag{12.34}$$

where we have assumed that $\rho_0 c = 407$ rayls.

In inspecting equation (12.34) we observe that the first two terms in the right-hand side of the equation represents L_{p1} under the conditions of a reverberant field near the wall in room 1. We also invoke equation (12.3) which expresses TL in terms of transmission coefficient τ, with the result that equation (12.34) simplifies to

$$L_{p2} = L_{p1} - TL + 10 \log \left(\frac{1}{4} + \frac{S_w}{R_2} \right) \tag{12.35}$$

From equation (12.35) one can estimate the sound pressure level L_{p2} near the wall in room 2, given the TL of the wall and the acoustic parameters of room 2. On the other hand, if we know the desired value of L_{p2}, (which is the usual case) we can rearrange equation (12.35) to find the necessary transmission loss of the wall as follows:

$$\text{TL} = L_{p1} - L_{p2} + 10\log\left(\frac{1}{4} + \frac{S_w}{R_2}\right) \qquad (12.36)$$

The above equation is valid in both English and metric units.

The term $L_{p1} - L_{p2}$ is referred to as the *noise reduction* denoted by the term NR. Equation (12.36) becomes

$$\text{NR} = \text{TL} - 10\log\left(\frac{1}{4} + \frac{S_w}{R_2}\right) \qquad (12.37)$$

We note if room 2 is entirely absorbent, that is, it supports no reverberation field or if the wall is an outside wall so that it radiates outdoors, the value of R_2 approaches an infinite value and equation (12.37) revises to

$$\text{NR} = \text{TL} + 6\,\text{dB} \qquad (12.38)$$

Example Problem 7

A 20 ft × 8 ft wall with a transmission loss of 32 dB separates two rooms. Room 1 contains a noise source that produces a reverberant field with a SPL = 110 dB near the wall. Room 2 has a room constant R_2 of 1350 ft^2. Determine the sound pressure level near the wall in room 2.

Solution

We apply equation (12.32) to find the value of L_{p2}:

$$L_{p2} = 110 - 32 + 10\log\left(\frac{1}{4} + \frac{20 \times 8\,\text{ft}^2}{1350\,\text{ft}^2}\right)\,\text{dB}$$

$$= 73.7\,\text{dB}$$

12.13   Sound Pressure Level at a Distance from a Wall

For the most part we wish to predict the sound pressure level at some distances from the wall. For appreciable distances the reverberant field will dominate over the direct field. In Figure 12.14, it is desired to find L_{p3} at a point located somewhat far from the wall, which results from the sound pressure level L_{p1} in room 1. But it should be noted that the difference $(L_{p1} - L_{p3})$ does not represent noise reduction, because region 3 in the chamber of room 2 is not directly contiguous to the wall.

An expression for L_{p3} will be derived on the basis that it consists almost entirely of the energy density in the reverberant field, that is, the energy density is given by

$$\delta_{r2} = \frac{4W_2}{R_2 c} = \delta_3 \qquad (12.39)$$

but since $p^2 = \rho_0 c^2 \delta$, equation (12.39) becomes

$$p_3^2 = \frac{4 W_2 \rho_0 c}{R_2} \tag{12.40}$$

Here p_3^2 represents the mean-square pressure in those regions of room 2 where the direct field emerging directly from the wall as a plane wave has already dispersed to the extent that the field can be considered a superposition of randomly reflected components, that is, it is essentially a reverberant field. Equation (12.28) can be used to eliminate W_2 in favor of W_1 in equation (12.40) which then becomes:

$$p_3^2 = \rho_0 c \frac{4 W_1}{R_1} \frac{S_w}{R_2} \tau$$

and because $\tau = W_2 / W_1$ and $L_{p3} = 10 \log(p^3 / 20 \,\mu\text{Pa})$, we will obtain

$$L_{p3} = L_{p1} + 10 \log \left(\frac{S_w}{R_2} \right) - \text{TL} \tag{12.41}$$

where $\rho_0 c = 407$ rayls, and L_{p3} denotes the sound pressure level at a distance sufficiently far from the wall that the direct field is negligible in comparison with the reverberant field.

Example Problem 8

Find the sound pressure level room 2 at a large distance from the wall in the Example Case 7.

Solution

Applying equation (12.41) yields

$$L_{p3} = L_{p1} + 10 \log \left(\frac{S_w}{R_2} \right) - \text{TL}$$

$$= 110 + 10 \log \left(\frac{20 \times 8}{1350} \right) - 32 = 68.7 \text{ dB}$$

12.14   Enclosures

Enclosures may be categorized as being either *full enclosures* or *partial enclosures*. These structures of varying sizes may enclose people or noise generating machinery. It is advisable to never enclose more volume than necessary. For example, an entire machine should not be enclosed if only one of its components (such as a gear box, motor, etc.) constitutes the principal noise source. In some cases, a partial enclosure may suffice to shield a worker from excessive exposure to noise. The walls of an enclosure should be constructed of materials that will provide isolation, absorption, and damping—all necessary for effective noise reduction. Moreover, any presence of cracks or leaks can radically reduce the noise reduction of an enclosure, so all mechanical, electrical, utility, and piping outlets must be thoroughly sealed. Access panels should fit tightly, and viewing windows should be constructed of double panes and be impervious to acoustical leakages. The interior of the enclosures should be lined with highly absorbing material so that the sound level does not build up from reflections, thereby decreasing the wall vibration and the resultant radiation of the noise.

Hoods constitute a special category of acoustical full enclosures designed specifically for the purpose of containing and absorbing excessive noise from a machine. A hood is designed to minimize leakages at its physical input and output, and access must be provided to allow periodic servicing. Usually hoods are sized so that the enclosed machine component does not occupy more than a third of the internal volume. The effectiveness of a hood is denoted by amount of *noise reduction* and *insertion loss*.

In the equilibrium condition, the total power W_1 radiated by a source is absorbed by the interior walls of the hood. In Figure 12.15, δ_1 and δ_2 respectively denote the energy densities of the interior and the exterior of the hood in the immediate vicinity of the enclosure wall. We assume the wall to be thin enough so that volumes

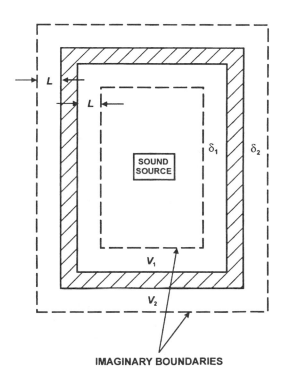

IMAGINARY BOUNDARIES

FIGURE 12.15. Plan view of a hood with energy densities δ_1 and δ_2 just inside and outside the hood.

V_1 and V_2 are nearly equal and denoted by V. As the time t required for an acoustic wave to travel distance L is given by L/c, the energy contained in volume V_1 may be approximated by $E_1 = W_1 L/c$, with the energy density now given by

$$\delta_1 = \frac{E_1}{V_1} = \frac{W_1 L}{Vc}$$

and similarly for the external side of the wall

$$\delta_2 = \frac{E_2}{V_2} = \frac{W_2 L}{Vc}$$

and because the transmission coefficient $\tau = W_2/W_1$

$$\delta_2 = \frac{\tau W_1 L}{Vc} = \tau \delta_1$$

Using the relationships $p_1^2 = \rho_0 c^2 \delta_1$ and $p_2^2 = \rho_0 c^2 \delta_2$, and the fact $L_p = 10\log(p/p_{ref})^2$ we readily obtain

$$L_{p1} = 10\log\left(\frac{\rho_0 c^2 \delta_1}{p_{ref}^2}\right) \quad \text{and} \quad L_{p2} = 10\log\left(\frac{\rho_0 c^2 \delta_2}{p_{ref}^2}\right)$$

Noise reduction NR is given by

$$\text{NR} = L_{p1} - L_{p2} = 10\log(1/\tau)$$

which we note is also the definition for transmission loss TL. Hence

$$\text{NR} = \text{TL}$$

which indicates that the noise reduction from just within the hood (located in a virtual free field) to a region very near the external hood wall is equal to the transmission loss of the wall alone. It should be realized here that the value of L_{p1} is supposed to be the value obtained with the hood in place, not the original *lower* value of L_{p1} measured in the vicinity of the noise source before the hood is placed over it. L_{p2} will be correspondingly higher, as the result of the hood insertion. As a result, it is more useful to know the insertion loss IL of the hood rather than the noise reduction. The insertion loss IL is given by

$$\text{IL} = L_{p0} - L_{p2} \tag{12.42}$$

where L_{p0} denotes the sound pressure level at a selected location without the hood, and L_{p2} is the sound pressure level at the same point with the hood enshrouding the sound source.

In deriving an approximate expression for insertion loss, we shall assume that the room is considerably larger than the hood. The sound pressure level L_{p0} at any location in the room may be found from equation (11.21)

$$L_{p0} = L_{w0} + 10\log\left(\frac{Q_0}{4\pi r^2} + \frac{4}{R}\right) \tag{12.43}$$

In a similar fashion we can express SPL at the same location with the hood in place by

$$L_{p2} = L_{w2} + 10 \log \left(\frac{Q_2}{4 \pi r^2} + \frac{4}{R} \right) \tag{12.44}$$

We need now to determine the total power W_2 emitted by the noise source and its hood acting as a single unit. Because the combination of noise source (usually a machine) and hood maintains a fairly equilibrium condition, the space average of the time-average energy density remains a constant value under the hood. The entire acoustic power emitted by the source is absorbed by the hood as losses or is radiated through the walls and from thence outside the hood. That amount of acoustic energy that passes through the hood is W_2, which can be approximated by equations (12.26) and (12.28):

$$W_2 = W_0 \left(\frac{S_h \alpha_h}{S \bar{\alpha}} \right) \bar{\tau} \tag{12.45}$$

where

W_2 = power radiate into the room by the hood

W_0 = acoustic power of the source

S_h = area of the walls and ceiling of the hood

α_h = absorption coefficient of the walls and ceiling

S = total area under the hood

$\bar{\alpha}$ = average value of sound absorption coefficient under the hood

$\bar{\tau}$ = average value of transmission coefficient for the hood, excluding the floor

We shall now simplify equation (12.45) by assuming that the total surface $S = S_h + S_f$, where S_f is the floor area, is such that $S_f \ll S_h$ so that $S \approx S_h$. With the system in equilibrium, we can set $\alpha_h = 1$, which means that all of the energy impinging on the walls of the hood becomes absorbed one way or the other. Equation (12.45) becomes

$$W_2 = W_0 \frac{\bar{\tau}}{\bar{\alpha}} \tag{12.46}$$

where $\bar{\tau} \leq \bar{\alpha} \leq 1$. The limits are established so that when $\bar{\alpha}$ approaches unity, the defining expression for τ is satisfied. But when $\bar{\alpha}$ approaches $\bar{\tau}$, nearly the entire acoustic power of the source is radiated outside the hood.

Setting $Q_0 = Q_2 = Q$ in equations (12.43) and (12.44) (i.e., the Q of the hood-source combination equals that of the source alone) and substituting into equation (12.42), we obtain

$$\text{IL} = L_{w0} - L_{w2} = 10 \log \left(\frac{W_0}{W_2} \right) \tag{12.47}$$

Inserting equation (12.46) into equation (12.47) yields

$$IL = 10 \log \left(\frac{\bar{\alpha}}{\bar{\tau}} \right) \qquad (12.48)$$

where $\bar{\tau} \leq \bar{\alpha} \leq 1$.

In actuality the average absorption coefficient $\bar{\alpha}$ has a lower nonzero limit as the result of air absorption inside the enclosure, viscous losses of waves near grazing incidence in the acoustical boundary layer inside the hood, and the change from adiabatic to isothermal compression in the immediate vicinity of the inner walls. According to Cremer, the latter two effects may be taken into account in terms of sound frequency f by the expression

$$\alpha_f = 1.8 \times 10^{-4} \sqrt{f}$$

The average excess air absorption coefficient $\bar{\alpha}_{ex}$ due to air absorption in a room or enclosure is

$$\alpha_{ex} = k \frac{4V}{S}$$

where k is an experimentally determined constant, V is the enclosed volume, and S is the room or enclosure area. Combining the last two expressions, we get a minimum $\bar{\alpha}_{min}$ expressed by

$$\bar{\alpha}_{min} = \alpha_{ex} + \alpha_f = \frac{4kV}{S} + 1.8 \times 10^{-4} \sqrt{f} \qquad (12.49)$$

which is also valid for reverberation rooms. Equation (12.49) helps to establish the upper limit of the reverberation time that can be achieved in an echo chamber.

The two limiting cases for equation (12. 48) are

$$IL = 0 \text{ dB} \quad \text{for } \bar{\alpha} = \bar{\tau}, \qquad IL = 10 \log(1/\tau) = TL \text{ dB} \quad \text{for } \bar{\alpha} = 1$$

The first case of $IL = 0$ obviously represents the worst case, and the second case where $IL = TL$ represents the best case for the insertion loss of the hood. This connotes that $\bar{\alpha}$ should be as near to unity as possible and $\bar{\tau}$ be made much less than unity for the most effective noise attenuation by a hood.

The above analysis is premised on the presence of high frequencies with a diffuse field inside the enclosure. For a more complete treatment that includes the effects of low frequencies, the reader is referred to I. L. Ver's article in the text edited by Snowdon and Ungar.

Example Problem 9

Determine the insertion loss of a hood with an average absorption coefficient of 0.30 and an average transmission coefficient of 0.002.

Solution

Applying equation (12.48)

$$IL = 10 \log \left(\frac{\bar{\alpha}}{\bar{\tau}} \right) = 10 \log \left(\frac{.30}{.002} \right) = 21.8 \approx 22 \text{ dB}$$

12.15   Small Enclosures

A small enclosure is one that fits closely around the noise source. If the noise source's geometry contains flat planes that are parallel to the walls of the enclosure, standing-wave resonances can occur at frequencies that are integer multiples of the half-wavelengths of the generated noise. This can render the enclosure useless from the viewpoint of noise attenuation unless the inside of the enclosure is lined with sound-absorption material at least one-quarter wavelength thick. If there is no sound absorption material inside the enclosure, situations can occur where the noise generated at resonant frequencies may actually be louder outside than would be the case without the enclosure!

12.16   Acoustic Barriers

The term *acoustic barrier* (or *noise barrier*) refers to an obstacle which interrupts the line of sight between a noise source and receiver, but does not enclose either source or receiver. An acoustic barrier may be in the form of a fence, a wall, a berm (a mound of earth), dense foliage, or a building between the noise source and the receiver. Noise attenuation occurs from the fact that noise transmission through the barrier is negligible in comparison with refracted noise, particularly if the barrier is solid, without holes or openings, and it is of sufficient mass. Because the sound reaches the receiver by an indirect path over the top of the barrier, the sound level will be less than the case would be if the sound had traveled the (shorter) direct path. The refraction phenomenon is highly dependent upon the frequency content of the sound. The calculations for barrier attenuation are based in part on Fresnel's work in optics.

In Figure 12.16 consider a room prior to inserting a barrier in the position shown. The mean square pressure at the sound receiver's location is given by

$$p_0^2 = W \rho_0 c \left(\frac{Q}{4 \pi r^2} + \frac{4}{R} \right)$$

where p_0^2 denotes the mean square pressure without the barrier. The sound pressure level is expressed by

$$L_{p0} = L_w + 10 \log \left(\frac{Q}{4 \pi r^2} + \frac{4}{R} \right)$$

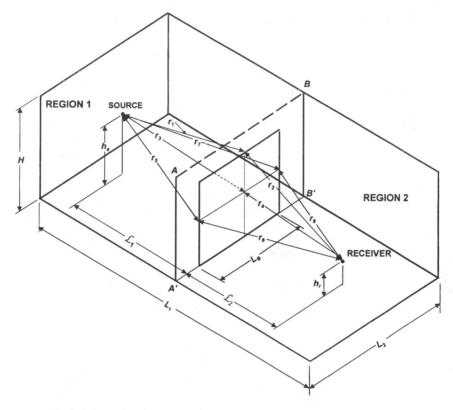

FIGURE 12.16. Schematic of a room with a barrier in position. Region 1 and region 2 are separated by a plane defined by $ABB'A'$.

where

$$L_{p0} = \text{SPL without the barrier}$$

$$L_w = \text{power level of the source}$$

$$Q = \text{directivity of the source}$$

$$R = \text{room constant without the barrier}$$

$$r = \text{distance between the source and the receiver}$$

Now let us insert the barrier as shown in Figure 12.16. The mean-square pressure p_2^2 at the receiver with the barrier installed is given by

$$p_2^2 = p_{r2}^2 + p_{d2}^2 \tag{12.50}$$

where the first term on the right-hand side of equation (12.50) is the mean-square pressure at the receiver due to the reverberant field, and the second term represents the mean-square pressure due to the diffracted field around the edges of the barrier. Recognizing that $L_{p2} = 10\log(p_2^2/p_{ref}^2)$, we can express the sound pressure level

L_{p2} at the receiver in terms of the reverberant and diffracted fields:

$$L_{p2} = 10 \log \left(\frac{p_{r2}^2 + p_{b2}^2}{p_{ref}^2} \right) \tag{12.51}$$

The barrier insertion loss IL is given in terms of SPL by

$$\text{IL} = L_{p0} - L_{p2} = 10 \log \left(\frac{p_0^2}{p_2^2} \right) \tag{12.52}$$

For the room of Figure 12.16, we shall assume an equilibrium condition and that the area of the barrier is considerably smaller than the planar cross section of the room, that is,

$$LH \gg L_b h \tag{12.53}$$

In this situation the reverberant field in the shadow zone of the barrier may be assumed to be the same with or without the barrier's presence. This reverberant field represents the minimum SPL in the shadow zone. The space average δ_r of the time average reverberant energy density in the room without the barrier is given by

$$\delta_r = \frac{4W}{Rc} = \frac{p_r^2}{\rho_0 c} \tag{12.54}$$

For the condition stated by equation (12.53), $\delta_r = \delta_{r1} = \delta_{r2}$, where the numerical subscripts denote the regions on each side of the barrier in Figure 12.15. Equation (12.54) can now be rewritten as

$$p_{r2}^2 = \frac{4W\rho_0 c}{R} \tag{12.55}$$

Here p_{r2}^2 denotes the mean-square pressure in the reverberant field of the shadow zone of the barrier. Our next step is to include the mean-square pressure p_{b2}^2 in area 2 at the location of the receiver due to the diffracted field from the edges of the barrier, and this is given by (see Moreland and Musa, 1972):

$$p_{b2}^2 = p_{d2}^2 \sum_{i=1}^{n} \frac{1}{3 + 10N_i} = p_{d2}^2 D \tag{12.56}$$

where p_{d2}^2 represents the mean-square pressure attributable to the direct field without the barrier, the Fresnel number N_i is defined by

$$N_i \equiv \frac{2d_i}{\lambda} \tag{12.57}$$

where

d_i = difference in direct path and diffracted path between the source and receiver

λ = wavelength of the sound

and D is the diffraction constant defined by[2]

$$D \equiv \sum_{i=1}^{n} \frac{1}{3 + 10N_i} \qquad (12.58)$$

In the case of Figure 12.16, the following path differences exist:

$$
\begin{aligned}
d_1 &= (r_1 + r_2) - (r_3 - r_4) \\
d_2 &= (r_5 + r_6) - (r_3 - r_4) \\
d_3 &= (r_7 + r_8) - (r_3 - r_4)
\end{aligned}
\qquad (12.59)
$$

In the case of rectangular barriers, three values of d_i, or $n = 3$, will usually suffice to describe the shadow zone of the barrier. Through equation (11.17) we can write for the mean-square pressure p_{d2}^2 due to the direct field as follows:

$$p_{d2}^2 = \frac{QW\rho_0 c}{4\pi r^2} \qquad (12.60)$$

with r constituting the direct length from the source to the receiver. Combining equations (12.56) and (12.60) yields

$$p_{b2}^2 = \frac{WQD\rho_0 c}{4\pi r^2} = \frac{WQ_B}{4\pi r^2}\rho_0 c \qquad (12.61)$$

where $Q_B \equiv Q_D$ is the effective directivity of the source toward the direction of the shadow zone. Inserting equations (12.61) and (12.55) into equation(12.50) results in

$$p_2^2 = W\rho_0 c \left(\frac{Q_B}{4\pi r^2} + \frac{4}{R} \right)$$

and the corresponding sound pressure is

$$L_{p2} = L_w + 10\log\left(\frac{Q_B}{4\pi r^2} + \frac{4}{R} \right) \qquad (12.62)$$

where

$$Q_B = Q \sum_{i=1}^{n} \frac{\lambda}{3\lambda + 20d_i} \qquad (12.63)$$

In English units, where distances are expressed in feet instead of meters, equation (12.62) becomes

$$L_{p2} = L_w + 10\log\left(\frac{Q_B}{4\pi r^2} + \frac{4}{R} \right) + 10 \qquad (12.64)$$

---

[2] In some literature the alternative definition for the diffraction coefficient is given by

$$D = \sum_{i=1}^{n} \frac{1}{3 + 20N_i}$$

For a rectangular barrier $n = 3$ in equation (12.63), and the required path differences are given by equation (12.59).

From equations (12.52) and (12.62) we obtain the barrier insertion loss IL:

$$IL = 10 \log \left(\frac{\frac{Q}{4\pi r^2} + \frac{4}{R}}{\frac{Q_B}{4\pi r^2} + \frac{4}{R}} \right) \tag{12.65}$$

The above equation applies to either the metric system or English system. It is interesting to note that if the barrier is located in an extremely reverberant environment, such as an echo chamber, the acoustic field reaches the receiver by rebounding unabated from the surfaces of the room to the degree that the effectiveness of the barrier in blocking the direct field becomes insignificant. Because $4/R \gg Q/(4\pi r^2)$ and $4/R \gg Q_B/(4\pi r^2)$ for a reverberant room, equation (12.65) will give a value of IL $= 0$ dB.

12.17   Barrier in a Free Field

The case of a barrier located outdoors or in an extremely acoustically absorbent room is of special interest. In a free field the room constant $R \approx \infty$, with the result that equation (12.65) becomes

$$IL = 10 \log \left(\frac{Q}{Q_B} \right) = 10 \log \left(\frac{Q}{QD} \right) = 10 \log \left(\frac{1}{D} \right) \tag{12.66}$$

where

$$D = \sum_{i=1}^{n} \frac{\lambda}{3\lambda + 20 d_i} \tag{12.67}$$

In treating the rectangular barrier in a free field or an acoustically high absorptivity chamber we expand equation (12.67) for $n = 3$ and insert the result into the rightmost term in equation (12.66) to yield

$$IL = -10 \log \left[\lambda \left(\frac{1}{3\lambda + 20 d_1} + \frac{1}{3\lambda + 20 d_2} + \frac{1}{3\lambda + 20 d_3} \right) \right] \tag{12.68}$$

If the barrier is of semi-infinite length, that is, $L_B \approx \infty$ in Figure 12.16, equation (12.68) reduces to

$$IL = -10 \log \left(\frac{\lambda}{3\lambda + 20 d_1} \right) \tag{12.69}$$

Example Problem 10

Consider the room of Figure 12.16 to be an anechoic chamber with dimensions 20 ft length $\times$ 15 ft width $\times$ 12 ft high. A sound source is located in the room 1.5 ft above the floor, 8 ft away from one of the shorter walls, in the center and

2 ft directly behind a 6-ft-wide × 5 ft-high rectangular barrier. The barrier itself has a very high transmission loss rating, and a receiver in the form of a SPL meter is located also in the center 2 ft above the floor and 6 ft away from the source. Determine the insertion loss of the barrier for the 1000-Hz octave band.

Solution

The path differences as given by equation (12.59) assume the following values:

$$d_1 = (r_1 + r_2) - (r_3 - r_4) = (4.03 + 4.98) - 6 = 3.01 \text{ ft}$$
$$d_2 = (r_5 + r_6) - (r_3 - r_4) = (3.61 + 4.98) - 6 = 2.59 \text{ ft}$$
$$d_3 = (r_7 + r_8) - (r_3 - r_4) = (3.61 + 4.98) - 6 = 2.59 \text{ ft}$$

The wavelength λ at 1000 Hz is found from

$$\lambda = \frac{c}{f} = 1128 \text{ ft/s} \div 1000 \text{ Hz} = 1.128 \text{ ft}$$

Eq (12.67) becomes:

$$D = \lambda \left(\frac{1}{3\lambda + 20d_1} + \frac{1}{3\lambda + 20d_2} + \frac{1}{3\lambda + 20d_3} \right)$$
$$= 1.128 \left[\frac{1}{3(1.128) + 20(3.01)} + \frac{1}{3(1.128) + 20(2.59)} + \frac{1}{3(1.128) + 20(2.59)} \right]$$
$$= 0.059$$

The insertion loss is then calculated:

$$\text{IL} = 10 \log \left(\frac{1}{D} \right) = 10 \log \left(\frac{1}{0.059} \right) = 12 \text{ dB}$$

12.18   Approximations for Barrier Insertion Loss

Consider the barrier of Figure 12.17. The Fresnel number N, given by equation (12.57), can be restated as

$$N = \frac{2(A + B - C)}{\lambda} = \frac{2(A + B - C)f}{c}$$

A number of researchers in the field developed and verified analytic models, with a view to applying the results of highway barriers. For a point source located behind an infinitely long solid wall or berm, the attenuation A' provided by a barrier are given by the following equations, where the arguments of tan and tanh

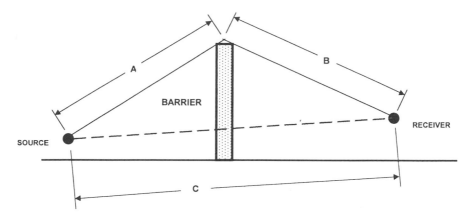

FIGURE 12.17. The geometry of a barrier used in the calculation of the Fresnel number.

are given in radians:

$$A' = 0 \qquad \text{for } N < -0.1916 - 0.0635b'$$

$$A' = 5(1 + 0.6b') + 20 \log \left(\frac{\sqrt{-2\pi N}}{\tan \sqrt{-2\pi N}} \right) \qquad \text{for } -0.1916 - 0.635b' \le N \le 0$$

$$A' = 5(1 + 0.6b') + 20 \log \left(\frac{\sqrt{2\pi N}}{\tan \sqrt{2\pi N}} \right) \qquad \text{for } 0 < N < 5.03$$

where $b' = 0$ for a wall and $b' = 1$ for a berm. The correction factor b' allows for the experimentally determined fact that berms tend to produce 3 dB more attenuation than walls of the same height. The above set of equations applies only to barriers that are perpendicular to a line between the source and the receiver. A detailed discussion of attenuation through scattering and diffraction is given by Pierce (1981).

References

Barry, T. M., and J. A. Reagan. December 1978. *FHWA Highway Noise Prediction Model.* U.S. Department of Transportation Report FHWA-RD-77 108.

Beranek, Leo L. 1971. *Noise and Vibration Control.* New York: McGraw-Hill; pp. 566–568.

Brüel & Kjær. 1980. *Measurements in Building Acoustics,* Nærum, Denmark; p. 19.

Cremer, L. 1961. *Statistische Raumakustik.* Stuttgart: Hitzel Verlag; Ch. 29.

Moreland, J. B., and R. S. Musa. 1972. *International Conference on Noise Control Engineering, October 4–6, 1972, Proceedings*; pp. 95–104.

Pallett, D. Wehrli, R., R. Kilmer, and T. Quindry. 1978. *Design Guide for Reducing Transportation Noise In and Around Buildings.* U S. National Bureau of Standards.

Pierce, A. D. 1981. *Acoustics: An Introduction to Its Physical Principles and Applications.* New York: McGraw-Hill.

Reynolds, D. D. 1981. *Engineering Principles of Acoustics.* Boston: Allyn and Bacon.

Ver, I. L. 1973. Reduction of Noise by Acoustic Enclosure. *Isolation of Mechanical Vibration, Impact, and Noise* (J. C. Snowdon and E. E. Ungar, eds.) ASME Design Engineering Conference, AMD-vol. 1; pp. 192–220.

Ver, I. L., and C. I. Homer. 1971. Interaction of sound waves with solid structures. *Noise Control* (L. L. Beranek, ed.). New York: McGraw-Hill; pp. 270–361.

Problems for Chapter 12

1. A room is subdivided in its middle by a concrete wall with a transmission loss of 52 dB. The room is 4 m high, and the division occurs across the entire width of 12.5 m. In the middle of the dividing wall there will be a door 2.5 m high × 1.0 m wide. The door's rated transmission loss is 28 dB. There is a crack underneath the door that will extend across the door's width and be 2.5 cm high.

   (a) Determine the effective transmission loss of this structure.

   (b) If you could reduce the crack to 0.30 cm (by lengthening the door) what will be the change in the overall transmission loss?

2. A 5.8 ft^2 sample of a building material is being tested by being placed in a window between a receiving room and a source room. There is no appreciable sound transmission except that through the sample. Average sound levels in the 1-kHz octave band are 93 dB in the source room and 68 dB on the receiving room. The equivalent absorption area of the receiving rooms is 18 ft^2. Estimate the transmission loss (TL) and the transmission coefficient (TC) for the sample at 1 kHz.

3. A 0.75 m^2 sample of building material is placed in a window between a receiving room and a source room. Sound transmission occurs only through the sample. The sound level in the source room is 89 dB and that for the receiving room, with 3.0 m^2 equivalent absorption, is 69 dB. Find the transmission loss and the transmission coefficient for this material sample.

4. A wall consists of the following: a 22-dB (transmission loss at 500 Hz) wood door which takes up 25 % of the exposed area, 2.2% airspace, and the remainder of the exposed wall area is a solid wall with 52 dB TL. Determine the TL of the composite wall at 500 Hz.

5. Repeat the above problem but with the airspace reduced to 0.2%.

6. Repeat Problem 4, using a 36-dB TL door. Also determine the benefit of reducing the airspace from 2.2% to 0.2%.

7. A room is subdivided in its middle by a concrete wall with a transmission loss of 62 dB. The room is 4.5 m high, and the division occurs across the entire width of 11.8 m. In the middle of the dividing wall there will be a door 2.5 m high × 1.0 m wide. The door's rated transmission loss is 29 dB. There

is a crack underneath the door that will extend across the door's width and be 2.6 cm high.

(a) Determine the effective transmission loss of this structure.

(b) If you could reduce the crack to 0.30 cm (by lengthening the door) what will the change in the overall transmission loss?

8. Develop an equation for transmission loss versus frequency for 4-mm glass in the mass-controlled region. Glass may be assumed to have a specific gravity of 2.6.

9. It is desired to increase the transmission loss of a panel in the mass-controlled region by 5 dB. Find the necessary change in thickness.

10. Find the transmission loss and transmission coefficient for 8-mm glass at the critical frequency. Assume a loss factor of 0.06.

11. Redo Problem 9 for 14-mm glass with a loss factor of 0.1.

12. Consider a double wall panel with airspace h (given in mm). One panel has 70% the surface mass of the other, and the sum of the surface masses is m_t (in kg/m^2). Plot the resonant frequency versus the product of m_t and h.

13. Determine the resonant frequency of a double-paneled partition constructed of 4 lb/ft^2 and 6 lb/ft^2 panels with 5.2-in. airspace.

14. A wall was measured for its transmission loss in one-third octaves beginning at 125 Hz. The values were: 25, 24, 30, 32, 39, 41, 41, 46. 47, 49, 47, 46, 48, 49, 45, and 46. Find the sound transmission class (STC).

15. Redo Problem 13 with the TL values in the four highest being 39, 40, 38, and 39.

16. Find the SIR for the wall of Problem 13.

17. A 6 m × 2.5 m wall with a transmission loss of 35 dB separates two rooms. A noise source in one room yields a reverberant field with a sound pressure level of 120 dB near the wall. The other room has a room constant $R = 130$ m^2. Predict the sound level pressure near the wall in the latter room.

18. Predict the insertion loss of a hood that has an average absorption coefficient of 0.4 and an transmission coefficient of 0.003.

19. Determine the insertion loss at 1000 Hz of an outdoor noise barrier with a noise source 1.2 ft above the ground, 12 ft away from the wall that is 8 ft tall and 20 ft wide. The SPL meter is located 3 ft above ground and is 6 ft away from the wall. Both source and receiver are equally far from the ends of the wall.

13
Criteria and Regulations for Noise Control

13.1 Introduction

Loss of hearing constitutes only one of the effects of sustained exposure to excessive noise levels. Noise can interfere with speech and sleep, and cause annoyance and other nonauditory effects. *Annoyance* is quite subjective in nature, rendering it difficult to quantify. Loud steady noise can be quite unpleasant, but impulse noise can be even more so. Moreover, impulse noise involves a greater risk of hearing damage. *Community response*, which can range anywhere from simple telephone calls to municipal authorities to massive public demonstrations of outrage, can be used as a measure of annoyance to citizens.

In this chapter we discuss these effects and describe the criteria and regulations that have been implemented over three decades for the purpose of controlling environmental noise. In a broader sense, the term *environmental noise* pertains to noise in the workplace and in the community. This pertains to noise sources arising from operation and use of industrial machinery, construction activities, surface and air transportation. We must also consider *recreational noise* which arises from the use of snowmobiles, drag strip racers, highly amplified stereo systems, and so on. A considerable number of regulations exists in a number of nations for the purpose of protecting hearing from overexposure to loud noise, providing salutary working conditions, and shielding communities from excessive environmental noise arising from the presence of manufacturing plants or construction activities and nearby surface and air transportation. The enactment and enforcement of these regulations motivates the control of noise levels in the workplace and in the community, not only for the purpose of promoting a proper auditory environment but also to avoid penalties that can be levied for uncorrected violations. The principal U.S. laws germane to the issue of noise control are:

- National Environmental Policy Act of 1969
- Noise Pollution and Abatement Act of 1970
- Occupational Safety and Health Act of 1970
- Noise Control Act of 1972

As the corollary to these acts, the Environmental Protection Agency (EPA), the

Department of Labor (DOL), and the Department of Transportation (DOT) have been designated as the principal federal agencies to issue and enforce noise control regulations.

Under the National Environmental Policy Act of 1969, any federally funded construction project requires the preparation and submission of an Environmental Impact Statement (EIS), which assesses the public impact of the noise that will be generated by the erection and operation of the completed project. The creation of the Office of Noise Abatement and Control (ONAC) as a branch of EPA was sanctioned by the Noise Pollution and Abatement Act of 1970. This office carried total responsibility for investigating the effect of noise on public health and welfare. The Occupational Safety and Health Act of 1970 set up the mechanism for enforcing safe working conditions, of which noise exposure criteria constitute a part. The Noise Control Act of 1972 probably contains the most important piece of federal legislation in regard to noisy environments. While EPA is given the primary responsibility for monitoring sound levels in the community, this legislation provides for a division of powers among the federal, state, and local governments.

13.2   Noise Control Act of 1972

With the passage of the Noise Control Act of 1972, the U.S. Congress established the national policy of promoting an environment that is free of excessive noise that would be deleterious to health, safety, and welfare. The Environmental Protection Agency (EPA) was charged with the responsibility of coordinating federal efforts in noise control research, identification of noise sources, and the promulgation of noise criteria and control technology, the establishment of noise emission standards for commercial products, and the development of low-noise emission products. This agency also bears the responsibility of evaluating the adequacy of Federal Aviation Administration (FAA) flight and operational noise controls and the adequacy of noise standards on new and existing aircraft (including recommendations on retrofitting and phaseout of existing aircraft). However, under the provisions of Section 611 of the Federal Aviation Act of 1958, the FAA retains the right to prescribe and amend standards for the measurement of aircraft noise and sonic boom, but the EPA can raise objections to FAA standards which, in the opinion of EPA, do not protect public health and welfare. Irreconcilable differences may be resolved by filing of a petition for review of action of the Administrator of the EPA or FAA only in the U.S. Court of Appeals for the District of Columbia Circuit. The EPA may subpoena witnesses and serve relevant papers to obtain information necessary to carry out the act.

Section 17 of the Noise Control Act also assigns the EPA the task of developing noise emission regulations for surface carriers engaged in interstate commerce by railroads, in consultation with the Secretary of Transportation. State and local governments are enjoined from adopting different standards except when rendered necessary by special local conditions, and then only with the permission of the EPA and the Secretary of Transportation. The provisions of Section 18 are nearly identical to those of Section 17 except they apply to motor carriers engaged in interstate commerce.

13.3   The Occupational Safety and Health Act of 1970

Under the directive of the Walsh–Healey Public Contract Act which was passed by Congress in 1969, the U.S. Department of Labor developed the first occupational noise exposure standard. In 1970, the Occupational Safety and Health Act was passed to apply the requirements of the standard to cover all workers engaged in interstate commerce. The Occupational Safety and Health Administration (OSHA) exists as an arm of the Department of Labor. The assigned task of this agency is to establish safety regulations (including those pertaining to noise levels) and enforce them.

According to OSHA regulations, the daily noise dose D in percent is given in terms of slow-response time-average A-weighted sound levels by

$$D = 100 \sum_{i=1}^{n} \frac{C_i}{T_i} \tag{13.1}$$

where C_i is the duration of exposure to a specific sound pressure level (SPL) and T_i is the allowable daily duration for exposure to noise at that value of SPL. Table 13.1 lists the A-weighted slow-response noise levels and their corresponding maximum daily exposure times T_i. Under OSHA regulations noise dosage exceeding 100 percent is not permitted. It will be noted in Table 13.1 that each 5 dB(A) increase in sound level cuts in half the allowable exposure time. For a time-average A-weighted average of 90 dB(A), the permitted exposure time is 8 hours. Elevation of this noise level by 5 dB(A) cuts the allowable exposure time to 4 hours. Exposure to noise levels above 115 dB(A) is not permitted. Moreover, for exposure to noise having a slow-response, time-averaged, A-weighted sound level of any value, the instantaneous peak sound pressure level may not exceed 140 dB. It should also be noted that the dosage levels are considered to be attributable to noise that actually reaches a worker's station at the ear level. If a worker moves from place to place

TABLE 13.1. OSHA Noise Exposure Limits, (OSHA, 1978).

SPL, dB(A) (slow-response)	Permissible Exposure, hr/day
90	8
92	6
95	4
97	3
100	2
102	1.5
105	1
110	0.5
115	0.25 or less

in the course of his or her occupational assignment, the dosage must be based on the noise exposure and the time spent at each station during the work period.

Example Problem 1

Sampling of a factory noise environment at a worker's station yielded the following time samplings over the course of a normal 8-hour workday: Compute the dosage and comment on the result.

Exposure level, dB(A)	Exposure period, hr
85	1.5
90	3.0
92	2.5
95	0.5
97	0.3
100	0.2

Solution:

Applying equation (13.1) and using Table 13.1, we obtain

$$D = 100 \times \left(\frac{3}{8} + \frac{2.5}{6} + \frac{0.5}{4} + \frac{0.3}{3} + \frac{0.2}{2} \right) = 112\%$$

This dosage of 112% violates OSHA regulations. However, if this 8-hour workday is cut down by 100/112 to 7.1 hours or less (assuming the noise level distribution is fairly consistent throughout the day), the daily dosage will not exceed 100%. It can also be arranged to move the worker to a region where the noise level averages 86 dB(A) or less (i.e., below the OSHA noise level range), for 1.16 hours in order to round out 8 hours daily on the job.

Annual audiology tests of all workers exposed to this environment are mandated if the noise level equals or exceeds 85 dB(A) over the course of a work day. Corrective measures must be taken when the dosage D exceeds unity. Aside from shortening time exposures to the offending noise, noise control measures described in the next chapter must be taken in order to decrease the sound-level pressure of noise sources. Hearing protection devices such as earplugs, earmuffs and special helmets are considered to be temporary stopgaps, and these do not necessarily remove the obligation of the work facility management to cut down on noise levels. OSHA inspectors have the discretion to levy fines for violations that are not removed after an initial visit to the facility.

13.4   Perception of Noise

Loudness is not the only characteristic of noise that determines the degree of annoyance. There are other factors, both acoustical and nonacoustical, that are important. In a classic series of laboratory studies conducted by Kryter and his associates (1959, 1963), human subjects rated sounds of equal duration according

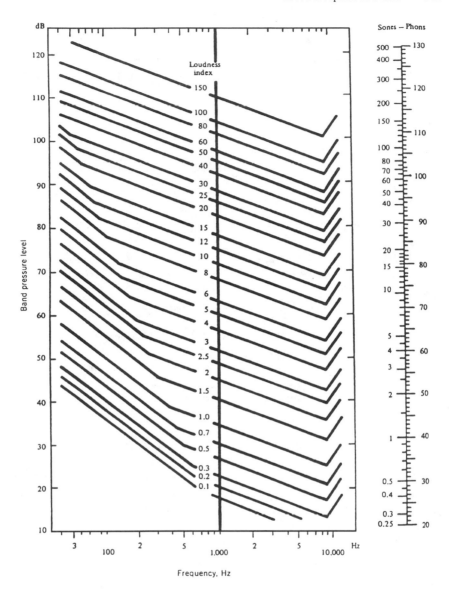

FIGURE 13.1. Equal loudness index curves.

to their noisiness, annoyance, or unacceptability. Through the use of octave bands of noise, Kryter with others established *equal noisiness contours*, which resemble those for equal loudness, but less acoustic energy is required at higher frequencies to produce equal noisiness, and more is needed at low frequencies.

The unit of noisiness index is the *noy N* (this term apparently derives from the second syllable of the word an*noy*). Figure 13.1 displays the equal loudness index curves. In order to determine the logarithmic measure of the perceived noise level (PNL), the following standardized procedural steps need to be taken:

1. Obtain and tabulate the octave band (or 1/3 octave band) sound pressure levels produced by the noise.
2. Use Figure 13.1 to calculate the noisiness index for each octave (or 1/3 octave) band. Then determine the total noisiness index from either one of the following two expressions:

$$N_t = N_{max} + 0.3\left(\sum_i N_i - N_{max}\right) \tag{13.2a}$$

$$N_t = N_{max} + 0.15\left(\sum_i N_i - N_{max}\right) \tag{13.2b}$$

Equation (13.2a) applies to one-octave bands and (13.2b) to 1/3 octave bands. For each octave (or 1/3 octave) band, N_{max} represents the maximum value of N_i found within that band.

3. The total perceived noisiness index N_t, which is summed over all frequency bands, is converted to the perceived noise level PNL (or L_{pn}) through the relationship

$$L_{PN} = 40 + (33.22)\log N_t \tag{13.3}$$

There has been some questions raised regarding the validity of this procedure because the listeners in the laboratory trials do not always seem to distinguish the difference between loudness and the noisiness and annoyance. But this procedure is now being used to evaluate single-event aircraft noise. The U.S. Federal Aviation Administration (FAA) adopted the PNL method to certify new aircraft.

13.5   Effective Perceived Noise Level (EPNL)

Noise of longer duration is obviously more annoying than noise of short duration. Moreover, if pure tones are present in the broadband noise system, the noise is judged to be even noisier than noise without such tones. In order to take into account the factors of duration and the presence of pure tones, the effective noise perception level (EPNL) has been defined as

$$L_{EPN} = L_{PN} + C + D \tag{13.4}$$

Here C is the correction factor for pure tones and D is the duration correction. The tone correction varies from 0 dB up to a maximum of 6.7 dB. The estimation of C entails a complex procedure (Edge and Cawthorn, 1977) that involves examination of the band spectra to detect any band whose level exceeds the level of the bands to either side. The duration correction D, expressed in decibels, which accounts for duration of the noise, may be either positive or negative, but it is usually negative. It is calculated from

$$D = 10\log_{10}\left(\sum_{k=0}^{d/0.5} \text{antilog}\frac{L_{PNT(k)}}{10}\right) - 13 - L_{PNT_{max}} \tag{13.5}$$

Here d represents the total length of the time elapsed when the noise begins to exceeds the background level to the moment when it falls back to the level of imperceptibility. The number 0.5 represents the increment index, that is, if the total duration d of the detectable sound is 5 s, then ten intervals are to be considered in the summation of equation (13.5). L_{PNT} is the tone-corrected value of L_{PN} (i.e., $L_{PNT} = L_{PN} + C$).

13.6   Indoor Noise Criteria

In order to render communication possible in both indoor and outdoor areas at work, it is necessary to minimize the speech interference arising from the background. The A-weighted sound level can be utilized to determine the acceptability of noise indoors, but it cannot give an indication as to what part of the frequency spectrum is responsible for interference. A number of noise-evaluation curves are available for rating the acceptability of noise in indoor situations. The most frequently used families of curves are the noise criterion (NC) curves, noise rating (NR) curves, room criterion (RC) curves, and balanced noise criterion (NCB) curves. These curves were developed in order to provide criteria to either determine the acceptable noise levels in buildings or to specify the acceptable noise in buildings.

Noise Criterion Curves

The NC curves of Figure 13.2 were the result of an exhaustive series of interviews with people in offices, industrial spaces, and public areas. It was found that the principle concern is the interference of noise with speech communication and listening to radio, television, and music. In order to find the NC rating of a particular area, the octave-band sound pressure levels of the noise are measured and plotted on the family of the NC curves of Figure 13.2. The highest curve penetrated by any octave-band and pressure level of the measured noise defines the NC value for the spectrum.

Example Problem 2

A noise generator in a room was found to yield 50 dB straight across the octave band from center frequencies 63 to 8000 Hz. What is the NC rating for that room?

Solution

Draw a line a horizontal line at the 50 dB level. The highest NC curve penetrated is 50. The NC value is therefore NC 50.

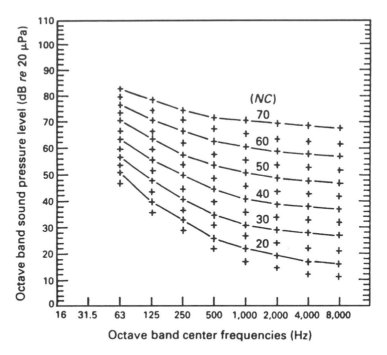

FIGURE 13.2. Noise Criterion (NC) curves.

Noise Rating Curves

The noise rating (NR) curves of Figure 13.3 are very similar to the NC curves. Their original purpose was to determine whether noise heard from industrial plants is acceptable at nearby apartments and houses. As with the NC curves, the noise spectrum is also measured at the affected locations and plotted in Figure 13.3. The NR curves differ in that they include corrections for time of day, fraction of the time the noise is heard and the type of neighborhood. In the range between 20 and 50 for NR or NC, there is little difference between results obtained from the two sets of calculated procedure.

Room Criterion Curves

The difficulty with NC curves is that they are not defined in the low-frequency range—that is, the 160 and 31.5 Hz octave bands—and complaints have been registered that they allow too much noise above the 2 kHz range. Accordingly, Blazier developed a set of room criterion curves on the basis of an extensive study conducted for ASHRAE of generally acceptable background spectra in 68 unoccupied offices. Most of the A-weighted sound pressure levels lay in the range of 40–50 dB. Blazier determined that the curve he derived from the measured data has a slope of approximately −5 dB/octave, and he developed a family of straight lines with this slope (cf. Figure 13.4). It was also established that intense

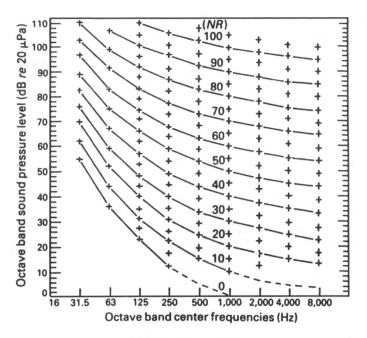

FIGURE 13.3. Noise Rating (NR) curves.

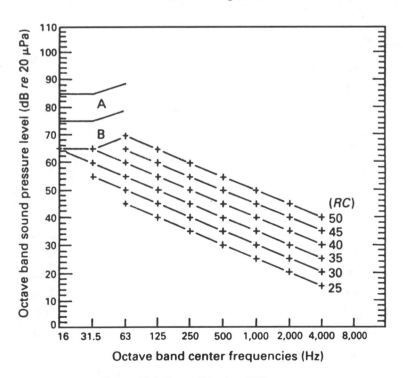

FIGURE 13.4. Room Criterion (RC) curves.

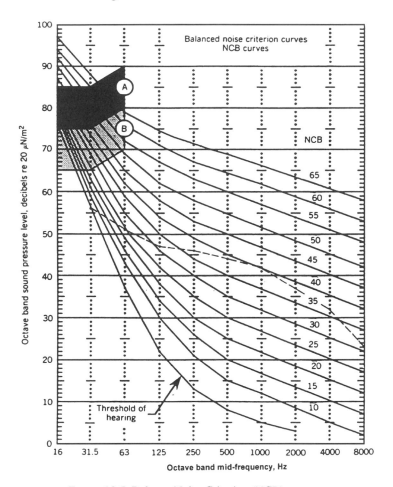

FIGURE 13.5. Balance Noise Criterion (NCB) curves.

low-frequency noise of 75 dB or more in region *A* Figure 13.5 is apt to cause mechanical vibrations and rattles in lightweight structures. Noise in region *B* has a low probability to generate such vibration. The RC value of a measured spectrum is defined as the arithmetic average of the sound levels at 500, 1000, and 2000 Hz. An evaluation of the low- and high-frequency content of the sound provides an additional evaluative parameter. These curves are based on measurements made with air conditioning noise only, so they are used mainly in the rating the noise of HVAC systems.

Balanced Criterion Curve

Beranek (1989) modified the NC curves by adding the 16- and 31.5-Hz octave bands and modifying the slope of the curves so that it became −3.33 dB/octave between 500 and 8000 Hz. He also incorporated regions *A* and *B* of the RC curves of Figure 13.4. The rating number of the resulting *balanced noise criterion (NCB) curve*, shown in Figure 13.5, represents the arithmetic average of the octave-band levels with center frequencies of 500, 1000, 2000, and 4000 Hz. Figure 13.5 is useful for rating air-conditioning noise in buildings. Table 3.2 (Beranek and Ver, 1992) lists the recommended categories of NCB curves for different uses of building interior space.

TABLE 13.2. Recommended NCB Curve Categories on Basis of Interior Use and Suggested Noise Criteria Range for Steady Background Noise.

Type of Space (and Acoustical Requirements)	NCB Curve[a]	Approximate L_A, dB
Broadcast and recording studios (distant microphone pickup used)	10	18
Concert halls, opera houses, and recital halls (for listening to faint musical sounds)	10–15	18–23
Large auditoriums, large drama theaters, and large churches (for very good listening conditions)	Not to exceed 20	28
Broadcast, television, and recording studios (close microphone pickup used only)	Not to exceed 25	33
Small auditoriums, small theaters, small churches, music rehearsal rooms, large meeting and conference rooms (for very good listening), or executive offices and conference rooms for 50 people (no amplification)	Not to exceed 30	38
Bedrooms, sleeping quarters, hospitals, residences, apartments, hotels, motels, etc. (for sleeping, resting, relaxing)	25–40	38–48
Private or semiprivate offices, small conference rooms, classrooms, libraries, etc. (for good listening conditions)	30–40	38–48
Living rooms and drawing rooms in dwellings (for conversing or listening to radio and television)	30–40	38–48
Large offices, reception areas, retail shops and stores, cafeterias, restaurants, etc. (for moderately good listening conditions)	35–45	43–53
Lobbies, laboratory work spaces, drafting and engineering rooms, general secretarial areas (for fair listening conditions)	40–50	48–58
Light maintenance shops, industrial plant control rooms, office and computer equipment rooms, kitchens and laundries (for moderately fair listening conditions)	45–55	53–63
Shops, garages, etc. (for just acceptable speech and telephone communication). Levels above NC or NCB 60 are not recommended for any office or communication situation.	50–60	58–68
For work spaces where speech or telephone communication is not required, but where there must be *no risk* of hearing damage.	55–70	63–78

[a] See Figure 13.5.

Example Problem 3

In Figure 3.5, the background noise spectrum from air conditioning is plotted as a dashed line. Find the NCB rating and comment on the noisiness of the equipment.

Solution

We note that this plot is tangent to the NCB-40 curve and no octave-band level exceeds this curve. This is a spectrum that might be acceptable for a business office, but is barely acceptable in a bedroom or a residential living room. According to Table 13.2, the NCB-40 rating is not at all acceptable in a concert hall, theater, or a house of worship.

13.7 Equivalent Sound Level and Day–Night Equivalent Sound Level

From Chapter 3, we repeat here the definition of equation (3.29) for the equivalent sound pressure level L_{eq} which is the A-weighted sound pressure level averaged over a measurement period T.

$$L_{eq} = 10 \log \left(\frac{1}{T} \int_0^T 10^{L/10} \, dt \right) = 10 \log \left(\frac{1}{N} \sum_{n=1}^N 10^{L_n/10} \right) \qquad (13.6)$$

This averaging time T can be anywhere from a few seconds to hours. L_{eq} can be readily measured through the use of a integrating sound-level meter. Because it takes into account both magnitude and duration, the equivalent sound level has proven to be a viable parameter for evaluating environmental noise from industry, railroads, and traffic. L_{eq} is found to correlate very well with the psychological effects of noise.

In order to account for different response of people to noise at night, the U.S. Environmental Agency developed the day–night equivalent level (DNL), as defined by equation (3.30), which we repeat here in modified form:

$$L_{dn} = 10 \log \frac{15 \left(10^{L_d/10} \right) + 9 \left(10^{(L_n+10)/10} \right)}{24} \qquad (13.7)$$

where L_d represents the 15-hr daytime A-weighted equivalent sound level (from 7:00 A.M. to 10:00 P.M.) and L_n is the 9-hr nighttime equivalent A-weighted sound level (from 10 P.M. to 7:00 A.M.). The nighttime value carries a penalty of 10 dB because noise at night is deemed to be much more disturbing than noise generated during the day. The use of the day–night equivalent level as a parameter is increasing in the United States and in some other nations for evaluating community noise and some cases of airport noise. The U.S. Federal Interagency Committee on Urban Noise (FICON) adopted DNL as the descriptor of environmental noise that affect residences.

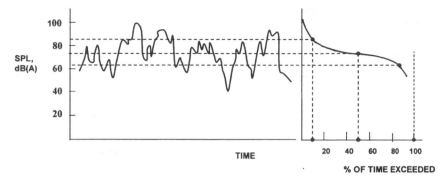

FIGURE 13.6. Example of a cumulative distribution and its percentile sound levels.

13.8   Percentile Sound Levels

Much of environmental noise tends to have a great deal of fluctuations in sound levels. There have been indications that unsteady noise, which occurs from noise sources such as passing aircraft or ground vehicles, is more disturbing than steady noise. In order to describe fluctuations in noise levels and the intermittent character-istics of some noises, percentile sound levels are used internationally as descriptors of traffic and community noise. The level L_n represents the sound level exceeded $n\%$ of the measurement time. For example, L_{20} represents the sound level ex-ceeded 20% of the time. In Figure 13.6 a cumulative distribution is given with examples of L_{10}, L_{50}, and L_{90} levels. In this example, $L_{10} = 85$ dB(A) denotes that 85 dB(A) is the level exceeded 10% of the time. $L_{50} = 75$ dB(A), and this is termed the *median* noise level, because half the time the fluctuating noise level is greater than this value and the other half of the time the noise is less.

In Japan, the median noise level is used to describe road traffic noise. Levels such as L_1 or L_{10} are used to describe the more intense short-duration noises. In Australia and in the United Kingdom L_{10} is used (over an 18-hr period, from 0600 to 2400 hours) to describe target values for new roads and for insulation regulations for new roads. High percentage levels such as L_{90} or L_{99} are usu-ally used to denote the minimum noise level or *residual* or *background noise* level.

13.9   Rating of Aircraft Noise

Aircraft noise is a cause of increasing concern in many nations. A considerable amount of effort has gone into developing a means for predicting and evaluating the annoyance caused by aircraft noise in communities. A 1995 survey (Gottlob) of rating measures revealed that 11 measures are in used in 16 countries surveyed. We discuss a few of these measures below.

Composite Noise Rating

The composite noise rating (CNR) traces its history as far back as the early 1950s. The CNR, originally used as a basic parameter, measures the *level rank* on the basis of a set of curves placed about 5 dB apart in the mid-frequency range, in nearly the same fashion as the NC and NR curves. The level rank was obtained by superposing the noise spectrum on the curves and determining the highest zone into which the spectrum protruded. The rank thus found was then modified to include algebraic corrections to reflect the spectrum characteristics, peak factor, repetitive nature, background noise level, time of day, adjustment to exposure, and even public relations factors. The realized value of CNR was associated with a range of community annoyance categories established by case histories, which can range from no annoyance through mild annoyance, varying degrees of complaints, threats of legal action, to downright vigorous community response.

The CNR had to be modified to meet the advent of jet aircraft (first military and then commercial) in the late 1950s. Instead of being assigned a level rank, the military aircraft noise was converted to an equivalent sound pressure level (SPL) in the 300–600 Hz range on the basis of a set of curves similar to the level-rank curves. The time-varying SPL was averaged and modified by corrections similar to those described in the preceding paragraph. When commercial jetcraft arrived on scene, the CNR was modified to use the perceived noise level (PNL) instead of SPL. The final version of CNR is of the form

$$\text{CNR} = \text{PNL}_{\max} + N + K$$

where $\text{PNL}_{\max}$ denotes the average maximum perceived noise level for individual aircraft flyover events (either landing or take-off) for a 24-hr period, N is a correction for the number of aircraft flyovers, and K is an arbitrary constant. The CNR methodology has proven to be useful for predicting community response to aircraft noise, ranging on a scale from *no reaction* to *vigorous community response*. Each ground point can be represented by

$$\text{CNR}_{ij} = \text{PNL}_{ij} + 10 \log (N_{d,ij} + 16.7 N_{n,ij}) - 12 \qquad (13.8)$$

where $N_{d,ij}$ and $N_{n,ij}$ represent the number of daytime and nighttime events, respectively, for each aircraft class i and flight path j. A penalty of 10 dB is imposed on nighttime flights, and the factor of 16.7 takes into account that there are fewer nighttime than daytime hours. The following expression gives the total CNR for the ground point:

$$\text{CNR} = 10 \log \sum_i \sum_j \text{antilog} \frac{\text{CNR}_{ij}}{10} \qquad (13.9)$$

But equation (13.9), the final version of CNR, does not correct for background noise, prior history, public relations, or other factors such as the occurrence of pure tones. We discussed here CNR even though it is no longer used, because this parameter serves as the precursor of currently used noise measures and descriptors such as NEF and NNI, which are described below.

Noise Exposure Forecast

The noise exposure forecast (NEF) resembles CNR, but it uses the effective perceived noise level EPNL instead of PNL (Bishop and Horojeff, 1967; Bishop, 1974), thus automatically incorporating the annoying effects of pure tones and the duration of flight events. NEF can be expressed as

$$\text{NEF} = \text{EPNL} + N + K$$

where EPNL designates the average effective perceived noise level for individual aircraft flyovers over a 24-hr period, N provides the correction for the number of flyover events, and K is an arbitrary constant. As was done for CNR in equation (13.8). NEF can be calculated for NEF_{ij} for a specific ground point:

$$\text{NEF}_{ij} = \text{EPNL}_{ij} + 10 \log (N_{d,ij} + 16.7 N_{n,ij}) - 88 \qquad (13.10)$$

Here EPNL_{ij} denotes the EPNL for aircraft class i and flight path j; and $N_{d,ij}$ and $N_{n,1}$ are the number of daytime and nighttime events, respectively, for each aircraft of class i and flight path j. In order to distinguish NEF values from CNR values, the constant 88 was selected for equation (3.10). Then

$$\text{NEF} = 10 \log \sum_i \sum_j \text{antilog} \frac{\text{NEF}_{ij}}{10} \qquad (13.11)$$

However, the use of NEF has been replaced by the day–night level L_{dn} in the United States. NEF is still being used in Canada and in a modified form in Australia.

Noise and Number Index

The noise and number index (NNI) is a rather subjective method for rating aircraft noise annoyance, developed and used in the United Kingdom. A survey conducted in 1961 of noise in the residential areas within a 10-mile radius of the London Heathrow Airport resulted in the creation of NNI which is defined by

$$\text{NNI} = \langle \text{PNL} \rangle_N + 15 \log_{10} N - 80 \qquad (13.12)$$

where $\langle \text{PNL} \rangle_N$ comprises the average peak noise level of all aircraft operating during a day and is given by

$$\langle \text{PNL} \rangle_N = 10 \log \left(\frac{1}{N} \sum_{n=1}^{N} \text{antilog} \frac{\text{PNL}}{10} \right) \qquad (13.13)$$

PNL denotes the peak perceived noise level generated by a single aircraft during the day, and N is the number of aircraft events over a 24-hr period. Because no annoyance apparently occurs at levels less than 80 PNdB a constant of 80 was introduced into equation (13.12) so that a zero value of NNI corresponds to no annoyance.

Another survey conducted in 1967 revealed that for the same noise exposure the reported annoyance was less than in 1961. This may be attributable to the fact that noise-sensitive people may have left the affected area, noise-insensitive people moved into the area, the residents adjusted to their environment or became apathetic to the noise, or the background noise arising from surface traffic, construction, or operation of industrial facilities had increased sufficiently to mask more of the aircraft noise.

The use of NNI was superseded in the United Kingdom in 1988 in favor of measurements based on the A-weighted L_{eq} [cf. equation (13.6)]. This parameter is averaged over the period from 0700 to 2300 hours (7 A.M. to 11 P.M.). The measure L_{dn} of equation (13.7) is not applied because nighttime flights are severely limited in the United Kingdom. Both Switzerland and Ireland still make use of NNI.

Equivalent Sound Levels

While some nations continue to use NEF or NNI as descriptors, it has been established that L_{eq} and L_{dn} are much simpler to measure and to evaluate because they seem to correlate well with subjective response. In addition to the United Kingdom which uses only L_{eq} over an 18-hr period (because nighttime flights are restricted), both Germany and Luxembourg have adopted the L_{dn} method with (0600–2200) day and (2200–0600) night. In the United States, as the result of the publication of EPA Report 550/9-74-004 in 1974 and similar documents, the use of CNR and NEF has been superseded by the day–night equivalent level DNL for rating of potential impacts of noise and for planning purposes and land usage near military and civilian airports. In fact, a directive (Part 256 of Title 32 of the Code of Federal regulations) was issued in 1977 by the U.S. Office of the Secretary of Defense that the day–night equivalent sound level DNL must be used as the basis for evaluating the impact of noise by air installations and that neither CNR nor NEF may be utilized.

13.10   Evaluation of Traffic Noise

Highway traffic noise probably impacts more people than any other source of outdoor noise. Consequently, many national, state, and local governments set requirements to assess the existing or potential noise impact of highways. In the United States, federal agencies are required by law to provide environmental impact statements (EIS) for proposed new roads and for any reconstruction of existing roads, where environmental impacts are likely to occur and federal funding is used to finance all or part of the project. The Federal Highway Administration (FHWA, not to be confused with the FHA, the Federal Housing Administration) delegates the responsibility for preparation of environmental impact statements to the individual departments of transportation of the affected states. Moreover, the U.S. Department of Housing and Urban Development (HUD) also requires that when a developer plans residential property with the aid of construction funds or a guarantee of such funds from HUD, the developer must assess the potential impact of transportation noise upon that development.

Assessments of traffic noise are usually made in terms of the overall A-weighted sound levels. Octave-band and 1/3-octave-band levels are generally used only for the purpose of developing vehicle noise abatement measures such as acoustic barriers. Traffic noise tends to vary greatly over time, so methods are required to deal with this variation and its resultant impact on people.

Time Period

The time period used as a basis by for noise evaluation FHWA is the 1-hr period when the traffic is at its heaviest. This period is called the *worst noise hour*. The Department of Housing and Urban Development and certain states, California, for example, require assessment over a 24-hr period. Two types of descriptors are in general use are *statistical descriptors* and *time-averaging descriptors*.

Statistical Descriptors. The 10-percentile-exceeded level L_{10} encompasses the acoustic magnitude of individual traffic noise events, such as passages of heavy trucks, and also the number of such events. Originally the FHWA noise abatement criteria specified only in terms of a 1-hr L_{10}. This descriptor is clearly inadequate for situations where (1) the hourly traffic rates are low, (2) vehicles are not evenly spaced along a road, and (3) the values of L_{10} could not be combined mathematically on the basis of calculations for separate events.

Time-Average Descriptors. Time-average descriptors are now widely used for assessing traffic. The most common such descriptor is the *1-hour average sound level* (abbreviated 1HL), which is essentially the A-weighted equivalent continuous sound level L_{eq} taken over a 1-hr period. This carries the advantage of (1) easy computation through the use of integrating meters, (2) assumptions regarding vehicle spacing are rendered unnecessary, and (3) the average levels for separate categories of sources may be combined readily. The principal disadvantage is that it can be extremely sensitive to isolated events having a high sound level, but which do not necessarily provoke a correspondingly high human response.

FHWA Assessment Procedures

The Federal Highway Administration (FHWA, 1976) requires that expected traffic impacts be determined and analyzed. Highway projects are classified by the FHWA into two categories: Type I, which is a proposed federal or federal-aided project to construct a new highway or to make major physical alterations to an existing road, and Type II, which is a project for noise abatement procedures that are added to an existing highway with no major alterations of the highway itself. All Type I projects are subject to FHWA regulations. Development and implementation of Type II projects are not mandatory, but a traffic noise analysis is mandated for eligibility of noise abatement measures for federal funding.

In the procedure for traffic noise analysis, the following steps are specified in FHWA regulations:

1. Identify activities and land uses that may be affected by the traffic noise. This can involve the use of aerial photographs, land-use maps, and highway plans in addition to the results of field reconnaissance. Also, determine regions of human activity and identify major sources of noise.

2. Predict future traffic noise levels through a method consistent with the FHWA Traffic Noise Prediction Model, employing noise emission levels that are either published in the FHWA regulations or determined by the agency using specified procedures.
3. Determine existing noise levels by actual measurements at sites that will be affected by the proposed highway. These ambient levels provide the basis for assessing impact and evaluating abatement feasibility.
4. Determine existing traffic noise impact. The predicted levels must be compared with the existing levels and with criteria based on land use. Table 13.3 lists the noise levels for different categories of land use. FHWA regulations describes two type of impacts on the land. The first type occurs when the predicted future levels "substantially exceed" the existing levels [but "substantially exceed" is not defined by FHWA, so different state transportation departments interpret the term differently, with minimum increases in A-weighted sound level ranging from 5 to 15 dB(A), with 10 dB(A) being a typical value]. The second type of impact occurs when the predicted future levels 'approach or exceed' the noise abatement criteria of Table 13.3. The criteria of Table 13.3 are listed according to the land-use activity and are in terms of 1-hr average sound levels (L_{1h}) or 1-hr 10-percentile-exceeded levels.
5. Evaluate abatement measures. In situations where severe impacts are identified, a state transportation agency needs to examine means of reducing substantially or eliminating the impact. FHWA regulations specify that primary consideration should be given to exterior impacts where frequent human use occurs and where a decreased level would be of benefit. The state agencies must also (a) consider the opinions of affected residents, (b) identify in the environmental reports the abatement measures likely to be incorporated into the project and impacts where no solutions are apparent, and (c) include the abatement measure in the approved project plans and specifications. Some abatement measures include traffic management (e.g., prohibition of certain vehicle types, time restrictions for certain vehicle types, speed limits, traffic control devices, etc.) and acquisition of property to serve as buffer zones to preempt development that would be adversely affected by traffic noise.

Highway Construction Noise

No specific quantitative rules or guidelines for limiting highway construction noise are provided by the FHWA, but some state agencies use the criteria of Table 13.3 as guidelines for assessing the impact of construction. A computer model (HICNOM) is available from FHWA to be used to predict construction noise and asses abatement measures.

Vehicle Noise

Major contributors to vehicle noise include the engine exhaust and air intake, engine radiation, fans and auxiliary equipment, and tires. To a lesser degree other

TABLE 13.3. Yearly Day–Night Average Sound Levels for Land-Use Compatability (*Source*: Federal Aviation Administration, 1985)*

Land Use	Yearly day–night average sound level (L_{dn}), dB					
	Below 65	65–70	70–75	75–80	80–85	Over 85
Residential:						
Residential other than mobile homes and transient lodgings	Y	N(1)	N(1)	N	N	N
Mobile home parks	Y	N	N	N	N	N
Transient lodgings	Y	N(1)	N(1)	N(1)	N	N
Public use:						
Schools	Y	N(1)	N(1)	N	N	N
Hospital and nursing homes	Y	25	30	N	N	N
Churches, auditoriums, and concert halls	Y	25	30	N	N	N
Governmental services	Y	Y	25	30	N	N
Transportation	Y	Y	Y(2)	Y(3)	Y(4)	Y(4)
Parking	Y	Y	Y(2)	Y(3)	Y(4)	N
Commercial use:						
Offices, business and professional	Y	Y	25	30	N	N
Wholesale and retail—building materials, hardware, and farm equipment	Y	Y	Y(2)	Y(3)	Y(4)	N
Retail trade—general	Y	Y	25	30	N	N
Utilities	Y	Y	Y(2)	Y(3)	Y(4)	N
Communication	Y	Y	25	30	N	N
Manufacturing and production:						
Manufacturing, general	Y	Y	Y(2)	Y(3)	Y(4)	N
Photographic and optical	Y	Y	25	30	N	N
Agriculture (except livestock) and forestry	Y	Y(6)	Y(7)	Y(8)	Y(8)	Y(8)
Livestock farming and breeding	Y	Y(6)	Y(7)	N	N	N
Mining and fishing, resource production and extraction	Y	Y	Y	Y	Y	Y
Recreational:						
Outdoor sports arenas and spectator sports	Y	Y(5)	Y(5)	N	N	N
Outdoor music shells, amphitheaters	Y	N	N	N	N	N
Nature exhibits and zoos	Y	Y	N	N	N	N
Amusements, parks, resorts and camps	Y	Y	Y	N	N	N
Golf courses, riding stables, and water recreation	Y	Y	25	30	N	N

Numbers in parenthese refer to notes.

* The designations contained in this table do not constitute a federal determination that any use of land covered by a program is acceptable or unacceptable under federal, state, or local law. The responsibility for determining the acceptable and permissible land uses and the relationship between specific noise contours rests with the local authorities. FAA determinations under Part 150 are not intended to substitute federally determined land uses for those determined to be appropriate by local authorities in response to locally determined needs and values in achieving noise-compatible land uses.

Key: Y(yes): Land use and related structures compatible without restrictions. N(no): Land use and related structures are not compatible and should be prohibited. NRL noise level reduction (outdoor to indoor) to be achieved through incorporation of noise attenuation into the design and construction of the structure. Land use and related structures are generally compatible; measures to achieve an NLR of 25, 30, or 35 dB must be incorporated into the design and construction of the structure.

noise sources are the transmission, rear axle, and aerodynamic noise due to the passage of the vehicle through air. The relative importance of each component depends on the vehicle type and condition, vehicle load (passenger and cargo), speed, acceleration, and highway grade and road surface condition. In order to aid in prediction of highway noise, FHWA conducted an exhaustive series of measurements of noise emission from, automobiles, trucks, buses, and motorcycles. The procedure for determining the noise emission levels entails the following steps:

1. A level open space free of large reflecting surfaces within 30 m (100 ft) of the vehicle path or microphone is identified. A 150° clear line-of-sight arc is required from the microphone position.
2. The surface of the ground should be free of snow and may be hard or soft. The roadway should be relatively level, smooth, dry concrete, or asphalt. There should be no gravel.
3. The background level from all sources except the vehicle in question should be at least 10 dB(A) lower than the level of the vehicle in question.
4. The microphone is situated 15 m (50 ft) from the centerline of the lane of travel.
5. The microphone is mounted 1.5 m (5 ft) above the roadway surface and not less than 1 m (3.5 ft) above the surface upon which the microphone stands. It should be oriented according to the manufacturer's specifications.
6. The vehicle in question should be traveling at steady speed without acceleration or deceleration.

Vehicles are grouped by FHWA into three classes:

1. Automobiles(A): All vehicles with two axles and four wheels, including automobiles designed for transportation of nine passengers or fewer, and light trucks. Generally, the gross vehicle weight (GVW) is less than 4500 kg (10,000 lb).
2. Medium Trucks (MT): All vehicles having two axles and six wheels, generally in the weight class 4500 kg (10,000 lb) $\leq$ GVW $\leq$ 12,000 kg (26,000 lb).

Cont. Table 13.3 Footnote

Notes: (1) Where the community determines that residential or school uses must be allowed, measures to achieve an outdoor to indoor noise level reduction (NLR) of at least 25 dB and 30 dB should be incorporated into building codes and be considered in individual approvals. Normal residential construction can be expected to provide an NLR of 20 dB; thus, the reduction requirements are often stated as 5, 10, or 15 dB over standard construction and normally assume mechanical ventilation and closed windows year-round. However, the use of NLR criteria will not eliminate outdoor noise problems. (2) Measures to achieve NLR 25 dB must be incorporated into the design and construction of portions of these buildings where the public is received, office areas, noise-sensitive areas, or where the normal noise level is low. (3) Measures to achieve NLR of 30 dB must be incorporated into the design and construction of portions of these buildings where the public is received, office areas, noise-sensitive areas, or where the normal noise level is low. (4) Measures to achieve NLR 35 dB must be incorporated into the design and construction of portions of these buildings where the public is received, office areas, noise-sensitive areas, or where the normal level is low. (5) Land use compatible provided special sound reinforcement systems are installed. (6) Residential buildings require an NLR of 25. (7) Residential buildings require an NLR of 30. (8) Residential buildings not permitted.

TABLE 13.4. Typical Octave-Band Sound Pressure Levels of Automobiles and Heavy Trucks, Measured at 1.2 m Above the Ground at a Distance of 15.2 m.

	Octave-band center frequency, Hz						A-weighted sound level, dB(A)
	125	250	500	1000	2000	4000	
Automobile Speed							
56 km/h (35 mph)	65	61	62	61	57	53	65
88 km/h (55 mph)	71	68	66	68	66	60	72
Heavy Truck Speed							
56 km/h (35 mph)	87	84.5	81.5	78	74.5	70.5	83.5
88 km/h (55 mph)	87.5	85	87.5	82.5	77	73.5	87.5

3. Heavy Trucks (HT): All vehicles having three or more axles, including three-axle buses and three-axle tractors with and without trailers. Generally GVW $\geq 12{,}000$ kg.(26,000 lb).

Table 13.4 shows the octave-band sound pressure levels and A-weighted sound levels typical of automobiles and trucks at two different speeds.

Measurements are made by vehicle type for each selected speed ± 5 km/h. For a given category of vehicles at a given speed, the *reference energy emission level* is given by

$$L_0 = L_{0\,\text{mean}} + 0.115 L_{0SD}^2$$

where $L_{0\,\text{mean}}$ is the arithmetic average emission level for the specific category and speed and L_{0SD} the standard deviation of that emission level.

Vehicle Noise Prediction According to the FHWA Model

On the basis of many measurements, FHWA developed a model for predicting highway noise (Barry and Reagan, 1978). We first compute the *reference mean level* $[(L_0)_E]_i$ for each vehicle type i. This level represents a speed-dependent value of the average (energywise) of the maximum passby levels measured at a reference distance of 15.2 m (50 ft) for a given vehicle type. The U.S. Federal reference mean emission levels are computed from:

$$\text{Automobile} : \{L_0\}_E = 38.1 \log (S) - 2.4 \text{ dB(A)} \qquad (13.14a)$$

$$\text{Medium trucks} : \{L_0\}_E = 33.9 \log (S) + 16.4 \text{ dB(A)} \qquad (13.14b)$$

$$\text{Heavy trucks} : \{L_0\}_E = 24.6 \log (S) + 38.5 \text{ dB(A)} \qquad (13.14c)$$

where S is the average operating speed in km/h.

Adjustments for Traffic Conditions

A number of adjustments have to be made to the reference mean emissions levels computed through the use of equations (13.14). These adjustments are for (a) the traffic density, (b) distances to areas of human activity, (c) finiteness of roadway segments, (d) presence of road gradients, (e) shielding by buildings, (f) shielding by rows of trees, and (g) barrier attenuation.

The traffic flow adjustment $(\Delta_{\text{traffic}})_i$ is rendered to account for the hourly flow of vehicle type i and to average the level over a 1-hr time period, as follows:

$$(\Delta_{\text{traffic}})_I = 10 \log (N_i d_0 / S_i) - 25 \text{ dB(A)} \qquad (13.15)$$

where

$$N_i = \text{hourly flow rate of vehicles of type } i, \text{ vehicles/hr}$$

$$d_0 = \text{the reference distance of 15.2 m}$$

$$S_i = \text{speed of the } i\text{th vehicle type, km/h}$$

Distance adjustment takes into consideration the type of ground cover (e.g., sound will attenuate more readily over a soft surface than over a hard parking lot pavement). Field data gathered by FHWA has shown an average rate of attenuation to be approximately 4.5 dB(A) per doubling of distance on a grassy surface and 3.0 dB(A) per doubling of distance over a paved surface. These two rates are used by FHWA to establish a ground cover parameter called an *alpha factor* (α). For hard sites, the alpha factor has a value of zero; for soft sites, it assumes the value 0.5. The distance adjustment Δ_{distance} can be generally expressed as

$$\Delta_{\text{distance}} = 10(1 + \alpha) \log (d_0/d)$$

where d is the perpendicular distance from the receiver to the center of the travel lane. Table 13.5 provides guidance on when to use propagation rates of 3.0 or 4.5.

The methodology described above assumes an infinitely long road. In analytical practice, highways are subdivided into a series of straight segments of finite length.

TABLE 13.5. Guidelines for Choosing Sound Propagation Rates.

Situation	Propagation rate, dB(A)
1. All situations in which the source or the receiver are located 3 m above the ground or whenever the line of sight[a] averages more than 3 m above the ground	3 ($\alpha = 0$)
2. All situations involving propagation over the top of a barrier 3 m or more in height.	3 ($\alpha = 0$)
3. Where the height of the line of sight is less than 3 m and a. There is a clear (unobstructed) view of the highway, the ground is hard, and there are no intervening structures, or	3 ($\alpha = 0$)
b. The view of the roadway is interrupted by isolated buildings, clumps of bushes, or scattered trees, or the intervening ground is soft or covered with vegetation.	4.5 ($\alpha = \frac{1}{2}$)

[a] The line of sight (L/S) is a direct line between the noise source and the observer.

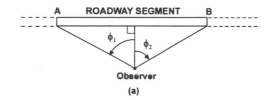

(a)

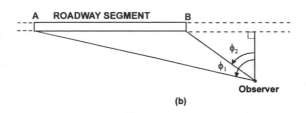

(b)

FIGURE 13.7. Subdivision of a road for noise analysis purposes with segment ends A and B defining the angles with respect to the normal from the observer to the road. In (a) an angle clockwise to the observer is positive and an angle counterclockwise is negative. In (b) the angles are measured from the normal line to the extension of the road segment.

The 1-hr average sound levels are computed separately for each segment, and then they are combined in the end. Such subdivision of the roadway for analytical purposes should be executed when (a) where the traffic volumes or speeds change (e.g., at an exit or entry ramp or a road fork), (b) where the ground cover changes significantly, (c) where a curved road is being analyzed as a series of straight segments, and (d) where the vertical gradient of the road changes. With reference to Figure 13.7, the *finite roadway segment adjustment* (which depends on the ground cover) is defined as

$$\Delta_{\text{segment}} = 10\log \int_{\phi_1}^{\phi_2} \frac{1}{\pi}(\cos\phi)^\alpha d\phi \qquad (13.16)$$

where ϕ_1 and ϕ_2 are angles in radians at the receiver, as shown in Figure 13.6; ϕ_1 is the angle to the left end of the segment, and ϕ_2 is the angle to the right end of the segment. If an angle is measured counterclockwise from the normal line, it is assigned a negative value; if measured clockwise, a positive value. For a hard site equation (3.16) reduces to

$$\Delta_{\text{segment}} = 10\log \frac{\phi_2 - \phi_1}{\pi} \qquad (13.17)$$

In the case of a soft site, equation (13.16) needs to be evaluated by numerical integration. For the general case of a soft site and an infinitely long roadway (where $\phi_1 = -\pi/2$ radians and $\phi_2 = +\pi/2$ radians), the adjustment is -1.2 dB(A).

It is commonly observed that large trucks become noisier as they travel uphill in lower gears. The FHWA recommends that the 1-hr average levels for heavy trucks be increased as a function of roadway gradient[1] in the following manner:

1. 0 to 2 percent: $\Delta_{grade} = 0 \, dB(A)$
2. 3 to 4 percent: $\Delta_{grade} = +2 \, dB(A)$
3. 5 to 6 percent: $\Delta_{grade} = +3 \, dB(A)$
4. Over 7 percent: $\Delta_{grade} = +5 \, dB(A)$

If there are present one or more rows of buildings between a study point and a road, an adjustment for the shielding, $\Delta_{shielding}$, by these buildings must be estimated. A rule of thumb can be used in the following manner: if 40–65 percent of the length of the first row of buildings is occupied by the building themselves, then subtract 3 dB(A) from the average sound level. If 65–90 percent of the length of the row is occupied by the buildings, subtract 3 dB(A) from the average sound level. And if the percentage exceeds 90 percent, the buildings may be treated as noise barriers. Each successive row of buildings adds an additional 1.5 dB(A) to this adjustment, up to a maximum reduction of 10 dB(A). Any excess ground attenuation stops at the first row.

A second type of shielding, also referred to as $\Delta_{shielding}$, can occur from the presence of trees between the road and the receiver. The trees must be dense enough to virtually disallow a direct view of the road. For a 30-m (100-ft) belt width, the FHWA suggests an adjustment of -5 dB(A). For an additional 30-m belt width there is an additional -5 dB(A) adjustment for a total of -10 dB(A). When this adjustment is used, both distance adjustment and a segment adjustment should be applied on the basis of the propagation rate of 3 dB(A) per doubling of distance from the line source.

Barrier attenuation adjustment $\Delta_{barrier}$ can be computed using the path-difference procedure described in the last chapter. According to the FHWA model, this should be computed by using an incoherent line source model separately for each vehicle type. The attenuation is computed for a series of paths, defined by the angles with the perpendicular source-to-receiver line, and the results are combined via numerical integration. This calculation is usually achieved with the use of a computer program such as the FWHA STAMINA program.

The total 1-hr average sound levels for each vehicle type can be obtained by summing up the various adjustments:

$$(L_{eq})_i = [(L_0)_E]_i + (\Delta_{traffic})_i + (\Delta_{distance})_i + (\Delta_{grade})_I + (\Delta_{shielding})_i + (\Delta_{barrier})_i$$

$$(13.18)$$

The 1-hr average levels for automobiles, medium trucks, and heavy trucks may now be combined to yield the total 1-hr average level.

---

[1] When a road changes its elevation n meters (or feet) for every 100 meters (or feet) horizontal travel, the road is said to have an n percent gradient.

Computer Programs

The FHWA noise prediction methodology has been incorporated into the following computer programs:

1. SNAP 1.1, intended for relatively simple geometries, can handle up to twelve roadways and one simple barrier. This program provides a detailed output and can provide separate results according to vehicle type.
2. STAMINA 2.0 and its derivative programs are used by most states, along with a noise barrier program called OPTIMA. Complex highway sites with many roadways, barriers and receivers can be analyzed with this program. Additional vehicle types can be specified in addition to the three classes of automobiles, medium trucks and heavy trucks. The output of the program can be chosen to include the 1-hr average sound level and 10-percentile-exceeded level for each receivers for the initial barrier segment elevations, the average sound levels for each individual road segment, and matrices of sound energy contributions from each barrier segment at each elevation for each receiver.
3. OPTIMA is used to design noise barriers on the basis of input information such as the type of material being used for each barrier. This program can be used to optimize the design on a cost effective basis.

13.11   Evaluation of Community Noise

Transportation noise, both surface and air, constitutes the dominant source of noise exposure in residential neighborhoods. Other sources can include noise emanating from industrial and commercial enterprises, rowdiness of individuals carousing in the streets, operation of lawnmowers, leafblowers, or snowblowers, chain saws or other gear, and local construction. Concern about noise pollution began to intensify when commercial jets began to appear in the skies in the late 1950s. There was little in the way standardization of community response survey methods, questionnaire items, and even noise measurements and analytical techniques throughout the 1960s and 1970s. But research was carried on during these years on annoyance, speech, and sleep-interfering properties of noise. Researchers have come up with a number of ways to assess community response. In addition to frequency-weighting schemes, some of the properties of noise thought to have relevance include tonality, impulsiveness, rise time, onset time, periodicity, time of day, and temporal variability. Dozens of physical measures of sound have been considered as predictors of annoyance caused by noise exposure (Pearsons and Bennett, 1974; Schultz, 1982). Even today there is no single purely physical metric that can function as a definite predictor of annoyance with noise exposure, and it may not even be possible to develop such a predictor.

Social Surveys

Perhaps the least ambiguous procedure to evaluate the prevalence of noice-induced annoyance in a community is through the means of a social survey.

Survey techniques, however, can range widely in the degree of their respective sophistication.

Empirical Dosage Response Relationship. Schultz (1978) documented quantitative dosage-response relationships through meta-analysis, which constitutes a major step toward the setting up of a standard method for predication of transportation noise effects. Schultz executed a "best-fit" third-order polynomial to a data set relating the day–night average sound level (DNL) to the degree of annoyance in communities. A simple quadratic-fit equation provides a purely empirical basis for predicting the prevalence of annoyance in communities:

$$\% \text{ highly annoyed} = 0.036L_{dn}^2 - 3.27L_{dn} + 79.14 \qquad (13.19)$$

However, equation (13.19) produces meaningless predictions when evaluated outside the range 45 dB $< L_{dn} < 85$ dB. The U.S. Federal Interagency Committee on Noise prefers a different logistic fit to a subset of data reported by Fidell, Barber, and Schultz, which resulted from disregarding the results of certain studies in which relatively low levels of noise were associated with a high degree of reported annoyance:

$$\% \text{ highly annoyed} = \frac{100}{1 + e^{11.13 - 0.141L_{dn}}} \qquad (13.20)$$

Equation (13.20) predicts somewhat lower levels of annoyance at lower noise exposure levels than the quadratic fit of equation (13.19). For example, at $L_{dn} = 65$ dB equation (13.20) yields a prevalence rate of 12.3% in contrast to 18.8% obtained through the use of (13.19).

Equal Energy Hypothesis. DNL has been adopted by many federal agencies as a convenient descriptor of long-term environmental noise, which soon enough became a predictor of annoyance. The DNL index is a time-weighted average (in effect, the average acoustic energy per second with arbitrary nighttime weighting), which is sensitive to the duration and magnitude of individual noise events and directly proportional, on an energy ($10 \log N$) basis, to the number of events.

Reliance on such an integrated energy metric is based on the "equal energy" hypothesis, which states the notion that the number, level and duration of noise events are fully interchangeable determinants of annoyance as long as their product (energy summation) remains equal. This quantification of noise exposure in terms of DNL for the purpose of predicting annoyance carries the implication that a person would be annoyed to the same degree by small numbers of very high-level noise events as by large numbers of lower-level noise and/or longer-duration noise events. The equal energy hypothesis has provided an adequate account for data on the prevalence of annoyance to sporadic (e.g., urban) noise in the range $55 < L_{dn} < 75$ dB, but the validity of the hypothesis falls off in extreme cases. For example, no community is likely to tolerate even infrequent operation of a noise source powerful enough to damage hearing or a very occasional shock wave from passing supersonic jet plane from a nearby military base.

13.12   Guidelines and Regulations in the United States, Canada, Europe, and Japan

The most well-established of noise effect guidelines in the United States are those promulgated by the Federal Interagency Committee on Noise (FICON). FICON is composed of a number of federal agencies with interests in environmental noise [e.g., FAA and Department of Defense (DoD)]. The guidelines and recommendations of FICON for "land-use compatibility" are couched in ranges of DNL. FICON considers noise exposure levels lower than $L_{dn} = 65$ dB to "be compatible with most residential land uses." However, FICON does confess its realization that this limit may be somewhat too high in highly rural areas where it would be more appropriate to characterize the effects of noise pollution not in acoustic terms but rather in terms of annoyance.

On the international scene, a major degree of consensus has evolved over the years as to what constitutes unacceptable levels of noise exposure. In the mid 1980s the Organization for Economic Cooperation and Development (OECD)[2] suggested a standard guideline value for average outdoor noise levels of 55 dB(A) for normal daytime hours in order to prevent significant interference with normal activities of local communities. According to OECD:

1. At 55–60 dB(A) noise creates annoyance.
2. At 60–65 dB(A) annoyance increases considerably.
3. Above 65 dB(A) constrained behavior patterns occur, symptomatic of serious damage by noise.

The World Health Organization (WHO) listed additional guidelines in 1996 for noise exposure in dwellings, schools, hospitals, concert halls, and so on. The exposure levels are given in terms of the time-averaged A-weighted sound level L_{eq}. For example a private bedroom should sustain a sound level no higher than $L_{eq} = 30$ dB(A) at night in order to allow undisturbed sleep; and the background noise level in a classroom should be no greater than $L_{eq} = 35$ dB(A) to facilitate the teaching processes.

Workplace Noise Exposure

In the United States the passage of the Noise Control Act of 1970 gave rise to OSHA regulations which are described in Section 9.4 above. In Europe, the Council of European Communities issued a *Council Directive 86/188/EEC* on May 12, 1986, which set the guidelines on the protection of workers from the risks related to exposure to noise at work. This directive does not prejudice the right of members of EU to introduce or apply even stricter provisions that reduce the permissible level

---

[2] While its name indicates its mission, the Organization for Economic Cooperation and Development also concerns itself with environmental matters. Its membership includes industrial nations from North America, Europe, and Asia.

of noise. For an 8-hr day, under the provisions of 86/188/EEC, a worker should not be exposed to more than A-weighted $L_{eq} = 85$ dB per 8-hr day. This value is determined from measurements made at the spatial positioning of a person's ears during work, preferably in that person's absence, using a technique that minimizes the effect on the sound field. If the microphone has to be located close to the person's body, appropriate adjustments should be made to determine an equivalent undisturbed field pressure. The daily personal noise exposure does not take into account of the effect of any personal ear protectors used. Should the allowable daily exposure be exceeded, the affected workers are to be notified, and procedures must be enacted to lower the exposure, which may includes periodic checks of the work environment, provision of ear protectors, and audiology tests. If the daily personal noise exposure exceeds 90 dB(A) or the maximum value of the unweighted instantaneous sound pressure is greater than 200 Pa, the employer must exert every effort to cut down this exposure.

Vehicle Noise Regulations

Exterior noise tests for road vehicles have been used in Japan since 1971. The current version of the tests are quite thorough and consists of three parts: (a) a fixed passby test at 7 m, 60% rated engine speed or 65 km/hr; (b) an acceleration passby test at 7.5 m; and (c) a stationary test behind the exhaust outlet at 20 m.

In these tests the microphone is placed 1.2 m above ground level. The noise limits for new highway vehicles are given in Table 13.6 which went into effect in 1998. Another 2 dB reduction in these noise limits are planned for 2002.

Table 13.7 lists the mandatory exterior limits for new highway vehicles sold in the European Union. The EU vehicle noise test procedure is prescribed in *Council Directives* 92/97 EEC and 81/334/EEC, presented in the *Official Journal of the European Communities*. Amendments were added for automatic transmission-equipped and high-powered vehicles. The test is based on ISO 362, which is

TABLE 13.6. Japanese Noise Limits for New Highway Vehicles.

Vehicle Description	GVWR (metric tons)	Power (kW)	Limit [dB(A)]
Passenger car	—	—	76
Light truck	<3.5	—	79
Medium truck, bus	<3.5	<150	81
Heavy bus	>3.5	>150	81
Heavy truck	>3.5	>150	81
Moped <50 cc			70
Motorcycle			
>50 cc < 125 cc			70
>125 cc < 250 cc			73
>250 cc			76

TABLE 13.7. European Union Highway Vehicle Noise Limits.

Vehicle Description	Unloaded Weight (metric tons)	Power (kW)	Noise Limit, dB(A)
Passenger car	—	—	74
Mini bus < 9 seats	<2	—	76
	>2 < 3.5	—	77
Bus > 9 seats	>3.5	<150	78
Light truck/van	<2	—	76
	<3.5	—	77
Medium truck/van	>3.5	<75	78
	>3.5	>75 < 150	78
Heavy trucks	>12	>150	80
Motorcycles			
≤80 cc			75
>80 ≤ 175 cc			77
>175 cc			80

substantially equivalent to SAEJ140 (cf. *Handbook of the Society of Automotive Engineers*). A 1-dB(A) tolerance is permitted in this test. The test and its derivatives are used widely throughout the world, including Japan [which does not have 1 dB(A) tolerance]. The same test site configuration is used for all classes of vehicles, including motorcycles, with the measurement distance set at 7.5 m.

Railroad Noise Regulations

In the United States, regulations for railroad noise are published in Section 40, Part 201, and Section 49, Part 210, of the *Code of Federal Regulations*. Test measurements are conducted within a cleared level area 30 m from the track center line. Locomotives are tested either in motion or stationary using remote load cells. Noise limits have been specified for locomotives manufactured after 1979:

> Stationary, idle throttle setting: 70 dB(A)
> Stationary, all other throttle settings: 87 dB(A)
> In motion :   90 dB(A)

For rail cars, the limits are:

> Moving at speeds 83 km/h (50 mph) or less: 88 dB(A)
> Moving at speeds greater than 83 km/h: 93 dB(A).

The test is to be conducted during dry weather without dust or powdery snow.

The EU has no exterior noise standards for railroad equipment because of intermingling with equipment from outside the EU and differences in rail gauges. Railroads have used, on their own, a number of noise standards such as the International Union of Railways ORE E82/RP4 and the ISO 3095: 1975.

Highway Construction and General Construction Noise

An interesting development in EU is the use of sound power as a noise measure, rather than sound pressure at a single point in space, for earth-moving and other construction equipment. Sound power measures the total noise emanating in all directions from a source. The fundamentals of sound power measurement are given in Chapter 9, Sections 9.12 et seq.; and details are given in the EU directives 95/27/EU, 89/514/EEC, 86/662/EEC, 84/533-536b/EEC, and 79/113/EEC. The distribution of measurement positions specified by 79/113/EEC is given for a hypothetical hemisphere enclosing a stationary piece of earth-moving equipment. Sound pressure is measured at the prescribed measurement points, which is converted to sound intensity, applying the far-field approximation and integrated to yield sound power. The sound power measurement would be more accurate if sound intensity measurements were used instead of sound pressure, In its current form the test only uses half of the measurement points displayed in Directive 79/113/EEC. A similar sound power test was developed for lawn mowers in 84/538/EEC, and amended by 88/181/EEC.

New noise limits came into effect at the end of 1996 for earth-moving machinery of net installed power less than 500 kW. Machines with power exceeding 500 kW operate in quarries and mines and so are considered to have a negligible effect on community noise. For the period between 1996 and 2001, the permissible sound power levels L_{WA}, in A-weighted decibels relative to 1 pW, are given by the following:

1. Tracked vehicles (except excavators):

$$L_{WA} = 87 + 11 \log P \qquad \text{for } L_{WA} \geq 107 \qquad (13.21)$$

2. Wheeled bulldozers, loaders, excavators-loaders:

$$L_{WA} = 85 + 11 \log P \qquad \text{for } L_{WA} \geq 104 \qquad (13.22)$$

3. Excavators:

$$L_{WA} = 83 + 11 \log P \qquad \text{for } L_{WA} \geq 96 \qquad (13.23)$$

where P is the net installed power of the vehicle in of the construction vehicle in kilowatts. Below the lower limit given above, the machine automatically passes the test. After 2001, the numerical A-weighted decibel values in equations (13.21–13.23), including the lower limits, are reduced by 3 dB(A). The coefficient 11 remains unchanged.

More recently, the European Commission of the EU has been laboring on a new directive regarding machinery used outdoors. This proposed statute, Directive 98/0029, covers about 60 types of machines (principally construction machinery, gardening equipment, and commercial vehicles); and it is intended to supersede similar directives now existing separately in the EU's member nations. The European Parliament is expected to adopt in early 2000 this proposed directive that lists allowable noise emission for each type of construction equipment. In Directive 98/0029, it also becomes the task of the member states to ensure that the equipment covered by the directive complies with the requirements when placed on the market or put into service in the member states. However, this directive does not affect the existing statutes that specifically protect workers by regulating the use of outdoor equipment. It is the responsibility of the manufacturer or its authorized representative established in the Community to ensure that the equipment being sold or used in the Community conforms with the provisions of the Directive.

Over the past two decades, a greater awareness of the impact of construction noise on the part of government agencies led to a series of codes and regulations for the control and mitigation of noise from construction sites. These acts generally cover (a) the erection, construction, alteration, repair, or maintenance of buildings, structures or roads; (b) the breaking up, opening, or boring under any road or adjacent land in connection with the construction, inspection, maintenance, or removal of public or private works; (c) piling, demolition, or dredging works; or (d) or any other work entailing engineering construction.

A major cornerstone in the development of effective construction noise control programs may very well be the Construction Noise Control Specification 721.560 developed by the Massachusetts Turnpike Authority for the Central Artery Tunnel (CA/T) Project in the Boston area. At the close of the twentieth century, this 12-year undertaking ranks as the largest infrastructure construction project in the United States. With this project's planned completion in 2004, the notorious Boston traffic bottleneck on U.S. Interstate Highway I-93 and to/from Logan International Airport should be alleviated, thus freeing up the City of Boston and the entire New England corridor to normal traffic flow. Aside from doubling Boston's highway capacity, this project will also modernize Boston's underground utilities and enable the city to achieve positive growth into the new millennium.

Construction over the 7-mile long project occurs 24 hours a day in various locations through the city. Construction equipment operate in close proximity to thousands of residential and commercial properties, in some cases as close as 10 feet away. Hundreds of construction units can be found functioning at any one time throughout the project. The range of equipment types used is wide, including cranes, slurry trenching machines, hydromills, hoe rams, pile drivers, jackhammers, dump trucks, concrete and chain saws, and gas and pneumatically powered hand tools. When the project is completed, more than 13 million cubic yards of excavated materials will have been removed and nearly 4 million cubic years of concrete will be poured.

In order to contain the acoustical impact of the project, while supporting and maintaining the progress of construction, the CA/T Project people developed a Policy Summary to establish an overall noise control program that includes (a) a commitment statement to minimize noise impact on abutting residents while sustaining construction progress; (b) a summary of the Project Noise Control

Specification criteria and components; (c) an expression of willingness to develop area-specific noise mitigation strategies tailored to particular community needs and sensitivities; (d) an approach and criteria for judging the worthiness of mitigation measures; and (e) a commitment to provide qualified noise experts in the field to oversee contractor compliance. Construction Noise Control Specification 721.560, adopted and enforced by the CA/T Project, is the most comprehensive and stringent noise code of any public works project in the United States.

In addition to the CA/T project, other outstanding sets of codes or regulations pertaining to construction noise (Raichel and Dallal, October 1999 and November 1999), include those by the Hong Kong and the Singapore governments as well as those developed for the I-15 project in Utah. Germany's "Blue Angel" seal program certifies noise-generating equipment that operate below specified noise levels.

References

Anonymous. 1997. *Low-Noise Construction Has a Future (Leises Bauen hat Zukunft): Noise Reduction with the "Blue Angel": (Lämminderung mit dem "Blauen Engel"*, 1st ed. Kalsruhe, Germany: Fachinformationszentrum, Gesellschaft für wissenschaftlish-technische.

Barry, T. M., and J. R. Reagan. 1978. *FHWA Highway Traffic Noise Prediction Model.* Report FHWA-RD-77-108, Washington, DC: Federal Highway Administration.

Beranek, Leo L. 1989. Balanced noise criterion (NCB) curves. *Journal of the Acoustical Society of America* 86: 650–664.

Beranek, Leo L. 1989. Applications of NCB noise criterion curves. *Noise Control Engineering Journal* 33: 45–56.

Blazier, W. E. 1981. Revised noise criteria for application in the acoustical design and rating of HVAC systems. *Noise Control Engineering* 16: 64–73.

Bowlby, W. (ed.). 1981. *Sound Procedures for Measuring Highway Noise: Final Report.* Report FHWA-DP-45-1. Washington, DC: Federal Highway Administration.

Bowlby, W., and L. F. Cohn. 1982. *Highway Construction Noise: Environmental Assessment and Abatement, vol. 4: User's Manual for FWHA Highway Construction Noise Computer Program*, HICNOM, Vanderbilt University Report VTR 81-2. Washington, DC: Federal Highway Administration.

Bowlby, W. 1980. *SNAP 1.1—A Revised Program and User's Manual for the FHWA Level 1 Highway Noise Traffic Noise Prediction Computer Program.* Report FHWA-DP-45-4. Arlington, VA: Federal Highway Administration.

Bowlby, W., J. Higgins, and J. Reagan (ed.). 1982. *Noise Barrier Cost Reduction Procedure, STAMINA 2.0/OPTIMA: User's Manual.* Report FHWA-DR-58-1. Washington, DC: Federal Highway Administration (based on Menge, C. W. 1981. *User's Manual: Barrier Cost Reduction Procedure, STAMINA 2.0 and OPTIMA*, Report 4686. Cambridge, MA: Bolt, Beranek and Newman).

Council of the European Communities. May 12, 1986. *Council Directive 86/188/EE on the Protection of Workers from the Risks Related to Exposure to Noise at Work.*

Department of Labor Occupational Noise Exposure Standard. May 29, 1971. *Code of Federal Regulations*, Title 29, Chapter XVII, Part 1910, Subpart G, 36 FR 10466.

Department of Labor Occupational Noise Exposure Standard. March 8, 1983. *Amended Code of Federal Regulations*, Title 29, Chapter XVII. Part 1910, 48 FR 9776–9785.

Edge, Jr., P. M., and J. M. Cawthorn. February 1977. *Selected Methods for Quantification of Community Exposure to Aircraft Noise*, NASA TN D-7977.

Environmental Protection Agency (EPA). July 1973. *Public Health and Welfare Criteria for Noise*. Washington, DC: EPA.

European Union Council. June 17, 1999. *Proposal for a Directive of the European Parliament and of the Council on the Approximation of the Laws of the Member States Relating to Noise Emission by Equipment Used Outdoors*. Interinstitutional File 98/0029. Brussels. European Council of EU.

Federal Aviation Administration (FAA). January 18, 1865 (revised). Part 150: Airport noise compatibility planning. *Federal Aviation Regulations*. Washington, DC: Federal Aviation Administration.

Federal Highway Administration. July 8, 1982. Procedures for abatement of highway traffic noise and construction noise. 23 C.F.R., Part 722. *Federal Register* 47: 29653–29657.

Federal Interagency Committee on Noise (FICON). 1992. *Final Report: Airport Noise Assessment Methodologies and Metrics*. Washington DC: FICON.

Federal Register, 40. May 28, 1975; p. 23105.

Fidell, S., D. Barber, and T. Schultz. 1991. Updating a dosage–effect relationship for the prevalence of annoyance due to general transportation noise. *Journal of the Acoustical Society of America* 89(1): 221–233.

Gottlob, D. 1995. Regulations for community noise. *Noise/News International* 3(4): 223 – 236.

Society of Automotive Engineers (SAE). 1991. *Handbook of the Society of Automotive Engineers*. Warrendale, PA: SAE.

Harris, Cyril M. (ed.). 1991. *Handbook of Acoustical Measurements and Noise Control*, 3rd ed. New York: McGraw-Hill; Chs. 23–26, 46–50

Hong Kong Environmental Protection Department. 1999. *A Concise Guide to the Noise Control Ordinance*. Hong Kong: EPA. Internet address: http://www.info.gov.hk.epdinkh/noise/book2/book20.htm.

Massachusetts Turnpike Authority, Central Artery (I-93)/Tunnel (I-90) Project. Revised July 28, 1998. *Construction Noise Control Specification 721.560*. Boston, MA: Commonwealth of Massachusetts.

Occupational Noise Exposure Standard. June 28, 1983. U.S. Department of Labor, Occupational Safety and Health Administration (OSHA), Code of Federal Regulations, Title 29, Part 1910, Section 1910.95 [29 CFR 1910.95], *Federal Register* 48: 29687–29698.

Olson. N. 1970. *Statistical Study of Traffic Noise*, Report APS-476. Ottawa: National Research Council of Canada, Division of Physics.

Pearsons, K. S., and R. Bennett, *Handbook of Noise Ratings*. Washington, DC: NASA, 1974.

Raichel, D. R., and Dallal, M. October 1999. Prediction and attenuation of noise resulting from construction activities in major cities. *Journal of the Acoustical Society of America* 106, 4, Pt. 2: 2261. Presented at the 138th Meeting of the Acoustical Society of America, November 4, 1999.

Raichel, D. R., and Dallal, M. November 29, 1999. *Final Report: Construction Noise*, Contract No. 99F88151, Subagreement No. 001 with the New York City Department of Design and Construction, vols. 1 and 2. New York: The Cooper Union Research Foundation.

Schultz, T. J. 1978. Synthesis of social surveys on noise annoyance. *Journal of the Acousti-cal Society of America* 64(2): 277–405.

Schultz, T. J. 1982. *Community Noise Rating.* New York: Elsevier.

Singapore Government. 1990. *The Environmental Public Health Act (Chapter 95) S466/90 (Control of Noise from Construction Sites) Regulations.*

Problems for Chapter 13

1. An outdoor survey of a neighborhood indicated averaged sound pressure levels at 40 dB(A) from 7 A.M. to 9 A.M.; 55 dB(A) from 9 A.M. to 3 P.M.; 59.5 dB(A) from 3 P.M. to 10 P.M.; and 39.8 dB(A) from 10 P.M. to 7 A.M. What is the day–night sound level? Is the value obtained acceptable for a residential area?

2. A factory environment was found to have an average of 88 dB for 2.5 hr, 90 dB for 1.6 hr, 92 dB for 1.4 hr, and 95 dB for 2.5 hr. Is the 8-hr exposure acceptable by OSHA standards? If not, what can be the maximum total daily working time, given the same histographic distribution?

3. A lathe in a machine shop was found to yield the following octave-band analytical results:

Center frequency, Hz	Octave band loudness, dB
63	60
125	70
250	68
500	75
1000	72
2000	75
4000	74
8000	58

   What is the RC rating for this machine shop?

4. Redo Problem 2 but find L_{eq} on the basis of the information given. If the 8-hr day is excessive exposure by the European Union standards, how much must that workday be cut down to in order to meet the daily exposure guidelines?

5. Examine the U.S. OSHA noise exposure regulations for workers and compare them with the counterpart European regulations. Which set of regulations is better for the worker? State the reasons for your choice.

6. Predict the mean emission levels for (a) an automobile going 135 km/h, (b) a medium truck moving at 120 km/h, and (c) a heavy truck traveling at 100 km/h.

7. Find the traffic flow adjustment for 3600 cars per hour moving at an average of 100 km/h for a observation point 15.2 m from the road.

8. A number of noises sources are located at different distances from a measurement point on a property line. The data consisting of the measured sound levels of each individual sources are given below:

Number of sources	1	2	3	1	2	4
L at 25 ft	80	75	82	90	84	87
Distance, ft	58	78	74	68	120	92

   (a) Determine the noise contribution of each source or group of sources.
   (b) Find the combined noise level at the measurement point on the property
       line.
9. Use the date of Problem 7 to compute L_{eq24} and L_{dn} if these sources run
   continuously at the same noise output levels.
10. Find the permissible sound power levels of the following highway construction
    vehicles under the European Union rules:
    (a) a nonexcavating tracking vehicle rated at 120 kW
    (b) a 200-kW bulldozer
    (c) a 100-kW excavator
    What will be these corresponding values after the year 2001?
11. Outline a noise code suitable for a suburban residential neighborhood. It should
    include the appropriate noise level limits, and specification of how and where
    measurements are to be taken.
12. Develop a noise code for a rural jurisdiction.
13. What are the problems that are applicable to urban areas? What are the special
    needs and goals with respect to noise levels in large cities?

14
Machinery Noise Control

14.1   Introduction

Workplace noise at high levels is detrimental to the welfare of workers. Not only do high sound levels affect hearing and hinder oral communication, they detract the employee from performing at peak capacity. In spite of possible economic drawbacks such as the cost of increased maintenance, the employer does have the moral obligation to provide a safe environment for both office and plant workers. Often when noise-attenuation measures are taken, some payback may accrue from the use of quieter machinery. For instance, when a diesel engine undergoes excessive vibration it becomes subject to severe stresses that can cause it to fail. Retuning of the engine so that it operates more smoothly lessens the stresses on its crankshaft or accessory parts and cuts down on its fuel consumption as well as the noise output. In many highly industrialized nations, such as the United States, Germany, France, the Scandinavian nations, and Japan, there are regulations that limit noise exposure levels. Some of these regulations were discussed in Chapter 13.

This chapter deals with industrial noise sources, predictions of their respective acoustic output, and means of attenuating noise in the workplace. Specific types of noise sources are considered, followed by descriptions of general methodologies of noise control which may or may not be machine specific.

14.2   Noise Sources in the Workplace

A considerable number of industrial machines and processes generate high levels noise which can cause physical and psychological stresses as well as considerable hearing loss. High-noise output machinery includes blowers, air nozzles, riveters, pneumatic chisels and hammers, diesel generators, chipping hammers, rock crushers, die casting machines, drop hammers, metal presses, power saws, grinders, ball mills, stamping machines, and so on. In addition, building accouterments, such as furnaces, air conditioning and ventilation (HVAC) systems and plumbing as well as office equipment, can roil the office environment and add to the cacophony of an industrial plant.

In setting up noise control measures, the first step is to identify the noise sources and to measure the sound power output. Ideally, it would be desirable to take a source into a well-defined environment such as a reverberation chamber and measure its sound power output, but many stationary sources cannot be moved. But the sound power output for immovable sources can still be estimated from sound pressure measurements made on a hemispherical or rectangular enveloping surface (cf. Chapter 9). The sound pressure level will be increased by the presence of background noise and room reverberation, so a correction factor for either/both background noise must be applied in such cases. For reasons which we shall see in the sections following, a spectrum analysis of noise from specific machines will often prove useful in tracing malfunctioning machinery parts so that they can be realigned or replaced.

Many noise problems in the workplace can be avoided by heeding the old adage that an ounce of prevention is worth a pound of cure. Prior to any purchase of machinery, the sound power output of each unit should be obtained beforehand, either directly from the vendor or by conducting actual measurements on an existing installation. Wise planning of the plant layout includes not only promoting production efficacy and personnel safety; it should always involve prediction of noise output of all equipment in normal operation. Adjustments can then be made at this stage in the choice of quieter equipment and/or incorporating noise reducing devices so that the planned facility will operate within the sound exposure limits mandated for workers.

14.3   Estimation of Noise Source Sound Power

A nondirectional point source in a free field will radiate sound uniformly and radially in all directions. Such a source L_W represents the true octave-band sound power level, with units of decibels based on the reference power 1 pW (10^{-12} W). Most pieces of machinery, however, are not point sources, nor do they radiate sound power uniformly. In planning of facilities, it is generally necessary to estimate the expected sound power for individual machines that will affect the environment. For certain machines, a sound power conversion factor F_n can be used to determine the output on the basis of the total power rating of the machine,

$$P = F_n \times P_m \tag{14.1}$$

where

$$P = \text{sound power of the machine, W}$$

$$P_m = \text{machine rated power, W}$$

The relationship of (14.1) applies to both mechanical and electrical machinery. Estimated conversion factors for a number of common machinery are listed in Table 14.1. It should be noted that the ranges are quite large for any machine.

TABLE 14.1. Power Conversion Factors for Some Common Noise Sources.

Noise Source	Conversion Factor F_n		
	Low	Midrange	High
Compressors, air (1–100 hp)	3×10^{-7}	5.3×10^{-7}	1×10^6
Gear trains	1.5×10^{-8}	5×10^{-7}	1.4×10^{-6}
Engines, diesels	2×10^{-7}	5×10^{-7}	2.5×10^{-6}
Motors, electric (1200 rpm)	1×10^{-8}	1×10^{-7}	3×10^{-7}
Pumps, >1600 rpm	3.5×10^{-6}	1.4×10^{-5}	5×10^{-5}
Pumps, <1600 rpm	1.1×10^{-6}	4.4×10^{-6}	1.6×10^{-5}
Turbines, gas	2×10^{-6}	5×10^{-6}	5×10^{-5}

Example Problem 1

Estimate the sound power of a "quiet" 100-hp electric motor operating at 1200 rpm.

Solution

From Table 14.1 $F_n = 1 \times 10^{-8}$. Applying equation (14.1), and because 1 hp = 746 W,

$$P = F_n \times P_m = (1 \times 10^{-8})(100\ \text{hp} \times 746\ \text{W/hp}) = 7.46 \times 10^{-4}\ \text{W}$$

Then

$$L_W = 10 \log \left(\frac{P}{10^{-12}} \right) = 10 \log \left(\frac{7.46 \times 10^{-4}}{10^{-12}} \right) = 89\ \text{dB}$$

14.4   Fan or Blower Noise

Fans (or blowers) are devices that use power-driven rotating impellers to move air. A fan has at least one inlet and at least one outlet. The rotating impeller imparts mechanical energy from the fan shaft to the airstream; the energy in the air exists as kinetic energy and air pressure. Different fan applications require fans with appropriate operating characteristics, including those of noise output. These characteristics are determined principally by the design of the rotating impeller.

Two principal types of fans—centrifugal fans and axial-flow fans—are normally used for central air-conditioning systems, industrial ventilation systems, and industrial processing applications. While many manufacturers make no distinction between fans and blowers, the latter term is more often used to describe high-pressure devices used to convey material, for example, a dust or a leaf blower. Centrifugal fans are low-pressure high-flow volume devices, while axial fans generally operate at higher pressures and tend to be noisier than centrifugal fans. The selection of the fan type, size and speed depends firstly on the performance necessary to

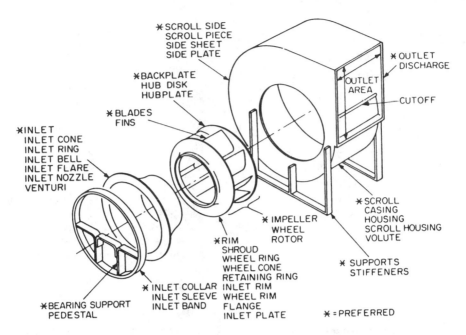

FIGURE 14.1. Exploded view of a centrifugal fan. (Harris, 1991, p. 41.2.)

move a given amount of air against a specified pressure, and the noise character-istics are then established on a secondary basis. Figures 14.1 and 14.2 show the general construction of a centrifugal fan and an axial-flow fan, respectively, along with listing of commonly used nomenclature.

Centrifugal fans come with a variety of blades. Three types are shown in Figure 14.3 Axial-flow fans divide into three main categories (vaneaxial, tubeax-ial, and propeller), as illustrated in Figure 14.4. In addition to flow and pressure requirements, fans are selected to meet environmental conditions, withstand cor-rosion, ease of maintenance, budget limits, and so on. The noise characteristics of various types of fans are fairly predictable, as they are not significantly altered by minor changes in the fan geometry.

Fan Noise Characteristics

Table 14.2 lists the broad-band noise characteristics of typical fan designs, The average specific sound power levels in eight octave bands are given for well-designed fans installed in well-designed systems. This data can be utilized to estimate fan noise at the design stage. In the selection process during the design stage, actual noise data should be obtained from the manufacturer. Octave-band noise levels should be used in calculations. Single-number ratings for fan noise should be avoided.

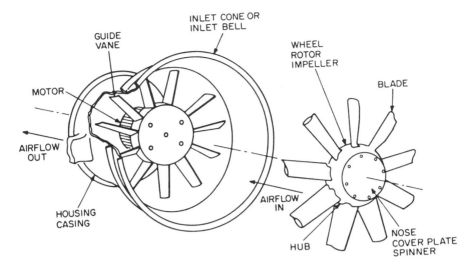

FIGURE 14.2. Components of an axial flow fan. (Harris, 1991, p. 41.2.)

When a blade passes over a given point, the air receives an impulse. The repetition rate of this impulse, termed the *blade frequency*, determines the fundamental tone that is produced by the blade. It can be predicted from

$$f_B = \frac{nN}{60} \qquad (14.2)$$

where

f_B = blade frequency, Hz

n = fan speed, number of revolutions per minute (rpm)

N = number of blades in the fan rotor

According to equation (11.23), the sound pressure level L_p from a specific source

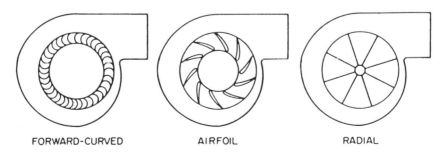

FIGURE 14.3. Three types of centrifugal fans. (Harris, 1991, p. 41.3.)

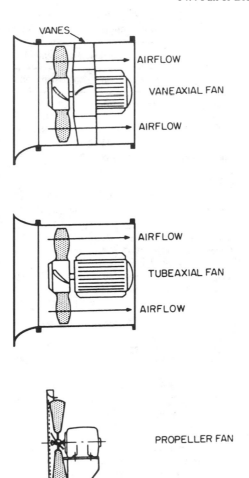

FIGURE 14.4. Three types of axial flow fans. (Harris, 1991, p. 41.3.)

in a room depends on room conditions as well as the sound power level L_W of that source. In order to predict the contribution of the fan noise to the room sound level, the data on sound power levels of the fan models under consideration should be provided by the manufacturer. However, the manufacturer has no control over the system design for the room, nor of the acoustical nature of the room and so cannot be responsible for the resultant noise level. If lower sound pressure levels are required than those generated by even properly designed fans, then it may be necessary to provide acoustic attenuators installed as an integral part of the fan assembly. Figure 14.5 shows a centrifugal fan with sound attenuators on both the inlet and the outlet, which is used in the supply system of a central station ventilating system. A sound attenuator is fitted at the outlet to lessen the flow of acoustic energy from the discharge of the air to the supply air ductwork.

TABLE 14.2. Relative Sound Power Generated by Different Types of Fans.

Fan type	Wheel size	63	125	250	500	1000	2000	4000	8000	BFI[a]
		\multicolumn{9}{Octave-Band Center Frequency, Hz}								
Centrifugal fans										
Airfoil or										
backward-curved										
or	Over 0.75 m	85	85	84	79	75	68	64	62	3
backward-inclined	Under 0.75 m	90	90	88	84	79	73	69	64	3
Radial fans:										
Low-pressure										
(4- to 10-in.	Over 1 m	101	92	88	84	82	77	74	71	7
static pressure)	Under 1 m	112	104	98	88	87	84	79	76	7
Medium-pressure										
(10- to 20-in.	Over 1 m	103	99	90	87	83	78	74	71	8
static pressure)	Under 1 m	113	108	96	93	91	86	82	79	8
High-pressure										
(20- to 60-in.	Over 1 m	106	103	98	93	91	89	86	83	8
static pressure)	Under 1 m	116	112	104	99	99	97	94	91	8
Forward-curved	All	98	98	88	81	81	76	71	66	2
Axial fans										
Vaneaxial:										
Hub ratio 0.3–0.4	All	94	88	88	93	92	90	83	79	6
Hub ratio 0.4–0.6	All	94	88	91	88	86	81	75	73	6
Hub ratio 0.6–0.8	All	98	97	96	96	94	92	88	85	6
Tubeaxial	Over 1 m	96	91	92	94	92	91	84	82	7
	Under 1 m	93	92	94	98	97	96	88	85	7
Propeller	All	93	96	103	101	100	97	91	87	5

[a] BFI = blade frequency increment.

Note: The data listed here are given in terms of specific sound power in dB re 1 $(10)^{-12}$ watt based on a volume flow of 1 m^3/s and a total pressure of 1 kPa. Equation (14.3) must be used to adjust for actual pressures and volume flow rates. To convert these values into English units, subtract 45 dB in all bands. The base for the English units is a total pressure of 1 in. water gauge and a volume flow rate of 1 ft^3/min. These values are those for total sound power radiated from the fan. To obtain the power levels at either the inlet or the outlet, subtract 3 dB from all bands. No change in blade frequency increment is to be made. In performing the calculations for sound power levels from this table, do not use a total pressure less than 0.125 kPa. For applications where the total pressure is lower than 0.125 kPa, simply use the value of 0.125 kPa.

Specific Sound Power Level

Specific sound power level is defined as the sound power level generated by a particular fan operating at an air flow rate of 1 m^3/s (2120 cfm) and at a pressure of 1 kPa (4.0 in. water gauge). Table 14.2 lists the relative sound power for a variety of fan types. In reducing all fan noise data to this common base, the concept of specific sound power level allows direct comparison to be made between the octave band levels of different types of fans. A blade frequency increment (BFI) is also listed in Table 14.2; this represents the number of decibels that must be added to the level of the octave band, which includes the blade frequency, in order to

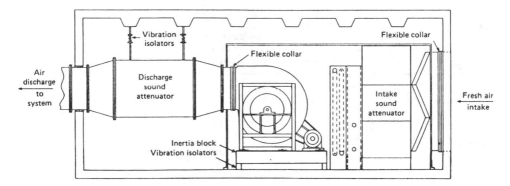

FIGURE 14.5. Centrifugal fan with sound attenuators at inlet and outlet. (Harris, 1991. p. 41.13)

account for the presence of such a tone. Also, a means is provided for estimating the noise level of fans under actual operating conditions by the procedure which consists of the following steps:

1. Select the fan type and obtain the specific power levels in octave bands from Table 14.2. These sound power levels are expressed in decibels re 1 picowatt.
2. Adjust the octave-band levels for the volume flow rate and the operating pressure by adding to each octave band one of the following values:

$$10 \log Q + 20 \log p_t \text{ dB} \qquad \text{(for metric units)} \qquad (14.3a)$$

or

$$10 \log Q + 20 \log pt - 45 \text{ dB} \qquad \text{(for English units)} \qquad (14.3b)$$

where

$$Q = \text{volume flow rate, m}^3\text{/s or cfm}$$

$$p_t = total\ pressure, \text{kPa or inches water gauge}$$

3. Account for the blade frequency component of the fan by adding the BFI for the fan type chosen to the octave-band level of that band which includes the blade frequency. The blade frequency is found from the use of equation (14.2)
4. The sum of the above equals the total sound power level of the radiation from the inlet *and* the outlet. Subtract 3 dB in each octave band to yield the sound power level of radiation from the inlet *or* the outlet.

Example Problem 2

Consider a radial forward-curved fan with 24 blades, having a rotor diameter of 0.8 m, and operating at 750 rpm with a volume flow rate of 18 m³/s and with a total pressure of 1.5 kPa. Find the total sound output power at the inlet.

TABLE 14.3. Calculation Results for Sound Power Level.

Procedural Step	Octave-Band Center Frequencies, Hz							
	63	125	250	500	1000	2000	4000	8000
1	98	98	88	81	81	76	71	66
2	16	16	16	16	16	16	16	16
3	0	0	2	0	0	0	0	0
4	-3	-3	-3	-3	-3	-3	-3	-3
Radiation from inlet	111	111	103	94	94	89	84	79

Solution

We use Table 4.2 and list the calculations at each step for each octave band level in Table 4.3. For step 2, using equation (14.3a)

$$10 \log Q + 20 \log p_t = 10 \log 18 + 20 \log 1.5 = 16$$

The blade frequency is from equation (14.2)

$$f_B = \frac{nN}{60} = \frac{750 \times 24}{60} = 300 \text{ Hz}$$

which lies in the octave band with the center frequency of 250 Hz. The BFI is 2 dB. Step 3 is now completed. Step 4 consists of subtracting 3 dB in each octave band to obtain the noise from the inlet alone.

Fan Laws

Fan laws can predict fan performance quite well over a wide range of size and speed. These laws are as follows:

$$Q_a = Q_b \left(\frac{d_a}{d_b}\right)^3 \left(\frac{n_a}{n_b}\right) \tag{14.4}$$

$$P_{ta} = P_{tb} \left(\frac{d_a}{d_b}\right)^2 \left(\frac{n_a}{n_b}\right)^2 \tag{14.5}$$

$$P_a = P_b \left(\frac{d_a}{d_b}\right)^5 \left(\frac{n_a}{n_b}\right)^3 \tag{14.6}$$

$$L_{wa} = L_{Wb} + 70 \log \left(\frac{d_a}{d_b}\right) + 50 \log \left(\frac{n_a}{n_b}\right) \tag{14.7}$$

where

$$Q = \text{volume flow rate, m}^3\text{/s}$$

$$p_t = \text{total pressure, kPa}$$

$$P = \text{fan power, kw}$$

$$L_W = \text{sound power level, dB re 1 pW}$$

$$d = \text{rotor diameter, m}$$

$$n = \text{rotor speed, rpm}$$

Subscript a denotes the parameters for the base curve performance conditions, and subscript b denotes the parameters for the desired performance conditions.

Equation (14.7) is less accurate than equations (14.4)–(14.6) for predicting performance characteristics, but it is sufficiently accurate for estimating purposes. These fan laws state mathematically that when two fans have similar design configurations, their performance curves are similar, and at the equivalent point of rating on each performance curve the efficiencies should be equal. In order to apply the fan laws, it is necessary to have the test data for one fan in the same design series. The applicability of the fan laws is restricted to cases where all linear dimensions of the larger or smaller fan are proportional to the fan for which there are test data.

14.5   Electric Motors and Transformers

Electric motors convert electrical power into mechanical power. Some of the fundamental noises occurring in electric motors are caused by rotational unbalance, rotor/stator interaction, and slot harmonics. The electromagnetic force between the armature and the field magnet, or rotor and stator, gives rise to vibration. Noise is also generated by the excitation of natural frequencies of the motor structure, air resonance chambers, and the movement of air itself. The intensity of the noise is typically a function of rotational speed and motor type.

As a means of estimating the noise output of a motor, the following expression can be used to obtain the total sound power level in the bands 500, 1000, 2000, and 4000 Hz:

$$L_w = 20 \log \text{hp} + 15 \log n + K_m \qquad \text{dB} \qquad (14.8)$$

where

$$\text{hp} = \text{rated horsepower (1–300 hp)}$$

$$n = \text{rated speed of motor, rpm}$$

$$K_m = \text{motor constant} = 13 \text{ dB}$$

Other more involved techniques have been developed, which entail different motor constants for each of the octave bands of interest (Magrab, 1975; Webb, 1976).

Transformers exist for the purpose of stepping up or stepping down voltages. Their changing magnetic field causes deformation of the transformer coil, occurring at the alternating (AC) current frequency and higher harmonics, especially at the twice the AC frequency. This results in the characteristic frequency of 120 Hz in the United States and 100 Hz wherever 50-Hz AC current is used.

14.6   Pumps and Plumbing Systems

Pump noise arises from both hydraulic and mechanical sources, namely, cavitation, pressure fluctuations in the fluid, impact of mechanical parts, imbalance, resonance, misalignment, and so on. The hydraulic causes are, however, the predominant noise generators. Pumps will generate even more noise if they are not operated at rated speed and discharge pressure, when the rate of compression is high or the inlet pressure is below atmospheric, or if the temperatures run too high. An unfortunate situation is created when the noise from the pump easily transmits through the fluid or piping to other system components.

Pumps generate two types of noise: discrete tones and broadband noise. The pump's fundamental frequency f_p is found from

$$f_p = \frac{n \times N_c}{60} \tag{14.9}$$

where

$n = $ pump rotational speed, rpm

$N_c = $ number of pump chamber pressure cycles per revolution

In large pumps the noise emission is the loudest at this fundamental frequency. As the pump size decreases, the frequency at which the maximum noise emission occurs increases, often to a frequency that constitutes a third or fourth harmonic of the fundamental. Above 3 kHz, the noise becomes more broadband, approaching an essentially flat spectrum. This is due to phenomena such as high-velocity flow and cavititation.

The total sound power level of pumps in the four octave bands of center frequencies 400, 1000, 2000, and 4000 kHz can be estimated from the following:

$$L_w = 10 \log hp + K_p \quad dB \tag{14.10}$$

where $K_p = $ pump constant which has the following values: 95 dB for the centrifugal type, 100 dB for the screw type, and 105 dB for reciprocating type. For rated speeds below 1600 rpm, 5 dB is to be subtracted for reciprocal pumps. The sound power in each of the four bands may be considered to be 6 dB less than the total sound power L_w computed from equation (14.10).

Controlling Noise in Plumbing Systems

Table 4.4 lists the sources of noise in a building's plumbing system and their likelihood of being annoying. Flow can be either *laminar* flow, that is, smoothly flowing, or *turbulent flow*, in which occurs an irregular, random motion of the fluid

TABLE 14.4. Different Types of Plumbing Noises. Their Means of Generation, and Annoyance Potentials. (From Harris, 1991, Ch. 44).

Plumbing System Component/Equipment	Generation Mechanism	Potential Annoyance
Piping runs:		
Couplings	Turbulence	Minimal
Elbows	Turbulence	Minimal
Tees	Turbulence	Minimal
Fixtures:		
Bar sink	Cavitation/turbulence/splash/ waste flow	Minimal
Bathtub	Cavitation/turbulence/splash/ waste flow	Very significant
Bidet	Cavitation/turbulence/waste flow	Nominal
Flushometer	Cavitation/turbulence	Significant
Hose pipe valves	Cavitation/turbulence	Nominal
Laundry tubs	Cavitation/turbulence/splash/ waste flow	Nominal
Pressure regulator	Cavitation/turbulence	Nominal
Shower	Cavitation/turbulence/splash/ waste flow	Very significant
Sink	Cavitation/turbulence/waste flow	Significant
Valves	Cavitation/turbulence	Significant
Water closet, tank stool	Cavitation/turbulence/splash/ waste flow	Very significant
Urinal	Cavitation/turbulence/splash/ waste flow	Nominal
Appliances:		
Dishwasher	Vibration/cavitation/spray/ water hammer	Very significant
Drinking fountain	Cavitation/turbulence	Minimal
Washing machine	Vibration/cavitation/impact/ motor/water hammer	Very significant
Waste disposal	Vibration/waste flow	Very significant
Water heater	Cavitation/turbulence	Minimal
Supply and waste pumps:		
Booster	Rotational flow/cavitation/ motor	Significant
Recirculation	Rotational flow/cavitation/ motor	Nominal
Sewage	Rotational flow/cavitation/ motor	Significant
Sump	Rotational flow/cavitation/ motor	Significant

particles. The influencing factor that determines whether a flow will be turbulent or laminar is the *Reynolds number Re*, a dimensionless parameter defined by

$$Re = \frac{\rho\, dv}{\mu}$$

where ρ is the density of the fluid, d is the internal pipe diameter, v is the flow velocity, and μ is the absolute viscosity of the fluid. For $Re < 2000$, the flow is laminar. For transition region $2000 < Re < 4000$ the flow may be either laminar or turbulent. Above $Re = 4000$, the flow will be turbulent. Noise generated by laminar flow tends to be quite low in intensity and is usually of no concern.

In most real plumbing systems, the velocities are sufficiently high to result in turbulent flow, which is a basic mechanism for noise generation within piping runs and fixtures of the plumbing system. A potentially great cause of noise is cavitation, which is the formation and subsequent collapse of cavities (bubbles) within the flow of water through and past a restriction in the flow. For cavitation to occur, a localized restriction or a projection must exist within the piping system, which ensues in localized high velocities and low pressures. The formation and sudden collapse of these bubbles result in extreme local pressure fluctuations, which can be detected as noise. Other water noises occur from splashing (i.e., impact of liquid striking a surface) and waste water flow (i.e., flow into drainpipes).

Much more serious is the sharp intense noise known as *water hammer*. It occurs when a steady flow in a liquid flow system is suddenly interrupted, for example, by a quick-action valve. When the fluid is in motion throughout a piping system, even at relatively low velocities, the momentum from this sudden interruption can be quite large. The sudden interruption of the flow creates an extremely sharp pressure rise that propagates as a shock wave upstream from the interrupting valve. The steep wavefront can be reflected numerous times back and forth throughout the various parts of the piping system until the energy is finally dissipated.

Noise control factors that must be dealt with in order to effect noise control include (a) water flow and piping characteristics, (b) radiation of sound to the building structure, (c) selection and mounting of fixtures, (d) isolation of pump systems, and (e) water hammer noise control.

Water pressure in a plumbing system influences the flow noise caused by pipe runs and water supply valves. According to typical building codes, water pressure should be maintained at at least 100 kPa (15 psi), but not more than 500 kPa (80 psi). For acceptable system performance, the supply pressure should be somewhere between 230 and 370 kPa (35 and 55 psi), with a preference toward the lower range in order to minimize noise. Flow noise radiation from pipes can be lessened by minimizing the number of pipe transitions (elbows, tees, Y-connections, and the like). This reduces the opportunity for the onset of turbulence and cavitation. Larger pipes are used in building design when noise control is given a higher priority. In the United States, 1/2-inch diameter piping is generally used in domestic plumbing systems, but it can be as high as 3/4-inch diameter to cut down on noise by as much as 3 to 5 dB.

Noise resulting from water flow in pipes may be transmitted to the rooms through which they pass, particularly if they are in direct contact with large radiation surfaces such as walls, ceilings, and floors. Isolation of these pipes from the building structure provides significant noise reduction. If the pipes are mounted with foam isolators, instead of being rigidly connected to the building structure, a considerable noise reduction of 10–12 dB may be obtained. Whenever piping passes through a structure (e.g., block, stud, joist, or plate) or is in contact with a wall or masonry, resilient materials such as neoprene or fiberglass should be used to provide isolation, It is vital to seal with a resilient caulking around the perimeter of all pipes, faucets, and spouts which penetrate through floors, walls, and shower stalls.

The impact of water hammer can rupture piping; thus, it will spring leaks, cause weakening of connections and produce damage to valves. Water-hammer pulsing associated with the use of washing machines and dishwashers can be partially damped by connecting these machines to the water supply with extra-long flexible hose. Figure 14.6 shows a schematic of a capped pipe which provides an air chamber for water hammer suppression. The length of the pipe ranges from 30 to 60 cm and may be of the same or larger diameter than that of line it serves. The volume of the air chamber, which serves as an air cushion, depends on the nominal pipe diameter, the branch line length, and the supply pressure. But if the air chamber becomes filled with water, it becomes ineffective. A petcock is

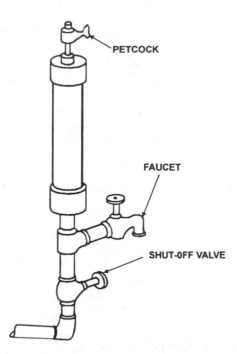

FIGURE 14.6. A capped pipe which serves to arrest water hammer.

provided along with a shut-off valve, so that the chamber may be drained of water and vented, thus reactivating the unit.

There are commercial devices, called *water-hammer arresters*, which are not subject to the limitation of capped pipes, because a metal diaphragm separates the water from the air. These devices are best placed near quick-acting valves and should also be installed at the termini of long pipe runs.

14.7   Air Compressors

Air compressors find wide use in industry, and they are identified as a major source of noise. Usually driven by a motor or a turbine, these devices are used to elevate the pressure of air or another gas. The noise emission characteristics are a function of the type of unit. In portable air compressors the driving engine is the major source of noise rather than the compressor itself. The second largest source is the cooling fan. Compressors are either rotary or reciprocating, A reciprocating compressor typically generates a strong low-frequency pulsating noise, the characteristics of which depend on the rotational speed and the number of cylinders. In centrifugal compressors the noise generated is a function of a number of parameters such as the interaction of the rotating and stationary vanes, the radial distance between impeller blades and diffuser vanes, the rotational speed, the number of stages, the inlet design, the horsepower input, turbulence, the molecular weight of the gas undergoing compression, and the mass flow.

As in the case of fans, the blade passage frequency constitutes an importance frequency component in certain types of compressors. In the diffuser-type machines, the *blade rate component*, arising from the movement of one set of blades past another, is of primary importance. The frequency of this component is found from

$$f_{BRC} = \frac{N_r \times N_s}{60 \, K_{BRC}} \times n \quad \text{Hz} \tag{14.11}$$

where

$$f_{BRC} = \text{blade rate component frequency, Hz}$$
$$N_r = \text{number of rotating blades}$$
$$N_s = \text{number of stationary blades}$$
$$K_{BRC} = \text{greatest common factor of } N_r \text{ and } N_s$$
$$n = \text{rotational speed, rpm}$$

Example Problem 3

Find the frequency of the blade rate component of a diffuser-type compressor with $N_r = 16$ and $N_s = 24$ and operating at a speed of 6000 rpm.

Solution

Using equation (14.11) we find that

$$f_{BRC} = \frac{N_r \times N_s}{60\, K_{BRC}} \times n = \frac{16 \times 24}{60 \times 8} \times 6000 = 4800 \text{ Hz}$$

If the blade-rate component frequency falls within the audio range, we should expect an increment of several decibels in the octave-band sound power level in which it occurs.

The total sound power level in the four octave bands with center frequencies of 500, 1000, 2000, and 4000 can be estimated roughly for both reciprocating and centrifugal compressors from the expression

$$L_w = 10 \log \text{hp} + K_c \qquad \text{dB} \qquad (14.12)$$

where K_c = air compressor constant = 86 dB for the 1–100 hp range. Equation (14.12) is very similar to equation (14.10) for pumps, and either expression will yield a straight line in a semi-log plot. Also, it is not unreasonable to estimate that the sound power is equally divided among the four octave bands. Thus, each band level is 6 dB below the total estimated from equation (14.12).

14.8   Gears

Internal combustion engines and electric motors generally operate at speeds of one to several thousand revolutions per minute. These high speeds help maximize the power-to-weight and power-to-initial cost ratios. Gearing and other speed reducers are applied when the driven machinery requires high torque and low speeds. Mechanical power, shaft speed, and torque are related as follows,

$$P_{kW} = 10^{-6}\, T\omega \qquad (14.13)$$

where

$$P_{kW} = \text{transmitted power, kW}$$

$$T = \text{torque, N} \cdot \text{mm}$$

$$\omega = \pi n/30 = \text{angular velocity, rad/s}$$

$$n = \text{shaft speed, rpm}$$

In English units, equation (14.13) becomes

$$P_{hp} = \frac{Tn}{63{,}025} \qquad (14.14)$$

where

$$P_{hp} = \text{transmitted power, hp}$$

$$T = \text{torque, lb-in}$$

$$n = \text{shaft speed, rpm}$$

It was observed by Hand (1982) that gear noise increases with speed at the rate of 6–8 dB per doubling of speed, It was also observed that an increase of 2.5–4 dB in gear noise occurs for each doubling of load. Thus, according to equation (14.13) or (14.14) a reduction in speed results in an increase in torque, if the transmitted power is to be sustained, and so the noise effect of speed reduction is somewhat offset by the increase in the torque.

Meshing Frequencies

The profile of most gear teeth is that of an involute curve. Force transmits through the driving gear to the driven gear along the line-of-action, which is fixed in space (excepting planetary gear trains). If the gears are ideal, perfectly fabricated, rigid, and transmitting constant torque, the power should be transmitted smoothly and without vibration or noise. Real gears, however, have tooth errors in spacing and tooth profile, and in some cases, an appreciable shaft eccentricity. Gear teeth do act elastically and flex slightly under load. Consequently, the driving gear teeth that are not in contact deflect ahead of their theoretical rigid-body positions, while the driven teeth that are not in contact lag behind their theoretical positions. This results in a rather abrupt transfer of force when each pair of teeth comes into contact, instantaneously accelerating the driven gear and decelerating the driving gear. The fundamental frequency of the noise and vibration is given by

$$f = \frac{nN}{60} \qquad (14.15)$$

where

f = the fundamental tooth meshing frequency, Hz

n = rotational speed of the gear in question

N = the number of teeth in the gear in question

Harmonics—the noise and vibration at integer multiples of the tooth meshing frequency—are usually present. The most significant contributions are usually made by the first two or three harmonics, $f_2 = 2f_1$, $f_3 = 3f_1$, and so on. A typical gear noise spectrum is displayed in Figure 14.7.

Tooth Error

If a single tooth is imperfectly cut, chipped, or damaged, then it will generate a noise impulse once every shaft revolution. The fundamental frequency of the noise or vibration due to tooth error is given by

$$f_{1E} = \frac{n}{60} \qquad (14.16)$$

where

f_{1E} = fundamental frequency due to tooth error, Hz

n = shaft speed

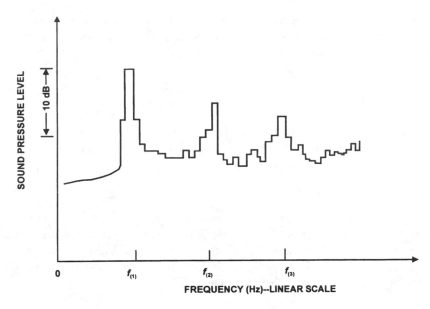

FIGURE 14.7. Gear noise spectrum.

Harmonics of tooth-error frequency may also occur. Moreover, if the shaft center-line is not straight or if a gear or bearing is not concentric with the shaft centerline, noise and vibration at the tooth-error frequency may result. These can result in *sideband frequencies* f_S that accompany the tooth meshing frequencies. These frequencies are found as follows

$$f_S = f_1 \pm f_{1E} \tag{14.17}$$

Gear Trains

Consider a pair of nonplanetary gears a and b in mesh. The tangential velocity V_t at the mesh is the same for both teeth. But the V_t of each tooth of gear i of radius r_i rotating at angular velocity ω_i is given by

$$V_t = r_i \times \omega \tag{14.18}$$

The number of teeth N_i on a gear i is directly proportional to the radius r_i of the gear, and $\omega_i = 2\pi n_i/60$. Because $r_a \times \omega_a = r_b \times \omega_b$ for gears a and b at their mesh point, we obtain

$$\frac{n_a}{n_b} = \frac{N_b}{N_a} \tag{14.19}$$

Thus, the ratio of rotational speeds is inversely proportional to the ratio of numbers of teeth: If both gears are external spur gears, the speed ratio is negative, that is, one gear will turn clockwise and the other counterclockwise. If one gear is an internal gear (i.e., a ring gear) the speed ratio is positive, meaning that both gears will rotate in the same angular direction. For a gear train comprised of several gears, the output to input speed ratio is given by

$$\frac{n_{\text{output}}}{n_{\text{input}}} = \prod_i \left(\frac{N_{\text{driving}}}{N_{\text{driven}}} \right)_i \tag{14.20}$$

An idler gear serves as both a driving gear and a driven gear. Idler gears must be included in determining the direction of rotation of shafts in a gear train.

Example Problem 4

In Figure 14.8 showing a gear train, gear 1 on the input shaft has 40 teeth and rotates at 3600 rpm. Gears 2, 3, and 4 have 90, 44, and 86 teeth, respectively. Narrow-band spectrum analysis of noise and vibration shows discrete tones and vibration energy peaks of 14, 27, 1227, 2400, 2455, and 4800. Establish the possible contributions to the noise and vibration at these frequencies.

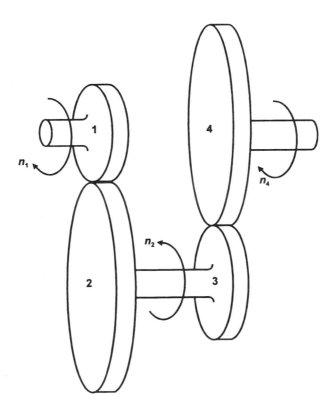

FIGURE 14.8. Gear train for Example Problem 4.

Solution

Gear speeds are determined from

$$\frac{n_2}{n_1} = \frac{N_1}{N_2}$$

from which we obtain

$$n_2 = n_3 = 3600 \times \frac{40}{90} = 1600 \text{ rpm}$$

From equation (14.20) we have

$$\frac{n_{\text{output}}}{n_{\text{input}}} = \frac{40 \times 44}{90 \times 86} = 0.227$$

and

$$n_{\text{output}} = 3600(0.227) = 818.7 \text{ rpm}$$

Fundamental tooth-error frequencies are given by

$$f_{1E} = \frac{n}{60}$$

$$= \frac{3600}{60} = 60 \text{ Hz for gear 1}$$

$$= \frac{1600}{60} = 26.7 \text{ Hz for gears 2 and 3}$$

$$= \frac{818.7}{60} = 13.6 \text{ Hz for gear 4}$$

Fundamental tooth-meshing frequencies are obtained from

$$f = \frac{nN}{60}$$

$$= 3600 \times \left(\frac{40}{60}\right) = 2400 \text{ Hz for gears 1}$$

$$= 1066 \times \left(\frac{46}{60}\right) = 1226.7 \text{ for gears 3 and 4}$$

If we compare the calculated results with the spectrum analysis, it seems that the 14- and 27-Hz discrete tones and vibration energy peaks correspond to the fundamental and the first harmonic of the tooth-error frequency for gear 4. This would indicate that the shaft straightness and the teeth of gear 4 should be investigated. The discrete tones of 1227 and 2455 Hz correspond to the fundamental and the first harmonic of the meshing frequencies of gears 3 and 4. Discrete tones at 2400 and 4800 Hz correspond to the fundamental and the first harmonic of the tooth meshing frequency of gears 1 and 2.

Contact Ratio (CR)

The contact ratio CR for a pair of gears is defined as the average number of pairs of teeth in contact. At least one pair must obviously be in contact at all times. Taking into consideration that tooth error, wear, shaft deflection, and machining tolerances affect the way gears work, a contact ratio of 1.2 is selected as the practicable minimum. When the contact ratio is near minimum value, contact commences at the tips of the driven teeth, and impact loads due to teeth meshing are high. A contact ratio of 2 or more usually results in better tooth load distribution and quieter operation, because two or more teeth will be in contact at all times.

Gears are selected to meet specified power and speed ratio requirements. These requirements affect the selection of gear material, tooth width, module or diametral pitch, number of teeth, and so on. If the number of teeth is increased for each gear in a train, the contact ratio will usually increase, with the result that the tooth meshing frequency and the accompanying noise level will decrease.

Helical Gears

When a pair of spur gear teeth begins to mesh, the contact occurs at once across the entire face width. In the case of a helical gear in which teeth entwine about an axial surface rather than cut straight across (Figure 14.9), the contact occurs gradually, with the contact beginning at a point and then extending across the tooth face. This results in reduced vibration and lower impact loads that produce tooth meshing frequency noise. Hence, helical gears on parallel shafts may be substituted for spur gear trains to reduce noise, and the actual number of teeth in contact will also be substantially increased. In order to ensure smooth and quiet operation, it is recommended that the thickness of the helical gear be 1.2–2 times the axial pitch (the distance between corresponding points on two adjacent teeth). Thrust (axial)

FIGURE 14.9. Helical gears. (Courtesy of Designatronics, Inc., New Hyde Park, NY.)

loads must be taken into consideration when specifying helical gears in the design of shafts and bearings. Well-made helical gears cost more than spur gears, but the payback in terms of smoother and quieter operation can more than offset the greater expense.

Other Aspects of Gear Noise Control

Gear production methods involve die casting, milling, drawing, extruding, stamping, and production from sintered metal. These methods tend to produce gears with significant tooth error to a more or lesser degree. Most gears produced by milling cutters also tend to have their tooth forms only approximate the correct configuration. In order to produce a precise tooth form, each milling cutter should be dedicated to only one diametral pitch or module and for a specific tooth number. In practice, a different milling cutter is used for each diametral pitch, but each cutter is used for a range of tooth numbers. Inaccuracies in cast, drawn or extruded gears result from shrinkage and other dimensional changes.

Precision methods of generating gear teeth usually result in cutting down noise due to tooth error. These include the computer-controlled generating rack cutter, the generating gear shaper cutter, and the generating hob. Precision gears are usually finished by grinding and shaving, and other finishing methods include burnishing, honing, and lapping.

Large gears benefit by damping to absorb vibration energy. Constrained layer damping, which sandwiches a layer of damping material between the gear web and a rigid steel plate, can reduce gear noise. If loads are low and the temperatures are not excessive for the material, fiber, plastic, fiberglass reinforced, and composite gears are used in such applications. These materials have low moduli of elasticity and high internal damping compared with steel. Shock loads from tooth meshing and tooth error are absorbed, thereby reducing noise.

Gear enclosures or housing can and should be designed to control noise. The enclosure should be isolated so as not to transmit vibration to adjacent structures; and provisions for adequate lubrication of the gears should be included. Resonances of the enclosure should not correspond to the tooth meshing or other excitation frequency; otherwise the housing will radiate a large amount of noise energy. This can be avoided by stiffening the enclosure structure to "tune out" its resonance to a higher frequency. The stiffeners themselves should be designed so that they have a low radiation efficiency. Advantage should be taken of the directivity patterns of noisy gear trains in the orientation of machinery, so that personnel noise exposure is minimized.

14.9   Journal Bearings

A journal bearing is the simplest type of bearing. It consists of portion of a shaft rotating inside a circular cylinder with a layer of lubrication separating the shaft and bearing surfaces. Hydrodynamic rotation depends on shaft rotation to pump a film of lubricant between the shaft and the bearing. If the lubricant viscosity and shaft rotational speed are adequate for the load on the project bearing area, then thick-film hydrodynamic lubrication will prevail. This ensures stable operation without metal-to-metal contact, as well as quiet operation. Starting, stopping, and load direction changing may result in temporary cessation of the lubricant film and in metal-to-metal contact. This calls for *hydrostatic* lubrication which introduces fluid (e.g., oil or air) to the bearing surface at a pressure sufficient to support the shaft, even when the bearing is stationary. A large increase in the noise level produced by a journal bearing is apt to indicate a failure of the lubrication system.

14.10   Ball and Roller Bearings.

An incipient failure of ball or roller bearing can be detected by noise and vibration measurements. For the most part, bearing life can be predicted, but premature failure can and does occur. The occurrence of discrete tones in the noise spectrum for an operating bearing may indicate manufacturing defects, and narrowband vibration spectra can be utilized to detect wear and defects.

A ball or roller bearing consists of the following elements: a set of balls or rollers are enclosed in a cage or separator sandwiched between an inner race (which is usually attached to the shaft) and the outer race (usually attached to the supporting stationary structure). When the shaft turns, the balls (or rollers) roll along the surface of the inner raceway, which allows for the rotational freedom of the shaft. The cage or separator follows the motion of the balls. The speed relationships are as follows:

$$\frac{n_O - n_C}{n_I - n_C} = -\frac{D_I}{D_O} \qquad (14.21a)$$

and

$$\frac{n_B - n_C}{n_I - n_C} = -\frac{D_I}{D_B} \qquad (14.21b)$$

where

$$n = \text{rotational speed, rpm}$$

$$D = \text{diameter, mm or in.}$$

and the subscripts denote the variables as follows:

$$I = \text{inner race}$$

$$O = \text{outer race}$$

$$B = \text{ball or roller}$$

$$C = \text{ball cage or separator}$$

The race diameters are measured at the point of contact with the balls or rollers.

Ball or Roller Bearings with Stationary Outer Races

In most bearing applications the outer race is stationary. Equation (14.21a) becomes

$$\frac{0 - n_C}{n_I - n_C} = -\frac{D_I}{D_O} \tag{14.22}$$

which yields the separator speed

$$n_C = \frac{n_I D_I}{D_O + D_I} \tag{14.23}$$

and the speed of the balls relative to the separator is given by

$$n_B^R = n_B - n_C = -\frac{(n_I - n_C)D_I}{D_B} = -\frac{n_I^R D_I}{D_B} \tag{14.24}$$

Here n_I^R is the speed of the inner race relative to separator speed. Equation (14.24) is obtained by eliminating n_B through the use of relation (14.21b).

Vibration and noise frequencies are related to the absolute speeds n, relative speeds n^R, and the number of balls or rollers v_B. For bearings with stationary outer races, the following fundamental frequencies are likely to arise:

1. Shaft imbalance

$$f = \frac{n_I}{60} \tag{14.25}$$

2. Outer race defect

$$f_O = \frac{n_O^R v_B}{60} = \frac{n_C v_B}{60} \tag{14.26}$$

3. Inner race defect

$$f_I = \frac{n_I^R v_B}{60} = \frac{(n_I - n_C)v_B}{60} \tag{14.27}$$

4. Defect or damage to one ball or roller

$$f_B = \frac{2n_B^R}{60} = \frac{n_B - n_C}{30} \tag{14.28}$$

5. Imbalance or damage in the separator or cage

$$f_C = \frac{n_C}{60} \tag{14.29}$$

With the appearance of the fundamentals as calculated above, harmonics $2f$, $3f$, ... are apt to occur.

Example Problem 5

Consider a ball bearing that uses twelve 8-mm-diameter balls. The inner and the outer diameters are $D_I = 28$ and $D_O = 44$ mm, measured at the ball contact points. The outer race is stationary and the inner race rotates at speed $n_I = 4500$ rpm clockwise. Determine (a) the rotational speeds and (b) the noise and vibration frequencies which will occur if there are defects and imbalance present.

Solution

It should be obvious that $D_I + 2D_B = D_O$

(a) Use equation (14.23) to find the separator speed

$$n_C = \frac{n_I D_I}{D_O + D_I} = 4500 \times \frac{28}{28 + 44} = 1750 \text{ rpm}$$

and, from equation (14.24), the speed of the balls relative to the separator is

$$n_B^R = n_B - n_C = -\frac{(n_I - n_C)D_I}{D_B} = -(4500 - 1750) \times \frac{28}{8}$$

$$= -9625 \text{ rpm (counterclockwise)}$$

(b) We find the fundamental frequencies as follows:

1. Shaft imbalance:

$$f = \frac{n_I}{60} = \frac{4500}{60} = 75 \text{ Hz}$$

2. Outer race defect:

$$f_O = \frac{n_O^R v_B}{60} = \frac{n_C v_B}{60} = \frac{1750 \times 12}{60} = 350 \text{ Hz}$$

3. Inner race defect:

$$f_I = \frac{(n_I - n_C)v_B}{60} = \frac{(4500 - 1750) \times 12}{60} = 550 \text{ Hz}$$

4. Defect or damage to one ball:

$$f_B = \frac{2n_B^R}{60} = \frac{n_B - n_C}{30} = -\frac{9625}{30} = 320.8 \text{ Hz}$$

Notice that the negative sign is ignored in this answer.

5. Imbalance or damage in the separator or cage

$$f_C = \frac{n_C}{60} = \frac{1750}{60} = 29.2 \text{ Hz}$$

Harmonics of all of the above frequencies are likely to all occur.

Ball or Roller Bearings with Nonstationary Outer Races

For some applications, the shaft attached to the inner race of a bearing remains stationary while the outer race rotates. In this case, $n_I = 0$ and we obtain the following speed relationships:

$$\frac{n_O - n_C}{0 - n_C} = -\frac{D_I}{D_O}$$

with carrier speed

$$n_C = \frac{n_O D_O}{D_O + D_I}$$

and speed of the balls with respect to the carrier

$$n_B^R = n_B - n_C = -\frac{(n_O - n_C)D_O}{D_B} = -\frac{n_O^R D_O}{D_B}$$

The frequencies that may occur when the outer race is nonstationary are as follows:

1. Shaft imbalance:

$$f = \frac{n_O}{60}$$

2. Outer race defect:

$$f_O = \frac{n_O^R v_B}{60} = \frac{(n_O - n_C)v_B}{60}$$

3. Inner race defect:

$$f_I = \frac{n_I^R v_B}{60} = \frac{n_C v_B}{60}$$

4. Defect or damage to one ball or roller:

$$f_B = \frac{2n_B^R}{60} = \frac{n_B - n_C}{30}$$

5. Imbalance or damage in the separator or cage:

$$f_C = \frac{n_C}{60}$$

Harmonics will also constitute additional components of the spectrum.

For any bearing fault, the noise and vibration spectra will be essentially tonal, that is, characteristic peaks will exist. A great deal of the noise energy will be concentrated in the bands that include the fundamental frequency and its first and second harmonics.

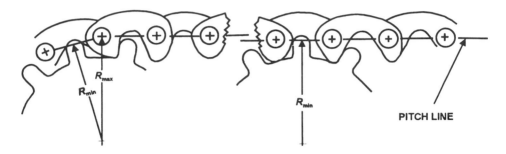

FIGURE 14.10. Roller chain.

14.11   Other Mechanical Drive Elements

Chain Drives

Figure 14.10 shows a roller chain meshing with toothed sprockets. The roller chain is constructed of side plates and pin and bushing joints designed to mesh with sprockets. The flexibility of the chain aids in limiting shock and vibratory forces, but the initial contact between the chain and sprockets can be noisy at high speeds. For transmission of high loads at high speeds, inverted tooth (or "silent") chains are often applied. Because the inverted-tooth chain engages with the sprockets with less impact force, it generally operates more quietly than roller chain.

The links are of finite length in either roller chain or inverted-tooth chain. While the chain links engage a sprocket, the pitch line (i.e., the centerline of the link pins) bobbles up and down due to the chordal action. In Figure 14.10, the center of a link pin is located at R_{max} from the center of the sprocket over which it moves. R_{max} is the maximum center of the pitch line from the sprocket center. As the link engages further, the pitch line moves nearer to the sprocket center and both pins of the link coincide with the pitch line. The new position is given by

$$R_{min} = R_{max} \cos\left(\frac{180°}{N_s}\right) \tag{14.30}$$

where N_S indicates the number of sprocket teeth. Roller chains usually have a minimum of 8 teeth and inverted-tooth chains, 17 teeth.

Consider a drive sprocket rotating at constant speed n_s. The pitch line velocity of the chain will change in direct proportion to its distance from the center of the sprocket, that is, $V = R \times (2\pi n_s)$. The fractional velocity change $V/\Delta V$ is given by

$$\frac{\Delta V}{V} = \frac{R_{man} - R_{mean}}{R_{mean}} = \frac{1 - \frac{1 + \cos(180°/N_s)}{2}}{\frac{1 + \cos(180°/N_s)}{2}}$$

$$= \frac{1 - \cos(180°/N_s)}{1 + \cos(180°/N_s)} \tag{14.31}$$

The angular velocity $(2\pi n_s)$ canceled itself out in the ratio of equation (14.31), but the speed ratio of a chain drive is given by

$$\frac{n_2}{n_1} = \frac{N_1}{N_2} \tag{14.32}$$

where n refers to the rotational speed (rpm) of the sprocket and N the number of sprocket teeth. Subscripts 1 and 2 denote the driver sprocket and the driven sprocket, respectively. Fundamental frequencies of noise and vibration can be due to the following:

1. Imbalance in the driver shaft or the driven shaft or damage to one sprocket tooth:

$$f_1 = \frac{n_1}{60} \text{ Hz} \qquad \text{and/or} \qquad f_2 = \frac{n_2}{60} \text{ Hz} \tag{14.33}$$

2. Damage to one chain link. In this case for a chain with N_L links, a defective link will strike each sprocket $n_1 N_1 / N_L$ times per minutes. Hence

$$f_{CL} = \frac{n_1 N_1}{30 N_L} \tag{14.34}$$

3. Tooth engagement and chordal action:

$$f_{TE/CA} = \frac{n_1 N_1}{30} \tag{14.35}$$

When the above fundamentals are present, harmonics are likely to occur. Noise levels and vibratory amplitudes due to tooth meshing and chordal action are functions of the sprocket configuration and also speeds and masses of the sprockets and chains. Also, if there are idler sprockets in the system, the values given in equations (14.33)–(14.35) will undergo change.

Example Problem 6

Given an 12-tooth drive sprocket rotating at 2400 rpm, a driven sprocket with 24 teeth and a chain with 60 links. Determine the velocity variation with respect to the mean velocity and the possible vibration and noise frequencies due to this chain drive.

Solution

We apply equation (14.31) to the smaller sprocket, that is, the driver, because the ratio is larger:

$$\frac{\Delta V}{V} = \frac{1 - \cos(180°/N_s)}{1 + \cos(180°/N_s)} = \frac{1 - \cos(180/12)}{1 + \cos(180/12)} = .0173$$

Here the velocity varies by nearly ±2% from the average. Equation (14.32) is used to find the speed of the driven sprocket's speed:

$$n_2 = n_1 \frac{N_1}{N_2} = 2400 \times \frac{12}{24} = 1200 \text{ rpm}$$

Equations (14.33)–(14.35) give us the following possible fundamental frequencies:

$$f_1 = \frac{n_1}{60} = \frac{2400}{60} = 40 \text{ Hz}, \qquad f_2 = \frac{n_2}{60} = \frac{1200}{60} = 20 \text{ Hz}$$

$$f_{CL} = \frac{n_1 N_1}{30 N_L} = \frac{2400 \times 12}{30 \times 60} = 16 \text{ Hz}, \qquad f_{TE/CA} = \frac{n_1 N_1}{30} = \frac{2400 \times 12}{30} = 960 \text{ Hz}$$

Belt Drives

Belt drives are often used instead of gears or chains in order to save on costs and provide some degree of noise control. The elasticity of the belt prevents shock loads in a driven machine from being transmitted back to the driver. Solid-borne noise and attendant vibration are reduced. Flat belts and V-belts mounted on smooth pulleys depend on friction to transmit power, so there must be adequate belt tension to prevent slipping as well as to oppose centrifugal effects. When suddenly loaded, these belts can slip, causing a squealing noise. This sometimes can be fixed by increasing the tension; but excessive tension can shorten the life of a belt and induce excessive bending moments in connected shafts. When precise speed ratios must be sustained, such as the camshaft of an automotive engine, toothed belts mounted on tooth pulleys can be utilized. Toothed belts can maintain timing and phase relationships just as well as meshing gears, but these belts isolate vibration and shock forces rather than transmitting them through the driving and driven elements.

Universal Joints

Universal joints are used when the relative position of a driving element changes with respect to a driven element, such as with an automobile transmission that is connected through a drive shaft to a rear axle. Flexible couplings and flexible shafts can also be used, but the former can accommodate only relatively slight misalignments, and the latter can handle large misalignments, but cannot handle large amounts of torque. In general, for a given rotational speed n, the frequency associated with noise and vibration is given by

$$f = \frac{n}{30} \text{ Hz}$$

14.12   Gas-Jet Noise

A most common and also worrisome noise source is the gas jet. This is also referred to as aerodynamic noise, and examples include blowdown nozzles, gas or oil burners, steam valves, pneumatic control discharge vents, aviation jet engines, and so on. An acoustically unmitigated steam valve of a large cooker in a major food processing plant can measure as much as 120 dB. Research on aerodynamically generated noise began in earnest in the early 1950s as the result of the appearance of the commercial jet engine, when it was realized that its mechanism of sound generation had to be understood better in order to effect noise control (Lighthill, 1952; 1978).

Prior to Lighthill's pioneering work, there had been even earlier studies made on aerodynamic noise in conjunction with efforts to reduce noise in axial fans. We can go back even further to find that the effect of jet streaming was mentioned in the earliest recorded references to sound. When the wind blew past the pillars of the ancient Greek temples, eerie discrete tones were produced. The Greeks adjudged these tones to be the voice of Aeolus, the god of the wind, and hence these tones are called *Aeolian tones*.

The mechanism of the Aeolian sound can be explained by visualizing the flow of air over a cylinder. At a given velocity, the downsteam flow behind the cylinder exhibits an oscillatory pattern, as vortices are shed alternatively on one side of the cylinder and then the other. The ensuing trail of eddies form what is referred to as a *Kàrmàn vortex street* which contains strong periodic components, resulting in a sound of nearly pure tonal quality. On the basis of empirical data, the frequency of the Aeolian tone can be predicted from

$$f_{\text{Aeolian}} = \frac{\alpha v}{d}$$

where

$$v = \text{velocity of air, m/s}$$

$$d = \text{diameter of cylinder, m}$$

$$\alpha = \text{Strouhal number, approximately } 0.2$$

Figure 14.11 displays the simplest example of a gas jet, in which is the high-velocity airflow emanating from a reservoir through a nozzle. The gas accelerates from virtually zero velocity in the reservoir to a peak velocity in the core at the exit of the nozzle. The peak velocity of the jet depends greatly on the pressure difference between the reservoir pressure p_r and external (ambient) pressure p_a. As the pressure ratio increases, the velocity of the gas at the discharge increases up to a point when the pressure ratio of 1.89 (for the case of the gas being air) is reached. When the flow velocity reaches the velocity of sound, any further increase

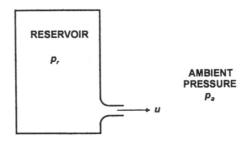

FIGURE 14.11. A simple gas jet.

of the reservoir pressure will not result in an increase of the velocity at the end of the straight duct. When the critical pressure ratio of 1.89 is reached, the nozzle is said to be *choked*. In order to increase the velocity further, the nozzle must increase in its cross-sectional area beyond the duct station where choking occurs.

In the frictionless, isentropic flow of an ideal gas from one point to another, the applicable energy equation describing this flow is

$$e + \frac{p}{\rho} + \frac{u^2}{2} = \text{const} \qquad \text{(W/kg)} \qquad (14.36)$$

where e is the internal energy of the gas, and u is the gas flow velocity. For the reservoir, $u = 0$ and pressure $p = p_r$. Also, from thermodynamics theory of an ideal gas,

$$e + \frac{p}{\rho} = \text{enthalpy} = c_p T$$

where c_p is the specific heat of the gas at constant pressure and T is the absolute temperature. Mach number M is defined by

$$M = \frac{u}{c} \qquad \text{(nondimensional characteristic)}$$

But propagation speed of sound c in an ideal gas is given by

$$c^2 = \gamma RT = \gamma \frac{p}{\rho}$$

Equation (14.36) becomes

$$e + \frac{p}{\rho} + \frac{u^2}{2} = c_p T + \frac{\gamma M^2 RT}{2} = c_p T \left(1 + \frac{\gamma M^2 R}{2 c_p}\right)$$

$$= c_p \frac{p}{R_\rho} \left(1 + \frac{\gamma M^2 (\gamma - 1)}{2\gamma}\right) = \text{const} \qquad (14.37)$$

We apply the fact that $R = c_p - c_v$ in an ideal gas, and the ratio of the specific heats $c_p/c_v = \gamma$. Selecting two flow stations, one at a point r inside the reservoir and the other at the point in the duct where choking occurs (Mach number $M = 1.0$), we obtain from equation (14.37):

$$\frac{p_r}{\rho_r} = \frac{p_{ch}}{\rho_{ch}} \left(1 + \frac{\gamma - 1}{2}\right) \quad \text{or} \quad \frac{p_r}{p_{cr}} = \frac{\rho_{ch}}{\rho_{ch}} \left(1 + \frac{\gamma - 1}{2}\right) \qquad (14.38)$$

In an isentropic flow, according to thermodynamic theory:

$$p\rho^{-\gamma} = \text{const}$$

Then equation (14.38) becomes

$$\frac{p_r}{p_{ch}} = \left(1 + \frac{\gamma - 1}{2}\right)^{\frac{\gamma}{\gamma - 1}} \qquad (14.39)$$

The ratio γ of specific heats is equal to 1.4 for air, so equation (14.39) yields the critical pressure ratio required for a shock to appear:

$$\frac{p_r}{p_{ch}} = \left(1 + \frac{1.4 - 1}{2}\right)^{1.4/1.4-1} = (1.2)^{3.5} = 1.89$$

Noise is generated from gas jets through the creation of fluctuating pressures from the turbulence and shearing stresses, as the high-velocity gas impacts with the ambient gas. In establishing the theory of aeroacoustics, the nonlinear effects of the momentum flux $\rho u_i u_j$ (i.e., the rate of transport of any momentum ρu_i across a unit area by any velocity component u_j) cannot be neglected as they were for linear acoustics. The momentum flux acts as a stress, because the rate of change of momentum constitutes a force. This momentum flux $\rho u_i u_j$ generates sound as a distribution of time-varying stresses. The forces between the air flow and its boundary radiate sound as distributed dipoles, and the stresses (which act on fluid elements with equal and opposite dipole-type forces) radiate as distributed quadrupoles.

The nature of the noise from jets cannot be predicted accurately, owing to the complex nature of the jet itself and uncertainties associated with turbulence, nozzle configuration, temperature vacillations, and so on. However, first-order estimates can be derived from empirical data obtained for the most part from experimentation in the aviation industry. The earliest measurements of jet noise demonstrated that intensity and noise power varied very closely with the eighth power of the jet exit velocity (Lighthill's eighth power law), and it is now generally agreed that the overall sound power P can be expressed as

$$P = \frac{K \rho_0 U^8 D^2}{c_0^5} \tag{14.40}$$

where K is a constant, with a value of 3–4; D is the jet diameter (in meters); U is the jet flow velocity (m/s); ρ_0 is the density of ambient air (kg/m^3); and c_0 is the ambient speed of sound (m/s). The factor $\rho_0 U^8 D^2 c_0^{-5}$ is often called *Lighthill's parameter*. Because the kinetic power of a jet is proportional to $\frac{1}{2}\rho_0 U^2 \cdot UD^2$, the fraction of the power converted into noise is the noise-generating efficiency η,

$$\eta \propto M^5 \tag{14.41}$$

where M $= U/c_0$, the Mach number of the flow referenced to the *ambient* speed of sound.

Aerodynamic noise can be modeled as monopoles, dipoles and quadrupoles. A jet pulse through a nozzle or discharge from HVAC ducts can be modeled as a monopole. In fans and compressors, the turbulent flow generally encounters rotor or stator blades, grids, and baffles; this type of flow can be modeled as a dipole. Quadrupole modeling applies to noise occurring from turbulent mixing in jets where there is no interaction with confining surfaces.

The velocity term U in equation (14.40) is the fluctuating velocity that varies throughout the jet stream. Consequently, U is not easily measured nor amenable to analytical calculation. But we can consider the average velocity V and assume that the size of the energy-bearing eddies are of the same order of magnitude as the jet diameter; and the total radiated acoustic power P is proportional to the kinetic energy of the jet flow. The total radiated power is simply a fraction of the total power discharged from the nozzle, that is,

$$P = \frac{\varepsilon M^5 \rho_0 V^3 A}{2} \text{ in watts} \qquad (14.42)$$

where

V = average flow velocity through the nozzle, m/s

M = Mach number of the flow $= V/c$, dimensionless

ρ_0 = density of ambient air

A = nozzle area

ε = constant of proportionality, in the order of 10^4

Equation (14.42) constitutes a first-order approximation that is applicable to many industrial situations where the average velocity of the jetstream lies in the range of $0.15c < V \le c$. The efficiency factor εM^5 is plotted from empirical data in Figure 14.12 over a range of Mach numbers. As a reflection of the uncertainty

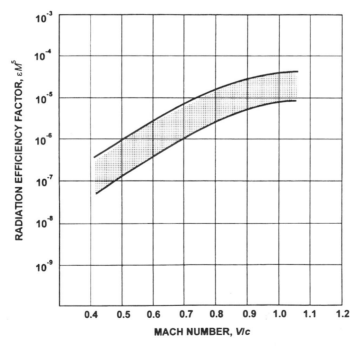

FIGURE 14.12. Efficiency factor εM^5 versus Mach number.

associated with turbulence, temperature, and so on, the efficiency factor is given as a range.

Example Problem 7

An air jet, operating through a choked 1-cm diameter nozzle, exhausts into the atmosphere. Determine the overall acoustical power and the sound power level L_W at 20°C.

Solution

Because the nozzle is choked, the average velocity of the air jet must be equal to the speed of sound at 344 m/s. The density of air at atmospheric pressure is 1.204 kg/m³ found from the ideal gas relation

$$\rho = \frac{p_a}{RT} = \frac{1.01325 \times 10^5}{287 \times (20 + 273.2)} = 1.204 \text{ kg/m}^3$$

and the area of the nozzle is $\pi D^2/4 = \pi(10^{-2})^2/4 = 7.85(10)^{-5}$ m². From Figure 14.12, the radiation efficiency factor εM^5 is found to be approximately 3×10^{-5}. Inserting into equation (14.42) we find

$$P = \frac{\varepsilon M^5 \rho_0 V^3 A}{2} = \frac{3.0 \times 10^{-5} \times 1.204 \times 344^3 \times 7.85 \times (10)^{-5}}{2} = 0.0577 \text{ W}$$

thus giving us the sound power level

$$L_W = 10 \log \left(\frac{0.0577}{10^{-12}} \right) = 107.6 \text{ dB}$$

From this last example, the acoustical power can be calculated to the first order on the basis of the nozzle diameters and exit velocities. However, the acoustical power of the air jet may have an accuracy of ±5 dB or thereabouts. But what if the gas jet is hot and extremely turbulent, as in the case of a gas burner? This situation can be resolved by applying a first-order correction for the temperature:

$$\text{correction due to temperature} = \Delta_T = 20 \log \left(\frac{T}{T_a} \right) \qquad (14.43)$$

where T and T_a are the absolute temperatures of the gas jet and the ambient air, respectively.

Example Problem 8

Consider the nozzle described in Example Problem 7 above. What would be the total radiated acoustical power level L_W if the temperature of the jet is to be raised from 20°C to 260°C?

Solution

Using equation (14.43),

$$\Delta_T = 20\log\left(\frac{T}{T_a}\right) = 20\log\left(\frac{(260+273.2)°\ K}{(20+273.2)°\ K}\right) = 5.2\ \text{dB}$$

The temperature-corrected sound power level becomes

$$L_W = 107.6 + 5.2 = 112.8\ \text{dB}$$

When the overall power of a jet is determined, the overall sound pressure at any location in the region surrounding the jet can be estimated through the use of the relationship between L_p and L_W:

$$L_p = L_W + 10\log\left(\frac{Q(\theta,\phi)}{4\pi r^2}\right) = L_W + 10\log[Q(\theta,\phi)] - 20\log r - 11$$

$$(14.44)$$

where $Q(\theta,\phi)$ is the directivity in three-dimensional space, and r is the distance from the source of L_W to the location where the value of L_p is desired. The directivity index DI is given by

$$DI = 10\log\left(\frac{Q(\theta,\phi)}{4\pi r^2}\right)$$

In many cases of interest, the jet can be generally regarded as a point source with typical directionality as shown in Figure 14.13. The parameter ϕ can be disregarded in axisymmetric flows.

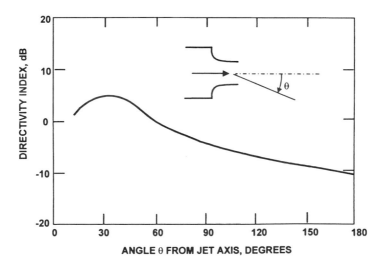

FIGURE 14.13. A fairly typical configuration for directivity DI_θ in a small subsonic jet. Note that the peak levels occur approximately in the angular range of 15° to 45° from the central axis of the jet.

Figure 14.13 illustrates that peak levels occur in the range of 15°–45° from the axis of the jet. The ordinate of Figure 14.13 giving the relative sound pressure level constitutes a directivity index DI_θ.

Example Problem 9

Find the radiated overall sound pressure level at a radial distance of 15 m from the nozzle of Example Problem 7 at angular positions of 0°, 45°, 90°, and 180°.

Solution

We make use of equation (14.44) and the results of Example Problem 7. The sound pressure level at $r = 15$ m is

$$L_p = L_W + 10 \log[Q(\theta, \phi)] - 20 \log r - 11 = 107.6 - 20 \log(15) - 11 + DI_\theta$$
$$= 73.1 + DI_\theta \text{ dB}$$

From Figure 14.13, the $DI_\theta = 0$ at $\theta = 0°$, and $L_p = 73.1 + 0 = 73.1$ dB. At 45°, $DI_{45°}$ is approximately +3 dB, and $L_p(45°) = 73.1 + 3 = 74.1$ dB. Similarly, for the other angles, we have

$$DB_{90°} = -5 \text{ dB}, \qquad L_p(90)° = 73.1 - 5 = 68.1 \text{ dB}$$
$$DI_{180°} = -10 \text{ dB}, \qquad L_P(180°) = 73.1 - 10 = 63.1 \text{ dB}$$

It becomes apparent from the above example that the jet has a strong directional character that must be accounted for in determining the sound level pressures.

Gas jets also manifest a strong frequency dependence. A first-order estimate for the peak frequency can be obtained for a given power level from the empirical relation

$$f_{peak} = \frac{St \times V}{D} \qquad (14.45)$$

where

$St =$ Strouhal number, a const $= 0.15$ for a wide variety of nozzle

diameters and operating conditions

$V =$ average exit velocity of the nozzle, m/s

$D =$ nozzle diameter, m

Example Problem 10

Given a 5-cm diameter nozzle with an exit velocity of 344 m/s, determine the peak frequency of the air blowdown.

Solution

Applying equation (14.45)

$$f_{\text{peak}} = \frac{0.15 \times 344 \text{ m/s}}{0.05 \text{ m}} = 1032 \text{ Hz}$$

Thus, the peak frequency should be expected to occur in the 1 kHz octave band.

On either side of the peak frequency, the octave band distribution of the acoustical power falls off. Experimental data indicates that the spectrum falls off at the average rate of -6 dB per octave above the peak frequency and -7 dB per octave below the peak frequency. The upshot is that the magnitude and the spectral character of gas jet noise can be only estimated roughly.

14.13   Gas Jet Noise Control

The major challenge in dealing with jet engine noise is to reduce the high noise levels with minimal impact on the thrust. The greatest progress came about by reducing the jet velocity while keeping the thrust constant because the sound power $\sim$ (thrust) $\times U^6/c_0^5$, according to the Lighthill equation (14.40). Early efforts were centered on modifying the nozzle shape to variations of the "cookie-cutter" forms or using multiple smaller nozzles. Some of these designs produced up to 10 dB noise reductions with a rather small loss of thrust. Newer by-pass and fan-jet engine designs entail much lower jet velocities, large streamlined center bodies, and annular jets that may be subdivided in any one of several ways.

In the case of simple high-velocity air jets in industrial environments, such as those used to power air tools, provide cooling or venting, parts ejection, and so on, a number of straightforward noise reduction measures can be applied. The basic steps include the following:

1. Reduction of the required air velocity by moving the nozzle closer to a part being ejected, while maintaining the same value of thrust.
2. Adding additional nozzles, reducing the required velocity but again sustaining the same thrust magnitude.
3. Installation of newer models of quieter diffusers and air shroud nozzles.
4. Interruption of air flow in sequence with ejection or blow-off timing.

Methods 1 and 2 above result in noise reduction through cutting down on the jet velocity. Reduction of the airstream velocity should be the first consideration of any noise reduction program. Usually, the only constraint is the preservation of the air-jet thrust. The thrust T_j of a jet is given by

$$T_j = \dot{m} \, V \, (\text{N}) \tag{14.46}$$

where

$$\dot{m} = \text{ mass flow rate, kg/s}$$

$$V = \text{ average jet velocity}$$

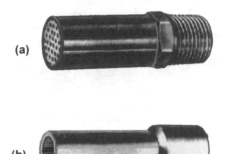

FIGURE 14.14. (a) Multijet diffuser and (b) restrictive diffuser nozzle.

Equation (14.46) indicates that thrust can be preserved if the mass flow is increased while reducing the jet velocity, Increasing the nozzle exit area and moving the nozzle closer to the ejection target will also provide considerable velocity reduction. Adding on two or more nozzles will also permit reduction in air-jet velocity and a corresponding noise level reduction.

Multiple-Jet and Restrictive Flow Silencer Nozzle

Figure 14.14(a) shows an example of a multijet diffuser. The noise reduction accrues mainly from the reduction in jet air core size, which also lessens the turbulence in the mixing regions. Also, the smaller inner jets flow along with the outer layer of high velocity air, thereby reducing the shearing action with the static ambient air. In the restrictive flow nozzle of Figure 14.14(b), the high-velocity core is minimized by the sintered metal restrictor, typically mounted into the nozzle exit. The flow velocity is reduced somewhat, with a corresponding drop in the amount of noise radiation. In both of these nozzle types, there will be some loss in the jet thrust, so additional nozzles may be required to keep up the same amount of the jet thrust.

Air Shroud Silencer Nozzles

In Figure 14.15, an air shroud silencer nozzle is shown with its air flow pattern. Here the noise reduction is achieved through bypassing some of the some of the primary air flow around and over the nozzle. The bypassed air lowers the velocity gradients between the primary jet and the ambient air, thus cutting down on the shearing action and the resultant radiation of noise. There is usually little change in the mass flow for this type of silencer and the jet thrust is generally preserved. A micrometer dial provides control of amount of bypassed air. The typical noise reduction through the use of air-shroud silencer nozzles is 10–20 dB in the critical high-frequency range of 2–8 kHz for small high-velocity jets.

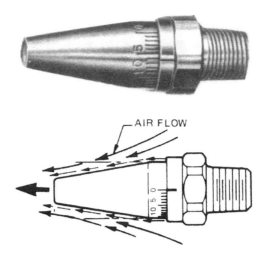

FIGURE 14.15. Air shroud silencer. The micrometer shown here adjusts the amount of air supply bypassing the nozzle instead of going through it. (Courtesy of ITW Vortec, Cincinnati, OH.)

Impingement Noise

Impingement noise occurs when a gas jet is brought close to and impinges upon a solid surface or object. A sharp increase in the noise level occurs, particularly in the range of higher frequencies. In addition, the impingement of the gas jet on the surface can bring about unsteady forces in the form of aerodynamic dipoles, which can be described as a pair of point sources of equal magnitudes separated by a small distance and oscillating with an angular phase difference of 180°. This is to say that these two point sources are out of phase—i.e., when one of the point sources is positive, the other is negative. These dipoles constitute the basic mathematical models used to describe the noise and radiation of many common noise sources including propellers, valves, loudspeakers, air duct diffusers, a number of musical instruments, and so on. There are also some directivity patterns present when the noise evolves from dipole sources.

Impingement noise, as with jet noise, is difficult to predict in the way of its amplitude and spectral characteristics of the noise-generating mechanism. From both analytical and experimental considerations, the radiated sound power for impinging subsonic jets depends in the first order on the fifth or sixth power of the flow velocity. Again, as with the free jet flow, even a slight reduction in the flow velocity can bring about appreciable reductions in impingement noise. If a jet flows over a sharp edge or discontinuity, even more noise is likely to be generated. Whistlelike edge tones will also occur. The periodic components and the impingement noise can be lessened by cutting down on the flow turbulence created by jet flows over a cavity or an obstruction. In these cases, the impingement noise can be lessened by avoiding or eliminating the presence of cavities and by redirecting the jet away from the edge.

Gaseous Flows in Pipes or Ducts

Velocities in pipes or ducts usually occur in the subsonic range, but where there are valves or vents present to control the flow, extremely high noise levels can occur. Noise levels have been measured as high as 140 dB downstream of reduction valves in large steam pipes. In many cases, pressure drops across a control or regulator valve are sufficiently large to choke the flow at the discharge, with the result that the flow of the gas jet is sonic or almost sonic with corresponding generation of high-intensity aerodynamic noise. This noise can propagate through the pipe walls into the immediate surroundings, and what is even worse, it can propagate almost unabated downstream with very little attenuation.

Because of the complexity of the noise source mechanism and the degree of uncertainty in transmission loss of the pipe and ducts, it becomes quite difficult to predict the magnitude of the aerodynamic noise. But some guidance can be derived from empirical data which can be used to establish first-order estimates. The turbulent mixing areas downstream of the valve constitute the principal region of noise generation. But if valves are encased in thick housings, the noise levels are typically 6–10 dB lower. The spectral character of the noise resembles that for high-velocity gas jets, that is, there are present peak levels in the range of 2–8 kHz. In short, it can be expected that wherever high-pressure steam and gas flows are regulated or discharged through valves, noise levels exceeding 100 dB will most likely occur. The valves will have relatively thick walls, so the piping system itself, downstream of the valve, is the primary source of externally radiated noise.

Three basic approaches can be considered in reducing the noise from the control valve regions, namely, (a) revising the dynamics of the flow, (b) introducing an in-line silencer to absorb acoustic energy, and (c) increasing the transmission loss in the pipe walls.

Of the three approaches just mentioned, altering the dynamics of the flow, which means actually reducing noise at the source, is probably the most preferred method. Changing of the dynamics entails reducing the flow velocity via multiple stages of pressure reductions or diffusion of the primary jet. Figure 14.16 shows multiple stages of pressure reduction with the use of expansion plates. The flow velocity is lessened sequentially in each expansion chamber. These plates also act as a diffuser, reducing turbulent mixing. As much as 20 dB noise reduction have been

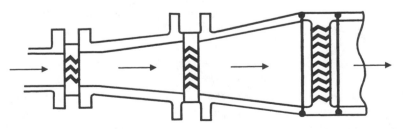

FIGURE 14.16. Multiple stages of pressure reduction in a throttling system with use of expansion plates.

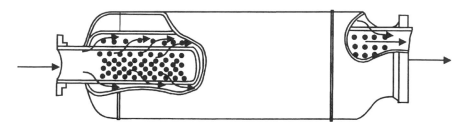

FIGURE 14.17. A cutaway view of an in-line silencer used to reduce aerodynamic noise in pipes. (Reproduced with permission of Fisher Controls International, Inc.)

achieved. A disadvantage of this setup is that a back pressure may be induced, which has the effect of impeding flows. The use of diffusers is another approach, in which the flow is diffused into smaller interacting jets,

In-line silencers, one of which is shown in Figure 14.17, basically consist of flow through ducts surrounded by absorptive materials separated from the flow by perforated metal sheets. A diffusive inlet may precede the tubular portion of the absorptive liner. The absorption materials typically consists of fiberglass or metal wool.

Finally, in the third approach, that of increasing the transmission loss in pipe walls, two methods can be used. One is to increase the pipe wall thickness and the other is to swath the pipe in acoustical absorption materials. However, no simple procedure exists for estimating the transmission loss through pipe walls. Piping standards are promulgated by mechanical engineering societies, which specify wall thicknesses for high-pressure, high-velocity flow installations. These thicknesses are specified with principal the aim of preventing ruptures and other failures of the pipes that are subject to high pressures. For larger pipes, the standard wall thickness is approximately $\frac{3}{8}$ in. A wall thickness greater than that required to meet stress requirements can be is selected so that more sound attenuation can be achieved from the presence of the thicker walls, in the range of 2–20 dB additional transmission loss.

However, a resonancelike condition can occur at the *ring frequency*, at which value the transmission loss virtually disappears. This is somewhat analogous to the coincidence frequency effect associated with barriers. This dip in transmission loss occurs in a pipe when a single wavelength of sound becomes equal to the nominal circumference of the pipe wall, a situation where the ring frequency f_r can be expressed mathematically as

$$ f_r = \frac{c_w}{\pi D_p} $$

where c_w is the longitudinal speed of sound in the pipe wall material in m/s; and D_p is the nominal diameter of the pipe (which can be considered the average of the inside and outside diameters of the pipe) in meters. For example, a 25-cm diameter steel pipe would have a ring frequency of 6598 Hz (because $c_w = 5182$ m/s for

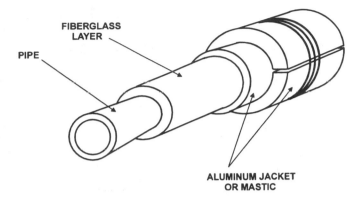

FIGURE 14.18. Lagging or jacketing of a pipe to minimize flow noise transmission.

steel). It would then be expected that the peak noise levels radiated from this pipe would occur in the vicinity of 6.6 kHz. Higher-order resonancelike conditions will also occur at $2f_r, 3f_r$, etc., but would be of little concern because those frequencies are extremely high.

Beside increasing wall thicknesses, another method for achieving noise attenuation in pipes is that of wrapping or *lagging* the pipes. A typical lagging setup is shown in Figure 14.18. A $2\frac{1}{2}$–8 cm layer of acoustically absorbent material (fiberglass, mineral wools, or polyurethane foam) is wrapped around the pipe wall. This absorbing layer, in turn, is sheathed in sheet metal, dense vinyl, or sheet lead. The outer dense layer is extremely important in providing a high level of noise reduction. Even more noise reduction can be obtained by applying an even denser outer layer and adding additional composite layers. The principal disadvantage of using lagging as the only noise reduction measure is that long lengths of piping would need treatment, so it would be advisable to give priority to noise source reduction (i.e., utilization of multiple pressure reduction stages or diffusers) and to use lagging as a secondary measure to bring the overall sound levels down to acceptable values.

14.14   Mufflers and Silencers

In the last section we discussed some of the elements of silencing air jet flow. In this section we examine further aspects of reducing the noise from air jet flow. The term *muffler* is commonly applied to the exhaust gas silencer for an internal combustion engine, and *silencer* usually denotes the noise suppressor installed in a duct or air intake. We use these terms interchangeably in this section because the same operational principle applies to both of them. Mufflers and silencers were developed to reduce noise energy while facilitating gas flow.

Both silencers and mufflers fall into one or the other of two categories: *dissipative* units and *reactive* units. Dissipative mufflers and silencers reduce noise energy by

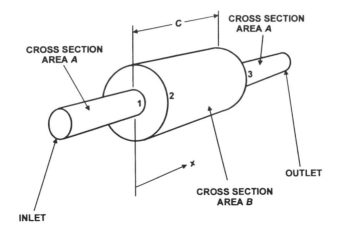

FIGURE 14.19. Expansion chamber of a reactive muffler.

employing sound-absorption materials which are flow resistive at frequencies in the audio range. Reactive mufflers and silencers reduce noise through destructive interference. Both reactive and dissipative principles may be combined into a single muffler or silencers in order to ensure effectiveness over a broad frequency range.

Reactive Mufflers and Silencers

Consider the basic reactive muffler in Figure 14.19. Sound waves become transmitted from left to right in the inlet pipe and reflections occur in the expansion chamber, causing destructive interference under the appropriate conditions. In this analysis (Davis et al., 1984), the following assumptions are made:

1. Sound wave pressure is small in comparison with absolute pressure in the expansion chamber.
2. No reflected waves occur in the tailpipe (i.e., the outlet pipe).
3. The expansion chamber walls do not transmit nor conduct sound.
4. Only plane waves exist.
5. Viscosity effects are negligible.

We denote the following subscripts in the description of incident and reflected waves: I for incident waves and R for reflected waves. The particle displacements ξ of the incident and reflected waves are described in complex format as follows:

$$\xi_I = A_I e^{i(\omega t - kx)} \tag{14.47}$$

and

$$\xi_R = A_R e^{i(\omega t + kx)} \tag{14.48}$$

Here A_I and A_R are complex amplitudes. Particle velocities are obtained by differentiating equations (14.47) and (14.48) with respect to time:

$$u_I = i\omega A_I e^{i(\omega t - kx)} \tag{14.49}$$

and

$$u_R = i\omega A_R e^{i(\omega t + kx)} \tag{14.50}$$

We recall that for a plane wave in the direct field, sound pressure is related to particle velocity by

$$p = \rho c u$$

which is used in equations (14.49) and (14.50) to obtain:

$$p_I = i\rho c\omega A_I e^{i(\omega t - kx)} \tag{14.51}$$

$$p_R = i\rho c\omega A_R e^{i(\omega t + kx)} \tag{14.52}$$

Let $x = 0$ designate the junction of the inlet pipe and expansion chamber. Pressure must be continuous at this junction, hence

$$(p_{I1} + p_{R1})_{x=0} = (p_{I2} + p_{R2})_{x=0} \tag{14.53}$$

Inserting equations (14.51) and (14.52) into (14.53) gives

$$A_{I1} + A_{R1} = A_{I2} + A_{R2} \tag{14.54}$$

Subscript 1 refers to the left of the junction and subscript 2 refers to the right. Continuity of flow requires that

$$A(u_{I1} - u_{R2})_{x=0} = B(u_{I2} - u_{R2})_{x=0}$$

Setting $B/A = m$, where A refers to the flow area of the inlet pipe and the outlet pipe and B the flow area of the expansion chamber, we have from continuity

$$A_{I1} - A_{R1} = m(A_{I2} - A_{R2}) \tag{14.55}$$

In Figure 14.19, C denotes the length of the expansion chamber. At the junction of the expansion chamber where $x = C$, pressure and flow continuity necessitates that

$$(p_{I2} + p_{R2})_{x=C} = p_3$$

from which

$$A_{I2}e^{-ikC} + A_{R2}e^{ikC} = A_3 \tag{14.56}$$

and

$$B(u_{I2} - u_{R2})_{x=C} = Au_3 \tag{14.57}$$

resulting in

$$m(A_{I2}e^{-ikC} + A_{R2}e^{ikC}) = A_3 \tag{14.58}$$

We now have five equations (14.54)–(14.58) and five unknowns, A_{I1}, A_{R1}, A_{I2}, A_{R2}, and A_3. These equations are solved simultaneously to correlate conditions at the inlet and outlet of the expansion chamber, resulting in the complex ratio

$$\frac{A_{I1}}{A_3} = \frac{(1+m)\left(1+\frac{1}{m}\right)e^{ikC} + (m-1)\left(\frac{1}{m}-1\right)e^{-ikC}}{4}$$

$$= \cos(kC) + \frac{i}{2}\left(m + \frac{1}{m}\right)\sin(kC) \tag{14.59}$$

In order to establish the transmission loss, the ratio of sound intensity at the expansion chamber inlet to transmitted sound intensity is needed:

$$\frac{I_{I1}}{I_3} = \frac{p^2_{rms(I1)}}{p^2_{rms(3)}} = \left|\frac{A^2_{I1}}{A^2_3}\right| = \cos^2(kC) + \frac{1}{4}\left(m + \frac{1}{m}\right)^2\sin^2(kC)$$

$$= 1 + \frac{1}{4}\left(m - \frac{1}{m}\right)^2\sin^2(kC) \tag{14.60}$$

Because transmission loss is a function of the ratio of power incident on the muffler to the power transmitted, that is,

$$TL = 10\log\left(\frac{W_{incident}}{W_{transmitted}}\right)$$

and if the inlet and outlet areas are equal, we obtain

$$TL = 10\log\left(\frac{W_{I1}}{W_3}\right) = 10\log\left(\frac{I_{I1}}{I_3}\right) = 10\log\left[1 + \frac{1}{4}\left(m - \frac{1}{m}\right)^2\sin^2(kC)\right]$$

$$\tag{14.61}$$

where $k = 2\pi/\lambda$ is the wave number. Equation (14.61) becomes invalid if any of the lateral dimensions of the expansion chamber exceeds 0.8λ.

Figure 14.20 displays a plot of the theoretical transmission loss versus C/λ for various area ratios. Equation (14.61) forecasts a transmission loss of zero when the argument of the sine function is 0, π, 2π, $2n\pi$, ... and a maximum transmission loss when the argument is $\pi/2$, $3\pi/2$, ..., $(2n+1)\pi/2$, ..., where $n = 0, 1, 2, 3$, etc. It then follows that the expansion chamber works best when length C constitutes an odd number of quarter-wavelengths, that is, $C = \lambda(2n+1)/4$ for maximal TL. But the expansion chamber becomes ineffective when the chamber length C measures out to an integer number of half-wavelengths, that is, $C = n\lambda$ (i.e., $\lambda/2$, λ, $3\lambda/2$, etc.). For extremely low frequencies, C/λ approaches zero, and the transmission loss likewise approaches zero.

Example Problem 11

A reactive muffler is to be designed to produce a transmission loss of 20 dB at 150 Hz, The inlet and outlet pipes are 60 mm in diameter and the average gas (air) temperature is 100°C. Determine the size of the expansion chamber.

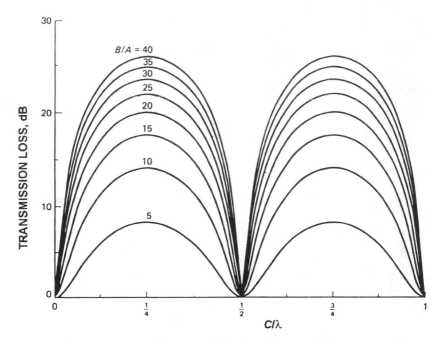

FIGURE 14.20. Transmission loss TL versus C/λ.

Solution

At 100°C, the velocity $c = 20.04\sqrt{T + 273.16} = 387.1$ m/s. The wavelength of the 150 Hz sound is given by $\lambda = c/f = 387.1/150 = 2.58$ m. For the shortest expansion chamber to produce maximum transmission loss, we examine the argument of the sine function,

$$kC = \frac{2\pi C}{\lambda} = \frac{\pi}{2} \text{ radians}$$

from which we get

$$C = \frac{\lambda}{4} = \frac{2.58}{4} = 0.645 \text{ m} = 645 \text{ mm}$$

From equation (14.61) the transmission loss is given by

$$TL = 10\log\left[1 + \frac{1}{4}\left(m - \frac{1}{m}\right)^2 \sin^2(kC)\right]$$

$$= 10\log\left[1 + \frac{1}{4}\left(m - \frac{1}{m}\right)^2 \sin^2\left(\frac{2\pi \times 0.645}{2.58}\right)\right]$$

$$= 10\log\left[1 + \frac{1}{4}\left(m - \frac{1}{m}\right)^2\right] \tag{14.62}$$

Let $[m - (1/m)]^2 = \Omega$. Then

$$\frac{A}{B}\left[\left(\frac{B}{A}\right)^2 - \frac{\Omega B}{A} - 1\right] = 0$$

where from (14.62)

$$\Omega = 2(10^{TL/10} - 1)^{1/2}$$

The solution to the quadratic expression inside the bracket is:

$$\frac{B}{A} = \frac{\Omega + \sqrt{\Omega^2 + 4}}{2}$$

Note the negative root carries no physical significance. For a transmission loss of 20, $\Omega = 19.90$ and

$$\frac{B}{A} = \frac{19.90 + \sqrt{19.90^2 + 4}}{2} = 19.95$$

The cross-sectional area A of the inlet pipe and the outlet pipe are each $\pi(60)^2/4 = 2827$ mm^2. The expansion chamber cross-sectional area is

$$B = A \times \frac{B}{A} = 2827 \times 19.95 = 56,399 \text{ mm}^2$$

If the expansion chamber has a cylindrical cross-section, its inside diameter d is found from

$$B = \frac{\pi d^2}{4}$$

Hence

$$d = \sqrt{\frac{4B}{\pi}} = \sqrt{\frac{4 \times 56,399}{\pi}} = 268 \text{ mm}$$

Area B represents the minimum cross-sectional area of the expansion chamber that is required to produce the 20-dB transmission loss.

Dissipative Mufflers and Silencers

In the preceding paragraphs, dissipative silencers were discussed. Lining the ducts of HVAC systems with sound-absorptive materials generally provide adequate noise attenuation. However, many industrial applications require a silencer that can provide a large amount of noise attenuation or insertion loss in a relatively small space. If the noise energy covers a narrow frequency range, reactive silencers constitute the better solution. Reactive silencers are preferred where the gas flows contain particles or other components that could contaminate sound-absorbing materials. Dissipative silencers are more effective both over a wider range of frequencies and in silencing fluctuating machinery noises. In a number of situations, in order to obtain the greatest noise reduction over a wider frequency range, both reactive and dissipative principles are combined into a single silencer.

14.15   Active Noise Control

Relatively recent advances in digital electronics have made it more economically feasible to apply *active noise control* in cutting down on noise in aircraft and automotive interiors, pumps, compressors, electric motors, transformers, and so on. Earlier active noise control techniques were first used in the study of noise in fan ducts. The advent of inexpensive digital processors enabled the conversion of analog audio frequency signals into digital form, then processing them through a digital filter and converting back into an analog signal with very little time delay. The basis of active noise control is to duplicate the noise that is the same but 180° out of phase, so that when the offending noise and the duplicated out-of-phase noise are combined, a cancellation will occur.

Figure 14.21 illustrates a simple single-channel active noise control system. Sound is detected by a microphone and processed though a digital filter imbedded in a special-purpose microprocessor prior to being fed into a loudspeaker that radiates the sound that is intended to interfere destructively with the "original" unwanted sound. The characteristics of the digital filter are designed to minimize the time-averaged signal at the error microphone on a continuous basis.

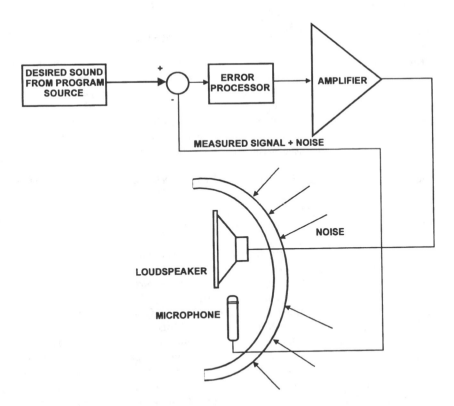

FIGURE 14.21. An example of an active noise control system for a communication headset.

The following factors, which may limit the effectiveness of active noise control, must be considered:

1. Continuous and reliable measurement, signal processing, and sound generation are required.
2. For maximum effectiveness the antinoise source must be near the noise source or near the receiver.
3. When changes occur in the relative position of the noise source and the observer, the effect could be one of sound reinforcement rather than cancellation. This consideration can limit many active-noise control systems to low-frequency sounds. However the presence of error detection microphones combined with digital filters that continuously adapt to changing conditions can mitigate the possibility of unintentional sound reinforcement.
4. Even under the most ideal real conditions, cancellation of sounds cannot be total, due to the statistical nature of the molecular motion in the sound propagation medium.

References

AMCA Standard 300. *Reverberant Room Method for Sound Testing of Fans*. Arlington Heights. IL: Air Movement and Control Association.

ASHRAE Standard 51. *Laboratory Method of Testing Fans for Rating*. Atlanta, GA: American Society of Heating, Refrigeration, and Air-Conditioning Engineers, Inc.

ASHRAE Handbook. Atlanta, GA: American Society of Heating, Refrigeration, and Air-Conditioning Engineers, Inc.

Davis, D. D., G. M. Stokes, D. Moore, and G. L. Stevens. 1984. Theoretical and experimental investigation of mufflers with comments on engine-exhaust muffler design. (Reprint, NACA Report 1192. 1954), *Noise Control*, M. J. Crocker (ed.). New York: Van Nostrand Reinhold.

Fuller, C. R., S. J. Elliott, and P. A Nelson. 1996. *Active Control of Vibration*. London: Academic Press.

Graham, J. Barrie, and Robert M. Hoover. 1991. Fan noise. *Handbook of Acoustical Measurements and Noise Control*, 3rd ed. Harris, Cyril M. (ed.). New York: McGraw-Hill; Ch. 24.

Hand, R. F. 1982. Accessory noise control. *Noise Control in Internal Combustion Engines*. Baxa, D. E. (ed.). New York: Wiley-Interscience; pp. 437–477.

Harris, Cyril M. (ed.). 1991. *Handbook of Acoustical Measurements and Noise Control*, 3rd edition. New York: McGraw-Hill; Ch. 34–45.

Hubbard, H. H. (ed.). 1995. *Aerodynamics of Flight Vehicles: Theory and Practice, vol. 1: Noise Sources, vol. 2: Noise Control*. Woodbury, NY: Acoustical Society of America. (A now classic compendium of theory and experimentation in the field of aeroacoustics, originally sponsored by the Aeroacoustics Technical Committee of the American Institute of Aeronautics and Astronautics (AIAA) and reprinted by the American Institute of Physics.)

Jones, Dylan M., and Donald E. Broadbent. 1991. Human performance and noise. *Handbook of Acoustical Measurements and Noise Control*, 3rd ed. Harris, Cyril M. (ed.). New York: McGraw-Hill; Ch. 41.

Lighthill, James M. 1952. On sound generated aerodynamically, I. general theory. *Proceedings of the Royal Society A* 211: 564–587. (A classic paper by the great master of modern aeroacoustics theory.)

Lighthill, James M. 1978. *Waves in Fluids*, Cambridge: Cambridge University Press. (A veritable classic.)

Magrab, E. B. 1975. *Environmental Noise Control*. New York: John Wiley and Sons, Inc.

Nelson. P. A., and S. J. Elliott. 1992. *Active Control of Sound*. London: Academic Press.

Nelson, P.A., and S. J. Elliott. 1997. Active noise control. Crocker, Malcolm J. (ed.). *Encyclopedia of Acoustics*, vol. 2. New York: John Wiley & Sons; Ch. 84, pp. 1025–1037.

Piraux, J., and B. Nayroles. 1980. A theoretical model for active noise attenuation in three-dimensional space. *Proceedings of Inter-Noise '80*. Miami, FL; pp. 703–708.

Webb, J. D. (ed.). *Noise Control in Industry*. Sudbury, Suffolk, Great Britain: Sound Research Laboratories, Ltd.

Problems for Chapter 14

1. A 6-blade fan rotates at 1150 rpm. Determine the frequencis of the noise that will emanate from the fan.

2. Find the sound power of a fairly efficient electric motor rated at 85 hp at 2400 rpm.

3. Determine the frequency of the blade rate component of a diffuser-type compressor with 24 blades in the rotor and 36 blades in the stator. The rotor rotates at 9000 rpm.

4. A hydraulic screw-type pump operates at 1200 rpm with 4 chamber pressure cycles per revolution. Find its fundamental frequency and the sound power output.

5. A radial forward-curved fan has 36 blades and a rotor diameter of 120 cm. It operates at 950 rpm with a air flow rate of 24 m^3/s under the effect of a total pressure of 1.8 kPa. What is the total sound power at the inlet?

6. Estimate the sound power of a "quiet" 100-hp electric motor operating at 1200 rpm.

7. A ball bearing has a stationary outer race. The inner and outer race diameters are 30 mm and 46 mm, respectively, measured at the point of ball contact. The inner race rotates at 5200 rpm. Determine the noise and vibration frequencies that can occur as the result of imbalance and defects. There are twelve balls in the bearing.

8. A ball-bearing is constructed as follows: it has twelve 15 mm balls, and the inner race diameter is 50 mm at the point of ball contact. The inner race has a rotational speed of 1800 rpm and the outer race remains stationary. Find
   (a) rotational speed of separator;
   (b) speed of the balls relative to the separator;
   (c) shaft imbalance frequency;
   (d) outer race defect frequency;
   (e) inner race defect frequency;

(f) frequency arising from damage to one ball;

(g) frequency attributable to imbalance in the separator.

9. A roller bearing contains ten 16-mm diameter rollers. The inner race diameter is 30 mm. The inner race rotates at 6000 rpm while the outer race remains stationary. Find:

(a) rotational speed of the separator;

(b) speed of the rollers relative to the separator;

(c) shaft imbalance frequency;

(d) outer race frequency;

(e) inner race frequency;

(f) frequency caused by a defect in one roller;

(g) frequency caused by imbalance of the separator.

10. In a nonplanetary gear transmission, the input shaft gear has 40 teeth and rotates at 1200 rpm. It meshes with a another shaft through a 30-tooth gear. Predict all of the fundamental frequencies that are likely to arise in this transmission.

11. The gears of a reverted gear train of Figure 14.8 carry the following specifications: Gear 1 (driver), 50 teeth; gear 2, 90 teeth; gear 3, 46 teeth; and output gear 4, 47 teeth. The input shaft rotates at 2400 rpm.

(a) Determine the rotational speeds of the other shafts.

(b) Compute each of the fundamental tooth-error and shaft frequencies and the first three harmonics.

(c) Find all of the fundamental tooth meshing frequencies and their first three harmonics.

(d) Determine the sideband frequencies.

12. It is specified that a pair of spur gears are to be used to reduce shaft speed from 2400 rpm to 1600 rpm. The driver has 18 teeth and a diametral pitch of 4. The pressure angle is $20°$ in the stub teeth.

(a) Predict the fundamental tooth-error frequencies.

(b) Determine the fundamental tooth meshing frequency.

(c) Find the contact ratio, and discuss the results in terms of noise.

13. Consider a chain drive system that contains a 15-tooth input sprocket that rotates at 2500 rpm. The chain has 90 links, and it is specified that the output sprocket rotate at 1500 rpm. Determine:

(a) the number of teeth needed for the output sprocket;

(b) the range of output speed if input speed remains constant;

(c) the probable frequencies of noise and vibration due to damage to one tooth or imbalance in either shaft;

(d) the probable frequencies due to damage to one link;

(e) the probable frequencies due to chordal action.

14. The input speed of a chain drive is to be reduced by one-half at the output. The 24-tooth input sprocket rotates at 1200 rpm. The chain contains 160 links.

(a) How many teeth are need in the output sprocket?

(b) For constant input speed, determine the range of output speed.

(c) Determine the likely frequencies of noise and vibration due to imbalance of either shaft or damage to one tooth.

(d) Find the possible frequencies due to damage to one link.

(e) Determine the possible frequencies due to chordal action.

15. A misalignment of $10°$ occurs in a Hooke-type universal joint. The input shaft rotates with a constant speed of 1200 rpm.

(a) Find the speed range of the output shaft.

(b) Determine the frequency resulting from vibration or noise.

16. A fairly simple $\frac{3}{8}$-diameter nozzle functions under an inlet pressure of 100 psi (gauge). Determine the sound level at a distance of 5 ft for an ambient air temperature of $80°F$.

17. A simple 25 mm nozzle operates at an inlet pressure of 600 kPa. Determine the sound pressure level at a distance of 1.0 m for an air temperature of $20°C$.

18. Fifty sound-level readings are taken at 10-s intervals. Find the percent exceeded noise levels L_{10}, L_{50}, and L_{90} for the following distribution:

Level [dB(A) ± .05]	Number of Readings
87	2
86	8
85	14
84	20
83	5
82	4
81	2

15
Underwater Acoustics

15.1 Sound Propagation in Water

Sound waves are absorbed and scattered in water to a much lesser degree than electromagnetic waves. Because of this property sound waves have proven to be particularly useful in detecting distant objects undersea by means of *sonar* (acronym for SOund NAvigation and Ranging). *Passive* sonar is strictly a "listening" device that detects sound radiation emitted (sometimes unintentionally) by a target. In *active* sonar, the process entails sending out a sound pulse and listening for a returning echo. The loudness of the echo hinges principally on the amount of energy absorption in the water and the degree of reflection from an intercepting surface. Some of the energy is scattered backward in a random fashion to the echo-ranging emitter, either by particles or inhomogeneities in the water, or by the ocean surface or sea bottom. This scattering results in a phenomenon referred to as *reverberation*. The sound directly reflected back from an obstacle (*target*)—such as a submarine or a whale—constitutes the *echo*.

Recognizable echoes have been mapped for schools of fish, dolphins, whales, patches of kelp and seaweed, sunken wrecks and pronounced irregularities in shallow depths. Certain water conditions give rise to echoes at very short ranges; and ocean swells also can generate echoes. Ship wakes and other types of bubble screens make effective targets as wells as icebergs. The most prominent use of sonar has been in subsurface warfare: submarines, surface vessels, and underwater mines are obviously the most important targets.

The reflection of the probing signals from targets to observation points (usually at signal sources) are evaluated in terms of *target strengths*, a quantitative measure of intensities of the echoes. Target strengths depend on a number of factors, viz. target size, shape, orientation with respect to the probe signal source, distance from the source to the target, and frequency of the probe signal. The intensity of the reflected sound is also a function of the intensity of the probe signal striking the target, the distance from the target to the point of echo measurement (which is usually at virtually the same location as the probe signal source), and the acoustic absorptivity of the target. The effects of these variants can be established only if the radii of curvature of the sound waves striking the target and returning to a receiver are of much greater magnitudes than those of the target's dimensions, that is, the probe signal waves should ideally be plane. In terms of ray acoustics the incident sound wave must be substantially parallel in the region of the target that they strike, and reflected sound rays should also be parallel over the area of the sonar receiver.

In the military use of sonar, the applicable spectrum range covers the ultralow frequency to the megahertz region. Acoustic mines detect the pressures below 1 Hz which are generated by moving ships. These mines detect the acoustic radiation and explode when the acoustic level reaches a certain level in their bandpass. Such mines can be destroyed harmlessly through the use of a minesweeper which is essentially a powerful signal source towed behind a mine-sweeping vessel.

In passive detection, the acoustic radiation of both water-surface and underwater vessels, are sensed by a hydrophone array mounted on the spy vessel or submerged at the bottom a long distance away. The receiving array must be directional in order to be able to locate the target through sensing of somewhat higher frequencies.

Modern echo-ranging (active) sonar consists of an elaborate array of equipment to send out signals in form of long, high-power pings in designated directions vertically and horizontally, and newer signal processing techniques present the echoed data to the observer. Transducer arrays are often enclosed in separate housings that are towed underwater behind the surface vessel, so that shallow thermal gradients[1] can be penetrated; and the sonar can probe in the stern (forward) direction, a procedure that cannot be achieved with a hull-mounted sonar.

Peaceful uses of sonar expanded greatly immediately after World War II. Originally developed for depth sounding, sonar is now being used to find fish, study fish migration, map ocean floors, locate underwater objects, transmit communications and telemetric data, serve as acoustic speedometers, act as position-marking beacons, and monitor well-head flow control devices for undersea oil wells. Passive sonar provides marine biology researchers the window for tracking sounds made by cetaceans (whales, dolphins, and porpoises).

15.2   Some Basic Concepts Pertaining to Underwater Sound

In an acoustic plane wave through a fluid medium of density ρ, the particle pressure p relates to the fluid particle velocity u as follows

$$p = \rho c u$$

The fluidic parameter ρc is called the *specific acoustic resistance*. Its value, on the average, for seawater is 1.5×10^5 g/cm^2-s or 1.5×10^6 kg/m$^2 \cdot$ s. For air,

---

[1] In the 1920s and 1930s, it was observed that good echoes were obtained with the early versions of shipboard echo-ranging equipment in the morning and poor or no results were obtained in the afternoon. This was due to the shifts in the seawater thermal gradients that caused sound to refract toward the sea bottom and thereby place a target in the "shadow zone."

$\rho c = 42$ g/cm$^2 \cdot$ s. From our previous chapters we recognize that this parameter can also assume a complex value in its role as the *specific acoustic impedance*.

In order to understand the importance of acoustic impedance, consider the fact that some sort of a piston must drive against a medium in order to generate acoustic energy. The medium presents a resistance to that drive; and in acoustic terminology, this resistance is the specific acoustic impedance. The conventional techniques applied in the design of (air) audio equipment, namely, loudspeakers and microphones, are not applicable to underwater sonar. Air is a very light substance, so the driving mechanism must produce a large displacement with very little force. In the case of underwater sound, however, it is necessary to provide a very large force to generate even a small displacement. This means that the sonar must possess a very large mechanical impedance, that is, a large ratio of the complex driving force to the complex velocity.

It is also evident that a propagating sound wave carries mechanical energy that includes the kinetic energy of the particles in motion and the potential energy of the stresses arising in the elastic medium. In the process of a wave propagating, a certain amount of energy per second crosses a unit area. This power per unit area describes the intensity of the wave. If a unit area is given an orientation with respect to reference coordinates, the intensity becomes a vector quantity represented by a Poynting vector normal to the unit area, in the same manner as in the theory of electromagnetic propagation. In a plane wave, the instantaneous intensity relates to the instantaneous acoustic pressure in the following way:

$$I = \frac{p^2}{\rho c}$$

For the cases entailing transient signals or those of signal distortions or target impingement occurring, it is more useful to use the concept of *energy flux density* of the acoustic wave, as defined by

$$E = \int_0^\infty I\,dt = \frac{1}{\rho c}\int_0^\infty p^2\,dt$$

The unit of intensity in underwater sound is the intensity of a plane wave having a root-mean-square pressure of 1 micropascal ($1\,\mu$Pa) or 10^{-5} dyne/cm^2. This amounts to 0.64×10^{-22} W·s/cm^2. This pressure of 1 μPa serves as the reference level for the definition of the decibel as applied to underwater acoustics, in contrast to the decibel referred to 20 μPa, that describes the sound level of acoustic propagation in air.

15.3   Speed of Sound in Seawater

The principal difference between the speed of sound in freshwater and seawater is that with the latter, salinity constitutes an additional factor besides pressure and temperature. In fact, the speed of sound in seawater varies with geographic location, water depth, season, and even time of the day. Sound velocity in a natural body of water was first measured in 1827 when Colladen and Sturm collaborated in

TABLE 15.1. Expressions for Sound Speed (Meters per Second) in Seawater as Functions of Temperature, Salinity, and Depth.

Expression	Limits[a]	Reference
$c = 1492.9 + 3(T - 10) - 6 \times 10^{-3}(T - 10)^2$ $- 4 \times 10^{-2}(T - 18)^2 + 1.2(S - 35)$ $- 10^{-2}(T - 18)(S - 35) + D/61$	$-2 \leq T \leq 24.5°$ $30 \leq S \leq 42$ $0 \leq D \leq 1,000$	Leroy
$c = 1449.2 + 4.6T - 5.5 \times 10^{-2}T^2$ $+ 2.9 \times 10^{-4}T^3 + (1.34 - 10^{-2}T)(S - 35)$ $+ 1.6 \times 10^{-2}D$	$0 \leq T \leq 35°$ $0 \leq S \leq 45$ $0 \leq D \leq 1,000$	Medwin
$c = 1448.96 + 4.591T - 5.304 \times 10^{-2}T^2$ $+ 2.374 \times 10^{-4}T^3 + 1.340(S - 35)$ $+ 1.630 \times 10^{-2}D + 1.675 \times 10^{-7}D^2$ $- 1.025 \times 10^{-2}T(S - 35) - 7.139 \times 10^{-13}TD^3$	$0 \leq T \leq 30°$ $30 \leq S \leq 40$ $0 \leq D \leq 8,000$	MacKenzie

[a] D = depth, in meters; S = salinity, in parts per thousand; T = temperature, in degrees Celsius.

striking a submerged bell in Lake Geneva and simultaneously setting off a charge of powder in the air (Wood, 1941). The intervals between the two events were timed across the lake, and they obtained a value of 1435 m/s at 8.1°C, which is amazingly close to the modern value. Subsequent measurements over the past forty decades entail direct measurements of speeds under carefully controlled conditions, and the speeds were mapped as functions of oceanographic parameters [Del Grosso (1952), Weissler and Del Grosso (1951), and Wilson (1960)].

Expressions have been derived in terms of three basic quantities: temperature, salinity, and pressure (hence depth). Save for the presence of contaminants such as biological organisms and air bubbles, no other physical properties were found to affect the speed of sound in seawater. The interdependence of these three parameters is not a simple one. Del Grosso (1974) developed an expression containing 19 terms each to 12 significant figures in the powers and cross-products of the three principal variables, and Lovett developed a simpler but still unwieldy expression. We list in Table 15.1 a number of simpler expressions that suffice for most practical work. Generally these expressions result in values with errors less than a few parts in ten thousand, or approximately 0.5 m/s.

The speed of sound in the sea increases with temperature, depth, and salinity. Table 15.2 gives the approximate coefficient for the rate of change in these

TABLE 15.2. Approximate Coefficients of Sound Velocity.

Parameter being varied	Coefficient[a]	Coefficient
Temperature (near 70°F)	$\frac{\Delta c/c}{\Delta T} = +0.001/°F$	$\frac{\Delta c}{\Delta T} = +5$ ft/s · °F
Salinity	$\frac{\Delta c/c}{\Delta S} = +0.0008/ppt$	$\frac{\Delta c}{\Delta S} = +4$ ft/s · ppt
Depth	$\frac{\Delta c/c}{\Delta D} = +3.4 \times 10^{-6}/ft$	$\frac{\Delta c}{\Delta D} = +0.016$ s^{-1}

[a] c = speed in ft/s; T = temperature in °F; S = salinity in parts per thousand (ppt); D = depth in ft.

principal parameters. In open deep water, salinity is found to have a rather small effect on the velocity.

15.4   Velocity Profiles in the Sea

The term *velocity profile* refers to the variation of the sound speed with depth; it is also called the *velocity–depth function*. Figure 15.1 shows a typical deep-sea velocity profile, which, in turn, can be subdivided into several layers. The *surface layer* lies just below the sea surface. The speed of sound in that layer is responsive to daily and local changes of heating, cooling, and action of the winds. The surface layer may consist of a mixed layer of isothermal water that is caused by action of the wind as it blows across the surface of the water. Sound becomes trapped in this mixed layer. On prolonged calm and sunny days, this mixed layer dissipates, to be replaced by water in which its temperature drops with increasing depth.

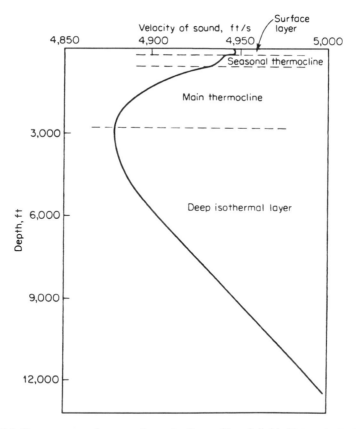

FIGURE 15.1. Deep-sea sound propagation velocity profile subdivided into principal layers.

The *seasonal thermocline* lies below the surface layer. The term *thermocline* denotes a layer in which the temperature varies with depth. The seasonal thermocline is usually characterized by a negative thermal or velocity gradient, meaning that the temperature and the speed of sound decreases with increasing depth; and it does vary with the seasons. During the summer and fall, when the ocean waters near the surface are warm, the seasonal thermocline is well defined and it becomes less so during the winter and spring and in the Arctic when it tends to become indistinguishable from the surface layer. Below the seasonal thermocline lies the *main thermocline*, which hardly varies throughout the seasons. It is this layer in the deep sea, that the temperature changes the most. Underneath the main thermocline, reaching down to the sea bottom is the deep isothermal layer having an almost constant temperature (generally about $3°-4°C$) in which the speed of sound *increases* with depth because of the effect of pressure on the sound speed. At the saddle point between the negative gradient of the main thermocline and the positive gradient of the deep layer, there occurs a region of velocity minimum toward which sound traveling at great depth tends to bend or where it becomes focused by refraction. In the more northern regions, the deep isothermal layer extends almost to the water surface. The region where this minimum occurs is called the *deep sound-channel axis*.

The existence and the thicknesses of these layers very according to latitude, season, time of day, and meteorological conditions. Figure 15.2(a) displays the

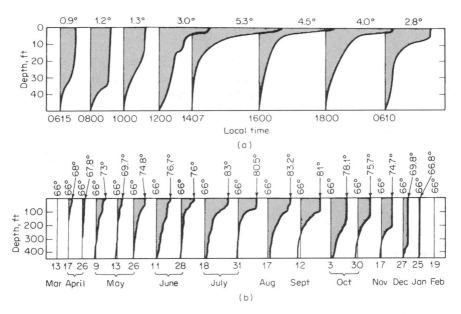

FIGURE 15.2. Diurnal and seasonal variation of a surface layer near Bermuda. In (a) temperature profiles at various times of the day show how surface temperature increases over the temperature at 50 ft depth. The temperature profiles for different parts of the year are given in (b). Temperatures and temperature differences are given in degrees Fahrenheit.

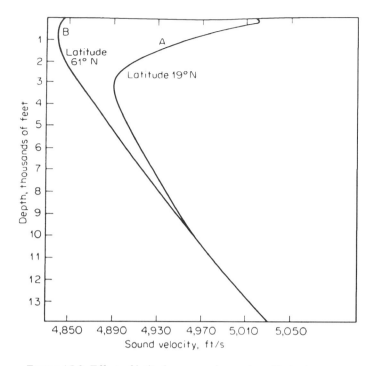

FIGURE 15.3. Effect of latitude on sound–speed profile in deep sea.

diurnal behavior of the surface layer near Bermuda. It demonstrates how the temperature profiles vary, when the surface waters of the sea warm up during the course of a sunny day and cool down during the night. These changes in temperatures affect considerably the transmission of sound from a surface-ship sonar, particularly in the afternoon when echo ranging tend to be poorest. Figure 15.2(b) illustrates a series of bathythermograms taken in the Bermuda area, showing how the seasonal thermocline evolves during the summer and autumn. The effect of latitude on sound–speed profile in the deep sea is shown in Figure 15.3 by profiles for two different locations in the North Atlantic at the same season of the year, At low latitudes (nearer the equator), the velocity minimum occurs at a depth of approximately 3,000 ft. At high latitudes the velocity minimum exists near the sea surdace, and the main and seasonal thermoclines show a tendency to disappear from the profile.

In shallow waters of the coastal regions and also on continental shelves, the velocity profiles tend to become far less clear-cut and rather unpredictable. The velocities tend be be influenced greatly by surface heating and cooling, changes in salinity, and the presence of water current. Nearby sources of freshwater tend to complicate these effects and contribute to the spatial and temporal instability of numerous gradient layers.

15.5 Underwater Transmission Loss

Transmission loss quantitatively refers to the weakening of sound between a point 1 yd from the source and a point someplace in the sea.[2] Let I_0 indicate the intensity at the reference point 1 yd from the "center" of the acoustic source (thus 10 log I_0 denotes the source level) and I_1 is the intensity at the point of interest. The transmission loss TL between the source and the point of interest is

$$TL = 10 \log \frac{I_0}{I_1} \text{ dB}$$

Time averaging is implied in the above definition. For short pulses, a transmission loss equivalent to that for continuous waves is given by the ratio of the energy flux density at 1 yd from the source E_0 to the energy flux density E_1 at the point 1 of interest, that is,

$$TL = 10 \log \frac{E_0}{E_1} \text{ dB}$$

If the metric units of distance are used so that the reference distance is 1 m, TL will be 0.78 dB less than for TL based on the reference distance of 1 yd.

Transmission loss can be subdivided into two types of losses: part of the losses is due to *spreading*, a geometric effect, and the remainder is attributable to *absorption losses* which represent conversion of acoustic energy into heat.

If a small source of sound is located in a homogeneous, infinite, lossless medium, the power generated by the source radiates outward uniformly in all directions. The total power radiating outward remains the same as its wave front expands as a spherical surface with an increasing radius (the radius increases at the rate of $c \times$ time t). Because power P is equal to intensity times the area, that is,

$$P = 4\pi I_1 r_1^2 = 4\pi I_2 r_2^2 = \cdots = \text{const}$$

and setting $r_1 = 1$ yd, the transmission loss to surface r_2 is

$$TL = 10 \log \frac{I_1}{I_2} = 10 \log \frac{r_2^2}{r_1^2} = 20 \log r_2$$

which can be recognized as the *inverse-square* law of spreading, also known as *spherical spreading*.

When the medium is bound by upper and lower parallel planes, the spreading is no longer spherical because the sound cannot cross the boundary plates. Beyond a certain range and as shown in Figure 15.4, the power radiated by the source spreads outward as a wavefront that constitutes a cylindrical surface represented

---

[2] The U.S. Navy still employs an odd melange of dimensional units in defining, say, source strength in terms of decibels (based on the metric reference of 1 μPa for sound pressure level or 1 pW for sound power level) and the distance from the source at 1 yd rather than the 1 m used elsewhere. Conversion to the entirely metric system requires the subtraction of 0.78 dB from the U.S. value.

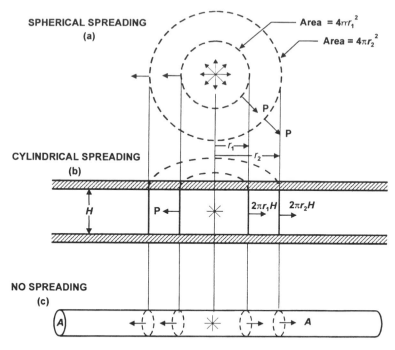

FIGURE 15.4. Spreading of acoustical energy: (a) in an infinite medium, (b) in a medium between two parallel plates, and (c) in a tube.

by an radius expanding at the rate of $c \times$ time t. The power P crossing the cylindrical surfaces at radius r is

$$P = 2\pi H I_1 r_1 = 2\pi H I_2 r_2 = \cdots = \text{const}$$

where H represents the distance between the two parallel boundary planes. Assigning $r_1 = 1$ yd, the transmission loss to r_2 is given by

$$\text{TL} = 10 \log \frac{I_0}{I_2} = 10 \log r_2$$

In this situation the *cylindrical* spreading occurs in the *inverse first power*. This sort of spreading occurs at moderate and long ranges when sound is entrapped within a sound channel in the sea.

We can also consider a *wave guide*, essentially a tube or pipe of constant cross section. This is a case where *no spreading* occurs. Beyond a certain range, the area over which the power is distributed remains constant; and the pressure intensity and the TL are independent of range.

A fourth type of spreading occurs when the signal from a pulsed source spreads out in time as the pulse propagates through the medium. The pulse becomes elongated by multipath propagation effects, which cause the pulse to smear out in time while it travels to a receptor. This type of effect is particularly evident in long-range propagation in deep-ocean sound channels. If the medium is infinite and the time stretching is proportional to the range traveled, the intensity falls off as the inverse cube of the range. In the case of a sound channel of Figure 15.4(b), time stretching that is proportional to the range causes the intensity to fall off as the inverse square of the range instead of the inverse first power. In summary the spreading laws may be listed as follows:

Propagation in	Type of spreading	Intensity varies as	TL, dB
Tube	None	r^0	0
Between parallel plates	Cylindrical	r^{-1}	$10 \log r$
Free field	Spherical	r^{-2}	$20 \log r$
Hypersphere	Free field with time stretching	r^{-3}	$30 \log r$

It should be noted that the hyperspherical case applies in a hypothetical sense to sonar theory.

Absorption varies with range in a manner different from the loss due to spreading. It occurs because acoustic energy becomes converted into heat; and this conversion embodies a true loss of energy within the propagation medium. Consider a plane wave passing through an absorbing medium. The fractional rate that the intensity of the wave decreases along distance x is proportional to the distance traveled, that is,

$$\frac{dI}{I} = -k \, dx \tag{15.1}$$

where k denotes the proportionality constant and the minus sign signifies that dI drops in the direction of increasing x. Integrating equation (15.1) between ranges r_1 and r_2, the intensity I_2 at r_2 is found from

$$I_2 = I_1 e^{-k(r_2 - r_1)} \tag{15.2}$$

Rewriting (15.2) yields:

$$10 \log I_2 - 10 \log I_1 = -10k(r_2 - r_1) \log_{10} e$$

Setting $\alpha = 10k \log_{10} e$, the change of the intensity level in dB is now expressed as

$$10 \log \left(\frac{I_2}{I_1} \right) = -\alpha(r_2 - r_1)$$

or

$$\alpha = \frac{10 \log(I_1/I_2)}{r_2 - r_1}$$

The quantity α is the logarithmic absorption coefficient and is usually expressed in decibels per kiloyard (dB/kyd) in the United States or decibels per kilometer (dB/km) in the metric system.

15.6 Parametric Variation of Absorption in Seawater

From actual measurements (Wilson and Leonard, among others) it became evident that the absorption of sound in seawater was unexpectedly higher than that of pure water, and it could not be attributable to scattering, refraction, or other effects of propagation in the natural environment. For example, the absorption in seawater in the frequency range 5–50 kHz was found to be approximately 30 times that in distilled water. Liebermann (1949) suggested that this excess absorption is attributable to the sort of chemical reaction that evolves under the influence if a sound wave and one of the dissolved salts in the sea.

The absorption of sound in seawater depends on three effects: one is the presence of *shear viscosity*, a classical effect studied by Lord Rayleigh, who derived the following expression for the absorption coefficient:

$$\alpha = \frac{16\pi^2 \mu_s}{3\rho c^3} f^2 \tag{15.3}$$

where

α = intensity absorption coefficient, cm^{-1}

μ_s = shear viscosity, poises (approximately 0.01 for water)

ρ = density, cg/cm^3 (1 for water)

c = sound propagation speed (approx. 150,000 cm/s)

f = frequency, Hz

According to equation (15.3) the value of α is $6.7 \times 10^{-11} f^2$ dB/kyd, but this amounts to only about one-third of the absorption actually measured in distilled water. The additional viscosity in pure water over that due to shear viscosity is attributed to another type of viscosity called *volume* or *bulk viscosity*, which is the result of a time lag for water molecules to "flow" in a expansive/compressive manner in reacting to acoustic signals. This viscosity effect adds to the shear viscosity, so the absorption coefficient incorporating both types of viscosity becomes:

$$\alpha = \frac{16\pi^2}{3\rho c^3} \left(\mu_s + \frac{3}{4}\mu_b \right) f^2 \tag{15.4}$$

where μ_b denotes the bulk viscosity. For water $\mu_b = 2.81\mu_s$.

Below 100 kHz the predominant reason for absorption in seawater is due to the phenomenon of ionic relaxation of the magnesium sulfate ($MgSO_4$) molecules present in the seawater. This association–dissociation process involves a relaxation time, an interval during which the $MgSO_4$ ions in the seawater solution dissociate under the impetus of the sound wave. Magnesium sulfate accounts for only 4.7% by weight of the total dissolved salts in seawater, but this particular salt was discovered to be the dominant absorptive factor in seawater, rather than the principal constituent NaCl (Leonard, Combs, and Skidmore).

The ionic relaxation mechanism causes a variation of the absorption with frequency, which is different from that in equation (15.4). On the basis of more than 30,000 measurements carried out at sea between 2 and 25 kHz out to ranges of 24 kyd, Schulkin and Marsh modified a frequency-dependency relation originally developed by Liebermann (1948). Their result was

$$\alpha = A\frac{Sf_tf^2}{f_T^2 + f^2} + B\frac{f^2}{f_T} \text{ dB/kyd} \tag{15.5}$$

where

S = salinity in parts per thousand (ppt)

A = .0186, a constant

B = .0268, another constant

f = frequency, kHz

f_T = temperature-dependent relaxation frequency

    = $21.9 \times 10^{6-1.520(T+273)}$

T = temperature in °C

But at frequencies <5 kHz, the attenuation coefficients measured at much higher values than those that would be obtained from equation (15.5), so it is apparent that there is an additional cause of attenuation other than the ionic relaxation of magnesium sulfate. A number of possible causes were advanced, but the currently accepted explanation for this excess attenuation is the boron–borate relaxation process discovered in the laboratory by Yeager et al. The boric acid [B(OH)$_3$] ionization process is not a simple one, because its mechanism seems to rely in a complicated way on the presence of other chemicals in the saltwater solution, particularly the pH or acidity factor in seawater.

The effect of depth on the absorption also has been well investigated theoretically and experimentally. For the range of hydrostatic pressure extant in the sea, the effect of pressure is to reduce the absorption coefficient by a factor $(1 - 6.54 \times 10^{-6}P)$, where P is the pressure expressed in terms of atmospheres. One atmosphere is the pressure equivalent of 33.9 ft (10.3 m) of water at 39°F (3.9°C). The absorption coefficient at depth d in the sea is found from

$$\alpha_d = \alpha_0(1 - 1.93 \times 10^{-5}d)$$

where d denotes the depth in feet and α_0 is the value of the attenuation coefficient at the surface of the water ($d = 0$). At a depth of 15,000 ft (nearly 4600 m), the absorption coefficient drops to 71 percent of its value at the surface. However, there has been a considerably larger depth dependence, by a factor of nearly two, observed in at-sea measurements (Bezdeck).

Figure 15.5 constitutes a summary of the processes of attenuation and absorption in the sea. The portion of the curve at the lower left of the plot is due to sound channel diffraction, caused by the fact that the deep sea is no longer an effective

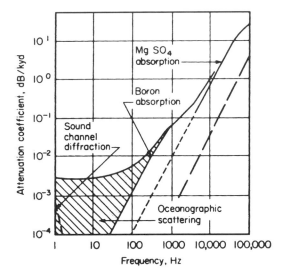

FIGURE 15.5. Attenuation and absorption processes in the sea.

duct for such long wavelengths, The dashed line at the right displays the effect of absorption due to viscosity.

15.7   Spherical Spreading Combined with Absorption

It has been found in propagation measurements made at sea that spherical spreading together with absorption provides a reasonable fit to the measured data under a surprisingly wide variety of conditions, even in situations where spherical spreading is not supposed to occur (e.g., trapping conditions in sound channels). When an approximation of the transmission loss is sufficient, the universal spherical-spreading law plus the loss due to absorption will serve as a useful working rule:

$$TL = 20 \log r + \alpha r \times 10^{-3} \qquad (15.6)$$

In equation (15.6) the first term denotes the spherical spreading and the second term the absorption effect, with 10^{-3} inserted to handle the fact that r is given in yards and α is expressed in dB/kyd.

15.8   Underwater Refraction

Refraction is the major factor in altering simple spherical spreading of sound in the ocean. As mentioned above, the factors affecting the sound propagation speed in seawater are temperature, depth, and salinity. Variations of salinity do occur, particularly at the mouths of large rivers where copious amounts of fresh water

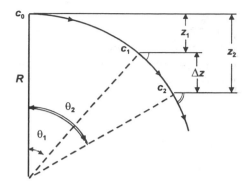

FIGURE 15.6. Diagram used for derivation of the relation between gradient G and the radius of curvature R of a sound ray.

intermingle with seawater, at the edges of large ocean currents such as the Gulf Stream, and in water near the surface, where rain, evaporation, and ice melting can impose maximal effects. Variations in speed of sound with depth is quite small (e.g., 0.1 percent over a 100 m depth). But variations in speed due to temperature changes are much greater and can fluctuate wildly, particularly near the surface.

When sound varies with ocean depth, the path of a ray through the medium can be determined by applying Snell's law ($\sin \phi / c =$ const). Because the rays of the greatest interest in the study of oceans are nearly horizontal, it is more usual to restate Snell's law as follows:

$$\frac{\cos \theta}{c} = \frac{1}{c_0} \tag{15.7}$$

where θ is the angle of refractive deflection made with the horizontal at a depth where the speed of sound is c, and c_0 is the speed at a depth (real or extrapolated) where the ray would become horizontal.

A complex profile of the propagation velocity versus depth such as that of Figure 15.1 can be simplified for analytical purposes by separating the profile into small enough segments so that the velocity gradient may be considered constant over its length. Advantage is taken of the fact that the path of a sound ray through a stratum of water over which the sound speed gradient G is a constant constitutes an arc of a circle whose center lies at a depth where sound speed extrapolates to zero.

In Figure 15.6 we consider a portion of the ray path with a radius of curvature R. It follows that $\Delta z = R(\cos \theta_1 - \cos \theta_2)$, and the gradient G is

$$G = \frac{c_2 - c_1}{\Delta z} \tag{15.8}$$

We can combine the last two equations with Snell's law of equation (15.7) which now yields

$$R = -\frac{c_0}{G} = -\frac{c}{G \cos \theta} \tag{15.9}$$

When G is constant and hence R is a constant, the path of the ray is therefore a circle. The center of curvature of the circle lies at the depth where $\theta = 90°$ which corresponds to $c = 0$. In the case illustrated in Figure 15.6, the speed gradient is negative, so R is positive. Otherwise, if the speed gradient were to be positive, R would become negative and the path would refract upward.

Once the radius of curvature of each segment of the path is established, the actual path can be traced through graphic or computerized means. Let the initial angle of deflection of the ray be designated θ_0, and use be made of the geometry of Figure 15.6 along with equation (15.9). The changes in range Δr and depth Δz are given by

$$\Delta r = \frac{1}{G}\frac{c_0}{\cos\theta_0}(\sin\theta_1 - \sin\theta_2) \tag{15.10}$$

$$\Delta z = \frac{1}{G}\frac{c_1}{\cos\theta_1}(\sin\theta_2 - \sin\theta_1) \tag{15.11}$$

Applying the small angle approximations ($\cos\theta \approx 1$, $\sin\theta \approx 0$, etc.) and eliminating θ from the last two equations yields a convenient approximate relationship between the range and depth increments along a ray for $|\theta|$ less than $20°$ and for $\Delta r \ll |c_1/G|$:

$$\Delta z = \tan\theta_0\Delta r - \frac{G}{2c_0}(\Delta r)^2 \tag{15.12}$$

15.9   Mixed Layer

Wave action can cause the water to mix in the surface layer, thus creating what is called a *mixed layer*. The positive sound–speed gradient in this layer entraps sound near the surface. After it is developed, the mixed layer tends to exist until the sun heats up the upper portion, decreasing the gradient. This heating effect engenders a negative gradient that leads to a downward refraction and the loss of sound from the layer. Because this occurs later during daytime, this effect became known as the *afternoon effect*. During the night, surface cooling and wave mixing permit this isothermal layer to reestablish itself.

A computer-produced ray diagram (the discontinuous form of the rays are due to the manner the velocity profile was subdivided in the computer program) for a source in a fairly typical mixed layer is shown in Figure 15.7. For the conditions for which the diagram was plotted, the ray leaving a source at $1.76°$ becomes horizontal at the base of the layer. Rays that leave the source at smaller angles stay entrapped in the layer, and rays that leave the source at greater angles are sent into the lower depths of the sea. A shadow zone is created beneath the mixed layer at ranged beyond the direct and near sound field. This zone is isonified by scattered sound from the sea surface and by diffusion of sound out of the channel, caused by the nature of the lower boundary.

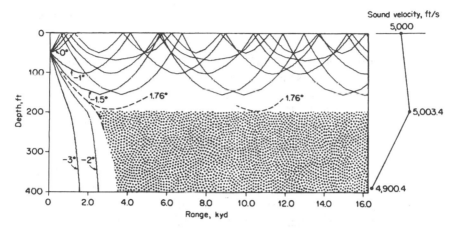

FIGURE 15.7. A computer-produced ray diagram for sound transmission from a 50-ft (15.24-m) source in a 200-ft (61-m) mixed layer.

15.10   Deep Sound Channel

In Section 15.4 reference was made to a region constituting the deep sound channel where the sound propagation speed reaches a minimum in the ocean depths. All rays originating near the axis of this channel and making small angles with the horizontal will return to the axis without reaching the ocean surface or bottom, thus remaining entrapped within that channel. Absorption of low frequencies in seawater tends to be quite small, so the low-frequency components of explosive charges detonated in this channel can travel tremendous distances, and they have been detected more than 3000 km away. The reception of these explosive signals by two or more well-separated hydrophone arrays can permit an accurate determination of the explosion's location by triangulation. Passive sonar is currently being used in deep sound channels to monitor activities in deep ocean.

15.11   Sonar Transducers and Their Properties

Underwater sound equipment were designed to detect and analyze underwater sound. They generally consist of a hydrophone array that consists of transducers that convert acoustic energy into electrical energy, and vice versa, and a signal processing system to analyze and display the signals aurally or visually. A transducer that accepts sound and converts it into electricity is called a *receiver* or *hydrophone*. A transducer that converts electrical energy into sound is called a *projector*. Some sonar systems use the same transducer to generate and receive sound. There are two principal types of transducers, according to their special properties of their activation materials. One type of transducer depends on *piezoelectricity* and its variant, *electrostriction*, and the other type functions on the principle of *magnetostriction*.

Certain crystalline substances, such as quartz, ammonium dihydrogen phosphate (ADP), and Rochelle salt, generate a charge between certain crystal surfaces when they are subject to pressure. Conversely, when a voltage is placed between is placed across their surfaces, a stress occurs in these crystals. Such materials are said to be *piezoelectric*. Electrostrictive materials are polycrystalline ceramics that produce the same effect and these have to be properly polarized in a strong electrostatic field. Barium titanate and lead zirconate titanate are examples of such materials.

A magnetostrictive material such as nickel exhibits the same effect as piezo-electricity, but it does so under the influence of a magnetic field rather than applied stresses. It changes its dimensions when it is subjected to a magnetic field, and, conversely, changes the magnetic field within and around it when it becomes stressed. In other words, when a properly designed nickel element is subjected to an oscillating magnetic field, a mechanical oscillation is produced that generates acoustic waves in water. Magnetostrictive materials are also polarized in order to avoid frequency doubling and to achieve a higher efficiency.

Piezoelectric and magnetostrictive types of transducers are more suitable than other kinds of transducers for use underwater due to better impedance matching with water. Because they are relatively inexpensive and can be readily fashioned into the desired shapes, ceramics materials are finding increasing applications as underwater devices. Other types of units now include the thin-film transducers [Hennin and Lewiner, 1978] and fiberoptic hydrophones [Hucaro et al., 1977].

Arrays

While single piezoelectric or magnetostrictive elements are normally used in hydrophones for research or measurement purposes, much of the applications of hydrophones entail hydrophone *arrays* which use a number of properly spaced elements. The following reasons exist for the use of arrays:

1. The array is more sensitive, as a number of elements will generate more voltage, if connected in series, or more current, if connected in parallel.
2. The array provides directivity that enables it to discriminate between sounds coming from different directions.
3. An improved signal-to-noise ratio over that of a single hydrophone is provided, because the array discriminates against isotropic or quasi-isotropic noise to favor a signal arriving from a direction that the array is pointing to.

Because of the above advantages, most practical applications of underwater sound make use of arrays. Moreover, the first and second benefits listed above also apply to projectors as well as to hydrophones.

Examples of a cylindrical and a plane array are shown in Figure 15.8. Spherical arrays have also been constructed for installation on submarines.

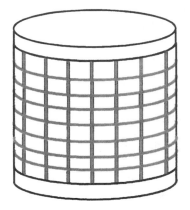 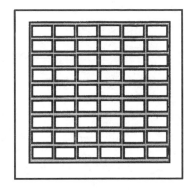

FIGURE 15.8. Cylindrical array and plane array of transducers.

Array Gain

The improvement in the signal-to-noise ratio (SNR), which results from the use of arraying of hydrophones is measured by a parameter called *array gain*, which is defined by

$$AG = 10\log \left(\frac{(S/N)_{\text{array}}}{(S/N)_{\text{single element}}} \right) \qquad (15.13)$$

The numerator of equation (15.13) denotes the SNR at the array terminals and the denominator represents the SNR of a single element of the array. It is assumed that all of the elements in the array are identical.

Transducer Response

The effectiveness of a hydrophone in converting sound into an electric signal is called the *response* of the hydrophone. It relates the generated voltage to the acoustic pressure of the sound field. The receiving response of a hydrophone is defined as the voltage produced across the terminals by a plane wave of unit acoustic pressure (the value before the introduction of the hydrophone into the sound field). The receiving response is usually expressed as the *open-circuit response* which is obtained when the hydrophone connects to an infinite impedance. The customary unit of the receiving response is the number of decibels relative to 1 V produced by an acoustic pressure of 1 μPa, and it is written as decibels re 1 V/μPa.

Example Problem 1

Find the voltage across the hydrophone terminals exposed to an acoustic rms pressure of 1 μPa if the receiving response is -80 decibels re 1 V.

Solution

We use the relationship

$$SPL = 20\log(p/1\,\mu Pa)$$

and because $SPL = -80$ dB,

$$p/1\,\mu Pa = antilog(-80/20),$$

yielding $p = 0.0001\,\mu Pa$. For the 1 μPa sound field, the voltage is .0001 V across the terminals.

For a projector, the *transmitting-current response* is the acoustic pressure produced at a point 1 m from the projector in the direction of the axis of its beam pattern by a unit current into the projector. This means that the transmitting response is stated in terms of the number of decibels relative to 1 μPa as measured at the reference distance, produced by 1 A of current into the electric terminals of the projector. While the transmitting response is usually referred to a reference distance of 1 m from the source, a correction of 0.78 dB must be added in order to convert the transmission response expressed in terms of a reference level of 1 yd instead of 1 m.

Example Problem 2

Predict the rms pressure at 1 m when the projector is driven with a current of 1 rms ampere, if the response is rate at 100 dB re 1 μPa/A (referred to 1 m). Express the response in terms of a reference distance of 1 yd instead of 1 m.

Solution

The corresponding rms acoustic pressure for 100 dB is $10^5\,\mu Pa$, which is produced by 1 rms ampere current. The corresponding transmitting response for 1 yd is $100 + 0.78 = 100.78$ dB.

15.12   The Sonar Equations

The purpose of the sonar equations, originally formulated during World War II and derivative of similar considerations in radar, is twofold: (a) prediction of the capabilities of existing sonar equipment with respect to detection probability or search rate, and (b) design of new equipment to meet a preestablished range of detection or actuation.

In our mathematical exposition of the sonar principle, we subscribe to the assumption that target strength is a function of the source and echo levels, respectively, as well as the transmission loss that occur in the echo-ranging process. The function of the sonar may be the detection of an underwater target, or it can be the homing of an acoustic torpedo at the instant when it begins to ascertain its target. Of a total signal energy received by a sensor, a portion may be desired, and is considered to be the *signal*. The balance of the acoustic energy is undesired and is termed the *background*. The background consists of *noise*, the basically steady state portion that is not attributable to the echo-ranging, and *reverberation* which represents the slowly decaying portion of the background caused by the return of the original acoustic output from scatterers dispersed in the sea. In the design of a sonar system, it is the objective to find ways of increasing the overall response of the system to the signal and to decrease the response of the system to the background, in other words, to increase the signal-to-background ratio.

A sonar system serves a practical purpose such as detection, classification (establishing the character of the target), torpedo hunting, communication, or fish finding. In each of these tasks there will be a specific signal-to-background ratio and a level of performance in successfully detecting targets with a minimum of "false alarms" which erroneously indicate the presence of a target when no target is present. If the signal increases sufficiently to equal the level of the background, the desired purpose will be achieved when the signal level equals the level of the background that just masks it, that is,

$$\text{signal level} = \text{background masking level} \qquad (15.14)$$

Masking does not mean that all of the background interferes with the signal. Only that portion that lies in the frequency band of the signal will cause masking, just as in psychoacoustics, where a broadband noise masks out a pure tone or narrow-band signal presented to the human ear.

The equality of (15.14) constitutes the one instant of the time when a target approaches or recedes from a receiver. At short ranges, the signal level from a target should handily exceed the background masking level, while the reverse will occur at long ranges. But it is at the instant of (15.14) when the sonar system just begins to perform its function which is of greatest interest to the sonar designer.

The source level that is defined in terms of intensity at 1 m (formerly 1 yd) was derived from physical concepts in order to express separately the effects on the signal strength of the echo, viz. (i) the size, shape and orientation of the target; (ii) the intensity of the source; and (iii) the range of the target. At long ranges, only the transmission loss depends on the range. At short ranges, target strength depends on that range as well as the size, shape, and orientation of the target. If the source is quite close to the target, different parts of the target are struck by sound of different intensities, or if the receiver is so close that the spreading of the sound reflected from the target to it is not the same as the spreading from a point source, the target strength term will depend on the range.

Active and Passive Equations

The sonar parameters are determined by the equipment, the medium, and the target. We denote the following parameters, which are stated in terms of dB relative to the standard reference intensity of a 1-μPa plane wave:

Equipment Parameters
    SL: projector source level
    NL: self-noise level (also called electronic noise)
    DI: directivity index
    DT: detection threshold
Medium Parameters
    TL: transmission loss
    RL: reverberation level
    NL: ambient noise level
Target Parameters
    TS: target strength
    SL: target source level

Note that two pairs of parameters are given the same symbol because they are identical. This set of parameters is not necessarily all-inclusive, nor is this set unique, for other parameters such as sound velocity or backscattering cross-section could be considered. The parameters chosen above are conventional ones applied in underwater technology.

In order to understand the significance of the above listed quantities, consider Figure 15.9 which illustrates a schematic of an echo-ranging process. A transducer operating as both sound source and receiver produces a source level of SL decibels at a unit distance (generally 1 m worldwide and 1 yd in the English system) on its axis. Let the axis of the sound source be properly aimed toward the target; the radiated sound will reach the target with a transmission loss TL, and the level of the signal reaching the target will be SL − TL. On reflection or scattering from the target with target strength TS, the reflected or the backscattered level will be SL − TL + TS (at a distance of 1 m from the acoustic center of the target in the direction back to the source). This reflected signal also becomes attenuated by the transmission loss TL as the result of its travel to the source. The level of the echo reaching the source thus becomes SL − 2TL + TS. Now if we consider the background noise and assume it to be isotropic noise rather than reverberation, the background level will simply be NL. But this level will be lessened by the directivity index DI of the transducer serving as a receiver, so the relative noise power at the transducer interface is NL − DI. Because the axis of the transducer points in the direction from which the echo is traveling, the relative echo power is unaffected by the transducer directivity, At the transducer terminals, the echo-to-noise ratio becomes

$$SL − 2TL + TS − (NL − DL)$$

Now let the sonar act as a detector, that is, it is to give an indication on some kind of display that an echoing target is present. When the input signal-to-noise ratio exceeds a specific detection threshold, thus meeting preset probability criteria, a relay can be activated to indicate on a display that a target is present. Otherwise, when the input signal-to-noise ratio falls below the detection threshold, the

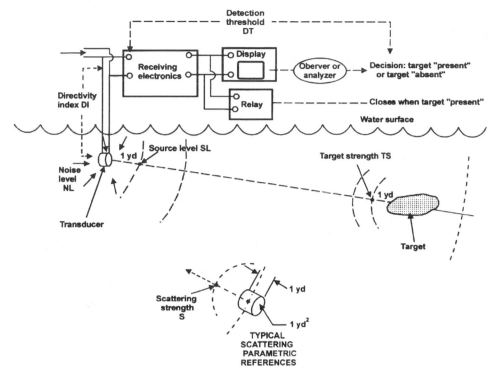

FIGURE 15.9. Schematic of echo ranging process.

indication will be that a target is absent. But when the target is just being detected, the signal-to-noise ratio equals the detection threshold, that is,

$$SL - 2\,TL + TS - (NL - DI) = DT \qquad (15.15)$$

Equation (15.15) characterizes the active-sonar equation as an equality in terms of the *detection threshold*. In recognizing that only a portion of the noise power lying above the detection threshold masks the echo, we could rearrange (15.15) as follows:

$$SL - 2\,TL + TS = NL - DI + DT \qquad (15.16)$$

Equation (15.16) places the echo level effects on the left-hand side and those pertaining to the noise-masking background level on the other side. Equation (15.16) constitutes the active-sonar equation for the *monostatic case*, one in which the source and the receiving hydrophone are coincident and the echo from the target travels back to the source. In some sonar applications, a *bistatic* arrangement is used—i.e., the source and the receiver are separate, and the two transmission losses to and from the target are not generally equal. In some sonars it is virtually impossible to resolve the receiving directivity index DI and detection threshold DT, so it becomes legitimate to combine these two terms as DI − DT as a single parameter describing the increase in signal-to-background ratio produced by the entire receiving system which includes the transducer, processing electronics, and the display.

What happens when the background consists of reverberation rather than noise? Instead of DI, which was defined in terms of an isotropic background and is now inappropriate, the (NL − DI) term in equation (15.16) is replaced by an *equivalent plane-wave reverberation level* RL observed at the hydrophone terminals. Equation (15.16) then becomes

$$SL - 2\,TL + TS = RL + DT \qquad (15.17)$$

The parameter DT will possess a value for reverberation that is different from the DT for noise.

In the passive or 'listening' situation, the target itself produces the signal that is detected, and one-way transmission rather than two-way transmission is involved. Target strength TS now becomes irrelevant, and the passive sonar equation becomes

$$SL - TL = NL - DI + DT \qquad (15.18)$$

Table 15.3 lists the parameters and their definitions in brief, while Table 15.4 provides the terminology of commonly used terms for describing sonar parameters. A number of the parameters in Table 15.3, namely, SL, TL, TS and the scattering

TABLE 15.3. Sonar Parameters

Parametric symbol	Reference	Definition
Source level, SL	1 yd from source on its acoustic axis	$10 \log$ (intensity of source/1 μPa)
Transmission loss, TL	1 yd from source and at target or receiver	$10 \log \left(\dfrac{\text{signal intensity at 1 yd}}{\text{signal intensity at target or receiver}} \right)$
Target strength, TS	1 yd from acoustic center of target	$10 \log \left(\dfrac{\text{echo intensity at 1 yd from target}}{\text{incident intensity}} \right)$
Noise level, NL	At hydrophone location	$10 \log \left(\dfrac{\text{noise intensity}}{1\,\mu\text{Pa}} \right)$
Receiving directivity index, DI	At hydrophone terminals	$10 \log \left(\dfrac{\text{noise power generated by an equivalent isotropic hydrophone}}{\text{noise power generated by actual hydrophone}} \right)$
Reverberation level, RL	At hydrophone terminals	$10 \log \left[\dfrac{\text{(reveberation power at hydrophone terminals)}}{\text{power corresponding to 1}\,\mu\text{Pa}} \right]$
Direction threshold, DT	At hydrophone terminals	$10 \log \left[\dfrac{\text{(signal power to just at perform a certain function)}}{\text{noise power at hydrophone terminals}} \right]$

Note that some of the parameters, viz. SL, TL, TS (and the scattering strength) in Table 15.1 use 1 yd as the reference distance. To obtain their values based on the reference distance of 1 m, 0.78 dB must be subtracted from the values of these parameters.

TABLE 15.4. Sonic Parameters: Terminology of Various Combinations.

Nomenclature	Parameter	Comments
Echo level	$SL - 2\,TL + TS$	Intensity of the echo measured in the water at the hydrophone
Noise-masking level	$NL - DI + DT$	Determines minimum
Reverberation-masking level	$RL + DT$	Detectable echo level
Echo excess	$SL - 2\,TL + TS - (NL - DI + DT)$	Detection just occurs, when echo excess just equals zero
Performance figure	$SL - (NL - DI)$	Difference between the source level and the NL measured at the hydrophone
Figure of merit	$SL - (NL - DI + DT)$	Equals the maximum allowable 2-way TL for TS = 0 for active sonar, or 1-way TL in passive sonars

strength (which determines RL) use 1 yd as the reference distance. To convert to 1 m reference, these quantities should be reduced by $0.78 = 20\log\,[1\ \text{meter} \times (39.37\ \text{in./m}) \times (1\ \text{yd/36 in.})]$. The attenuation coefficient commonly expressed in dB/kiloyard should be multiplied by 1.094 to convert the coefficient to dB/km. In the United States, it is generally more convenient to find the range first in kiloyards and then to divide by the factor 1.094 to express the range in kilometers.

15.13   Noise, Echo, and Reverberation Levels

The above sonar equations constitute a statement of equality between the *signal*, which is the desired portion (the echo or noise from the target) of the acoustic field, and the undesired portion, that is, the *background* of noise and reverberation. This equality holds true at only *one range*; and at all the other ranges, the equality will no longer exist. This fact is demonstrated in Figure 15.10 in which the curves of the echo level, noise masking level, and the reverberation-masking level are displayed as functions of range. The echo and reverberation levels drop off with increasing range, but the noise remains fairly constant. The echo level curve falls off more rapidly with range than does the reverberation masking curve, and intersects it at the reverberation-limited range r_r. This curve will also meet the noise-masking level curve at range r_n. If the reverberation is high, the echo will be of lesser value, and the range may be considered to be *reverberation-limited*. But if the noise-masking level occurs at the level represented by the dashed line in Figure 15.10, the echoes will then fade away into a background of noise rather than reverberation. The new noise-limited r'_n will become less than the reverberation-limited range r_n, and the range will hence be *noise-limited*. Both ranges can be established by the use of the appropriate sonar equation.

It is necessary for a sonar designer or sonar operator to know whether a sonar will be noise-limited or reverberation-limited. Generally speaking, the curves for

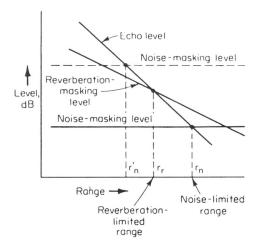

FIGURE 15.10. Effect of range on echo level, noise masking, and reverberation masking.

echo and reverberation will not occur as straight lines because sound propagation and distribution of echo-yielding scatterers add to the complexity of the situation. In the design of a sonar system for specific purposes, these curves should be created from the best information available for the conditions that are most likely to be encountered.

A convenient graphical method of solving the sonar equations for passive sonars is the SORAP, the acronym for "sonar overlay range prediction." Figure 15.11 consists of two plots that are overlaid on each other. The solid lines constitute an

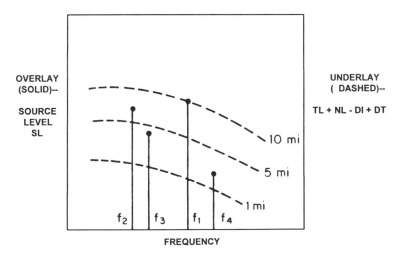

FIGURE 15.11. SORAP (Sonar Overlay RAnge Prediction) graphical method for solving the passive sonar equation.

overlay of a plot of SL vs frequency for a specific passive target or class of targets. The underlay denoted by dashed lines is a plot of the sum of the parameters $(TL + NL - DI + DT)$ for a specific passive sonar at a number of different ranges. The range and frequency at which the target can be discerned can be determined by inspection of the plots. For example, the target will be first detected at a range of 10 miles according to the line component at frequency f_1. But suppose it requires that three spectral lines must appear on the display before a detection is called; the range would then be reduced to 4 miles, and the lines at frequencies f_1, f_2 and f_3 would be displayed. This procedure helps to distinguish the target parameter SL from the equipment and medium parameters at the location where it is used, while it accommodates a wide range of frequencies. Targets can therefore be compared for the same locations, or locations can be compared for the same targets, and so on.

15.14   Transient Form of Sonar Equations and Pulses

So far, the sonar equations have been expressed in terms of the average acoustic power per unit area or *intensity* of the sound radiated by the source or received from the target. But the time interval implied by the terminology "average" can yield unreliable results in situations where short transient sources exist or whenever severe distortion is incurred in sound propagation in the medium in the course of scattering from the target.

We can adopt a more general approach by writing the sonar equations in terms of energy flux density, which is defined as the acoustic energy per unit area of the wavefront. Consider a plane acoustic wave that has a time-dependent pressure $p(t)$. The energy flux density of the wave is given by

$$E = \frac{1}{\rho c} \int_0^\infty p^2(t)\, dt \tag{15.19}$$

Because the intensity is the rms pressure of the wave divided by the acoustic impedance ρc, averaged over a time interval T, that is,

$$I = \frac{1}{T} \int_0^T \frac{p^2(t)}{\rho c}\, dt$$

it follows that

$$I = \frac{E}{T} \tag{15.20}$$

The time interval T represents the duration over which the energy flux density of the sound wave is to be averaged to yield the intensity. In the case of long-pulsed active sensors, this time interval equals the duration of the emitted pulse and very nearly equals the duration of the echo. But for short transient sonars, the interval T becomes rather ambiguous, and the duration of the echo can be considerably different from the duration of the transient emitted by the source.

Urick (1962) demonstrated that under these conditions that the intensity form of the sonar equations can be applied, providing the source level is defined as:

$$SL = 10 \log E - 10 \log \tau_e \qquad (15.21)$$

where E is the energy flux density of the source referred to 1 yd and measured in units of the energy flux density of a 1-μPa plane wave taken over an interval of 1 s, and τ_e represents the duration of the echoes expressed in seconds for an active sonar. As an example, E can be determined for explosives by measurements for a given charge weight, depth where the explosion occurs, and type of explosive.

Consider a pulsed sonar emitting a rectangular pulse of constant source level SL' over a time interval τ_0. The energy density of a pulse equals the average intensity times duration,

$$10 \log E = SL' + 10 \log \tau_0 \qquad (15.22)$$

Combining equations (15.21) and (15.22), the effective source level SL to be used accordingly in sonar equations is

$$SL = SL' + 10 \log \frac{\tau_0}{\tau_e} \qquad (15.23)$$

where τ_0 denotes the duration of the emitted pulse of source level SL' and τ_e the duration of the echo. For long-pulsed sonars, τ_0 is equal to τ_e, and therefore SL = SL'. In the case of short-pulsed sonars, $\tau_0 > \tau_e$, and thus the effective source level SL is less than actual source level SL' by the amount $10 \log(\tau_0/\tau_e)$. This effect of time-stretching is depicted in Figure 15.12. A short pulse of duration τ_0 at source level SL' is replaced in a sonar calculation by a longer pulse of duration τ_e at a lower source level SL. The energy flux density source levels must be the same for these two source levels, that is,

$$SL + 10 \log \tau_e = SL' + 10 \log \tau_0$$

which is merely a rearrangement of equation (15.23). A pulse emitted by the source stretches out in time and becomes reduced in level by the multipath effect

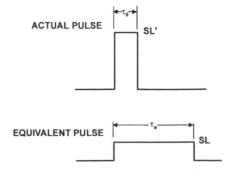

FIGURE 15.12. Time stretching: equivalent source level in short-pulse sonar.

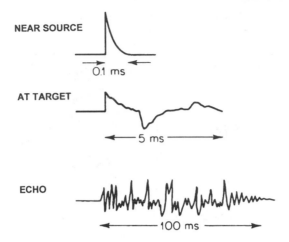

NEAR SOURCE

0.1 ms

AT TARGET

5 ms

ECHO

100 ms

FIGURE 15.13. Distortion of the pressure resulting from an explosive pulse upon arrival at an extended target. The return echo in the vicinity of the source is also shown here.

of propagation and by the mechanism of target reflection. The appropriate values of the other sonar parameters in the sonar equations, such as target strength TL and transmission loss TL, are those applying to long-pulse of CW conditions, in which the effects of multipaths in the medium and target reflection are summed up and accounted for.

In active short-pulse sonars, the echo duration τ_e becomes in itself an important parameter. In Figure 15.13, a pulse is shown as a short exponential transient at the source, as a distorted pulse at the target, and as an echo received back near the source. A shock wave pulse, for example, from an underwater explosion occurs as an exponential pulse. It may have an initial duration of 1/10 of a millisecond, and it can become distorted into an echo that is 1,000 times as long.

The duration of the echo can be subdivided into three components: τ_0, which is the duration of the emitted pulse near the source; τ_m, the additional time needed for two-way propagation in the underwater; and τ_r, the additional duration assessed by the extension in range of the target. Hence

$$\tau_e = \tau_0 + \tau_m + \tau_r \qquad (15.24)$$

Typical values of these three components of the echo duration for different circumstances are listed in Table 15.5.

Example Problem 3

Estimate the echo time duration of an explosive echo from a beam aspect submarine patrolling shallow water.

TABLE 15.5. Examples of Echo Duration.

Component	Representative Values, ms
Duration of the emitted pulse at short ranges	Explosives: 0.1 Sonar: 100
Duration produced by multiple paths	Deep water: 1 Shallow water: 100
Duration produced by a submarine target	Beam aspect: 10 Bow–stern aspect: 100

Solution

From Table 15.5 and equation (15.24), the time duration is $0.1 + 100 + 10 = 110.2$ ms

15.15   Overview of the Sonar Equations

A summarization of the sonar equations may be listed as follows:

Active sonars (monostatic)

Noise background

$$SL - 2\,TL + TS = NL - DI + DT \tag{15.25}$$

Reverberation background,

$$SL - 2\,TL + TS = RL + DT_R \tag{15.26}$$

Passive Sonars:

$$SL - TL = NL - DI + DT_N \tag{15.27}$$

The detection threshold DT differs quantitatively for reverberation and for noise and so it carries subscripts to denote the difference.

15.16   Shortcomings of the Sonar Equations

The sonar equations expressed in terms of intensities may not always be complete for certain types of sonars. For example, the short pulse sonars need the addition of the echo duration to account for time-stretching caused by multipath propagation. It must also be realized that the sea is an ever-moving medium that contains inhomogeneities of different sorts, with irregular boundaries with the topmost boundary on the move. Multipath propagation tends to predominate, because so many of the sonar parameters fluctuate erratically with time. Other irregularities may occur because of internal changes in measurement equipment and possible reconfiguration of the platform on which the equipment is mounted. In essence, the "solutions" proffered by the sonar equations really constitute a "best estimate" time average of what is really a stochastic problem. Thus, precise calculations to, say, the nearest tenth of a decibel are exercises in futility, and the solutions must be considered "best guesses" or "most probable" values.

15.17   Theoretical Target Strength of a Sphere

Consider a sphere of radius r as the subject target. As we shall see, the effect of the pulse length of the probe signals constitutes a very important factor. No account is taken of the wave nature of sound, that is, the effects of interference, diffraction, and phase differences are ignored here. Let I_0 denote the intensity of the incident sound wave striking the target and I_r the intensity of the reflected sound signal measured at some particular point. With all the other factors remaining constant, we can write

$$I_0 \propto I_r$$

The intensity I_r of the reflection is a function of the target orientation and the location of its measurement. Because I_r is usually measured at the probe signal point, mathematical treatment becomes considerably simplified. The inverse square law, which holds for large, not small, distances is expressed as

$$I_r = \frac{KI_0}{r^2} \tag{15.28}$$

where K is a constant dependent upon the size, shape and orientation of the target. Equation (15.28) does not apply to explosive sounds. For the incident signal

$$I_0 = \frac{F}{r^2} \tag{15.29}$$

where F is the intensity of the projected sound 1 m away from the source. Here it is tacitly assumed that

$$r \gg D$$

where D is the order of magnitude of the size of the source. Combining equations (15.28) and (15.29)

$$I_r = \frac{KF}{r^4} \tag{15.30}$$

For an ideal medium, according to equation (15.30), the intensity of an echo is inversely proportional to the fourth power of the range, providing the echo is measured in the source location and the range is much larger than the dimensions of either the target or the source.

Reexpressing equation (15.30) in logarithms, we obtain

$$10 \log I_r = 10 \log K + 10 \log F - 40 \log r \tag{15.31}$$

In the course of a signal traveling to a target and the echo reflecting back to the signal source, some attenuation in the signal intensity must occur. This drop in intensity in each direction has been defined as the transmission loss TL which should equal 20 log r in the idealized case of equation (15.31). The total transmission loss is therefore 2 TL, that is, $= 40 \log r$, which represents the attenuation of the signal traveling to and from the target. It should be recalled that losses do not occur solely to the effect of the inverse square law; they are also attributable to absorption and scattering in seawater, bending by temperature gradients (with consequent focusing and spreading out). Therefore, the actual transmission loss TL does not equal 20 log r alone, but also includes other effects of attenuation. The function 2 TL represents a more general situation than 40 log r in equation (15.31) which then is recast as:

$$10 \log I_r = 10 \log K + 10 \log F - 2 \text{ TL} \tag{15.32}$$

As mentioned earlier, because of the great variance of oceanographic conditions the quantity 2 TL can be established only through careful measurements. Setting target strength

$$\text{TS} = 10 \log K$$

and the echo

$$\text{E} = 10 \log I_r$$

and source level

$$\text{SL} = 10 \log F$$

we obtain the first sonar equation of Table 15.4, which constitutes the fundamental definition of signal strength:

$$\text{TS} = \text{E} - \text{SL} + 2 \text{ TL} \tag{15.33}$$

As it entails only directly measurable quantities equation (15.33) is particularly useful in the computation of target strengths from data measured at sea.

A sphere presents a perfectly symmetric target. The echoes it returns to a sound source are completely independent of its own orientation, and it is for this reason that spheres make convenient experimental targets in echo-ranging measurements. In a simple derivation, we consider a plane wave of intensity I_0 striking a sphere of radius a and cross sectional area πa^2. The total sound energy intercepted by the sphere is thus $\pi a^2 I_0$ in the ideal case of perfect reflection. Now let us assume uniform reflectivity in all directions. At a distance r from the sphere's center the acoustic energy will be spread uniformly over the surface of a sphere of radius r or over the surface area $4\pi r^2$. Because the intensity I_r of the reflected sound equals the total energy $\pi a^2 I_0$ reflected by the target sphere per unit time divided by $4\pi r^2$ over which it is distributed, then at a distance r from the sphere's center

$$I_r = \frac{\pi a^2 I_0}{4\pi r^2} = \frac{a^2}{4r^2} I_0$$

But from equation (15.28), $I_r = KI_0/r^2$, where r represents the distance from the target to the measurement point,

$$K = a^2/4$$

and so

$$\text{TS} = 10 \log K = 20 \log(a/2).$$

References

Bezdek, H. H. 1973. Pressure dependence of sound attenuation in the Pacific Ocean. *Journal of the Acoustical Society of America* 53: 782.

Browning, D., M. Fechner, and R. Mellon. 1981. Regional dependence of very low-frequency sound attenuation in the deep sound channel. *Naval Underwater Systems Center Technical Document 6561.*

Del Grosso, V. A. 1952. The velocity of sound in sear water at zero depth. *U.S. Naval Research Laboratory Report 4002.*

Del Grosso, V. A. 1974. New equations for the speed of sound in natural waters with comparisons to other equations. *Journal of the Acoustical Society of America* 56: 1084.

Hennin, C., and J. Lewiner. 1978. Condenser electret hydrophones. *Journal of the Acoustical Society of America* 63: 279.

Hucaro, J. A. et al. 1977. Fiber-optic hydrophone. *Journal of the Acoustical Society of America* 62: 1302.

Leonard, R. W., P. C. Combs, and I. R. Skidmore. 1949. Attenuation of sound in synthetic sea water. *Journal of the Acoustical Society of America* 21: 63.

Leroy, C. C. 1969. Development of simple equations for accurate and more realistic calculation of the speed of sound in sea water. *Journal of the Acoustical Society of America* 46: 216.

Liebermann, L. N. 1948. Origin of sound absorption in water and in sea water. *Journal of the Acoustical Society of America* 20: 868.

Liebermann, L. N. 1949. Sound propagation in chemically active media. *Physics Review* 76: 1520.

Medwin, Herman. 1975. Speed of sound in water for realistic parameters. *Journal of the Acoustical Society of America* 58: 1318.

Mackenzie, K. V. 1981. Nine-term equation for sound speed in the oceans. *Journal of the Acoustical Society of America* 70: 807.

National Defense Research Committee. 1946. Application of oceanography to subsurface warfare, *Div. 6 Summary Tech. Rep.*, vol. 6A; Figures 17, 32.

Schulkin, M., and H. W. Marsh. 1962. Absorption of sound in sea water. *Journal of the British IRE* 25: 293. Also: 1962, "Sound Absorption in Sea Water." *Journal of the Acoustical Society of America* 34: 864.

U.S. Navy. 1975. An Interim Report on the Sound Velocity Distribution in the North Atlantic Ocean. *U.S. Navy Oceangoing Office Tech. Report 171.*

Urick, Robert J. 1962. Generalized form of the sonar equations. *Journal of the Acoustical Society of America* 34: 547.

Urick, Robert J. 1983. *Principles of Underwater Sound*, 3rd ed. New York: McGraw-Hill. (A classic text in the field by one of the leading researchers.)

Weissler, A., and V. A Del Grosso. 1951. The velocity of sound in sea water. *Journal of the Acoustical Society of America* 23: 219.

Wilson, W. A. 1960. Speed of sound in sea water as a function of temperature, pressure, and salinity. *Journal of the Acoustical Society of America* 32: 219. Also see extensions and revisions in *Journal of the Acoustical Society of America* (1960), 32: 1357 and (1962) 34: 866.

Wilson, O. B., and O. B. Leonard. 1954. Measurements of sound absorption in aqueous salt solutions by a resonator method. *Journal of the Acoustical Society of America* 26: 223.

Wood, A. B. 1941. *A Textbook of Sound*. New York: Macmillan; p. 261.

Yeager, E., F. H. Fisher, J. Miceli, and R. Bressel. 1973. Origin of the low-frequency sound absorption in sea water. *Journal of the Acoustical Society of America* 53: 1705.

16
Ultrasonics

16.1 Introduction

As a subcategory of acoustics, *ultrasonics* deals with acoustics beyond the audio frequency limit of 20 kHz. Although ultrasonics have been employed for most of the twentieth century, the tempo of new and improved applications has reached virtually explosive proportions only in the past few years, particularly in medical diagnostics and therapeutics.

Applications of ultrasonics fall into two categories—*low intensity* and *high intensity*. Low-intensity applications carry the purpose of simply transmitting energy through a medium in order to obtain information about the medium or to convey information through the medium. Nondestructive testing, medical diagnostics, acoustical holography, and measurements of elastic properties of materials fall into this category. Even marine applications are included in this category, despite the large energy input into operating sonar submarine detectors, depth sounders, echo ranging processors, and communication devices.[1]

High-intensity applications deliberately affect the propagation medium or its contents. Uses of high intensities include medical therapy and surgery, atomization of liquids, machining of materials, cleaning, welding of plastics and metals, disruption of biological cells, and homogenization of materials.

Human beings are not alone in the use of ultrasonics, and even in this respect they have been preceded by thousands of years by other species in the animal kingdom. Certain animals are capable of generating and detecting ultrasonic signals in order to locate and identify food, navigate their way through their environment, and detect danger. In fact, the study of these animals has helped, and is still helping, scientists to develop and improve techniques in the application of ultrasonic energy.

Bats are known to emit pulses in the 30–120 kHz range, and it has been hypothesized that the bats judge range by sensing the time delay between an emitted pulse and the echo. Small bats can fly at full speed through barriers constructed

---

[1] Emsinger suggested that a third category not based on intensity be designated to specifically cover underwater applications.

of 0.4 mm vertical wiring spaced only one wingspan apart. They are capable of catching small insects at the rate of one every 10 s for as long as a half-hour. Their ability to discriminate between objects such as food and raindrops or foliage can be described as being nothing less than phenomenal; and yet when a large number of bats fly in close proximity to each other in a potentially confusing background of ultrasonic noise, they continue to locate prey and avoid collisions with one another.

The *Noctillio* bat of Trinidad catches small fish by dipping its feet below the surface of the water, after emitting a series of repetitive pulses. It has been conjectured that the characteristic ripples created by the fish are being detected by the bat rather than echoes from an object beneath the water surface. This is particularly remarkable in view of the fact the sound must penetrate a barrier with a very high reflection coefficient. As with bats, the echolocation of porpoises appear to be unaffected by the presence of interfering noises or jamming.

Moths, a prime target of bats, use ultrasonics for self-defense. When a moth detects a sonic pulse from a bat it immediately takes evasive action through zig-zagging in its flight and executing power dives. Moths can detect bats as far away as 13 m, and their ears can detect the cries of approaching bats at this distance, but when the bats are moving away their ears stop registering at about this distance. It was also observed by Roeder and Treat that when a bat makes a straight-on approach, it was observed to emit an uneven, sporadic signal; this was interpreted as an indication that the bat was counter-maneuvering by varying the intensity of its sonar pings.

Cetaceans constitute a group of sea mammals which includes whales, dolphins, and porpoises. They are extremely intelligent as well as beautiful creatures, and are of the greatest interest to acousticians. Porpoise sounds have been described as whistling, barking, rasping, repetitive clicks or pulses, and mewing. The cetaceans emit signals for the purpose of echolocation and communication with one another. Signals as high as 170 kHz have been observed in the clicks of porpoises, which vary in repetition rate from five clicks to several hundred per second. Extensive observations are being conducted to observe the emission patterns of different cetacean species, which vary from location to location at different times of the year even with the same group that has been tracked on an almost continuous basis.

At least two species of birds, *Steatornia* and *Collocalla*, are known to be echolocators. Ornithologists have reported that calls by various birds may be comprehended only by members of their own local flocks. Gulls or crows will always respond to calls from members of their own flocks, but they may or may not respond to distress or assembly calls from gulls or crows from another region.

The hearing of dogs extend well beyond the frequency range of human hearing. The "silent" dog whistles that generate ultrasonic output can thus be used to summon a dog. Although dogs are generally endowed with especially keen senses of smell, hearing, and sight, there is no evidence that they make use of echolocation.

Certain effects become more evident in wave propagation through a medium as the acoustic signal extends into the ultrasonic range. The attenuation of the signal's amplitude occurs not only because of the spreading of the wave front, the conversion of the acoustical energy into heat, and scattering from irregular surfaces—the phenomena of relaxation add to the attenuation. The process of relaxation, which represents the lag between the introduction of a perturbation and the adjustment of the molecular energy distribution to the perturbation, requires a finite time; and the energy interchange approaches equilibrium in an exponential fashion. Considerable information regarding the nature of matter can be derived from the study of relaxation phenomena.

High-intensity ultrasound can result in energy absorption that yields a considerable amount of heat, to the extent that glass or steel can be melted quickly. Ultrasonic waves can also generate stresses, resulting in cavitation in fluids. Cavitation is also capable of producing free chemical radicals, thus fostering specific chemical reactions. The stresses produced in the cavitation process are sufficiently concentrated to erode even extremely sturdy materials. Cavitation also provides the mechanism of ultrasonic cleaning.

16.2   Relaxation Processes

Relaxation entails the molecular interactions in gases and liquids. These interactions have an effect on absorption and velocity dispersion both of which depend on frequency (and on pressure, in the case of gases). Chemical reactions also entail relaxation processes on their own, but they will not be considered in this chapter, except for the effect of ultrasound on reaction rates.

To better understand the phenomenon of relaxation, let us consider an ideal gas made up of diatomic molecules. The individual molecules move translationwise in three principal directions in a nonquantitized fashion, that is, any translational energy is allowable. In addition, the molecules rotate about three perpendicular axes (actually two, since one axis has zero moment of inertia, hence zero rotational energy), and the molecules also vibrate along the direction of the bond joining the atoms. These molecules collide with one another in translation motion, exchanging energy between themselves. A single collision is usually sufficient to transfer translational energy from molecule to another, but a certain period of time is needed to randomize energy associated with excessive velocity in a particular direction. This amount of time is referred to as the *translational relaxation time* (Herzfeld and Litovitz), and it is given by

$$\tau_{tr} = \frac{\eta}{p} = 1.25\tau_c$$

where τ_{tr} is the translational relaxation time, p is the gas pressure, η is the viscosity of the gas, and τ_c is the interval between collisions. As the gas pressure is lowered, the rate of collisions decreases in the same proportion.

Unlike the case of translational motion, rotation and vibration are quantized. When a collision occurs, a change in rotational or vibrational state will occur only when the change of energy of another state is sufficient to permit at least one quantum jump. For rotational energy transfer the spacing between energy levels is given by $2(J + 1)B$, where J denotes the rotational quantum number (which must be an integer), $B = \hbar^2/2I$, I is the effective moment of inertia, and $\hbar$ is the

Planck constant $1.055(10)^{-34}$ J · s. Let us assume that J is the most probable value according to the Boltzmann distribution. The value of $2(J + 1)B$ for a typical molecule (for example, O_2) in temperature units is about 1 K. This indicates that in a gas above 1 K all collisions will have enough translational energy to engender multiple changes, with the result that rotation rapidly equilibrates with translation. Hydrogen, however, constitutes an exception, because it has much larger rotational energy level spacing due to its small moment of inertia. According to Winter and Hill (1967), as many as 350 collisions may be necessary to transfer a quantum of rotational energy in the hydrogen molecule. As a pressure of 1 atm, this corresponds to a relaxation time of 2×10^{-8} s. It should be understood that a specific collision either does or does not transfer a quantum of rotational energy in the hydrogen molecule. The 350-collision average indicates that only one of the 350 collisions possesses the proper geometry and energy to execute a transfer of one quantum of rotational energy. The number of collisions necessary, on the average, to engender the transfer of one quantum of rotational energy is termed the *collision number Z*. Where rotational energy is entailed, the collision number is written as Z_{rot}. The inverse of this dimensionless parameter represents the probability of transferring a quantum in a collision, symbolized by P_{rot}. Rotational energy levels are spaced unevenly, that is, a $1 \rightarrow 2$ transition should be more probable than a $2 \rightarrow 3$ transition. The probabilities of these events occurring are distinguished by the symbols $P_{rot}^{1 \rightarrow 2}$ or $P_{rot}^{2 \rightarrow 3}$.

As a rule, the probability for transferring a quantum of energy through a collision drops off rapidly with the size of the quantum transferred. Because vibrational energy levels are much more widely spaced than rotational energy levels, the vibrational relaxation times are considerably longer than rotational relaxation times. Vibrational levels in a single vibrational modes are virtually evenly spaced, so energy can therefore be exchanged between levels (e.g., the vibrational quantum number goes up in one molecule and goes down in another) with hardly any energy exchanged between vibration and translational. Thus, the vibration-to-vibration exchanges occurs very rapidly. The amount of time it takes energy to transfer between translation and the lowest lying vibrational level determines the vibrational relaxation time.

Because this energy level varies greatly for different molecules, the probabilities of vibrational energy transfer during a collision also varies greatly. In the case of N_1 molecules colliding with N_2 molecules, Z_{10} (the number of collisions needed to transfer energy from the lowest vibrational energy to translation) is approximately 1.5×10^{11}, so the relaxation time is close to 15 s (Zuckerwar and Griffin, 1980). Larger molecules possess very closely spaced vibrational energy levels so fewer collisions are required to transfer a quantum of first level vibrational energy into translational energy. For example, a relatively large molecule such as C_2H_6 need to undergo about 100 collisions to execute this type of transfer.

Let us consider the case of a vibrational state that is excited to an energy E_v (the subscript v denotes vibration) that is greater than energy $E_v(T_{tr})$ that would exist in a Boltzmann equilibrium with translation. This excess vibrational energy will equilibrate with translational energy, in accordance with the following standard expression for relaxation:

$$\frac{dE_v}{dt} = \frac{1}{\tau}[E_v - E_v(T_{tr})] \tag{16.1}$$

This reversion to equilibrium occurs through individual molecular collisions in which energy transfers result.

Let k_{10} be defined as the rate at which the molecules descend from the first excited state to the ground state, owing to collisions at 1 atm pressure. This is simply the collision frequency M multiplied by the probability of energy transfer $P^{1 \to 0}$ multiplied by the mole fraction x_1 of the molecules in the first excited state. The reverse process will also occur, that is, some molecules in the ground state will become excited at a rate k_{01}. In equilibrium both rates are equal, and we can write:

$$k_{10}x_1 = k_{01}x_0$$

The energy is quickly shared from the first excited level of the vibrational mode with high-level modes through vibrational exchanges. On the basis of quantum mechanics governing the probabilities of energy exchanges between vibrational levels of a harmonic oscillator, Landau and Teller derived the following expression:

$$-\frac{dE_v}{dt} = k_{10}\left(1 - e^{-h\nu/kT}\right)[E_v - E_v(T_{tr})] \tag{16.2}$$

where h is the Planck constant (6.626×10^{-34} J·s), ν the vibrational frequency of the relaxing mode, k is the Boltzmann constant (1.38×10^{-16} ergs K^{-1}), and T is the absolute temperature in K. Through comparison with equation (16.1), the relaxation time τ for equation (16.2) is found to be

$$\tau = \frac{1}{k_{10}\left(1 - e^{-h\nu/kT}\right)} \tag{16.3}$$

Relaxation time and ultrasonic absorption and dispersion are interlinked. The relaxation process causes the specific heat of the gas to be frequency-dependent. The specific heat of a simple gas can be traced to translational, vibratory, and rotational contributions. Let us now consider the situation where any acoustically imposed temperature variation is followed by both translational and rotational energy equilibrating extremely quickly. The specific heat C' is deemed independent of temperature, or the temperature never deviates appreciably from the equilibrium value T_0; and we can write

$$[E_v - E_v(T_{tr})] = C'(T' - T_0) \tag{16.4}$$

and insert this last expression into equation (16.1) to yield

$$\frac{dT'}{dt} = \frac{1}{\tau}(T' - T_{tr}) \tag{16.5}$$

Now consider the case when the translational temperature is suddenly raised at time $t = 0$ from T_0 to a value T_1. Two cases can be ascertained: The external temperature is kept constant at T_1 after $t = 0$, the energy seeping from the outside into the internal (rotational and vibrational) degrees of freedom. The solution of equation (16.5) is

$$T' = T_1 + (T_0 - T_1)e^{-t/\tau}$$

But if the external energy is not constant but is lessened by the amount flowing into the internal degrees of freedom, we have

$$0 = \bar{C}(T_{tr} - T_1) + C'(T' - T_0) \tag{16.6}$$

where $\bar{C}$ refers to the specific heat arising from external degrees of freedom and C' denotes the specific heat belonging to the internal degrees. Equation (16.6) is rewritten as follows:

$$\begin{aligned} 0 &= \bar{C}(T_{tr} - T') + \bar{C}(T' - T_1) + C'(T' - T_0) \\ &= \bar{C}(T_{tr} - T') + C(T' - T_0) \end{aligned} \tag{16.7}$$

where

$$T_2 = \frac{\bar{C}}{C}T_1 + \frac{C'}{C}T_0$$

where C is the total specific heat and T_2 is the final equilibrium temperature. We eliminate $(T_{tr} - T')$ from equation (16.5) with the use of equation (16.7) to obtain

$$-\frac{dT'}{dt} = \frac{C}{\bar{C}}\frac{1}{\tau}(T' - T_2)$$

which yields the solution

$$T' = T_2 + (T_0 - T_2)e^{-t/\tau'}$$

The apparent relaxation time is

$$\tau' = \frac{\bar{C}}{C}\tau = \frac{C - C'}{C}\tau$$

Equation (16.5) can be written as

$$-\tau\frac{dT'}{dt} = T' - T_0 - (T_{tr} - T_0)$$

or

$$T' - T_0 + \tau\frac{d}{dt}(T' - T_0) = T_{tr} - T_0 \tag{16.8}$$

If $T_{tr} - T_0$ is periodic in time, that is, it is proportional to $e^{i\omega t}$, $T' - T_0$ will likewise be proportional to $e^{i\omega t}$. After the transient dies out, then we have

$$T_{tr} - T_0 = (1 + i\omega t)(T' - T_0)$$

or

$$T' - T_0 = \frac{T_{tr} - T_0}{1 + i\omega \tau}$$

which can also be rewritten as

$$\frac{dT'}{dT_{tr}} = \frac{1}{1 + i\omega t}$$

The effective specific heat can be established from

$$dE = (C_v)_{eff} \, dT_{tr} = C_v \, dT_{tr} + C' \, dT' = \left((\bar{C}_v)_v + C'\frac{dT'}{dT_{tr}}\right) dT_{tr}$$

and

$$(C_v)_{eff} = \bar{C}_v + \frac{C'}{1 + i\omega\tau} = C_v - \frac{C'i\omega\tau}{1 + i\omega t} \tag{16.9}$$

The acoustic propagation constant k can be written in the form

$$\frac{k^2}{\omega^2} = \left(\frac{1}{c} - \frac{i\alpha}{\omega}\right)^2 = \frac{\rho_0\kappa_T}{\gamma_{eff}}$$

where c represents the acoustic velocity, α is attenuation coefficient. ρ_0 is the equilibrium density, κ_T is the compressibility of the gas, and γ_{eff} is given by

$$\gamma_{eff} \equiv \frac{(C_v)_{eff} + R}{(C_v)_{eff}}$$

where R is the gas constant. In the case of this simple single relaxation, for $\alpha/\omega \ll 1$

$$\lambda a = \pi\left(\frac{c}{c_0}\right)^2 \frac{\varepsilon\omega\tau}{1 + (\omega\tau_s)^2} \tag{16.10}$$

and

$$\left(\frac{c_0}{c}\right)^2 = 1 - \frac{\varepsilon(\omega\tau_s)^2}{1 + (\omega\tau_s)^2} \tag{16.11}$$

where

$$\varepsilon = \frac{c_\infty^2 - c_0^2}{c_\infty^2}$$

c_0 = speed of sound for $\omega\tau_s \ll 1$

c_∞ = speed of sound at frequencies $\gg$ relaxation frequency

λ = wavelength

The adiabatic relaxation time τ_s is related to the isothermal relaxation time τ as follows:

$$\tau_s = \frac{C_v + R}{C_v^\infty + R}\tau$$

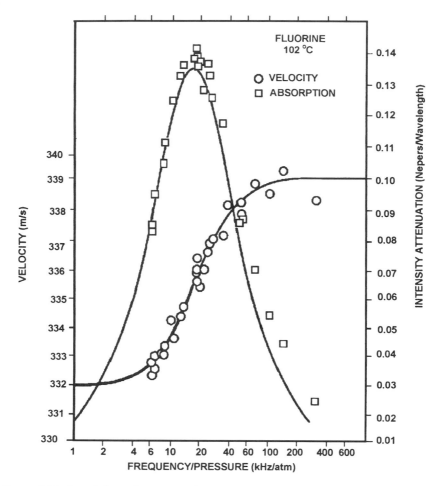

FIGURE 16.1. Sound absorption per wavelength and velocity dispersion in fluorine at 102°C.

The frequency at which the maximum absorption per wave length occurs is called the *relaxation frequency*, symbolized by f_r. It is related to the adiabatic absorption time τ_s as follows:

$$f_r = \frac{1}{2\pi\tau_s} \frac{c_\infty}{c_0}$$

We take as an example the gas Fl_2 at 102°C in Figure 16.1 which shows curves for absorption per wavelength and velocity dispersion due to a single relaxation process (Shields, 1962). Measured values are also plotted for the purpose of comparing with theory.

In the case of polyatomic gases or mixtures of relaxing diatomic gases, the relaxing modes can be coupled by vibration-to-vibration exchanges. These multiple relaxation processes follow the general behavior given by equations (16.10) and

(16.11), but the magnitude of the absorption and dispersion and the associated relaxation frequencies assume a different connotation. For this case of the multiple relaxing internal energy modes, equations (16.10) and (16.11) are changed into the following formats:

$$\alpha\lambda\left(\frac{c}{c^\infty}\right) = -\pi \sum_j \frac{\delta_j k_s / k_s^\infty}{1 + (\omega\tau_{s,j})^2} \tag{16.12}$$

$$\left(\frac{c}{c^\infty}\right)^2 = 1 + \sum_j \frac{\delta_j k_s / k_s^\infty}{1 + (\omega\tau_{s,j})^2} \tag{16.13}$$

where $\delta_j k_s / k_s^\infty$ is a relaxation adiabatic compressibility (which has a negative value) and j denotes that there may be more than a single relaxation process entailed. In such complex cases, $\tau_{s,j}$ and $\delta_j k_s / k_s^\infty$ can no longer be associated respectively with a single transfer reaction and relaxing energy of a specific mode, because the various modes and reaction pathways are coupled. So the summations in equations (16.12) and (16.13) should cover all eigenvalues of the energy transfer matrix, which also accounts for all reactions. These two equations constitute the standard equations for calculating sound absorption in moist air as function of frequency and temperature. Equations (16.12) and (16.13) can also be used in the reverse manner, where the measured values of absorption and velocities can be used to derive the transition rates. But when the number of relaxing modes increases, the number of possible relaxation paths multiplies very rapidly, thus limiting this procedure to only a few special cases.

While relaxation processes similar to the energy exchanges in gases occur in liquids, there are important differences that arise from the greater density of molecules inherent in liquids and the consequential multibody interactions. The concept of a rate equation is less applicable, but the existence of relaxation time as a measure of the time for a system to revert to an equilibrium state in sustaining a perturbation remains valid.

In a few cases such as CS_2 and a number of organic liquids, the relaxation mechanism appears to be the same as for gases, that is, the internal energy of the individual molecules is excited by "collisions." This type of liquids is called *Kneser liquids*, and they generally have a positive temperature coefficient of absorption. With other liquids, the molecules bond temporarily to form large groups that reconfigure themselves when an ultrasound wave passes through. Such figurative relaxations tend to be very rapid, and there is the possibility of the frequency dependence of absorption and dispersion, which results in a distribution of relaxation times. Such liquids are termed *associated liquids*; water is a prime example of such a liquid.

Chemical reactions complicate matters even more: in a reversible chemical reaction with heat of reaction ΔH, ΔH plays a role in the relaxation equations in the same manner as ΔE for vibrational relaxation. Because chemical reactions increase the possibility of the number density of the molecules changing, additional relaxation absorption and dispersion tend to occur.

16.3   Cavitation

The phenomenon of cavitation, the rupture of liquids, is readily observed in boiling water, turbines, hydrofoils, and in seawater in the vicinity of a ship's rotating propeller. It occurs in those regions of liquids that are subject to high-amplitude, rapidly vacillating pressures. Cavitation also occurs in a liquid irradiated with high-energy ultrasound.

Consider a small volume of liquid through which sound travels travel. During the negative half of the pressure cycle the liquid undergoes a tensile stress, and during the positive half the liquid undergoes compression. Bubbles entrapped in the liquid will expand and contract alternatively. When the pressure amplitude is sufficiently great and the initial radius of the bubble is less than a critical value R_0 given by

$$R_0 = \frac{1}{\omega} \sqrt{3\gamma \frac{\left(p_0 + \frac{2T_{st}}{R_0}\right)}{\rho}} \qquad (16.14)$$

the bubble collapses suddenly during the compression phase. In equation (16.14), the symbols used are defined as follows:

ω = angular frequency of the signal

p_0 = hydrostatic pressure in the liquid

γ = ratio of the principal specific heats of the gas in the bubble

T_{st} = surface tension at the surface of the bubble

This sudden collapse of bubbles constitutes the phenomenon of cavitation and it can result in the very sudden release of a comparatively large amount of energy. The severity of this cavitation, as measured by the amount of the energy released, depends on the value of the ratio R_m / R_0, where R_m denotes the radius of the bubble when it has expanded to its maximum size. Obviously this ratio depends on the magnitude of the acoustic pressure amplitude, that is, the acoustic intensity.

The presence of bubbles facilitates the onset of cavitation, but cavitation can also occur in gas-free liquids when the acoustic pressures exceed the hydrostatic pressure in the liquid. During a part of the negative phase of the pressure cycle, the liquid is in a state of tension. This causes the forces of cohesion between neighboring molecules to become opposed, and voids are formed at weak points in the structure of the liquid. These voids expand and then collapse in the same manner as gas-filled bubbles. The cavities produced in this fashion contain only the vapor of the liquid. Cavitation in a gasless liquid can be induced by introducing defects in the structure of the liquid by adding impurities or by bombarding the liquid with neutrons.

A hissing noise often accompanies the onset of cavitation. This noise is referred to as *cavitation noise*. Th minimum intensity or pressure amplitude required to establish cavitation is termed the *threshold of cavitation*. Figure 16.2 displays how the threshold intensity varies with both aerated and gas-free water. The threshold

(SBSL). When a sufficiently strong acoustic field propagates through a liquid, placing it under dynamic stress, preexisting microscopic inhomogeneities serve as nucleation sites for liquid rupture. Most liquids such as water have thousands of potential nucleation sites per milliliter, so a cavitation field can harbor many bubbles over extended space. This cavitation, if sufficiently intense, will produce sonoluminescence of the MBSL type. It was more recently discovered that under certain conditions, a single, stable oscillating gas bubble can be forced into large amplitude pulsations that it produces sonoluminescence emission on each and every cycle (Gaitan and Crum, 1992; Gaitan, et al., 1992). This is the SBSL-type of sonoluminescence.

Sonochemistry deals with high-energy chemical reactions that occur during ultrasonic irradiation of liquids. The chemical effects of ultrasound do not result from direct molecular interactions, but occur principally from the effects of acoustic cavitation. Cavitation provides the means of concentrating the diffuse energy of sound, with bubble collapse producing intense, local heating and high pressures that are extremely transient. Among the clouds of cavitating bubbles, the highly localized hot spots have temperatures of roughly 5000 K, pressures exceeding 2000 atm and heating and cooling rates greater than 10^7 K/s. Ultrasonics can serve as useful chemical tool, as its chemical effects are diverse and it can provide dramatic improvements in both stoichiometric and catalytic reactions. In a number of cases, ultrasonic irradiation can increase reactivity by a millionfold.

The chemical effects can be categorized into three areas: (a) homogeneous sonochemistry of liquids, (b) heterogeneous sonochemistry of liquid–liquid or liquid–solid systems, and (c) sonocatalysis (which constitutes an overlap of the first two categories). Chemical reactions have generally not been observed in the ultrasonic irradiation of solids and solid–gas systems.

16.4 Phonons

In quantum mechanics, energy states are considered to occur only at discrete levels or *eigenstates*, not at any arbitrary values. In the analytical treatment of crystalline solids, the concept of phonons, often referred to as a "quantitized sound waves," is used to represent the effects of a transition between the eigenstates of a system of coupled quantum-mechanical oscillators. Phonons generally apply to discrete strictly linear systems, while "classical" sound waves derive from continuous, intrinsically nonlinear systems within the limits of small amplitudes. While phonons can occur in all states of matter, they are most easily discerned in crystalline solids. In 1819, Dulong and Petit discovered the first evidence for phonons in solids when they observed that the specific heat of a solid is twice that of the corresponding gas. This finding suggested the fact that solids have a way of storing potential energy, in addition to the kinetic energy that is so apparent in gases. Einstein first proposed an acoustic theory of the specific heat of solids by assuming that the kinetic energy and potential energy arose from atoms oscillating about the equilibrium positions in the crystalline lattices. Applying the Planck quantum

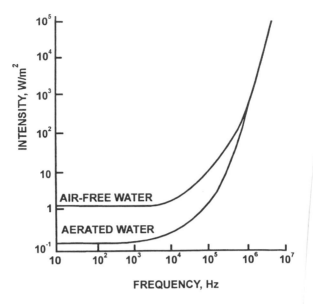

FIGURE 16.2. Variation of threshold intensity with frequency in aerated and at room temperature (20°C).

intensity is obviously considerably greater for gas-free water than f water. This parameter remains fairly constant up to about 10 kHz, the a steady increase up to about 50 kHz and a more pronounced expon beyond 50 kHz.

Generally speaking, the threshold intensity usually increases v pressure and decreases with increasing temperature. But a numbe to this rule exist (Hunter and Bolt, p. 234). The threshold intensity c time of exposure to sound is increased. This is the result of a time the acoustic excitation and the onset of cavitation. For pulsed wave intensity reduces in value as the pulse length is increased to an upp which it becomes independent of pulse length. This frequency-d limit would be of the order to 20 ms for a frequency of 20 kHz.

The amount of energy released by cavitation depends on the bubble growth and collapse of the bubbles. This energy shou surface tension at the bubble interface and lessen with the vapo liquid. Water has a comparatively high surface tensions, so it can b medium for cavitation. It can be made even more effective by the alcohol—this results in an appreciable increase in vapor pressure a decrease in the surface tension, but the former effect outweigh

Weak emission of light has been observed in cavitation. Th known as *sonoluminescence*. It was first observed in water in and Schultes. Two separate forms of sonoluminescence are multiple-bubble sonoluminescence (MBSL) and single-bubble

theory, Einstein related the energy to the frequency, but he had made the simplifying assumption that each atom oscillated independently of the other, so his formula for the specific heat was therefore incorrect. Peter Debye in 1912 correctly inferred that these atomic oscillators are coupled, and later Max Born, Theodore von Kármán, and Moses Blackman refined the theory to the extent of matching the experimental results of the temperature dependence of the specific heat of solids. Debye described the role of phonons in his explanation of thermal conductivity, and in that same year (1912) Frederick Landemond correlated lattice vibrations to thermal expansion and melting of solids, both of which are attributable to the nonlinearities of forces between atoms in solids.

Consider a one-dimensional array of masses m_j ($j = 0, \ldots, N + 1$) interconnected with ideal springs of stiffness s_j ($j = 0, \ldots, N$). A spring s_j interconnects mass m_j and mass m_{j+1}. Let ξ_j denote the displacement of mass m_j. The displacements ξ_0 and ξ_{N+1} provide the boundary conditions at each end of the system. Within the boundaries, the motion of each mass is assumed to follow Hooke's law of linear elasticity and Newton's law:

$$m_j \frac{d^2\xi_j}{dt^2} = -s_{j-1}(\xi_j - \xi_{j-1}) + s_j(\xi_{j+1} - \xi_j) \qquad (16.15)$$

The motion can be assumed amenable to Fourier analysis, so that it assumes a time dependence $e^{i\omega t}$. The left term of equation (16.15) then becomes $-m_j \omega^2 \xi_j$. The solution to this equation, through the application of the Bloch theorem for the normal modes of a coupled system, is a linear combination of the normal modes:

$$\xi_j = \sum_k \left(e^{ik(ja)} X_k + e^{-ik(ja)} X_{-k}\right) e^{i\omega_k t}$$

where

k = eigenvalues of the system determined by the boundary conditions

a = periodic spacing, or the "stretch" of the "springs" connecting the masses

$$X_k = \frac{\xi_1 - \xi_0 e^{-ika}}{2i \sin(ka)}$$

The values of ξ_0, ξ_1, X_k and the restrictions on k are established by boundary and initial conditions. For example, consider a system clamped at the ends, that is, $x_0 = 0$, $x_{N+1} = 0$. Then

$$\xi_j = \sin[k(ja)]\left(\frac{\xi_1}{\sin(ka)}\right)$$

The eigenvalue k is quantized with

$$k = \frac{n}{N+1}\frac{\pi}{a} \qquad \text{with } n = 1, 2, \ldots, N$$

In an infinite system or in a system with periodic boundary conditions, it is readily established that $X_k = 0$ pr $X_{-k} = 0$. If we apply a coordinate system with the mass m_0 located at its origin, then the location of the jth mass is $x = ja$. and we have

$$\xi_k(x) = e^{ikx} X_k \tag{16.16}$$

Equation (16.16) represents the customary Bloch wave result. The subscript k was added in equation (16.16) to serve as a label for the normal mode. The solution is complete as a linear combination of the normal modes, that is,

$$\xi_j = \sum_k \left(e^{ik(ja)} X_k + e^{-ik(ja)} X_{-k} \right) e^{i\omega_k t} \tag{16.17}$$

The normal modes are orthogonal, so the normalization constants can be obtained for any boundary conditions, whether they be clamped, periodic, and so on. The total energy E of the system can be found from

$$E = m \sum_k |X_k|^2 \omega_k^2$$

Coupled Quantum Particles

In dealing with harmonically coupled particles it is more expeditious to use a Lagrangian formulation rather than Newtonian formation of equation (16.15). The Lagragian for such a set of connected particles is

$$L = \frac{1}{2} \sum_j m_j \left(\frac{d\xi_j}{dt} \right)^2 - \frac{1}{2} \sum_j s_j (\xi_{j+1} - \xi_j)^2$$

With the canonical momenta $p_j = m_j (d\xi j/dt)$, the Hamiltonian becomes

$$H = \sum_j p_j \left(\frac{d\xi_j}{dt} \right) - L$$

The system is quantitized with the commutation relations

$$[\xi_j, p_{j'}] = i\hbar \, \delta_{j,j} \tag{16.18}$$

where $\delta_{j,j}$ represents the Dirac delta function which equals unity when $j = j'$ and is zero when $j \neq j'$. We then assume a periodic system and set $m_j = m$ and $s = s_j$, and also make use of equation (16.17) without the coefficient $e^{-i\omega_k t}$. We obtain the Hamiltonian

$$H = \frac{1}{2m} \sum_k P_k P_{-k} + \frac{1}{2} m \sum \omega_k^2 X_k X_{-k}$$

where

$$P_k = m \frac{dX_{-k}}{dt}, \qquad \omega_k = \omega_0 \sin\left(\frac{ka}{2} \right)$$

The above Hamiltonian could be used to construct a Schrödinger wave equation for a field $\psi(X)$, where X represents a point in the $2N$-dimensional X_k space and

$$P_k = -i\hbar\left(\frac{\partial\psi}{\partial X_k}\right)$$

Another approach is to construct the properties of the eigenfuctions and eigenvalues through the use of the commutation relations of equation (16.18) for x_j and p_j. Defining

$$a_k = \sqrt{\frac{m}{2\hbar\,\omega_k}}\left(\omega_k X_k + \frac{i}{m}P_{-k}\right)$$

and

$$a_k^* = a_{-k}^*, \qquad \omega_k = -\omega_{-k}$$

the Hamiltonian assumes the form

$$H = \sum_k \hbar\,\omega_k\left(N_k + \frac{1}{2}\right)$$

where

$$N_k = a_k^* a_k$$

A general state of the system is constructed from a superposition of the eigenstates:

$$\psi = \sum_k \sum_n C_{kn}\psi_{kn}$$

where $|C_{kn}|^2$ represents the probability that the system is in the state ψ_{kn}. From the customary quantum mechanical relation $E\psi = H\psi$, the expectation value of the total energy of the system is

$$\langle\psi|E|\psi\rangle = \sum_k \sum_n |C_{nk}|^2 \hbar\,\omega_k\left(n_k + \frac{1}{2}\right)$$

The expectation value of the square of the momentum operator is found from

$$\langle\psi|P^2|\psi\rangle = m\sum_k\langle\psi|X_k|\psi\rangle\omega_k^2 = \sum\sum |C_{nk}|^2 \hbar\,\omega_k n_k = \langle\psi|E|\psi\rangle - E_0$$

where E_0 is the zero-point energy.

From the expectation state of the position operator, a state ψ_{nk} can be coupled only to the state $\psi_{(n+1)k}$ or $\psi_{(n-1)k}$. Consequently the time dependence will behave as $2\cos(\omega_k t)$. The term phonon can be defined in the following manner: a phonon is emitted or absorbed when a system of harmonically coupled quantum-mechanical particles executes a transition from a state ψ_{nk} to a state $\psi_{(n-1)k}$ (which results in emission) or $\psi_{(n+1)k}$ (which results in absorption).

The absorption of ultrasonic waves in solids are attributable to a number of different causes, each one of which is characteristic of the physical properties of the material concerned. They can be classified as (a) losses characteristic of poly-crystalline solids, (b) absorption due to lattice imperfections, (c) absorption on ferromagnetic and ferroelectric materials, (d) absorption due to electron–phonon interactions, (e) absorption due to phonon–phonon interactions, and (f) absorp-tion due to other possible causes. It is also interesting to note that a rapid decrease in attenuation occurs at the critical temperature for superconductivity. This varia-tion of attenuation with temperature has been explained in a Noble prize–winning paper by Bardeen, Cooper, and Schrieffer through the B.C.S. theory (so-called after the initial of their surnames) which predicts a temperature-dependent energy gap 2ε wide around the Fermi level at the critical temperatures of T_c and less. As the temperature is reduced, the gap increases toward a maximum at zero absolute temperature, where the predicted value of ε is equal to $1.75 \times \kappa T_c$, where κ is the Boltzmann constant. In the realm of $l_e > \lambda/2\pi$. the B.C.S. theory predicts that:

$$\frac{\alpha_s}{\alpha_n} = \frac{2}{e^{\varepsilon/\kappa T} + 1}$$

where l_e denoted the mean free path of an electron, α_s and α_n represent the values of absorption in the superconducting and normal state, respectively, at absolute temperature T. This variation was confirmed experimentally by Morse (1959) and Bohm for indium at 28.5 MHz. Gibbons and Benton measured the velocities of longitudinal waves in both normal and superconducting tin; they found a very small reduction in velocity (about 1/500,000th) for the superconducting state. Application of a magnetic field to a metal at low temperatures affects the mean free path and thus affects acoustic attenuation. For $l_e < \lambda/2\pi$, a decrease in the attenuation with increasing magnetic field H has been observed experimentally (Sternberg).

16.5   Transducers

A *transducer* is a device that generates and receives (i.e., it senses) sound waves. The transducer essentially functions as an energy converter. That is, it converts acoustical energy to or from other forms of energy (e.g. electrical, mechanical, or thermal). A transducer is said to be *reversible* if it can convert in either direction. Most high-intensity ultrasonic generators in use are basically crystal oscillators or magnetostrictive devices.

Transducers fall into the following categories: (1) crystal oscillators that operate on the piezoelectric effect, which is reversible; (2) magnetostrictive devices based on the magnetostriction phenomenon which is also reversible; (3) mechanical gen-erators and receivers, which includes whistles and sirens acting as generators and also radiometers and Rayleigh discs serving as receivers; (4) electromagnetic trans-ducers which operate on the same principle as the customary audio loudspeaker, but can function only in the lower range of ultrasonic frequencies; (5) miscellaneous types such as chemical, thermal, and optical transducers. In addition there are ul-trahigh frequency transducers that work in the megahertz and gigahertz ranges.

In this chapter we concentrate on the first two categories that constitute a major portion of the transducer types currently in principal use.

Ultrasonic receivers fall into two categories, namely, receivers that terminate ultrasonic paths of propagation and receivers that serve as probes. Receivers that terminate ultrasonic paths have cross-sectional dimensions extending over several wavelengths, with the result that the presence of such a transducer will materially affect the acoustic field mainly through reflections. The purpose of an ultrasonic probe is to gauge the characteristics of an acoustic field, so its dimensions must be sufficiently small so as not to affect the field. A probe diameter is typically only about one-tenth of the wavelength.

Piezoelectric Crystals

In 1880, the Curie brothers discovered that when a crystal having one or more polar axes or lacking axisymmetry is subjected to mechanical stress, an electrical potential difference occurs. Consider a segment of such a crystal, in the form of a slab or a disk, that is cut with its parallel surface running normal to a polar axis. When this segment undergoes a mechanical stress, equal and opposite electric charges arise on the parallel surfaces. The magnitude of the charge density (i.e., dielectric polarization) is directly proportional to the applied stress, provided the applied stresses does not strain the crystal beyond its elastic limit.

The opposite effect, predicted by Lippmann in 1881 and verified experimentally by the Curie brothers the same year, occurs when an electric field is applied in the direction of a polar axis, causing a mechanical strain in the crystal segment. The amount of strain is directly proportional to the intensity of the applied electric field. From the viewpoint of the principle of conservation of energy, the piezoelectric effect and its converse may be deemed to be equal and opposite. Such effects occur in crystals such as quartz (a member of the trigonal system, as shown in Figure 16.3), Rochelle salt and lithium sulfate.

Quartz is very commonly applied for ultrasonic generation. A quartz crystal is shown in Figure 16.3, with a hexagonal cross section normal to the nonpolar optic axis, denoted by the z-axis. The axes joining opposite edges are designated as x-axes, and the associated axes, which are perpendicular to these and joining opposite faces are termed y-axes. The x- and y-axes are *polar axes*, and slabs cut with their faces perpendicular to them manifest the piezoelectric effect. Crystals that are cut with their faces perpendicular to an x-axis or y-axis are termed x-*cut* and y-*cut* crystals, respectively. The x-cut crystals are generally utilized to propagate compression waves, and the y-cut crystals are applied to generate shear waves.

Now consider an x-cut crystal in the form of a rectangular prism shown in Figure 16.3. Applying an electric field along the x-axis yields a compression in that direction, while an expansion occurs simultaneously along the y-direction. If the direction of the field is reversed, an expansion occurs along the x-axis with an associated compression along the y-axis. No strain, however, occurs along the z-axis. If a pair of surfaces normal to either of the polar axes (x- and y-axes) are coated with a conductive material to form electrodes, small-amplitude oscillations

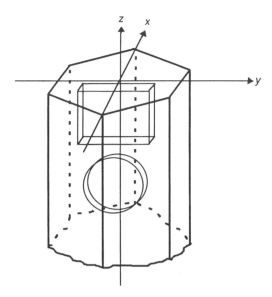

FIGURE 16.3. A hexagonal quartz crystal with x-cut rectangular and circular plates.

will result when an alternating voltage of frequency f is applied across them. When the frequency f equals one of the natural frequencies of mechanical vibration for a particular axis, the response amplitude jumps to a considerably higher value. Crystals are generally operated at resonant frequencies for either "length" or "thickness" vibrations, as denoted by the resonance occurring in the direction parallel with or normal to the radiating surfaces, respectively. The natural frequency for mechanical vibrations is proportional to the inverse of the dimension along which they occur, so it becomes obvious the lower frequencies are generated by "length" vibrations along the direction of the longer dimension while the higher frequencies are produced by "thickness vibrations" along the direction of the smaller dimension.

Maximum acoustic intensities are obviously obtained by operating at the fundamental natural frequencies. But material constraints in crystals may necessitate the use of higher harmonics to obtain higher frequencies. For example, an x-cut quartz plate can be only 0.15 mm thick in order to generate a fundamental "thickness" mode for 20 MHz. Such a quartz plate is extremely brittle and it can shatter under the impetus of a exceedingly high applied voltage, or its dielectric properties may break down. To avoid this situation, it is customary to use thicker slabs of crystals with lower-resonance frequencies and operate at one of the upper harmonics. An example is the vibration of a 1-cm thick quartz crystal at its 191st harmonic to generate 55 MHz ultrasound.

The piezoelectric effect occurs only when opposite charges appear on the electrodes, and for that reason, only odd harmonics can be generated. At the nth harmonic, the thickness of the crystal is divided into n equal segments with

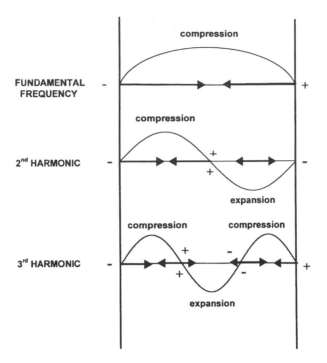

FIGURE 16.4. Crystal divided into segments.

compressions and expansions alternating in adjacent sections, as illustrated in Figure 16.4. For even harmonics in the nth mode, compressions occur in $n/2$ segments and expansions occur in the other $n/2$ segments, with the result no net strain exists in the crystal. When n is odd, the $(n-1)/2$ compressions offset the same number of expansions, leaving either a compression or an expansion in the remaining segment.

The Electrostrictive Effect

The electrostrictive effect, which is the electrical analog of the magnetostrictive effect discussed in a later section, occurs in all dielectrics, but it is not a very pronounced phenomenon in most materials except for a certain class of dielectrics. The effect is much more apparent in this class, namely, the *ferroelectrics*. An electric field applied along a given direction produces a mechanical strain. The magnitude of the strain is proportional to the square of the strength of the applied electric field and is therefore independent of the sense of the field. A positive strain may thus result for both positive and negative values of the excitation field. For a sinusoidally varying electric field, the waveform of the strain assumes that of a rectified but unsmoothed alternating current, and its frequency is twice that of the applied field.

It is possible to obtain a sinusoidal variation in the strain, and this is done by permanently *polarizing* the transducer, namely, one that has magnetostrictive properties. The transducer is heated to a temperature above the Curie point, causing the magnetostrictive effect to vanish, and then it is cooled slowly in a strong direct field oriented in the direction along which it is intended to apply the exciting field. If the exciting field is kept small compared with the initial polarizing field, the strain should vary sinusoidally at the frequency of the exciting field. Because a polarized ferroelectric transducer appears to manifest the same effect as a piezoelectric transducer, it has been mislabeled as being "piezoelectric." Among the principal ferroelectrics, barium titanate, lead metaniobate and lead zirconate titanate are greatly used for electrostrictive applications. To construct this type of transducer, many small crystallites of ferroelectric substances are bonded together to form a ceramic of the appropriate shape. Because such materials are polycrystalline, they can be considered as being isotropic and thus do not have to be cut along specific axes. This renders possible the construction of a concave transducer so that the ultrasonic radiation can be focused without the need for an auxiliary lens system.

Fundamental Piezoelectric Relationships

For a given temperature, consider a piezoelectric element having cross-sectional area A, thickness t, and with electrodes attached to the opposite faces. A voltage V is applied across the electrodes to generate an electric field $E = V/t$, and a constant tensile stress σ is applied to the surfaces. Within the elastic limits, the resultant mechanical strain s relates to the stress as follows

$$s = a\sigma + bE \tag{16.19}$$

and we also have

$$D = cE + d\sigma \tag{16.20}$$

where D represents the electric displacement and a, b, c, and d are coefficients defined below.

Let us now short circuit the electrodes so that $E = 0$. Equation (16.20) now reads as

$$D = d\sigma$$

We note here that the electric displacement D equals the dielectric polarization P, or the charge per unit area. Hence

$$P = d\sigma \tag{16.21}$$

under short-circuit conditions. The coefficient d constitutes the *piezoelectric strain constant*, which is defined as the charge density output per unit applied stress under the conditions of short-circuited electrodes. Now if the stress σ is reduced to zero, equation (16.19) modifies to

$$\sigma = bE$$

The principle of conservation of energy dictates that $b = d$, resulting in

$$\sigma = dE \tag{16.22}$$

for the no-load condition. The coefficient d may also be described as the mechanical strain produced by a unit applied field under the conditions of no loading and is expressed either in units of coulomb/newton or meter/volt. Equation (16.19) becomes altered to

$$s = a\sigma + dE \tag{16.23}$$

If no piezoelectric effect is present, the term d vanishes from equations (16.20) and (16.23), which yields the familiar relationships:

$$s = a\sigma = \sigma/Y \tag{16.24}$$

and

$$D = \varepsilon E \tag{16.25}$$

where Y is the elastic constant (or Young's modulus) for the material and ε is the corresponding electrical permittivity.

Under short-circuit conditions for the crystal, equations (16.21) and (16.24) lead to

$$P = es \quad \text{(for short-circuit conditions)} \tag{16.26}$$

where $e = d/Y$

When a compressive stress is applied to the crystal, equation (16.23) becomes

$$s = -a\sigma + dE \tag{16.27}$$

When the crystal is clamped so that the strain remains zero when stress is applied, we see from equations (16.26) and (16.27) that

$$\sigma = eE \quad \text{(for the constraint } s = 0) \tag{16.28}$$

where e is the piezoelectric stress constant which is expressed on coulomb/m^2 or newton/volt-m.

It must be realized at this stage that the piezoelectric phenomenon is a three-dimensional one. Not only we have to consider the changes in voltage, stress, strain, and dielectric polarization in the thickness direction of the crystal, we must also take into account the effects in any direction. A stress applied to a solid in a given direction may be resolved into six components, three tensile stresses σ_x, σ_y, and σ_z along the principal axes x, y, and z, respectively, and three shear stresses τ_{yz}, τ_{xz}, and τ_{yz} about axes x, y, and z. In the notation for shear, the subscripts indicate the action plane of the shear—thus yz denotes a shear in the yz-plane acting about the x-axis. Also we note that $\tau_{yz} = \tau_{zy}$, and so on. The corresponding components of strain are ε_x, ε_y, ε_z, ε_{yz}, ε_{xz}, and ε_{xy}. In general, the following stress–strain relationship of equation (3.6) can be generalized as

$$\sigma_{ji} = c_{jk}\varepsilon_{ij} \tag{16.29}$$

TABLE 16.1. Values of the Adiabatic Elastic Constants for Quartz $\times 10^{10}$ dyne/cm^2 or $\times 10^9$ newton/m^2.

$$c_{11} = 87.5$$
$$c_{33} = 107.7$$
$$c_{44} = 57.3$$
$$c_{12} = 7.662$$
$$c_{13} = 15.1$$
$$c_{14} = 17.2$$

where c_{jk} denotes the elastic modulus or stifness coefficient. This yields 36 values of c_{jk}:

$$\left.\begin{array}{l}
\sigma_{xx} = \sigma_x = c_{11}\varepsilon_x + c_{12}\varepsilon_y + c_{13}\varepsilon_z + c_{14}\varepsilon_{yz} + c_{15}\varepsilon_{xz} + c_{16}\varepsilon_{xy} \\
\sigma_{yy} = \sigma_y = c_{21}\varepsilon_x + c_{22}\varepsilon_y + c_{23}\varepsilon_z + c_{24}\varepsilon_{yz} + c_{25}\varepsilon_{xz} + c_{26}\varepsilon_{xy} \\
\sigma_{xx} = \sigma_z = c_{31}\varepsilon_x + c_{32}\varepsilon_y + c_{33}\varepsilon_z + c_{34}\varepsilon_{yz} + c_{35}\varepsilon_{xz} + c_{36}\varepsilon_{xy} \\
\tau_{yz} = c_{41}\varepsilon_x + c_{42}\varepsilon_y + c_{43}\varepsilon_z + c_{44}\varepsilon_{yz} + c_{45}\varepsilon_{xz} + c_{46}\varepsilon_{xy} \\
\tau_{xz} = c_{51}\varepsilon_x + c_{52}\varepsilon_y + c_{53}\varepsilon_z + c_{54}\varepsilon_{yz} + c_{55}\varepsilon_{xz} + c_{56}\varepsilon_{xy} \\
\tau_{xy} = c_{61}\varepsilon_x + c_{62}\varepsilon_y + c_{63}\varepsilon_z + c_{64}\varepsilon_{yz} + c_{65}\varepsilon_{xz} + c_{66}\varepsilon_{xy}
\end{array}\right\} \quad (16.30)$$

Because $c_{mn} = c_{nm}$, the number of these constants is reduced from 36 to 21. Symmetry of the axes will lessen this number even further. In the case of quartz, only six elastic constants are independent of one another, and the values of c_{mn} can be expressed as the following matrix:

$$\begin{vmatrix}
c_{11} & c_{12} & c_{13} & c_{14} & 0 & 0 \\
c_{13} & c_{11} & c_{13} & -c_{14} & 0 & 0 \\
c_{13} & c_{13} & c_{23} & 0 & 0 & 0 \\
c_{14} & -c_{14} & 0 & c_{14} & 0 & 0 \\
0 & 0 & 0 & 0 & c_{44} & c_{14} \\
0 & 0 & 0 & 0 & c_{14} & \frac{c_{11}-c_{12}}{2}
\end{vmatrix}$$

The values of these constants for adiabatic conditions are given in Table 16.1.

The Dynamics of Piezoelectric Transducers

A body undergoing forced vibrations can be considered analogous to an electric circuit that is activated by an electromotive force, with the current i corresponding to body velocity u and the voltage V to the applied force F. In terms of the strain s, velocity $u = l(ds/dt)$; and the current is expressed as $i = (dQ/dt) = A(dP/dt)$. Here l is the body length, Q is the electric charge, and A is the cross-sectional area. We invoke equation (16.26) to obtain

$$\frac{dP}{dt} = e\frac{ds}{dt}$$

which gives

$$i = \frac{Ae}{l}u = \alpha_T u \tag{16.31}$$

where the transformation factor $\alpha_T = Ae/l$, which constitutes a characteristic constant for a specific transducer. Because P has three components and S has six components, we can write equation (16.31) in the general matrix form

$$\mathbf{i} = \mathbf{a_T u}$$

From equation (16.28) we have

$$F = \frac{Ae}{l}V = \alpha_T V$$

or

$$\mathbf{F} = \mathbf{a_T V}$$

The mechanical compliance C_m is analogous to the electrical capacitance C, that is,

$$C_m = \frac{sl}{F} = \frac{sl}{\sigma A} = \frac{Yl}{A}$$

When an applied force F causes a strain s, mechanical energy W_m becomes stored in the transducer, according to

$$W_m = \frac{1}{2}Fsl = \frac{1}{2}F^2 C_m = \frac{1}{2}\alpha_T^2 V^2 C_m = \frac{1}{2}CV^2 \tag{16.32}$$

where

$$C = \alpha_T^2 C_m$$

The electrical capacitance C_e between the electrodes of the transducer follows the relationship

$$C_e = \frac{\varepsilon A}{l} \tag{16.33}$$

The corresponding electrical energy W_e is equal to $\frac{1}{2}C_e V^2$. From equations (16.32) and (16.33) the ratio of mechanical energy stored in a piezoelectric transducer to the electrical energy provided to it is given by

$$\frac{W_m}{W_e} = \frac{C}{C_e} = \alpha_T^2 \frac{C_m}{C_e} = k_e^2$$

Here the *electromechanical coupling factor* k_e^2 constitutes a measure of the efficiency of the transducer.

The Q Factor

The Q factor of either a mechanical or an electrical system determines the contour of the frequency response curve for that system. A low value of Q results in a resonance spreading over a wide frequency band. At higher values of Q, a resonance will be confined to a considerably narrower frequency band. Two Q factors exist in a transducer, one mechanical and the other electrical, denoted by Q_m and Q_e, respectively. The mechanical Q factor is defined by

$$Q_m = \frac{m\omega_r}{R_m}$$

where ω_r represents the resonance frequency of the transducer, m its mass, and R_m the mechanical resistance. In the simplest case for the radiating transducer surface A, the mechanical resistance is given by

$$R_m = \frac{\sigma A}{u} = \rho c A$$

in terms of the stress σ, radiating surface velocity u, material density ρ, and sound propagation speed c. For the electrical Q factor, electrical capacitance C_e between transducer electrodes must be taken into consideration. At resonant frequencies the only effective mechanical impedance is R_m. Thus

$$Q_e \cong C_e \omega_r R = C_e \omega_r \frac{R_m}{\alpha_T^2} = \frac{\pi^2/2k_e^2}{Q_m}$$

Magnetostrictive Transducers

Magnetostriction occurs in ferromagnetic materials and certain nonmetals which are termed ferrites. When a magnetic field is applied, a bar of ferromagnetic or ferrimagnetic material undergoes a change in length. Conversely, a mechanical stress applied to the bar will cause a change in intensity of magnetization. The former effect was discovered by Joule in 1847 and the converse effect by Villari in 1868.

Magnetostriction occurs prominently in materials such as iron, nickel, and cobalt. Whether there occurs an increase or decrease in length fully depends on the nature of the material as well as on the strength of the applied magnetic field. The change in length does not depend on the direction of the magnetic field. The magnitude of the strain varying as a function of the applied magnetic field is shown in Figure 16.5 for four different materials, viz. cast cobalt, permendar, nickel, and iron. Figure 16.6 shows strain varying as a function of magnetic polarization. The magnetostrictive effect generally decreases with a rise in temperature and disappears altogether at the Curie temperature.

When a sinusoidally varying magnetic field is applied in the direction of the axis of a bar of ferromagnetic material, the bar will oscillate at double the frequency

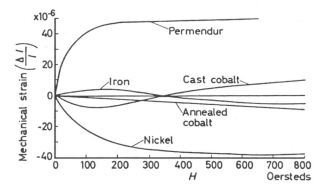

FIGURE 16.5. Mechanical strain as a function of magnetic field.

of the applied field. In accordance with the associated curve in Figure 16.6, a decrease in length occurs when a field is applied to nickel, regardless of the sense of the field. A negative strain occurs every half-cycle. The waveform of the strain occurs as a rectified sine curve, with the result that unwanted harmonics may be generated. A purely sinusoidal wave corresponding to the frequency of the applied field, along with a markedly increased energy output, will be obtained if the bar is polarized. This is achieved by simultaneously applying the alternating field and a direct magnetic field of sufficiently high intensity for the value of the resultant field to remain above zero.

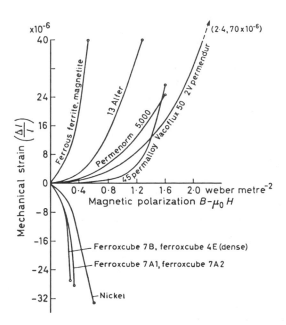

FIGURE 16.6. Mechanical strain as a function of magnetic polarization.

The maximum output for magnetostriction occurs by operating at the fundamental frequency f_r of the bar, given by

$$f_r = \frac{1}{2L}\sqrt{\frac{Y}{\rho}} \tag{16.34}$$

where Y is the Young's modulus for the bar material, ρ is the density of the material, and L is the length of the bar. The term $\sqrt{Y/\rho}$ in equation (16.33) also happens to be the propagation velocity of sound in the material. At resonant frequencies the mechanical strains reaches the order of 10^{-4} rather than magnitudes in the order of 10^{-6} which prevail in operating at nonresonant frequencies.

The Physics of Magnetostrictive Transducers

Magnetrostriction theory is highly analogous to that of piezoelectricity, but account in this case is taken of the polarizing field H_0. We now consider a ferromagntic rod undergoing polarization throughout its length with a magnetic field H_0, with B_0 denoting the associated flux density. The resultant strain ε_0 is directly proportional to the square of the flux density, that is,

$$\varepsilon_0 = C B_0^2 \tag{16.35}$$

where C is a constant. It is seen from equation (16.35) that the sign of the resultant strain is independent of the direction of the field. Now if we apply an exciting magnetic field of strength H that is appreciably less than H_0, with an associated flux density B, we can write

$$B = \mu_i H = \Delta B_0 \ll B_0 \tag{16.36}$$

where we have denoted μ_i as the incremental magnetic permeability. For a state of constant stress, we have for the resulting strain by differentiation of equation (16.35):

$$\varepsilon = \Delta \varepsilon_0$$
$$\Delta \varepsilon = 2C B_0 \Delta B_0$$

or:

$$\varepsilon = 2C B_0 B = \beta \mu_i H \tag{16.37}$$

Here $\beta = 2C B_0$ constitutes the *magnetostrictive strain coefficient* (given in units of m²/weber) which applies to small strains. Equation (16.37) is analogous to equation (16.22) for no-load conditions. When no alternating field is applied the value of the strain ε is given by Hooke's law

$$\varepsilon = \sigma/Y \tag{16.38}$$

in terms of stress σ and Young's modulus Y. We can now obtain the analogue to equation (16.23) by using H instead of the electric field F, Y instead of $1/a$, and

$\beta\mu_i$ instead of d. This gives us

$$\varepsilon = \frac{\sigma}{Y} + \beta\mu_i H \tag{16.39}$$

for a rod undergoing simultaneously a tensile stress σ and a magnetic field H. The analogy extended to equation (16.20) yields

$$B = \sigma\beta\mu_i + \mu_i H \tag{16.40}$$

Clamping the rod causes strain ε to be zero in equation (16.39), yielding

$$\sigma = Y\beta\mu_i H = \Lambda B$$

where σ denotes the compressive stress and $\Lambda = Y\beta$ represents the *magnetostrictive stress constant* (given in units of newton/weber). The reciprocal of Λ, preferred by some authors, is called the *piezomagnetic constant* (units of weber/newton).

16.6   Transducer Arrays

A single-element ultrasound transducer tends to radiate a rather narrow beam or receive signals over a narrow spatial range. In order to cover a wider area through a process called *scanning*, and, in many instances, to emit more powerful signals than is possible with a single element, a specially arranged group of transducers—or *arrays*—are used to extend the versatility of transducers.

Array transducers are also used to focus an acoustic beam. Variable delays are applied across the transducer aperture. The delays are controlled electronically in a sequential or phased array, and can be changed instantaneously to focus the beam on different areas.

With linear-array transducers, which are far more versatile than piston transducers, the electronic scanning entails no moving parts, and the focal point can be changed readily to any position in the scanning plane. A broad variety of scan formats can be generated, and received echoes can be processed for other applications such as dynamic receive focusing, correction for phase aberrations, and synthetic aperture imaging. The principal disadvantages of linear arrays obviously lie in the greater complexity and increased costs of the transducers and scanners. In order to ensure high quality imaging, many (as high as 128 and on the increase) identical array elements are required. Each array element tends to be less than 1 mm on one side and is connected to its own transmitter and receiving electronics.

Phased Arrays: Focusing and Steering

We examine how a phase-array transducer can focus and steer an acoustic beam along a specified direction. An ultrasound image is created by repeating the scanning process more than a hundred times to probe a two-dimensional (2D) or a

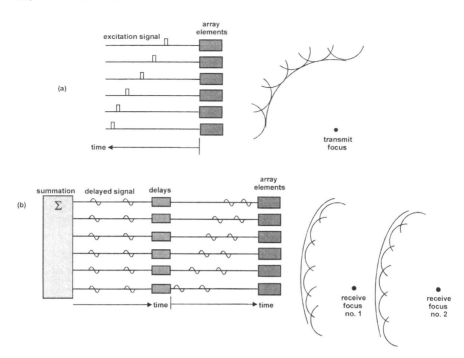

FIGURE 16.7. The phased array illustrated above provides for steering and focusing of an ultrasonic beam. In (a) the six-element linear array is shown in the transmit mode. In the receive mode of (b), dynamic focusing allows the scanner focus to track the returning echoes.

three-dimensional (3D) locale of the medium. In Figure 16.7(a), simple six-element array is shown focusing the transmitted beam. Each array element may be considered a point source which radiates a spherical wavefront into the medium. Because the topmost element is the furthest away from the focus in this example, it is activated first. The other elements are excited progressively at the right time so that the acoustic signals from all the elements reach the focal point simultaneously. According to Huygens' principle, the resultant acoustic signal constitutes the sum of the signals that came from the source. The contributions from each element add in-phase at the focal point to yield a peak in the acoustic signal. Elsewhere, some of the contributions add out-of-phase, lessening the signal relative to the peak.

In receiving an ultrasound echo, the phased array works in reverse. An echo is shown in Figure 16.7(b) originating from receive focus 1. The echo is incident on each array element at a different time interval. The received signals undergo electronic delay so that they add in phase for an echo originating at a focal point. Echoes originating elsewhere have some of their signals adding out of phase, thereby reducing the receive signal relative to the peak at focus.

In the receive mode, a dynamic adjustment can be made with a focal point so that it coincides with the range of returning echoes. After the transmission of a pulse, the initial echoes return from targets nearest the transducer. The scanner therefore focuses the phased array on these targets, located at focus point 1 in Figure 16.7(b). As echoes return from the more distant targets, the scanner focuses at a greater depth (e.g., focus point 2). Focal zones are achieved with adequate depth of field so that the targets always remain in focus in receive mode. This is the *dynamic receive focusing* process.

Array-Element Configurations

The dynamic receive focusing process is repeated many times to form an ultrasonic image in the scan of a 2D or 3D region of tissue. In defining the 2D image, the *scanning plane* is the azimuth dimension; the *elevation dimension* is normal to the azimuth scanning plane. In *linear sequential arrays*, as many as 512 elements constitute a sequential linear array in currently available scanners. A subaperture containing as many as 128 elements is selected to function at a given time.

In Figure 16.8(a), the scanning lines are directed perpendicular to the transducer face; the acoustic beams are focused but not steered. The advantage is that elements have high sensitivity when a beam is directed straight out. The disadvantage is that the field of view is limited to the rectangular region directly facing the transducer. Linear-array transducers also require a large footprint to obtain a sufficient wide field of view. Another type of array configuration is that of the *curvilinear array*. Because of its convex shape [Figure 16.8(b)], the curvilinear (or convex) array scans a wider field of view than does a linear-array configuration. The curvilinear array operates in the same manner as the linear array in that the scan lines are directed perpendicular to the transducer face. In the *linear phased array* of Figure 16.8(c), which may contain as many as 128 elements, each element is used to emit and receive each line of data. In Figure 16.8(c), the scanner steers the beam through a sector-shaped region in the azimuth plane. These phased arrays can scan a region considerably wider than the footprint of the transducer, thus rendering them suitable for scanning through acoustically restricted windows. This is ideal for use in cardiac imaging, where the transducer must scan through a small window in order to avoid obstruction by ribs and lungs.

The *1.5D array* is structurally similar to a 2D array, but operates as 1D. The 1.5D array consists of elements along both the azimuth and elevation directions. Dynamic focusing and phase correction can be implemented in both dimensions to enhance image quality. Steering is not possible in the elevation dimension since a 1.5D array contained a fairly limited number of elements in elevation (usually 3–9 elements). Figure 16.8(d) shows a B-scan conducted with a 1.5D phased array. It is also possible to use linear sequential scanning with 1.5D arrays. In the *2D phased array*, a large number of elements are employed in both the azimuth and elevation. This permits focusing and steering of the acoustic beam along both dimensions. With the application of parallel receive processing, a 2D array can scan is pyramidal volume in real time to yield a volumetric image, as illustrated in Figure 16.8(e).

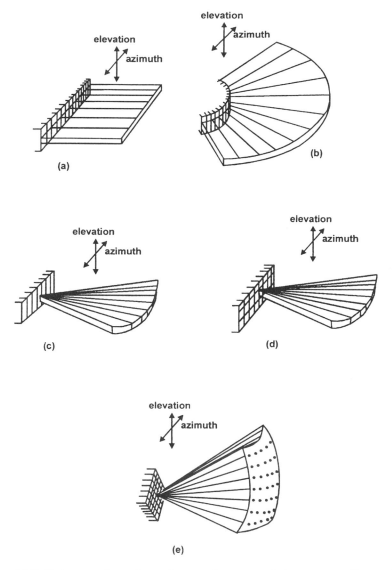

FIGURE 16.8. Various configurations of array elements and the corresponding regions scanned by the acoustic beam: (a) sequential linear array scanning a rectangular region; (b) curvilinear array scanning a sectored region; (c) linear-phased array sweeping a sectored region, (d) 1.5D array scanning a sectored region; and (e) 2D array sweeping a pyramidal region.

Linear-Array Transducer Specification

In the design or selection of an ultrasound transducer, a number of compromises are entailed. The ideal transducers have high sensitivity (SNR, signal-to-noise ratio), excellent spatial resolution, and freedom from spurious signals. Additionally, an individual array element should possess wide angular response in steering dimensions, low cross-coupling with another element, and an electrical impedance matching the transmitter.

16.7 Industrial Applications of Ultrasound

Ultrasonic Cleaning

Both cavitation and the agitation of the fluid by the waves are entailed in the process of ultrasonic cleaning. At lower frequencies, cavitation acts as the principal agent, but at high frequencies the cleaning effect is attributable mainly to agitation. Most cleaning applications are executed in the frequency range of 20–40 kHz, where cavitation effects occur more strongly. Either piezoelectric or magnetostrictive sources are used. The workpiece being cleaned is immersed in a tank containing a liquid selected on the basis of its susceptibility to cavitation, its detergent properties, ability to degrease, and so on. Trichlroethylene and cyclohexane are among the more satisfactory fluids used for ultrasonic cleaning.

Ultrasound cleaning lends itself to continuous processing in which a series of workpieces can be transported on a conveyor belt through a series of processes in separate tanks as shown in Figure 16.9. Ultrasonic cleaning supersedes other usual methods of cleaning, particularly when these methods are ineffective and liable to cause damage. Applications include the removal of lapping paste from lenses without scratching after grinding, the flushout of grease and machining particulates from small crevices in engine components, removal of blood and

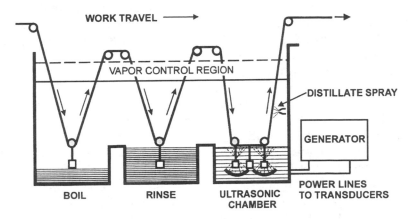

FIGURE 16.9. Conveyor belt for ultrasonic cleaning.

other organic material from surgical instruments after use, and so on. Very delicate parts that can be damaged by cavitation are cleaned by wave agitation at much higher frequencies, from 100 kHz to 1 MHz.

Flaw Detection and Thickness Measurements

A method of nondestructive testing, the pulse technique, is used extensively to determine the propagation constants of solids, particularly in the MHz frequency range. This method consists of sending a short train of sound waves through a medium to a receiver. In the transmission mode of the pulse technique, the receiver is placed at a measured distance from the source. In the echo mode, a reversible transducer acts as both source and receiver, with a reflector used to reflect the pulses. The speed of sound in a medium can be determined from the time of travel of the pulse over a given length of acoustic path. Longitudinal waves are generally used. In gauging the thickness of a specimen, advantage is taken of the fact that a beam of pulses will reflect from the specimen surface opposite to the side of the reversible transducer.

The use of the single pulse method for flaw detection is fairly straightforward when the specimen has two parallel surfaces and the defect is linear and roughly parallel to these surfaces, but not too close to a surface or another defect. If the pulse is followed on an oscilloscope, and there is no defect present, then two peaks, say A and B, will appear on the screen. A represents the instant of the transmission of the pulse, and B that of its return after a simple echo. Peak B is referred to as the bottom echo. When a defect is present, a discontinuity of the characteristic impedance and some, or possibly all, of the sound energy is prematurely reflected back to the transducer. Another peak will then occur between A and B. The distance AC indicates the depth where the flaw exists, and the height of the peak C determines the extent of the defect.

Figure 16.10 illustrates a schematic of a longitudinal wave probe used to detect flaws. A crystal transducer is normally used, and it is encased in a suitable housing. The crystal is mounted for heavy damping, which results in the propagation of small pulses to allow for greater accuracy in locating defects and better resolution of neighboring defects from one another. The transducer is protected by a plastic cover that is coupled to the crystal with oil so as to prevent wear by friction between the surfaces of the crystal and the material being tested. It would be ideal for the characteristic impedances of the transducer material, the material of the protective cover, and the oil should be similar. Sometimes it is more desirable to apply the immersion method, whereby both the specimen and the probed are immersed in water, with a probe at a fixed distance from the upper surface of the sample. An additional peak that represents the echo from the interface between the immersion liquid and the sample will appear on the oscilloscope screen.

When a defect is not parallel with a surface it a preferable to use an angled probe, which consists of a transducer mounted on a wedge. The enables the sound waves to be incident normally to the defect, and a greater degree of sensitivity is

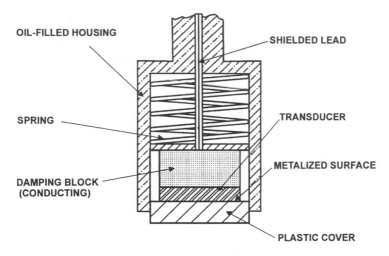

OIL-FILLED HOUSING

SHIELDED LEAD

SPRING

TRANSDUCER

METALIZED SURFACE

DAMPING BLOCK
(CONDUCTING)

PLASTIC COVER

FIGURE 16.10. Longitudinal wave probe for detecting flaws.

thus achieved. Using *variable-angled probes* makes it easier to better gauge the direction of defects that might otherwise remained undetected. For samples having irregular shapes, two probes are used, one acting as a sender and the other as a receiver in order to locate the defects. Care should be taken that coupling between the transducers should involve the medium and nothing else.

Flaws, such as those that occur in butt welds oriented at right angles to the surface, are commonly detected through the procedure of *forward scanning*. Here a beam of transverse waves is propagated, after refraction at the boundary, in the medium at a shallow angle to the surface. In Figure 16.11(a) the transducer is shown mounted on a plexiglass wedge and the longitudinal waves are directed to the surface with an angle of incidence greater than the first critical angle (this is the angle of incidence sufficiently large that the refracted ray is directed along the boundary rather than into the medium). The figure shows that the wedge is shaped in such a manner that longitudinal waves that are reflected at the surface become totally absorbed by subsequent reflections. In Figure 16.11(b) a similar probe is positioned at a suitable location to receive the waves from the defect after a reflection at the base of the specimen. In a given substance, the transverse wave velocity is generally about half that of the longitudinal wave velocity, so the sensitivity of this methodology is twice that for longitudinal procedures.

Surface defects can be discerned through the means of *surface waves*. They are produced by a probe similar to that shown in Figure 16.11(a), but the incident longitudinal waves are directed to the surface at the second critical angle where the transverse raves are refracted at an angle of 90 degrees (i.e., along the boundary surface). Laminar defects that exist just below the surface, which are hard to detect by normal longitudinal wave methods, can be located by Lamb waves (Worlton, 1957). According to Lamb, a solid plate can resonate at an infinite number of

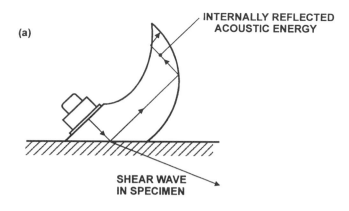

(a)

INTERNALLY REFLECTED
ACOUSTIC ENERGY

SHEAR WAVE
IN SPECIMEN

(b)

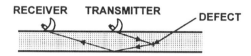

RECEIVER   TRANSMITTER   DEFECT

FIGURE 16.11. Transducer mounted on plexiglass wedge and setup of transverse wave probes for locating a defect in a specimen.

frequencies. The portion of specimen between the surface and a lamination close to it forms such a plate. If surface waves are directed toward this plate, it will resonate and generate a signal that can show up on an oscilloscope screen.

Determination of Propagation Velocity and Attenuation Through an Interferometer

The interferometer is a continuous-wave device that can accurately measure velocity and attenuation in liquids and gases that can sustained standing waves. It consists of a fluid column that contains a fixed, air-backed piezoelectric transducer at one end and a moveable rigid reflector at the other end. A fixed frequency is selected. The reflector is moved with respect to the transducer by a micrometer adjustment mechanism. As the reflector moves, the reflected waves becomes periodically in and out of phase with the transmitted waves, as the result of the corresponding constructive and destructive interference. The effect of the interference on the crystals influences the load impedance detected by electronic system. The load current in the electronic amplifier fluctuates accordingly. The wavelength of the sound is established by the distance the micrometer moves the reflector over one cycle of load current fluctuation, with the distance between two successive maxima being equal to half-wavelength $\lambda/2$.

Optical interference methods also have been used in this fashion to accurately measure the wavelengths of standing waves at high frequencies (near 1.0 MHz or above). An accuracy of 0.05% is typical for the interferometer, which depends on the quality of micrometer readings, the parallelism between transducer and reflector surfaces, and the accuracy of the frequency determination. The velocity of sound is found by simply multiplying the frequency by the wavelength. The attenuation is found from the decay of the maxima of the periodic amplitude plotted as the function of the distance x between the transmitter and the reflector increases.

Ultrasonic Delay Lines

Delay lines are used to store electrical signals for finite time periods, These are used in computers to store information to be extracted for a later stage of calculation. A method for generating the delay is to convert those signals into ultrasonic waves which then travel through a material to be reconverted into their original forms. The simplest ultrasonic delay line is a crystal transducer radiating into a column of liquid, such a mercury, that terminates in a reflector. An adjustment of the delay time can be effected by changing the position of the reflector relative to the crystal. As liquid delay devices are not always convenient to use, solid delay lines are more common. For delay times of a few microseconds, only a few centimeters of a solid rod or block is sufficient. The delay time may be doubled by using shear waves instead of longitudinal waves. If even longer delay times are required, say in the order of a few milliseconds, very long acoustic paths are necessary. These can be achieved by using materials in the form of polygons in which large numbers of multiple reflecions can occur. The solid delay line has the disadvantage that the delay time usually cannot be varied. However, the magnetostrictive delay line, illustrated in Figure 16.12. can be varied in length and unwanted reflections are avoided by coating the ends of the rods with grease, which completely absorbs the sound waves. The line may be either a wire, a rod, or some sort of ribbon of ferromagnetic material such as nickel. An electrical signal applied to coil A induces through a magnetostricive effect a sound pulse in the line. The pulse travels along the line and induces an electrical signal in coil B by the reverse magnetostrictive effect. The permanent magnets C and D provide the requisite polarization.

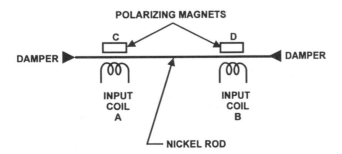

FIGURE 16.12. Magnetostrictive delay line.

Measuring Thicknesses Through the Pulse Technique

When the pulse technique is used for gauging thicknesses, better results are achieved through the use of a variable delay line. Two pulses are generated simultaneously. One pulse is sent through the sample and the other through a delay line, which can be a length of nickel wire or a column of liquid terminating in a reflector. The latter pulse is indicated on the oscilloscope by a trace following that representing the pulse passing through the sample. The delay line is adjusted in length by means of a micrometer device until the two traces on the screen coincide positionwise. The thickness of the specimen is derived from the predetermined calibration of the delay line.

A major advantage of using ultrasound for thickness measurement is that access to only one surface is needed. This is especially useful in measuring the extent of corrosion in infrastructures such as viaducts, sewers, gas pipes, and chemical conduits. The thicknesses of ship hulls can be monitored at sea without resorting to taking sample borings, an expensive and tedious process that is not 100 percent effective. In the livestock industry, ultrasonic thickness measurement is used to measure the amount of fat on the bodies of live animals.

Mechanical Stress Measurements

When a solid material undergoes a change in mechanical stress, changes also occur in its elastic moduli and hence in its acoustic velocities (Shahbender, 1971). This method can also determine the variations of stress in real time. In order to render possible the use of this procedure, a calibration curve is necessary to provide the reference data, namely, plots of velocity (in terms of percentage) in a direction as function of stress applied normal to that direction. Figure 16.13 displays typical calibration curves for longitudinal waves and also for shear waves

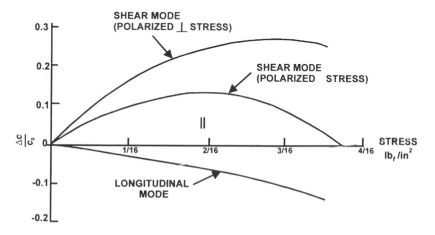

FIGURE 16.13. Calibration curves for longitudinal waves.

polarized perpendicular to direction of the stress and also polarized parallel to the direction of the stress. It can be inferred that the variations of the two components of shear velocity with time are different. Thus, at any given time the corresponding wave vectors will be out of phase with each other. This phase difference ϕ is given by

$$\phi = \frac{L\omega}{c_0}\left[\left(\frac{\Delta c}{c_0}\right)_n - \left(\frac{\Delta c}{c_0}\right)_p\right]$$

where ω denotes the angular frequency, L is the acoustic path length, c_0 the shear wave velocity in the unstressed medium; and $(\Delta c/c_0)_n$ and $(\Delta c/c_0)_p$ represent the fractional shear velocity for polarizations in directions normal to and parallel with the direction of stress, respectively. The value of L is found simply from longitudinal wave measurements while ϕ is obtained by the means of a suitable phase-shift network. This methodology is also applied to determine third-order elastic constants.

The Ultrasonic Flowmeter

The Doppler principle constitutes the operating basis of the ultrasonic flowmeter. Two reversible transducers are submerged in the liquid along the line of flow. One transducer acts as a signal source of ultrasonic pulses and the other acts as a receiver. At short regular intervals the roles of the transducers are reversed, so that the source becomes the receiver and the receiver becomes the source. The wave velocities are $c + u$ along the direction of the flow and $c - u$ in the opposite direction, where c represents the propagation velocity of sound in the fluid and u the velocity of the streamline flow of the liquid.

A number of techniques have been used to compare the upstream and downstream propagation rates. A "*sing-around*" method uses a pulse generator to produce a short train of ultrasonic waves. The received signal is amplified and used to retrigger the pulse generator. If we neglect delay times due to the electronic system and the distance the pulse travels beyond the fluid stream, the difference between the downstream and upstream pulse repetition rates is

$$f_1 - f_2 = \frac{2u}{d}$$

where d represents the distance between the two transducers.

Another type of ultrasonic flowmeter is based on the deflection of an ultrasonic beam by the fluid flow (Dalke and Welkowitz, 1960). In Figure 16.14 a transmitting transducer located on one side of the fluid conduit emits a continuous signal into the fluid stream. A split transducer on the other side of the pipe determines the amount of beam deflection. A differential amplifier is used to determine the difference in the outputs of the two receiving transducers. If there is no flow, the beam falls midway between the two receiving sections, the two sections generate equal voltages, and the output from the differential amplifier will be zero. When fluid flow occurs, the

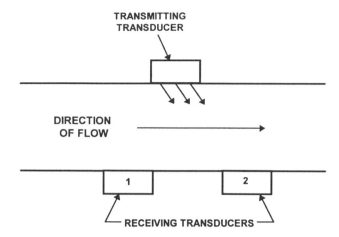

FIGURE 16.14. Schematic of a beam-deflection ultrasonic flowmeter.

beam shifts in the direction to the flow by an amount corresponding to the flow speed, and the outputs from the two sections differ. The difference voltage then corresponds to the rate of flow. Assuming a constant flow rate across the pipe, the deflection ϕ of the beam is computed from

$$\phi = \tan^{-1}\left(\frac{u}{c}\right)$$

Ultrasonic meters are being used to measure flow rates of rivers, nuclear reactor heat-exchanger fluids, gas-containing emulsion, slurries, corrosive liquids, and wind velocities. These measurement devices bear the advantage that they introduce negligible pressure loss in a system and they are economical to operate and can handle a wide range of flow rates, pipe diameters, and pressures.

Resonance Method of Measuring Sound Propagation Speed

The resonance method is similar to the interferometer method for measuring velocity, but it can be applied to solids as well as to fluids. This involves the used of a fixed transducer and a fixed reflector or two transducers spaced a fixed distance apart. The transducer is driven to sweep through a range of frequencies to determine successive resonances. In a nondispersive medium, the difference between two successive resonant frequencies equals the fundamental resonant frequency of the medium, that is,

$$c = 2 \ell \Delta f$$

where ℓ is the distance between the transducer and the reflecting surface and Δf denotes the difference between successive resonant frequencies.

Motion and Fire Sensing

One of the few ultrasonic applications in open air is that of the motion and fire sensor, which is restricted to the lower kHz range, where attenuation is not very much. A magnetostrictive transducer placed at some point in a room emits pulses in all directions. The reflected signals from the walls and furniture are eventually picked up by a receiver, from which a constant indication is generated. Any variation in the sound field, caused by an intruder or a change in temperature, gives rise to a change in this indication, which triggers an alarm. One version of this device is a light switch that goes on automatically when someone enters an empty room and goes off automatically when nobody is present in the room after a predetermined period of time.

Working of Metals and Plastics

Ultrasound have been successfully used in the treatment and working of metals and plastics. As a molten metal is cooled, bubbles should be removed before solidification sets in. Otherwise defects will occur. Irradiation of the melt with ultrasound sets the fluid particles in motion so that bubbles tend to coalesce and, when sufficiently large, tend to rise to the surface. As cooling progresses, crystals begin to form at the solidification temperature, The crystal growth depends on the rate of cooling and the presence of specific impurities in the metal. The high energy of the ultrasound-induced cavitation causes the crystals to be break up as they form. The solid that is finally formed has a much finer grain structure than it would have if cooling had taken place with the melt undisturbed.

Ultrasonic machining is one of the first industrial applications of high-intensity ultrasound. In one technique a slurry consisting of abrasive particles suspended in a low-viscosity liquid flows over the end of a tool shaped in the desired geometry of the impression to be made. The axial motion of the tool tip and the ensuing cavitation impart high accelerations to the abrasive particles and thus erode away the material from the workpiece. This method is suitable for machining of brittle ceramic materials as well as of powder-metal components. But in the latter application, there has been evidence of highly accelerated electrolytic erosion which adversely affects the tool.

Ultrasonic drills do not have transducers in immediate contact with the workpiece. An intermediate material is necessary, and this must help in matching the impedance. In order to achieve maximum velocity amplitude in the tool, a tapered rod which acts as an acoustic transformer in the same manner as a loudspeaker horn is used. The rod is made of a material that has a characteristic impedance matching that of the transducer to which it is rigidly fixed. It is exactly one wavelength long in order to achieve maximum transfer efficiency and clamped at a distance of one-fourth wavelength from the transducer.

The soldering of metals without any flux can be achieved through an ultrasonic soldering iron illustrated in Figure 16.15. The soldering iron is similar in design to an electronic drill, but the tool at the end of the drill is replaced by the electrically

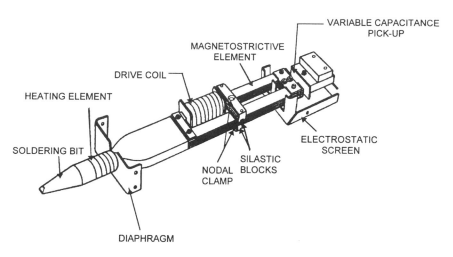

VARIABLE CAPACITANCE
PICK-UP

MAGNETOSTRICTIVE
ELEMENT

DRIVE COIL

HEATING ELEMENT

SOLDERING BIT

ELECTROSTATIC
SCREEN

SILASTIC
BLOCKS

NODAL
CLAMP

DIAPHRAGM

FIGURE 16.15. Ultrasound soldering iron.

heated bit. The vibration of the bit produces cavitation in the solder, thus effectively cleaning the surface of the workpiece and removing any oxide coating, Until fluxes were developed for soldering aluminum, ultrasound had provided the only effective method of soldering this metal. Ultrasound soldering is now being extensively used for soldering joints in miniaturized printed circuit boards.

Ultrasonic welding is effected when two materials are pressed together in intimate contact and the ultrasound vibrations produce shearing stresses at the interface, thereby generating a great deal of heat. This phenomenon lends itself well to produce spot and seam welds in metals such as 304 and 321 stainless steel, aluminum, brass, copper, zirconium, titanium, gold, molybdenum, 0.5% titanium, and platinum. This type of welding can be applied to very small wires to seam welding of metal plates to 0.5 cm in thickness. Both similar and dissimilar metals can be welded. While the formation of a thin molten film in the interface seems to be the primary mechanism of bonding, there is some evidence of solid-state bonding instigated by diffusion in the welding zone at low amplitudes and high clamping pressure.

Bonding of thermoplastic materials[2] through ultrasound is even easier than for metals, because the necessary equipment is smaller and requires appreciably less power. Ultrasonic welding of thermoplastic materials has gained wide acceptance, particularly in situations where thick materials are joined together and the toxicity of adhesives must be avoided. In addition to bonding plastic parts, the same equipment can be used to insert metal parts in plastic pieces. A hole slightly smaller than the metal is drilled or incorporated in a molding part and the metal is driven

---

[2] Plastics fall into two categories: thermoplastic and thermosetting. The former plastic type will soften at sufficiently elevated temperatures and reset upon cooling. Thermosetting plastics retain their rigidity at elevated temperatures.

into the hole ultrasonically. During the insertion the melted plastic encapsulates the metal piece and fills flutes, threads, undercuts, and so on.

Ultrasonic Viscometer

The ideal liquid should not support a shear stress, but the fact is that liquids do have viscosity that gives rise to shear waves. A viscoelastic liquid combining the attributes of both fluid and solid behavior (which produces shear stresses) is described by

$$-\frac{\partial \varepsilon}{\partial y} = \frac{1}{\eta} p_y + \frac{1}{G} p_y \tag{16.41}$$

where p_y represents a variable shear stress; ε is the fluid particle displacement; η is the viscosity coefficient; and G is the shear coefficient.

The solution to equation (16.41) is:

$$p_y = -\left(\frac{\partial \varepsilon}{\partial y}\right)_0 \left(1 - e^{t/(\eta/G)}\right) \tag{16.42}$$

Equation (16.42) indicates that a periodically varying shear produces a relaxation process characterized by a time constant $\tau = \eta/G$. The associated relaxation frequency is given by

$$f_0 = \frac{1}{2\pi \tau} = \frac{G}{2\pi \eta}$$

From equation (16.42) the attenuation of shear waves in a viscous liquid at a given frequency decreases with increasing viscosity. The damping of a vibrating shear wave transducer submerged in the liquid us a function of the coefficient of viscosity of the liquid. One way of measuring this quantity is to apply a pulse to a Y-cut crystal or to a torsionally vibrating rod immersed in the liquid so that it vibrates freely with damped harmonic motion. When the amplitude of the vibration drops to a predetermined level, another pulse is generated. The rate of pulse repetition increases with damping and hence decreases with increase in viscosity. The device is calibrated by using liquids having known coefficients of viscosity.

16.8   Ultrasound Imaging

The application of ultrasound to imaging processes is extremely important in industry and in medicine. Because imaging entails low-intensity ultrasound energy, it provides a valuable nondestructive testing technique. Ultrasonic imaging, which may be defined as any technique of providing a visible display of the intensity and phase distributions in an acoustic field, falls into a number of categories: (a) the electronic-acoustic imaging, (b) B-scanning most commonly used in medical diagnosis, (c) C-scanning widely used in nondestructive testing and inspection flat and cylindrical surface, (d) liquid-surface-levitation presentations, (e) liquid crystal display (LCD), photographic or similar display, (f) light-refraction methods, and (g) acoustical holography.

Electron-Acoustic Image Converters

The concept of an electron-image converter originated with the Russian scientist S. Ya. Sokolov (1939) who envisioned a device, similar to a video camera tube, in which the photosensitive element is replaced by a pressure-sensitive piezoelectric plate. Secondary electrons are given off when the plate is struck by a scanning beam of electrons. An ultrasound field aimed at the plate influences the electrical potentials on the faces of the plate. These potentials are proportional to the impressed acoustic pressure, with the result that the electrical potentials modulate the secondary emission of electrons. The secondary emission that result from the impingement of an electron beam is a function of the velocity of the primary electrons and also of the plate material. The ratio of the number of secondary electrons leaving the plate to the number of electrons impinging on the plate is called the *secondary emission ratio*. When the primary-electron voltage is increased from zero, the secondary emission rate also increases from zero, passes through a maximum and then decreases. With some piezoelectric materials, the maximum ratio exceeds unity and there are two velocities, one above and one below the point of maximum secondary emission, at which the ratio becomes unity. In other materials, the maximum secondary emission ratio never exceeds unity.

Smythe et al. (1963) improved on the Sokolov concept by developing an ultrasonic-imaging camera that could be operated in either an amplitude-sensitive mode or a phase-sensitive mode. In the amplitude-sensitive mode, a relatively low-voltage (160–200 V) scanning beam is used. The secondary-emission ratio is less than zero, which allows the potential of the surface of the (quartz) piezoelectric plate to nearly equal that if the electron gun cathode, in absence of ultrasound. After a scan without ultrasound, the ultrasound is activated and the plate is scanned again. The electrons are distributed over the surface of the plate in direct ratio to the piezoelectric voltage present at each point of the surface. A corresponding image charge forms on an anode located externally to the tube, and the video image is obtained directly from the anode charges. The ultrasound is then turned off and the native charge on the piezoelectric surface is removed by inundating the surface with gas ions, a process which occurs in about 0.2 s. High voltage scanning can also be applied to remove the negative charge, followed by a low-voltage scanning to restore the surface to the cathode potential, with the ultrasound turned off. This procedure removes the need for ion recharging and allows the use of a high-vacuum tube.

For the phase-sensitive mode of operation, higher-voltage (600 V) beams are used in conjunction with an auxiliary grid on which the secondary-emission current is collected. The inner quartz surface is stabilized at approximately the same voltage as the auxiliary grid. The piezoelcctric charge in the quartz plate tends toward

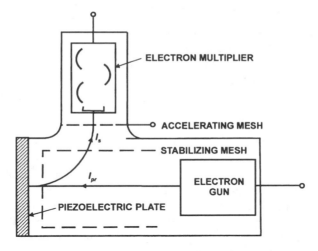

FIGURE 16.16. Jacobs device for ultrasonic scanning.

neutralization and, consequently, the current to the quartz plate and to the collector grid fluctuates at the frequency of the ultrasonic waves. The output pressure at each point on the plate is therefore proportional to the incident ultrasonic pressure.

Other methods have been developed to reproduce the image on the piezoelectric plate in order to circumvent problems posed by high-velocity scanning-beam stabilization of an insulating surface. One method entails the use of a photoemissive surface deposited on the piezoelectric surface, with the scanning being executed by a moving beam of light. Mechanical methods also were developed, and they proved to be appreciably more sensitive than scanning with an electron beam. The surface of a piezoelectric plate subjected to an ultrasonic field is scanned mechanically using a capacitive, noncontacting electrode or by sliding a small electrode over the surface itself.

Jacobs (1962) introduced the use of electron multipliers in the camera tube of ultrasound image converters based on Sokolov's method. A schematic diagram of Jacobs's device in shown in Figure 16.16. The construction provides shielding for the low-voltage circuits and virtually eliminates ground loops. The secondary-emission beams produced by the scanning beam are attracted toward a positively charged electrode in the electron-multiplier unit. The electron multiplier amplifies the signal currents by about 100,000 times greater than the threshold values before they are fed onto the associated amplifiers. Thus, the electron multiplier serves as a wideband amplifier with good noise characteristics. Its output constitutes the video signal that is processed through a conventional closed-circuit television system.

Two sealed ultrasonic camera tubes were developed by Jacobs: one version incorporates quartz crystals permanently sealed to the end of the tube and the features interchangeable crystals which permit operation over the frequency range from 1 to 15 MHz. Color also has been introduced to increase the sensitivity of the ultrasonic image converter, particularly in view of the fact that the human eye is more sensitive to changes in color than to differential changes in display brightness. The color display indicates both relative amplitude and phase of the ultrasonic signal undergoing analysis. In North America, the video method used is the National Television System Committee (NTSC) color broadcasting standard; and the operating frequency is 3.58 MHz.

Acoustic Lenses

Acoustic lenses are necessary in the use of electron-acoustic image converters. An acoustic lens can (a) increase the sensitivity of an imaging system through energy concentration and (b) provide coverage of a larger area by concentrating the image on the receiving piezoelectric element of the image converter in almost the same fashion an optical lens reduces a larger picture to a smaller area. For optimal performance, the velocity of sound in the lens material must differ considerably from that of the surrounding media, and the reflection of energy at the boundary between the lens and the surrounding should be minimal. The latter condition is fulfilled automatically when the acoustic impedances of the lens materials and the surrounding medium match each other. Liquid lenses of carbon tetrachloride or chloroform have the same acoustic impedance as water, but their toxicity generally preclude their use in industry. Plastic lenses also have been developed, but the sound propagation velocity in these materials exceeds that of water and they present an impedance mismatch between the lenses and water, but not to the degree of rendering them useless. The relatively high absorption of plastics limits their use to frequencies less than 15 MHz. Metallic lenses, which can be made of aluminum and other metals, possess low absorption characteristics, so they can be used at frequencies exceeding 15 MHz, but the impedance mismatch between metals and water essentially prevent their effective use.

The velocity of sound in solids is higher than in liquids, so solid concave lenses are convergent and solid convex lenses are divergent—unlike the case of light traveling in a vacuum or a gas through optical lenses.

Schlieren Imaging

Schlieren imaging has been used for many years as a tool to visualize sound fields. Advantage is taken of the physical fact that pressure gradients in an ultrasonic wave cause density gradients in the medium. When a light beam passes through these gradients it becomes refracted. The refracted light is used in a schlieren apparatus to produce an image of the sound field. Either one of two methods can be used to produce the sound field image: (a) interruption of the refracted part of the beam in order to remove it from the beam and focusing the remainder of the field onto an image detector (e.g., a ground glass screen or camera film), or (b) focusing the refracted rays on the image plane and eliminating the remainder of the beam of light from the image. The result of the first method is a dark image on a light background, and that of the second method is a light image on a dark background.

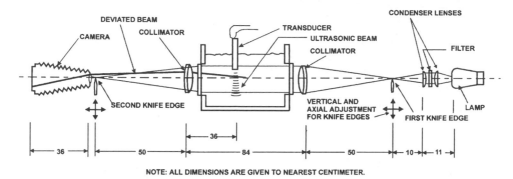

NOTE: ALL DIMENSIONS ARE GIVEN TO NEAREST CENTIMETER.

FIGURE 16.17. Schlieren device for direct viewing in nondestructive testing.

A schlieren device for direct viewing during nondestructive testing is schematically illustrated in Figure 16.17. The image is viewed directly on a ground plate glass screen. In the right-hand side of Figure 16.17, the lamp (e.g., zirconium arc), condenser lenses, filter, first knife edge, and first collimeter are mounted on a single 1-m optical bench. The second collimeter, second knife edge, and camera system are mounted on another 1-m optical benth. The two benches themselves are positioned on a 25 cm wide, $5\frac{3}{4}$-m long steel girder. The water tank is mounted apart from the optical system so that any movement or changes in the tank will not cause the optical system to go out of alignment.

The optical system must be mechanically-isolated from all the other parts of the apparatus. When the device is being used, the image of the light source is centered on the first knife edge, and then the collimating lenses, tank windows, second knife edge, camera lenses, and viewing screen are centered in the light beam. Centering of the optics must be precise to ensure maximum sensitivity and uniform field. Through careful alignment and focusing, the field can be made to go from light to gray to dark uniformly by adjusting the second knife edge further into the light beam until it intercepts it completely. In the range of gray settings, convection currents both in air and water become clearly visible. The ultrasonic field is best rendered visible when the second knife edge is set to intercept all of the main beam, thus allowing only light refracted in the ultrasonic field to pass and illuminate the viewing plane. The optical system may also be adjusted for bright-field operation, in which situation the ultrasonic perturbations appear as dark shadows.

Color schlieren photography is useful in the study of ultrasonic waves and shock waves. Color can indicate the various pressure levels in an ultrasonic field. One method of color schlieren photography uses a spectroscopic prism between a slit located at the position of the first knife in Figure 16.17 and the first collimating lens. This method produces colors ranging from red or from blue to green. Another method developed by Waddell and Waddell produces a complete color spectrum. The Waddells eliminated the spectroscopic prism and used a color-filter matrix in place of the second knife edge and a vertical slit in place of the first knife edge. Illumination was provided by a high-pressure mercury arc lamp. The color matrix consisted of three filters which represent the primary colors.

Liquid Crystal Imaging

Liquid crystals manifest properties of solid crystals that are not apparent in ordinary liquids. When a stress is placed on a liquid crystal, its optical properties changes. A certain class of liquid crystals known as *nematic crystals* are used to indicate the presence of an ultrasonic field with sensitivity equal to that of schlieren systems, high resolution, large-area capability, and handling ease.

Ultrasonic Holography

Holography is a form of three-dimensional imaging that was conceived and developed by Dennis Gabor who received the Nobel Prize in physics for his efforts. Gabor applied his discovery to electron microscopy to overcome the problem of correcting spherical aberration of electronic lenses. The principle of holography is this: A diffraction diagram of an object is taken with coherent illumination and a coherent background is added to the diffracted wave. A photograph so taken will contain the full information on the changes sustained by the illuminating wave in traversing the object. The object can be reconstructed from this diagram by removing the object and illuminating the photograph by the coherent background alone. The wave emerging from the photograph will contain a reconstruction of the original wave which seems to issue from the object, A hologram, therefore, is a recording or a photograph of two or more coherent waves. If one recorded wave is from an illuminated object and another is a reference wave, simply illuminating the hologram with the reference wave reconstructs twin images of the original object, thus giving the illusion of three dimensions.

The wave used in the reconstruction does not have to be the original, and this allows the use of ultrasonic waves and the subsequent reconstruction of the image using light from a laser. But the size of the image changes in proportion to the ratio of the wavelength of the reconstructing wave to the wavelength of the original illuminating wave. There are a number of methods of making acoustical holograms. One method in Figure 16.18 uses liquid levitation, where the acoustic waves from below the surface of a liquid forms an ultrasonic image at the surface of the liquid that can be rendered visible on a photographic plate. The reference beam may be obtained either by reflecting a portion of the irradiating beam onto the surface or by generating a separate wave through a second transducer. The height of the bulge of the liquid surface due to the impetus of an acoustic wave is critical; it should be small compared to the wavelength of the light.

16.9   Medical Uses of Ultrasound

Ultrasound use in the medical arts can be classified as being diagnostic or therapeutic. While the diagnostic procedures involving ultrasound have been in use for a number of years, ultrasonic therapeutics constitute a newer, rapidly growing domain. The ultrasound frequencies used in medical applications range from

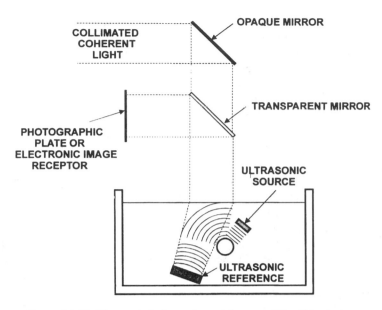

FIGURE 16.18. Ultrasonic holography making use of liquid levitation.

approximately 25 kHz used in dental plaque removal to the megahertz range which is required for medical imaging. The merits of ultrasound diagnostics include safety, convenience, and capability of detecting medical conditions to which X-rays and other means of diagnosis are insensitive. Because ultrasound is much safer than X-rays, it is used in fetal monitoring, detection of aneurysms, and echocardiography. The production of heat in the body through ultrasound is applied for its therapeutic value. The selectively greater absorption of ultrasound in cancerous tissues has proven its usefulness in hyperthermic treatment of cancer. Emerging developments include the use of ultrasonic waves to perform noninvasive or "bloodless" surgery, stop internal bleeding in trauma patients, and control delivery of drugs or other compounds. Enormous advances in electronic miniaturization are resulting in small handheld diagnostic units.

Diagnostic Uses of Ultrasound

Diagnostic medical applications are based on the imaging procedures described in the last section of this chapter for material inspection. One diagnostic technique is based on the pulse method, and the second diagnostic technique is based on the Doppler effect where the reflected wave is shifted in frequency from that of the incident wave impinging on a moving target. In the reflection or pulse-type equipment, A-scan, B-scan, or a combination of these two methods are utilized to present data on an oscilloscope display or process the data for permanent record. The A-scan presents echo amplitude and distance, and it is used principally in echoencephalography for the detection of midline shifts traceable to tumors or concussions. It has also been applied in obstetrics, gynecology, and ophthalmology in conjunction with the B-scan techniques. In B-scanning, the radar/sonar techniques of data processing are applied to synthesize the reflected signals into a pattern on the oscilloscope display that corresponds to a cross-section of the region scanned lying in a plane parallel to the direction of beam propagation.

The position of the probe is synchronized with the sweep of one of the axes of the oscilloscope, and the echo amplitude appears as a spot of a certain intensity at a position on the screen corresponding to the position of the plane causing the echo. The TM-mode (or M-mode) is a diagnostic ultrasound representation of temporal changes in echoes in which the depth of echo-producing interfaces is displayed along one axis and time (T) is displayed along the second axis, thus recording motion of the interfaces toward and away from the transducer.

In order to resolve small structural details, the transmitted pulse should be as short as possible, which means that the transducer must be highly damped. In order to promote good coupling between the transducer and the body, a film of oil or grease is applied at the selected spot, and care must be taken that the contacting film is free of air bubbles. In immersion procedures (such as that for kidney stone pulverization), coupling is through a bath of liquid, usually water. The impedance match between water and soft tissue is good, and little energy is lost in irradiating soft tissue. But the match between water and bone is poor and also the attenuation in bone is high. The result in echoencephalography is that considerable energy is lost, where almost immediate contact is made with the skull bone.

Ultrasound imaging is being extensively used to study cardiac functions. One use of the A-scan technique is to monitor for early signs of rejection following a heart transplant. As the heart fills at the onset of rejection, the muscle walls swell and stiffen. Measurements are made using a 2-cm diameter, 2.25 MHz transducer at a pulse repetition rate of 1000/s. Echo indications from the anterior wall and the posterior wall supply the measurement information, because the distance between the two walls indicates overall heart size. Echocardiography is useful in diagnosing pericardial effusion (escape of fluid from a rupture in the pericardium) because the echo received from the posterior wall of the heart is split when the transducer is located on the anterior chest surface. Echocardiography is also useful in assessing the degree of stenosis (narrowing of opening) in mitral valves. The transducer is aimed at the anterior mitral leaflet and the echo signal can be recorded on a strip chart so that an upward movement of the recorder pen corresponds to movement toward the transducer and a downward movement corresponds to movement away from the transducer. The slope of the tracing indicates the velocity of motion. The degree of stenosis affects the blood flow rate through the opening. A number of symptoms can be discerned with the use of this method, for example, rigidity or calcifcation of the mitral valve is indicated by decrease in the total amplitude of the anterior mitral leaflet between the closed position during ventricle systole and the position of maximum opening in early diastole.

Intercardiac scanning is used to obtain plan-position (C-scan) displays for the interior of the heart. A tiny probe is inserted into the right atrium through the external jugular vein or the femoral vein. An advanced version of the probe consists of many elements so that scanning can be achieved by sector techniques in which the positions of points on the recorded image are correlated with beam direction. The data from the transducer can be processed by computer and displayed. A-scanning may be combined with C-scanning to provide 3D information. The motion of selected regions of the heart can even be viewed on a video display, particular with the use of focusing transducers which can provide high-resolution images of the heart.

A tomographic method of observing interior structures of the heart three-dimensionally is based on a stereoscopic display of 2D images. The ultrasonic device functions in synchronism with the cardiac cycle to obtain phase-specific tomograms, which then can be displayed on a storage CRT. Tomographic systems are also used to investigate other internal organs such as right ventricle, atrium, and kidney cysts that have irregular shapes.

Accurate diagnosis is rendered possible in the field of ophthalmology through the use of ultrasound to diagnose conditions existing in the soft tissues of the orbit of the light-opaque portions of the eye. Focused transducers are used with a frequency typically being 15 MHz. This method can outline tumors and detached retinas, measuring the length of the axis of the eye, and also detect foreign bodies close to the posterior eye wall. One instrument combines a diagnostic transducer for locating foreign bodies with a surgical instrument for removing an object enables rapid removal of foreign bodies from the eye by directing the surgical tool to the object with least damage to the eye.

In the field of neurology, ultrasonic echoencephalography provides an immediate means of detecting lateral shifts in the midline septum caused by tumors or concussion. It is notable that every accident ambulance in Japan is outfitted with echoencephalographic equipment in order to identify victims with possible subdural hemorrhage so that they may be transported directly to special neurological units for treatment. Posttraumatic intercranial hemorrhage and skull and brain trauma can be quickly diagnosed and lesions located rapidly without discomfort to the patient. Ultrasonic pulses are transmitted through the temples. In the A-scan mode, echo indications from the midline and the opposite temple areas of the skull are presented on an oscilloscope screen. A shift in the midline is readily discernible. B-scans which could provide more information are more effective with small children and infants because their skulls are soft and have attenuation coefficients lower than those in adults.

The ultrasonic B-scan technique for examining the abdomen is useful for detecting pelvic tumors, hydatiform moles, cysts, and fibroids. It is also used to diagnose pregnancy at 6 weeks (counted from the first day of the last menstrual period) and afterwards. The obstetrician can follow the development of the fetus throughout the pregnancy, including size, maturity, and positioning of the placenta. The presence or twins or multiple pregnancy is also revealed. This avoids the need for X-rays and the attendant danger of irradiation to mother and child. Fetal death can be confirmed much earlier by ultrasound than it can by radiography.

Ultrasonic diagnosis by echo methods extends to all parts of the body. Air and other gases have a much lower impedance than either liquid or solids, so air cavities produce distinct echoes. Stationary air embolisms can be so identified because they are located in area usually filled with fluids. Moving embolisms can be identified by Doppler methods. Gallstones have an acoustic impedance equivalent to that of bone and so can provide good ultrasonic echoes. The ultrasonic B-scan technique has proven effective in diagnosing thyroid disorders. Disorders detected ultrasonically include neoplastic lesions including cystic nodules, solid adenoma and carcinoma, nonneoplastic lesions, subacute thyroidiis, and chronic thyroiditis.

The Doppler method takes advantage of the fact that a shift in frequency occurs when a ultrasonic wave is reflected from a moving target. Also any variation in fluid motion that causes a beam of ultrasound to be deflected causes a Doppler shift in frequency. One application of the Doppler shift principle is for detection of fetal blood flow and heartbeat, which can provide a mother the exciting assurance that her unborn baby is alive by listening to its heartbeat through a set of headphones connected to the detecting equipment. Blood flow is measured through intact blood vessels through the use of the Doppler principle with the carrier frequency ranging from 2 to 20 Mhz. The Doppler frequency signal is calibrated in terms of velocity, according to the pitch of the Doppler signal.

Doppler-type diagnostic units are used to determine the severity of atherosclerosis, locating congenital heart defects, and continuously monitoring fetal heartbeat during birth.

Safety of Ultrasonic Diagnosis

Extensive studies on the safety of ultrasound in Japan and elsewhere has led to issuance of Japanese industrial standards (JIS) for diagnostic ultrasound devices. These standards place limits on various diagnostic procedures:

1. Ultrasonic Doppler fetal diagnostic equipment—10 mW/cm^2 or less.
2. Manual scanning B-mode ultrasonic diagnostic equipment—10 mW/cm^2 or less for each probe.
3. Electronic linear scanning B-mode ultrasonic diagnostic equipment—10 mW/cm^2 or less in a single aperture.
4. A-mode ultrasonic diagnostic equipment—100 mW/cm^2 or less. This standard is limited to diagnosis of the adult head and is not for pregnancy where B-mode is the principal technique used. This higher permissible level is due to attenuation through the adult skull bone.
5. M-mode ultrasonic diagnostic equipment—40 mW/cm^2 or less. The M-mode is used in clinical diagnosis of the heart. For a combination of the M-mode with the B-mode, the intensity os limited by the B-mode standard—i.e. 10 mW/cm^2.

Therapeutic Uses of Ultrasound

We may very well be witnessing at this time only the beginning of the use of ultrasound for therapeutic purposes. Some techniques, such as the use of 25-kHz ultrasound combined with a water jet to remove plaque from teeth and the cleaning of dental and medical tools with ultrasound, have been well established for a number of years. Ultrasonic nebulizers of pharmaceuticals occurs without producing destructive temperature levels. Athletic centers and sports medicine specialists make use of ultrasound devices to heat sore muscles. Newer techniques are arriving on the market or are still in the testing stages. One example is the use of ultrasound in catheters to ream out arteriosclerositic deposits in arteries that is still very much in the experimental stage.

One therapeutic use of ultrasound already in widespread clinical use; *extracorporeal shock wave lithotripsy*, has completely changed the treatment of kidney stones. Kidney stones are calcified particles that tend to block the urinary tract. In this type of treatment, the patient is immersed in water to equalize as much as possible the acoustic impedances between the transducer and the patient's body. A focused, high-pressure ultrasonic pulse is directed through the water and into the patient's torso to break the stone into small pieces. The pulverized material can now pass out of the body unhindered. Lithotripsy causes very little damage to kidney tissue.

A promising procedure for therapeutic ultrasound is the laser-guided ablative acoustic surgery, in which sound supersedes the scalpel in destroying benign or malignant tissues. The ultrasound focused by a specially shaped set of transducers converges inside the body to create a region of intense heat that can destroy tumor cells. The spot of destruction is so small that a boundary of only six cells lies between the destroyed tissue and completely unharmed tissue, which connotes a precision far beyond any current method of surgical incision.

Acoustical surgery offers a potentially better means of treating cancerous tumors because it does not require an anesthetic, can be administered in a single treatment, and causes no observable side effects. In a Phase I clinical trial at Marsden Hospital in London, focused sound waves destructed parts of liver, kidney, and prostate tumors in 23 patients. In the next phase the researchers will attempt to fully destroy tumors in the liver and prostate. Also under testing is the Sonablade™ system by Focus Surgery of Indianapolis, IN, which incorporates proprietary transducer technology in a transrectal probe that provides imaging for tissue targeting and high-intensity focused ultrasound (HIFU) for tissue ablation. After the operator defines the area of periutheral tissue to be ablated, the treatment process begins under computer control, where the focus of the dual-function transducer is electromechanically stepped through the designated volume of the tissue. HIFU results in thermally induced coagulative necrosis only in the intraprostatic tissue encompassed by the focal volume, which affects intervening tissues. Confirmation of targeting accuracy is provided through continuously updated images. The necrotic tissue either is sloughed during urination or is reabsorbed along with the cessation of patient symptoms.

Ultrasound can also be used to stop internal bleeding through an effect called *acoustic hemostatis*. With sufficient power, ultrasonic pulses can elevate the body temperature at selected sites from $37°C$ to between $70°C$ and $90°C$ in an extremely short time, less than one second. This causes the tissue to undergo a series of phase transitions, and the protein-based bodily fluids and blood coagulate as the result of the proteins undergoing cross-linking (a process similar to cooking an egg).

A research team at University of Washington's Applied Physics Laboratory is investigating the use of ultrasound to stop internal bleeding during surgery and for treating trauma cases. The present method of stemming bleeding in delicate organs, such as the liver, pancreas, or kidney, is through cauterization on the surface with ion or microwave systems. The focused ultrasound waves, however, can penetrate deeply into the organ and 'cook' the tissue in a layer as thin as 1 mm. It follows that trauma patients could be treated without the need for a sterile environment of an operating room and without the danger of infection that accompanies conventional surgery. To date, success has been achieved by the University of Washington group in identifying patients with internal bleeding and in the use of HIFU in the operating room to stop bleeding in the organs and vessels of animals.

Advances in magnetic resonance imaging (MRI) and diagnostic ultrasound imaging will allow these two areas to be combined. Manufacturers of MRI equipment are developing models that combine therapeutic functions such as ultrasound with imaging processes. One company (Therus) is combining ultrasound diagnostic imaging with a separate therapeutic ultrasound capability to produce a relatively small, portable system that can be carried by paramedics and rescue workers for use at the site of disaster.

A newer method of targeted drug delivery is that of *sonophoresis*, which uses sound waves instead of needles to inject drugs such as insulin and interferon through the skin. The high-frequency waves open tiny holes in cell membranes, thus rendering the cells temporarily permeable in localized regions and allowing better penetration of the drug into the blood vessels below the skin. This results in greater effectiveness of the drug, lessens the dosage requirements and toxicity, and allows for more precise localization of drug delivery. While the mechanisms by which ultrasound augments these effects are only partially understood, it is known that ultrasound produces biophysical reactions yielding hydroxyl radicals that in turn affect cell membranes.

A system developed by Ekos Corporation of Bothell, WA, to dissolve life-threatening blood clots constitutes an early application of ultra-sound drug delivery. The Ekos device injects thrombolytic drugs through a cather using low-energy, localized ultrasound to the target site in the body. Thrombolytics, which are generally administered to break up clots, are dramatically more effective with the use of ultrasound because the sound waves help to concentrate the drugs at the site of the clot. Ekos's device and similar products are meant to provide treatment of cardiovascular obstructive diseases (e.g., stroke and arterial and deep-vein thrombosis). In these type of ailments, rapid response is critical, and therapy necessitates applications of massive doses of clot-dissolving drugs over 72 hours. Ultrasound may also be used with antiresrenosis agents which are intended to prevent coronary arteries from reclosing after angioplasty, a procedure in which a balloon-tipped catheter is inserted into a clogged vessel and the balloon is inflated to open up the vessel.

Another, long-term application is the use of ultrasound to deliver insulin through the skin for treating diabetes. It may be even possible through ultrasound to penetrate the blood–brain barrier, which insulates the brain from foreign substances and also prevents many drugs from reaching diseased tissues there, so that the effects of chemotherapy can be enhanced.

Higher frequencies of ultrasound can allow the transfer of large molecules such as DNA to migrate among cells, as the result of *sonoporation*, which enhances the porous effect on cell membranes by induction of ultrasonic shock waves through a lithotripter. One pharmaceutical company (ImaRx) developed a gene-delivering system for improving gene expression, that is, revising a gene's genetic code to make a specific protein. The ImaRx system employs acoustically direven microbubbles, which carry the gene and fluorinated compounds that serve as markers to detect the DNA. The ultrasound assists in delivering the DNA to targeted areas in the body and in tracking its progress through the use of contrasting agents injected with the microbubbles of genetic material.

References

Atchley, A. A., and L. A. Crum. 1988. Acoustic cavitation and bubble dynamics. In: *Ultrasound: Its Chemical, Physical, and Biological Effects* (K. S. Suslink, ed.). New York: VCH Publishers; pp. 1–64.

Bardeen, J. B., L. N. Cooper, and J. R. Schrieffer. 1957. Theory of superconductivity. *Physics Review* 108: 1175

Blake, J. R., J. M. Boulton-Stone, and N. H. Thomas (eds.). 1994. *Bubble Dynamics and Interface Phenomena*. Dordrecht: Kluver.

Blitz, Jack. 1967. *Fundamentals of Ultrasonics*, 2nd ed. New York: Plenum Press. (A bit dated, but this remains a very clear exposition of the ultrasonic field).

Brown, B., and J. E. Goodman. 1965. *High Intensity Ultrasonics*. London: Iliffe.

Crum, L. A. 1994. Sonoluminescence. *Physics Today*. September: 22–29.

Dalke, H. E., and W. J. Welkowitz. 1960. *Journal of the Instrument Society of America* 7(10): 60–63.

Ensminger, Dale. 1988. *Ultrasonics: Fundamentals, Technology, Applications*, 2nd ed. New York: Marcel Dekker. (A comprehensive, well-done modern text giving an overview of the field of ultrasonics.)

Frenzel, H., and Z. Schultes. 1934. *Physical Chemistry* 27B: 421.

Gaitan, D. F., and L. A. Crum. 1990. *Frontiers of Nonlinear Acoustics* (M. Hamilton and D. T. Blackstock, eds.). New York: Elsevier Applied Science; pp. 49–463.

Gaitan, D. F., L. A. Crum, A. Roy, and C. C. Church. 1992. *Journal of the Acoustical Society of America* 91: 3166.

Herzfeld, Karl F., and Theodore A. Litovitz. 1959. *Absorption and Dispersion of Ultrasonic Waves*. New York: Academic Press. (A classic reference on ultrasonic phenomena.)

Hunter, T. F., and R. G. Bolt. 1955. *Sonics*. New York: John Wiley & Sons.

Jacobs, J. E. 1962. *Proceeding on Physics and Nondestructive Testing*, October 2–4. San Antonio, TX: Southwest Research Institute; pp. 59–74.

Jacob, J. E., K. Reimann, and L. Buss. 1968. *Materials Evaluation* 26(8): 155–158.

Joyner, Claude R (ed.). 1974. *Ultrasound in the Diagnosis of Cardiovascular-Pulmonary Disease*. Chicago: Year Book Medical Publishers, Inc. (Written by clinicians primarily for clinicians.)

Landau, L., and E. Teller. 1936. Zur Theorie der Schaledispersion. *Physik Zeitschrift Sowjet-Union* 10, p. 34.

Leighton, T. G. 1994. *The Acoustic Bubble*. London: Academic Press.

Maeda, K., and M. Ide. 1986. *IEEE Transactions: Ultrasonics, Ferroelectrics, and Frequency Control*, UFFX-33(2); pp. 179–185.

Mason, Warren P. (ed.). 1964. *Physical Acoustics: Principles and Methods*, vol. 1, part A. New York: Academic Press; Chs. 3–5.

R. W. Morse. 1959. *Progress on Cryogenics* (K. Mendelssohn, ed.), vol. I. London: Heywood,

Neppiras, E. A. 1973. The prestressed piezoelectric sandwich transducer. *Proceedings of the International Ultrasonics Conference*; pp. 295–30.

Ouellette, Jennifer. 1998. New ultrasound therapies emerge. *The Industrial Physicist*. September: 30–34.

Shahbender, R. A. 1961. *Transactions of I. R. E.*, UE-8, p. 19.

Shields, F. D. 1962. Thermal relaxation in fluorine. *Journal of the Acoustical Society of America* 34(3): 271.

Smyth, C. N., F. Y. Poynton, and J. F. Sayers. 1953. *Proceedings of IEEE* 110(1): 16–23.

Sokolov, S. Ya. July 1937. U. S. Patent No. 2,164,185.

Sternberg, M. S. 1958. *Physics Review* 110: 772.

Suslick, Kenneth S., and Lawrence A. Crum. 1997. Sonochemistry and sonoluminescence. *Encyclopedia of Acoustics* (Malcolm J. Crocker, ed.), vol 1. New York: John Wiley & Sons; Ch. 26, pp. 71–281.

Waddell, J. H., and J. W. Waddell. 1970. *Research and Development* 30(32).

Winter, T. G., and Hill, G. L. 1967. High-temperature ultrasonic measurements of rotational relaxation in hydrogen, deuterium, nitrogen, and oxygen. *Journal of the Acoustical Society of America* 42(4): 848.

Worlton, D. C. 1957. *Nondestructive Testing* 15: 218.

Zuckerwar, A. J., and W. A. Griffin. 1980. Resonant tube for measurement of sound absorption of gases at low-frequency/pressure ratios. *Journal of the Acoustical Society of America* 68(1): 218.

17
Music and Musical Instruments

17.1 Introduction

From time immemorial music has impacted humanity in many ways. In moments of sadness, music provides solace; in happier times, enhanced exhilaration; during stressful periods, a greater sense of calm intertwined with an intensified feeling of purpose; and when diversion is needed, entertainment.

Music reached its greatest heights through the evolution of primitive contraptions into more elegant instruments and the emergence of great composers such as Monteverdi, Vivaldi, Bach, Handel, Haydn, Mozart, Beethoven, Verdi, and, in more modern times, Rimsky-Korsakov, Stravinsky, Mahler, Ravel, Gershwin, Schoenberg, and Ellington.

Music has given rise to a whole slew of industries: the manufacture and marketing of instruments; staging of performances which can range from solo appearances to a lavish operatic production or a frenzied rock concert, on stage or through electronic transmission (radio, television, the Internet); and distribution of recorded media (tape, CD or DVD) and playback equipment.

In surveying the musical scene, we must realize the manner in which music influences people cannot possibly be adequately gauged and this constitutes a situation that provides a fertile field of research in psychology and anthropology, to say nothing of musicology and music theory. In the medical field, classical music serves as a valuable tool in psychotherapy. The field of *musical acoustics* is an extremely broad interdisciplinary field, a field that deals with the production of musical sound, and the transmission of musical sound to the listener (Rossing, 1997). The study of music from a physical approach provides the opportunity to bridge the gap between art and science.

In this chapter the structure of music is examined from an acoustical approach and the principles of generating musical tones with instruments, which also includes the human voice, are outlined. Many but not all musical instruments are described herein, as well as the makeup of orchestras and bands. A brief overview is also given here of the means of recording and reproduction of sound, with some emphasis on the design and construction of loudspeakers and headphones and on the psychoacoustics of multichannel playback.

17.2   Musical Notation

As pointed out by Olson, music can be memorized and passed from one person to another by the direct conveyance of the sound, but this does not constitute a satisfactory, nor efficient method of communicating music to performers and preserving the music for future performances. Accordingly, musical notations were developed to use symbols on paper to denote frequency, duration, quality, intensity, and other tonal characteristics. A human can typically distinguish 1400 discrete frequencies, but in the equally tempered musical scale there are only 120 discrete tones ranging from 16 Hz to 16,000 Hz. Pitch is an attribute of aural sensation and depends on the frequency of the sound. Musical tones are assigned specific values, allowing for specific frequency values and leaving out the 'in-between' values. A principal reason that we can identify a musical instrument is that the musical instruments are essentially resonant instruments and therefore exhibit a response only to certain frequencies. These resonant frequencies are fixed and cannot be altered, except for certain instruments such as members of the violin family and the trombone. Moreover, when same notes are played on different instruments, the overtones differentiate one instrument from another. With the relatively small number of fundamental frequencies designated in Western music, matters are greatly simplified in designating the discrete frequency characteristics of tones.

In the five-line *staff* of Figure 17.1, the pitch of a tone is denoted by placing notes, ○, ♩, ♪, on the lines and in the spaces between the lines. The *pitch range* of a set of lines is designated by the *clef* (♭ or ♮) which is placed at the left of the staff. The most common clefs are the *treble* or *G clef* and the *bass* or *F clef*, both of which are shown in Figure 17.1. The notes are designated alphabetically from A to G. The interval in pitch between two notes with the same letter is the *octave*. As explained earlier in this text, the two sounds separated by an octave have a fundamental frequency ratio of 2. The pitch interval between adjacent notes (i.e., between a note on a line and a note in the adjacent space, as designated by successive letters) is a whole tone in the equally tempered scale. Pitches that are higher and below the staff are designated by notes written upon and between short lines called *leger lines*, as shown in Figure 17.1. The number of leger lines can theoretically be extended without limit. The sign 8va above the staff denotes that all tones are to be played an octave higher than their placement and, conversely, the sign 8va (or 8va basso) placed below the staff indicates that all tones are to be played an octave lower. This is shown in Figure 17.2.

There is also another clef—the *movable* or *C clef*—which was meant to accommodate music instruments with extended range (e.g., the bassoon, cello, or viola). In older musical manuscripts there are C-clefs for the soprano, alto, and tenor parts, Figure 17.3 shows the three different symbols to indicate the C-clef. The C-clef is placed on the middle C line. The old C-clef positions for the soprano, alto, and tenor are given in Figure 17.4. The position of the C clef always corresponds to the middle C.

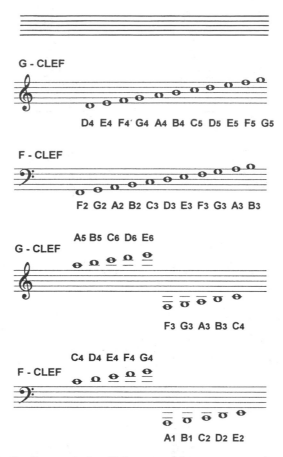

FIGURE 17.1. The five-line musical staff, the notes of the most common bass and treble clefs and the leger lines (the lines above and below the staff) for the treble and bass clefs.

A note can be moved up in its pitch a half step or *semitone*; this is designated a *sharp*. The sharp is designated by the natural note preceded by the ♯ sign, as shown in Figure 17.5. A note can also be moved down a semitone, thus rendering that note a flat with the symbol ♭, which is also shown in Figure 17.5. A natural designation ♮ nullifies a sharp or a flat, returning the note to normal. A note may be moved up by a whole step by a double sharp designation ♯ ♯ or ×. A note may be moved down a whole note by a double flat ♭♭. A standard system for the identification of tones used in music is given in line 1 of Figure 17.6 (Young, 1939). The other seven lines represent the various systems for identifying the musical tones without the benefit of using the staff, but the system represented by the first line is the most logical one to use and understand. The reference standard frequency C_0 is 16.352 Hz which just about constitutes the lowest frequency that a human ear can detect. It is the custom

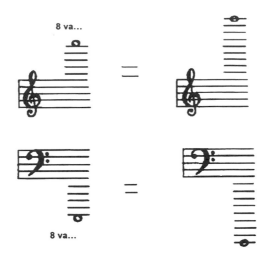

FIGURE 17.2. The position of 8va above the staff indicates all notes are sounded an octave higher. When 8va is located below the staff, the notes are to be sounded an octave lower.

FIGURE 17.3. Three symbols used to indicate the C, or movable, clef.

SOPRANO CLEF   ALTO CLEF   TENOR CLEF

FIGURE 17.4. The positions of the C-clef for the soprano, alto, and tenor clefs.

FIGURE 17.5. Sharp, flat, and natural designations. The sharp raises a pitch by a semitone, the flat lowers a pitch by a semitone, and the natural nullifies a sharp or flat to restore a note to normal.

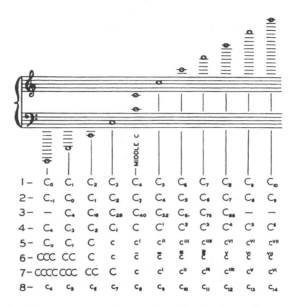

FIGURE 17.6. Eight systems used for tone identification in music (Young, 1939).

to consider C as the point to count whole octaves. Figure 17.7 displays the frequencies of the notes in equally tempered scale in the key of C from 16 Hz to 16 kHz.

17.3   Duration of Musical Notes

The duration of a musical tone is the length of time assigned to it in the musical composition. Figure 17.8 displays the symbols used to indicate duration. While the pitch of a tone is given by its position on the staff, its length is assigned by the choice of one of the symbols of Figure 17.8. However, the magnitude (i.e., its duration) of a tone is not rigidly fixed and it may vary from composition to composition. But nevertheless, in a particular composition the duration of each tone is kept in proportion to the magnitude of a whole note.

In the traditional musical notation a vertical bar is drawn across the staff. The time interval between two vertical bars in a staff is called a *measure*, or the less precise but more commonly used *bar*. The time intervals of all measures within a composition are usually equal. If two whole notes constitute a measure, then an equivalent measure will need four half notes, or eight quarter notes, or any combination that adds up to two whole notes in time interval. A double bar that consists of two vertical bars across the staff denotes the end of a division, movement, or an entire composition.

To indicate periods of silence in a composition, one or more *rest* symbols of Figure 17.9 are used to indicate the duration of the silence. The duration of a whole rest is equal to that of a whole note, the duration of a half rest is equal to that of a half note, and so on.

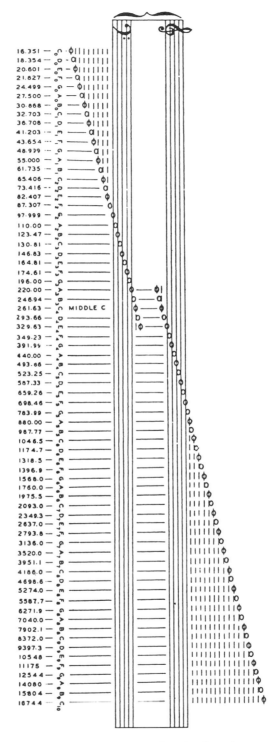

FIGURE 17.7. The frequencies of the notes in the C scale of equal temperament from 16 Hz to 16 kHz.

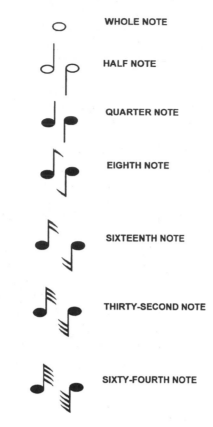

WHOLE NOTE

HALF NOTE

QUARTER NOTE

EIGHTH NOTE

SIXTEENTH NOTE

THIRTY-SECOND NOTE

SIXTY-FOURTH NOTE

FIGURE 17.8. Note values which indicate duration.

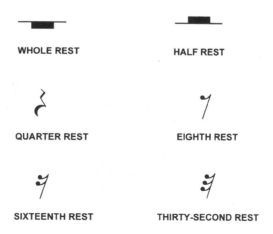

WHOLE REST

HALF REST

QUARTER REST

EIGHTH REST

SIXTEENTH REST

THIRTY-SECOND REST

SIXTY-FOURTH REST

FIGURE 17.9. Rest symbols.

The duration of a tone represented by a note or a rest of a certain denomination can be modified by the addition of a dot to the note. The effect of the dot is to lengthen the duration of the preceding note by half as much, that is, a whole note becomes equal in duration to a whole note plus a half note, a dotted half note equals the duration of a half note plus a quarter note.

There is no absolute time interval standard for the duration of a tone represented by a note, and it generally depends on the performer's interpretation of the music with respect to its tempo. Some compositions carry an indication of the setting of a *metronome* for a quarter note. A metronome is a mechanical device that consists of a pendulum activated by a clock-type of mechanism driven by a spring motor. At the extremities of the pendulum swing an audible tick is produced. The interval between ticks can be adjusted by moving a bob on the pendulum arm: the further the bob is located from the fulcrum, the longer the duration between ticks, and vice versa. The pendulum itself is graduated in ticks per minute. The numbers usually indicate the number of ticks per minute, the interval between ticks usually specified as that of a quarter note or, in some cases, half notes. Modern versions of metronomes use electronic means to generate ticks. The metronome setting defines the rate of movement or tempo of the music. Instead of metronome settings, the composer may specify one of a number of terms to designate tempos. Commonly used terms to describe tempos are as follows:

Largo: Slow tempo
Andante: Moderately slow tempo
Moderato: Moderate tempo
Allegro: Moderately quick tempo
Vivo: Rapid tempo
Presto: Very rapid tempo

17.4   Time Signature Notation

A musical selection's time signature is specified at the beginning of the staff by a fraction, as illustrated in Figure 17.10. Common time signatures include 2/4, 3/4, 4/4, and 6/8. The denominator indicates the unit of measure (i.e., the note used to define a pulse). The numerator stands for the number of these units or their equivalents included in a measure (i.e., the interval between two vertical lines across the staff).

In the upper portion of Figure 17.11 for 2/4 time, each measure contains one half note, or two quarter notes or four eighth notes. Each measure contains two beats, so when a musician plays the count is one, two. In 2/4 time a stressed pulse is followed by a relaxed pulse, a sequence used for marches. In the 3/4 time of Figure 17.11, each measure equals three quarter notes or 1 half note plus a quarter note, and so on. Each measure carries three beats, and usually there is one stressed pulse followed by two relaxed pulses, which yields a time used for waltzes. In the 4/4 (or common) time, each measure contains the equivalent of four quarter notes,

FIGURE 17.10. Time signatures for 2/4, 3/4, C or 4/4, 6/8, and 9/16 time.

with the performer counting one, two, three, four. In 4/4 time, a stressed pulse is followed by three relaxed pulses, and this time is used for dances. In 6/8 time each measure contains six eighth notes or a combination of notes equaling the same duration. There are then six beats to each measure, and in 6/8 time the stressed pulses are one and four of six beats.

Listeners mentally arrange the regular repetition of sounds into groups of stressed and relaxed pulses. These groups are called meters. The meter is assigned by the numerator of the time signature, and the most common are 2, 3, 4, 6, 9, and 12. Each measure contains a certain number of beats or pulses according to the meter. Meters are classified in terms of the numerators of the time signatures in the following manner:

1. *Duple meter*. Two beats comprise each measure, with the first beat stressed and the second beat relaxed. Example signatures 2/2 and 2/4 time.
2. *Triple meter*. Three beats occur in each measure, with the first one stressed and the following two relaxed. Example time signatures are 3/8, 3/4, and 4/8.
3. *Quadruple meter*. Four beats occur in each measure, with the first beat stressed and the remainder relaxed. Occasionally the third beat carries a secondary stress. Examples include 4/2, 4/4, and 4/8 time signatures.
4. *Sextuple meter*. Six beats occur each measure, with the first and fourth beats stressed. The 6/8 time signature is such an example.

Rhythm is the repetition of accents in equal intervals of time.

FIGURE 17.11. Notes and beats for 2/4 and 3/4 times.

17.5   Key Notation

The keynote defines the note with which any given scale begins. The tonic is the keynote of the scale, whether the latter is a major or a minor scale. Many short compositions are written in one key only, but more elaborate musical pieces may shift from one key to another. The key signature of a musical piece is denoted by the number and arrangement of flats and sharps following the clef sign at the beginning of each staff, or it may appear only once at the beginning. Figure 17.12 shows some of the most common key signatures for different major and minor keys.

Major and minor keys play a role in determining the mood of music. In earlier times, a key may have been selected by a composer because a number of wind instruments were able to play only in certain keys. Certain desired effects may be achieved better on more flexible instruments in a specific key than another key.

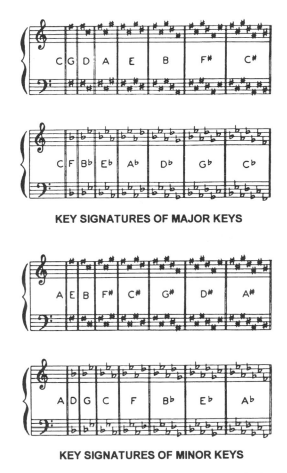

FIGURE 17.12. Key signatures for number of major and minor keys.

As Machlin pointed out, romantic composers developed affinities for certain keys, for example, Mendelsohn preferred E-minor, Chopin leaned toward C-sharp, and Wagner made use of D-flat major for majestic effects.

Whether it starts with C, D, E, or any other tone, a major scale follows the same arrangement of whole and half steps. Such an arrangement is known as a *mode*. All major scales typify the arrangement of whole and half steps.

The minor mode serves as a foil to the major. The principal difference from the major is that its third degree is flattened. For example, in the scale of C, the third degree is E♭ rather than E. In a natural minor scale, the sixth and seventh steps are also flatted (i.e., C-D-E♭-F-G-A♭-B♭-C). The minor differs considerably from the major in coloring and mood. It should not be inferred that the minor is deemed "inferior"—the nomenclature simply refers to the fact that the interval C–E♭ is smaller (hence minor, the Latin word) than the corresponding interval in the major scale.

If a mode is not specified, the major is implied. For example a *Minuet in G* indicates the G-major. The minor is always specified (e.g., Mozart's *Symphony No. 40 in G minor*).

To classical composers the tonal qualities of the minor key assumes a more somber aspect (e.g., the funeral music of Beethoven and Mendelsohn), while the triumphal portions of symphonies and chorales are generally played in major keys. Also, the minor mode carries a certain exotic tinge to Western ears, and thus in the popular view was associated with oriental and Eastern European music. This was reflected in such works as Mozart's *Turkish Rondo*, a number of Hungarian-style works by Schubert, Liszt, and Brahms, the main theme of Rimsky-Korsakov's *Scheherazade*, and other musical pieces that passed for exotica.

17.6   Loudness Notation

Loudness depends on the intensity of the musical signal. While loudness can be measured objectively with the use of a sound level meter, a conductor or a musician depends on his or her own sense of subjectivity to obtain the proper intensity or range of intensities. The common notations and abbreviations for loudness are as follows:

Pianisisssimo (*ppp*): softly as possible
Pianissimo (*pp*): very soft
Piano (*p*): soft
Mezzo piano (*mp*): half soft
Mezzo forte (*mf*): half loud
Forte (*f*): loud
Fortissimo (*ff*): very loud
Fortisissimo (*fff*): extremely loud

Loudness can vary in musical passages. An increase in loudness can be indicated by the term *crescendo* or the abbreviation *cres* or the sign ◁ . A crescendo connotes a gradual increase in the intensity of the music. A decrescendo is a decrease in loudness and is thus the converse of a crescendo. It is denoted by the word *decrescendo* (also *diminuendo*) or the abbreviation *decresc* or the symbol ⟩ .

17.7   Harmony and Discord

Consider two tuning forks being sounded together. Let us keep the pitch of one fork fixed at 261 Hz, while the pitch of the other begins at 262 Hz and is raised gradually. As the pitch rises, beats can be heard, due to the difference in the frequencies, for a time and then can no longer be discerned. The sound of the combined tones starts out by sounding pleasant to the ear and then it becomes gradually more unpleasant. The unpleasantness reaches a maximum at about 23 beats per second, and then begins to abate. This unpleasantness or *discord* declines only slightly, and the discord remains at a fairly uniform level until the octave-marking value of 522 Hz is reached, at which point the unpleasantness disappears.

If this experiment is repeated with violin strings, radically different results will be obtained. The discord does not stay at a uniform level, but fluctuates erratically. It almost vanishes at the interval of a major third, and again at the intervals of the fifth and octave. At the precise points at which the minimums of the unpleasantness occur, the frequency ratios of the variable to the fixed tone are found to have the values: 5/4, 4/3, 3/2, and 2/1.

It has been observed that tones sound well together when the ratio of their frequencies can be expressed in terms of small numbers. The smaller the numbers, the better is the consonance. Table 17.1 lists the intervals in order of increasing dissonance.

The further away from small numbers, the more we encroach into the realm of discord. Pythagoras knew this fact more than 2500 years ago when he associated consonance with the ratios of small numbers. The premise of the Pythagorean doctrine, "all nature consists of harmony arising out of number," may be somewhat

TABLE 17.1. Interval Nomenclature and Frequency Ratios.

Interval	Frequency ratio	Largest Integer Occurring in the Ratio
Unison	1:1	1
Octave	2:1	2
Fifth	3:2	3
Fourth	4:3	4
Major Third	5:4	5
Major Sixth	5:3	5
Minor Third	6:5	6
Minor Sixth	8:5	8
Second	9:8	9

simplistic, but the Chinese philosophers in Confucius's time also regarded small numbers 1, 2, 3, and 4 as the source of all perfection.

The Swiss mathematician Leonhard Euler adopted the psychological approach in declaring that the human mind takes pleasure in law and order, particularly in natural phenomena. His theory of harmony is this: the smaller the numbers required to express the ratio of two frequencies the easier it is to find this law and order, thus making it more pleasant to hear the combined sounds. Euler went as far as to propose a definitive measure of the dissonance of a chord. His idea was to express the frequency ratio of a specific chord by the smallest number possible and then to find the common denominator for these frequencies. For example, the frequency ratio of the common chord CEG c' is 4:5:6:8. The least common denominator is 120, because it is the smallest number of which 4, 5, 6, and 8 are all factors. But this theory falls apart when the same denominator is assigned to the chord of the seventh CEFGB (frequency ratios 8:10:12:15) which turned out to be far more unpleasant to listen to.

17.8   Musical Instruments

Music instruments fall into four categories: string, wind, percussion, and electrical instruments. A string instrument may have its strings struck, bowed, or plucked. Wind instruments can be subclassified as single mechanical reed, double mechanical reed, lip reed, air reed, and vocal-cord reed. Percussion instruments are classified as being either definite pitch or indefinite pitch. The advent of electronics has given rise to a whole new class of instruments, such as synthesizers which can effectively replicate the sounds of conventional strings, winds and percussion instruments as well as generate unusual sounds not heard from any other instruments. Even personal computers can function as musical instruments provided they are equipped with special sound boards and speakers and they are programmed to simulate various types of instruments. Table 17.2 lists a number of musical instruments and their respective classifications.

17.9   Strings

Strings

Possibly the oldest method of creating music is a vibrating string under tension, which is capable of producing a full range of overtones that are harmonics of the fundamental. As was explained in Chapter 4, the presence and the amplitudes of these harmonics depend on the manner by which the string is excited (i.e., by plucking, striking, or bowing) and where the excitation is applied. Because the string projects a small area, it is not an efficient producer of sound as it is not by itself capable of moving very much air. It is for this reason that strings are coupled to a large multiresonant surface (or *soundboard*) to increase the sound output.

TABLE 17.2. Classification of Musical Instruments.

String Instruments

Plucked Strings	Bowed Strings	Struck Strings
Lyre	Violin	Piano
Lute	Viola	Dulcimer
Harp	Violoncello	
Zither	Double bass	
Guitar		
Ukulele		
Mandolin		
Banjo		
Sitar		
Harpsichord		

Wind Instruments

Air Reed	Single Mechanical Reed	Double Mechanical Reed	Lip Reed
Whistle	Free-reed organ	Oboe	Bugle
Flue Organ Pipe	Reed organ pipe	English horn	Trumpet
Recorder	Accordion	Oboe d'amore	Cornet
Flageolet	Harmonica	Bassoon	French horn
Ocarina	Clarinet	Contra bassoon	Trombone
Flute	Bass clarinet	Sarrusophone	Bass trombone
Piccolo	Saxophone (soprano,		Tuba
Fife	alto, tenor, and bass)		
	Bagpipe		

Organ (combination air reed and mechanical reed)

Percussion Instruments

Definite Pitch		Indefinite Pitch	
Tuning fork	Celesta	Snare or side drum	Triangle
Xylophone	Kettledrums (tympani)	Military drum	Steel drum
Marimba	Bell	Bass drum	Cymbals
Chimes	Carillon	Gong	Tambourine
Glockenspiel		Castanets	

Electrical or Electronic Instruments

Siren	Electric piano	Electrical carillon	Computer (specially
Automobile horn	Electric guitar	Synthesizer	configured)
Electric organ	Music Box	Metronome	

All stringed musical instruments make use of a soundboard or a combination of a ported hollow body and a soundboard to couple the string to the air. The larger the radiating surface, the greater the acoustic impedance, and this provides the increased coupling to enhance the sound output of the strings. Soundboards exhibit the complex modes of vibrations that have been described in Chapter 6 which deals with membranes and plates. In a number of string instruments such as the lute, lyre, zither, guitar, mandolin, and the violin family, a hollow body with portholes is coupled to the strings. The vibration imparted by the strings to the body exterior produces radiation in a fashion similar to that of the soundboard. The hollow body with a hole(s) coupled to the outside air comprises a Helmholtz resonator, and the fundamental frequency can be established from the theory of Section 7.11. The dimensions of the resonator may equal or be larger than the wavelength of the sound in air for some of the tones or overtones produced in the instrument, so the hollow body with its holes can manifest other resonant frequencies in addition to the fundamental resonant frequency. A hollow body is a quite complex resonant system.

Some examples of a plucked-string instrument include the lyre, lute, and harp, all of which are illustrated in Figure 17.13.

The lyre traces its origin to ancient Greece, and it consists of a frame, finger board, and a hollow body with sound holes. The lyre is played by plucking the

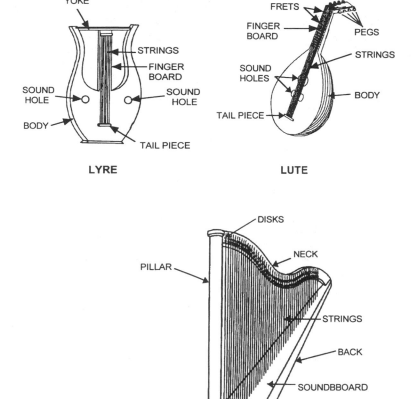

FIGURE 17.13. Plucked string instruments: lyre, lute, and harp.

strings with fingers, and the length of the string (and hence the resonant frequency) is varied by pressing the finger against the finger board. The hollow body serves as a soundboard to increase the sound output of the string.

The lute, developed more than 1000 years ago, is the precursor of the instruments of the guitar class. It is comprised of a pear-shaped hollow body, a neck with frets (projecting ridges across the neck), and a head equipped with pegs to tune the strings.

The zither, similar to the lyre in appearance but having its soundboard underneath the entire length of the strings, consists of two sets of strings stretched across a flat hollow body featuring a large round hole. One set of steel strings which pass over a fretted finger board is used to play the melody while a set of gut strings is used for accompaniment. The modern zither now consists of 32 strings, of which four are assigned to the fretted finger board. Each of the strings can be tuned by turning pegs of pins at one end of the instrument. The plucking of the zither occurs in the following way: A ring-type plectrum is used on the thumb of the right hand to play the melody. The left hand is used to stop the melody strings by pressing the strings against the frets of the finger board. The spacing of the frets are such that the sounds produced by stopping the strings on any two adjacent frets are one semitone apart. The first, second, and third fingers of the right hand play the accompaniment. Zithers are available in three sizes, called bass, bow, and concert types. The open melody strings of the concert zither are tuned to C_3, G_3, D_4, and A_4. The accompaniment strings provide the fundamental-frequency range of C_2 to $A\flat_4$.

The harp, often portrayed as being the instrument of angels, consists of strings stretched vertically upon a triangular frame and connected to a soundboard constituting the lower leg of the triangular frame. The soundboard is quite small, and so the strings are not highly damped and the sound from each string persists for a rather long time, yielding a rather mellow tone. The modern harp is usually provided with seven pedals which actuate a transposing mechanism for shortening the strings in two stages.

The mechanism for shortening the strings is illustrated in Figure 17.14. In panel A of the figure the string is positioned at its maximum length. Depressing the pedal halfway causes disk 1 to rotate along with the pins attached thereto, so the string is shortened by a semitone as shown in panel B. When the pedal is depressed all the way, disk 2 is rotated along with its attached pin, thus shortening the string even further so that the string is a whole tone higher, as shown in panel C. Each pedal operates on strings with notes having the same letter notation, that is, the C pedal controls all the C strings, the D pedal all the D strings, and so on. The harp is generally tuned in the key of C flat. The pillar of the harp serves to bear the stresses produced by the stretched strings and to serve as a housing for the rods connected to the pedals and for the transposing mechanism of Figure 17.14. The overall height of the standard harp is approximately 173 cm. The harp is played either by plucking the strings with the fingers or by sliding the fingers over the strings in a manner referred to as a *glissando*. There are 44 strings, usually made of gut (some bass strings may be constructed of silk overlaid with metal wire for greater density); portable models have as few as 30 strings.

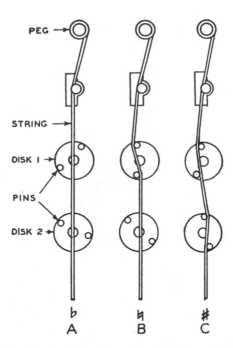

PEG →

STRING →

DISK 1 →

PINS

DISK 2 →

$\flat$ $\natural$ $\sharp$
A   B   C

FIGURE 17.14. Mechanism for shortening strings in the harp for transposing tones.

Figure 17.15 shows views of a ukulele, guitar, mandolin, and banjo. The ukulele is a small version of the guitar. The former consists of four strings, tuned to A_4, D_4, $F_4^{\sharp}$, and B_4, stretched between a combined bridge and tailpiece attached to the top flat surface of the body and the end of a fretted finger board. The body itself consists of two flat surfaces fastened together by a contoured panel at their outside edges. The bottom of the body is attached mechanically to the top by a post, and the cavity of the body, coupled to the external atmosphere, acts as a resonator. The body of the ukulele may be made of wood or steel or plastic. The 61-cm long instrument is played by strumming the strings with the fingers, and the resonant frequency of each string is varied by the fingers pressing against the frets which are spaced so the sounds produced by stopping a string on any two adjacent frets are one semitone apart. The guitar is similarly constructed except it is larger than the ukulele by nearly twice the length, and it consists of six strings, tuned to E_2, A_2, D_3, G_3, B_3, and E_4. The guitar is played by either plucking with fingers or with a pick or plectrum (a flat piece of metal or plastic) held firmly between the thumb and the first finger. The adjacent frets also result in notes that are one semitone apart.

The body of a mandolin consists of a flat top attached to a hollow semiellipsoidal body. This combination of the body and the body cavity coupled to the external air through a hole constitutes a complex resonator. A bridge mounted at the center of the flat surface of the body couples the vibrating strings to the body.

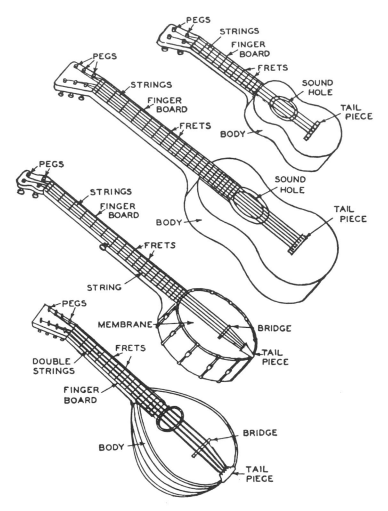

FIGURE 17.15. Additional plucked-string instruments from top to bottom: ukulele, guitar, banjo, and mandolin. (From Olson, 1967.)

The construction of a banjo differs from other string instruments in that its body consists of a skin membrane stretched over one end of a truncated cylinder, thus making the body drumlike. The other end of the cylinder is open. The bridge is supported by the stretched skin and it couples the strings to the stretched membrane, which, in turn, provides a large resonant area that is coupled to the air. Four long strings and a short string are stretched over the bridge between the tailpiece and the finger board. The relatively long neck is fretted; the short string is referred to as the *melody string*. A more modern version called the *tenor banjo* is equipped with four strings of equal length, and it has supplanted the older model with the one short string. The four open strings are tuned to C_3, G_3, D_4, and A_4. This instrument can be played by either plucking the strings with the fingers or with a pick or plectrum, that can be made of a flat piece of tortoise shell. The note of a string and hence the resonant frequency can be varied by pressing it against the frets. The spacing between a pair of adjacent frets correspond to a difference of one semitone. The overall length of a banjo is approximately 86 cm.

Over the past two decades, Europeans and Americans have become familiar with structured musical compositions called *ragas* through concert performances by Ravi Shankar who performed them on the *sitar* which is northern India's predominant string instrument. The sitar's seven main strings are tuned in fourths, fifths, and octaves to approximately $F_3^\#$, $C_2^\#$, $G_2^\#$, $G_3^\#$, $C_3^\#$, $C_4^\#$, and $C_5^\#$. In addition there are 11 sympathetic strings tuned to the notes of the raga. The inharmonicities are quite small; the high curved frets permit the player to execute with vibrato and glissando. The curved bridge allows for both amplitude and frequency modulation from the rolling and sliding of the string.

The harpsichord (or cembalo) and its older cousin, the clavichord, both of which resemble shrunken baby grand pianos in their respective configurations, trace their common origin as far back as the twelfth century. The harpsichord was the mainstay of chamber music during the baroque and classical period until the advent of the more versatile and louder pianoforte, the immediate precursor of the modern piano. Because so many excellent examples survive to this day, the art of constructing harpsichords has been revived so that modern audiences can today enjoy the music that has been composed expressly for this medium. During the twentieth century, a number of excellent performers revived audience interest in the instrument and the works written for it, among them the great Wanda Landowska and in this decade Igor Kipnis (the son of the great Ukrainian operatic basso, Alexander Kipnis).

Figure 17.16(a) illustrates the structure of a harpsichord, which consists of a large number of steel strings stretched over a rather triangular steel frame. The keyboard ranges about $4\frac{1}{2}$ octaves from A_1 to F_6, but different versions of harpsichords have been built to cover both larger and smaller ranges. The strings are excited by being plucked by a key-actuated mechanism shown in Figure 17.16(b). The key is coupled through a level system to a short plectrum of leather, fiber, or tortoise shell which plucks the string that deflects to let the plectrum slip past. When the key is released, the plectrum, which is attached to a short spring-loaded level, slips back under the string. A damping pad also mounted on the jack stops the sound.

Bowed string instruments are played by exciting the strings with a bow. Four modern instruments of this type, in order of increasing size, are the violin, viola, violoncello, and double bass (cf. Table 17.3). Figure 17.17 illustrates the comparative sizes of these instruments. Details and nomenclature of the violin's components (the nomenclature also applies to the other three instruments) are shown in Figure 17.18. The two smaller instruments, the violin and the viola, are generally played with the tailpiece end tucked to one side under the chin. The fingerboard is cradled in the nook formed between the (somewhat) outstretched thumb and the first finger of the left hand, and the fingers of that hand vary the notes of the strings by pressing them against the fingerboard surface. The larger violoncello and the

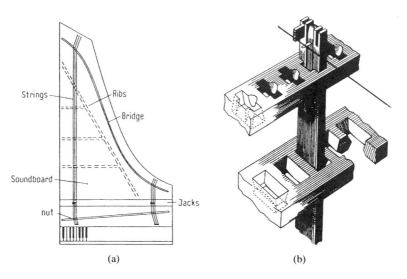

(a)   (b)

FIGURE 17.16. The harpsichord: (a) its construction and (b) the string-plucking mechanism of a harpsichord.

double bass, being much larger instruments, are played held in a tilted, almost vertical position toward the seated players' bodies, with monopods elevating the instruments from the floor.

The bow for any of these four instruments consists of horsetail hair stretched between the two ends of a thin wood, one end constituting the head and the other point of attachment being a movable frog that is connected at the other end of the bow to a screw inside the bow wood (cf. Figure 17.18). The screw can be turned to move the frog, thus adjusting the tension of the stretched horsehair. The horsehair is rubbed with rosin to provide friction between the bow and the strings. When the bow is drawn across the strings, the string vibrates as the result of its being dragged with the bow and then springing back under the impetus of the restoring force. This action of the string's drag and spring-back repeats constantly with the continued sweep of the bow over the string. The driving force occurs as a sawtooth shape, which corresponds to the fundamental and its harmonics. The constant excitation by the bow and the resultant waveform paired with the multiresonant properties of

TABLE 17.3. Typical Characteristics of Standard Bowed-String Instruments.

Bowed-String Instrument	Notes of Tuned Open Strings	Overall Length of Instrument, cm	Overall Length of Bow, cm	Range, Number of Octaves
Violin	G_3, D_4, A_4, E_5	60	75	>4
Viola	C_3, G_3, D_4, A_4	66	75	>4
Violoncello	C_2, G_2, D_3, A_3	69	71	3
Contrabass	E_1, A_1, D_2, G_2	198	66	3

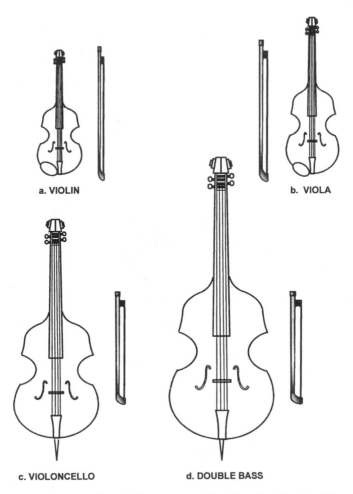

a. VIOLIN   b. VIOLA

c. VIOLONCELLO   d. DOUBLE BASS

FIGURE 17.17. Modern bowed-string instruments: (a) violin, (b) viola, (c) violoncello, and (d) double bass.

the strings yield a sound output rich in harmonics. Within certain limits, the amplitude of the string and its output sound are proportional to the pressure on the bow.

These instruments developed in Italy during the sixteenth and seventeenth centuries required extremely sophisticated skills to construct. Their quality of craftsmanship reached a peak during the eighteenth century in Cremona, Italy, particularly under the skilled hands of Antonio Stradivari (1644–1737) and Guiseppe Guarneri del Gesù (1698–1744). Because of the complexity of their construction which affects the quality of their tones, the violin family has been the object of much acoustical research. Savart, Helmholtz, and the Nobel Laureate C. V. Raman (1888–1970) contributed to the understanding of the generation of sound with these instruments. In more recent times, considerable work has been conducted

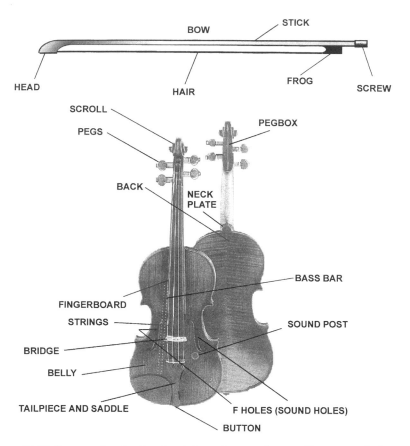

FIGURE 17.18. Construction details of a violin and nomenclature of its structural elements.

in Germany through the efforts of Werner Lottomoser, Jürgen Meyer, and Frieder Eggers. Especially noteworthy is the work of Lothar Cremer (1905–   ) and his colleagues, which culminated in Cremer's classic text *The Physics of Violins*. In the United States, Frederick Saunders (1875–1963), best known for his work in spectroscopy, investigated many violins, making many acoustical comparisons between the old and the new. He, Carleen Hutchins, John Schelleng, and Robert Fryxell established the Catgut Society, an organization (now based in Upper Montclair, NJ) that promotes research on the acoustics of the violin family. This organization was responsible for the development of the violin octet, an ensemble of eight specially scaled new violin family instruments (Hutchins, 1967).

Violin makers (or *luthiers*) are especially concerned with the vibration of the free top and back plates. They test these plates by tapping and listening to tones. Modern technology is now being used to visually observe mode shapes (Figure 17.19) through the use of holographic interferometry. The finished violin's vibrational modes are considerably different from those of the free plates forming the top and bottom of their bodies.

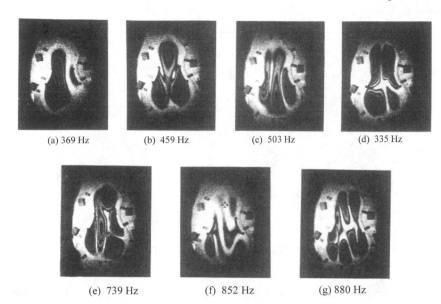

(a) 369 Hz   (b) 459 Hz   (c) 503 Hz   (d) 335 Hz

(e) 739 Hz   (f) 852 Hz   (g) 880 Hz

FIGURE 17.19. Holographic interferograms of a treble viol top plate (Ågren and Stetson, 1972).

The principal modern struck-string instrument is the piano which is available in several models, ranging in size from the more modest upright or spinet piano to the concert grand piano. The heart of the piano consists of a large number of steel strings stretched on a metal frame. The strings couple through a bridge to a large soundboard. The strings are activated by being struck by hammers which are connected to keys forming a keyboard. Pressing down on a key actuates the hammer, which in turn strikes the string. The conventional piano is equipped with 88 keys, and the piano covers a wide frequency range of more than seven octaves, from A_0 to C_8 (27.5 to 4,186 Hz).

Figure 17.20 illustrates a schematic of the piano mechanism for a grand piano. The strings stretch from the pin block across the bridge to the hitch-pin rail at the other end. When a key is pressed downward, the damper rises and the hammer impacts against the string, causing it to vibrate. The string's vibrations are transmitted to the soundboard through the bridge. The hammer rebounds, remaining about $1\frac{1}{4}$ cm from the string as long as key remains pressed. When the key is depressed the damping pad does not engage the string. When the key is released the damping pad engages the string to speed up the decay of the sounding note.

The largest version of the piano, the concert grand, has 243 strings which vary in length from about 200 cm at the bass end to approximately 5 cm at the treble end. In this group there are eight single strings wrapped with one or two layers of wires, five pairs of strings also wrapped, seven sets of 3 wrapped strings, and 68 sets of 3 unwrapped steel strings. The smaller pianos may contain fewer strings, but they still play the same number of notes. A small grand piano may carry 226 strings. The arrangement of the strings is such that the bass strings may overlay the

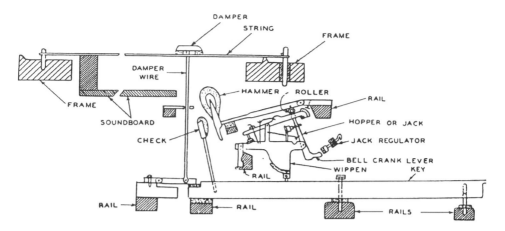

FIGURE 17.20. Schematic of the mechanism of a grand piano. (From Olson, 1967.)

middle strings, so that they can function nearer the soundboard. The soundboard itself is usually made of spruce and it is up to 1 cm thick; it acts as the principal source of radiated sound just as the top plate of a violin does. Because the tension forces in the strings are so high, the frames are fabricated of cast iron, which also provides dimensional stability necessary to maintain the state of tune.

Three pedals are provided on the conventional piano. The right, or *sustaining*, pedal removes all dampers from the strings so that the strings become damped only by the soundboards and end supports. The center pedal, or bass sustaining pedal, removes the dampers from all the bass strings. The left pedal, or the soft pedal, reduces the sound output by lessening the length of the stroke of the hammers or by shifting the hammers so that fewer strings are struck or by permitting the dampers to act.

The dulcimer, considered by some to be the forerunner of the piano, consists of a large number of strings stretched over a frame mounted in an oblong box. These strings pass over bridges which are coupled to a soundboard. The oblong box is mounted on legs, causing the instrument to resemble a square piano without keys. The instrument is played by striking the strings with two hammers, one in each hand. Dampers controlled by a foot pedal are provided. The range of the dulcimer is from D_2 to E_6.

17.10   Wind Instruments

A wind musical instrument is a device that generates sound by (a) blowing a jet stream of air across some type of opening, as in whistles, flutes, piccolos, fifes, and a flue organ pipe; or (b) by buzzing of lips (acting as reeds) in a bugle, French horn, trumpet, tuba, or trombone; or (c) by vibrating a reed (or a double reed) through the means of air flow in an accordion, clarinet, saxophone, oboe, bagpipe, English horn, bassoon, sarrusophone, and the human voice.

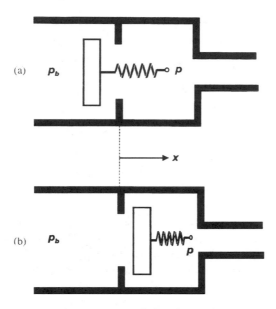

FIGURE 17.21. Two types of vibrating reed generator.

The reed instruments fall into two subcategories: one category entails instruments in which air pressure tends to force the reed valve open (e.g., the human larynx, buzzing lips of brass instruments, and harmonium reeds); the other category includes instruments in which the air pressure forces the reed valve to close (e.g., clarinets, oboes and similar woodwinds, and organ reed pipes). The first category tends to act as a sound generator over a relatively narrow fundamental frequency range, just above the fundamental frequency of the reed. The other category serves as a sound generator over a wider range of frequencies, just below the resonant frequency of the reed—but some type of coupling to a pipe resonator must be provided.

Figure 17.21 illustrates two possible configurations of a vibrating reed generator. The reed generator is really a pressure-controlled device for cutting off and reinstating air flow at selected frequencies. In both cases of Figure 17.21, the blow air pressure p_b (gauge pressure, relative to the atmospheric pressure) is applied from the left. If $p_b > 0$ (i.e., the blow pressure is above atmospheric pressure) the valve in (a) is forced closed by the positive pressure and the valve in (b) is forced open. The opposite situation occurs when $p_b < 0$. In general terms, the motion of a reed can be described by (Fletcher and Rossing, pp. 350–355):

$$m_r\left[\frac{d^2x}{dt^2} + \frac{1}{Q_r}\omega_r\frac{dx}{dt} + \omega_r^2(x - x_0)\right] = \gamma_{gr}(p_b - p) + \gamma_{be}\frac{U^2}{|x|^2}$$

where

$$m_r = \text{mass of reed}$$

$$Q_r = \text{quality factor of reed resonance}$$

x = position of valve relative to its seat

x_0 = equilibrium position of valve

ω_r = angular frequency of reed

p_b = blow pressure

p = acoustic pressure

γ_{g_r} = geometric factor for the exposed reed faces

γ_{be} = Bernoulli force factor based on internal flow
in the narrow part of valve gap

U = flow velocity

In the *air-reed instruments*, a steady stream of air does the activation by flipping in and out of the pipe or cavity at the resonant frequency of the system, as shown in Figure 17.22. This steady stream of air thus becomes an alternating flow, and because of the nonlinear nature of the exciting force, a number of the resonant elements in the system are excited, thus yielding a series of overtones that add to the fundamental tone. The whistle and flue organ pipe each consists of a cavity, a closed or open pipe coupled to an air reed which is activated by a steady air stream. The air stream entering the pipe vacillates between moving to the inside of the pipe and moving to outside of the pipe. Upon entering the pipe, the air stream compresses the air in front of it, as depicted in Figure 17.22 at $\theta = 0°$. The pressure inside the pipe builds up to the equilibrium point, and no more air will

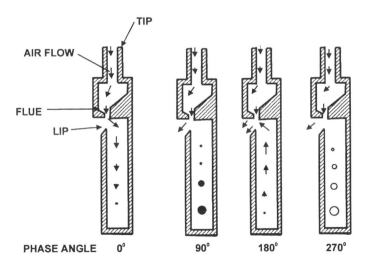

FIGURE 17.22. The mechanism of the flue pipe organ, an air reed instrument. The magnitude of the pressures are indicated by the diameter of the circles. A dark circle denotes a pressure above atmospheric, and a white circle indicates a pressure below atmospheric. The direction of the arrows represents the direction of air flow, and the magnitude of particle velocities are indicated by the length of arrows.

then enter the pipe, as indicated in Figure 17.22 at $\theta = 90°$. The excess pressure will now direct the incoming air stream to the outside at $\theta = 180°$. This results in the excess pressure being relieved and a rarefaction owing to the inertia of the outgoing air, as shown for $\theta = 270°$.

The decreased pressure pulls in the air to renew the cycle at $0°$. This cycle, consisting of the four phases, repeats itself at the resonant frequency of the system. The frequency of the complete cycle occurs at the resonant frequency of the closed pipe. Odd harmonics are also produced, because the closed pipe resonates at these frequencies as well as the fundamental; this action steers the airstream vacillating between the interior and the exterior. In the case of the open pipe, both odd and even harmonics are produced. Sectional views of the open-flue pipe (made of metal) and a stopped-flue pipe are shown in Figure 17.23. A whistle, such as that used by police officers and referees of athletic contests, operates on the same principal as the organ flue pipe except that the resonating chamber is a Helmholtz resonator, which is essentially a chamber with a narrow neck. The resonance of a whistle can be found from application of equation (7.66), on the basis of the volume of the

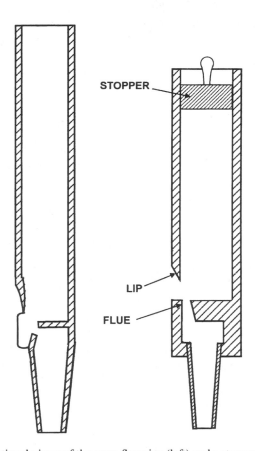

FIGURE 17.23. Sectional views of the open flue pipe (left) and a stopped-flue pipe (right).

whistle chamber and the area of the sound radiating hole coupling the chamber to the outside air. The overtones due to resonances within the small chamber occur at relatively high frequencies and are effectively suppressed by the inertance of the sound-radiating hole, thus resulting in a nearly pure tone of the whistle. A calliope is a group of whistles with frequencies corresponding to the notes of a musical scale. Each whistle is controlled by a valve connected to a key, which is part of a keyboard similar to that of a piano. Either steam or compressed air is used to actuate the whistles.

The recorder and the flageolet can be categorized as instruments of the whistle class. Each is equipped with a mouthpiece, a fipple hole, and a cylindrical tube bearing a set of finger holes. The recorder has eight finger holes plus one thumb hole on the opposite side of the tube to alter the resonant frequency of the air column. Different types of recorders have been constructed ranging in length from 30 cm to almost 900 cm. The flageolet features a set of four finger holes plus two thumb holes. The ocarina's construction differs from that of a flageolet in that its resonating system is a cavity-and-hole combination, not a pipe. The resonator is coupled to the air through several holes, including the fipple hole. The resonance frequency is increased as the number of holes is increased, because the inertance decreases with the number of holes. Ocarinas, which may be made of metal, ceramic or plastic, cover about a range of an octave and a half.

The flute illustrated in Figure 17.24 consists of two cylindrical segments joined by a conical tube. One end of the flute is open and the other is closed. The embouchure (blowhole) is located a short distance from the closed end. The holes are controlled by closing or opening them in order to vary the resonant frequencies corresponding to the musical scale, either by the fingers directly or through actuation of keys. This system of keys and connecting shafts and levels constitute a mechanism that actuates valves which are basically disks that cover finger holes. Springs are utilized to keep the valves in the unactuated position, which may be either open or closed. This system renders it possible to open or close finger holes that are too far apart or are too large to be stopped by the fingers alone.

The sound of the flute is generated in the following manner: in Figure 17.25 an air stream from the lips impinges upon the embouchure of the flute. Resonant frequencies are generated by the air stream slipping back and forth between entering the flute body and flowing past the embouchure. A stream of air enters the

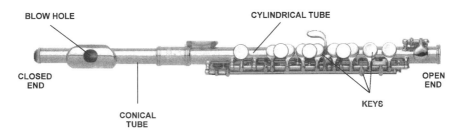

FIGURE 17.24. The modern flute.

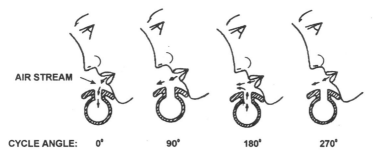

AIR STREAM

CYCLE ANGLE: 0°   90°   180°   270°

FIGURE 17.25. The action of the flute. The direction of the arrows indicate air flow direction.

blowhole ($\vartheta = 0°$), causing a pressure to build up to the extent that it stops air from entering ($\vartheta = 90°$). The excess pressure ($\vartheta = 180°$) then forces the air out of the blowhole until it stops leaving the blowhole ($\vartheta = 270°$). The cycle then repeats itself.

The piccolo can be considered a smaller version of the flute that operates one octave higher. The operational principle of the piccolo is fundamentally the same as that of the flute, but the fundamental range is from D_5 to B_7 instead of the range from C_4 to C_7 of the latter instrument. The fife is a somewhat simpler instrument, consisting of a tube of metal or wood that is closed at one end and open at the other. It is equipped with six finger holes distributed along the tube and a blowhole near the closed end.

Mechanical reed instruments are those that use a steady air stream to actuate a mechanical reed to throttle the air flow at the resonant frequency of the reed and the associated acoustical system. These instruments may be subdivided into two categories: single- and double-reed instruments.

Single-reed instruments use a single mechanical reed to throttle a steady air stream to generate a musical sound. A free-reed instrument is one that radiates its sound directly into the air. Modern examples of the free-reed instrument are the free-reed organ (harmonium), accordion, and harmonica. In other instruments (e.g., reed organ pipe, clarinet, saxophone or bagpipe), the reed couples to a resonant air column. The action of the reed is that of a free-end (cantilever) vibrating bar described in Chapter 5. The mechanical reed in its quiescent state is positioned so that it forms an opening with the structure of the instrument as shown in Figure 17.26. The velocity of the air moving past the reed through the opening results in a lower pressure on the flow side, due to the Bernoulli effect,[1] which causes the reed to flex toward the lower pressure, causing the opening to become

---

[1] Bernoulli's principle states that the dynamic pressure of a nonviscous fluid flow will drop with an increase in the velocity of the flow and vice-versa according to the relation

$$p + \frac{1}{2}\rho u^2 = \text{const}$$

where p is the air pressure, ρ is the air density, and u is the air velocity.

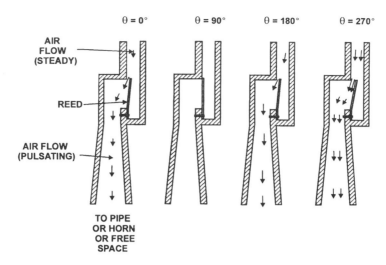

FIGURE 17.26. Reed position and particle velocities (indicated by arrows) in a mechanical reed instrument for a complete cycle. (After Olson, 1967.)

smaller. The air flow then becomes reduced by the constricted passage, which causes the pressure on the flow side to increase, and the reed springs back to its original position. It then moves beyond its original position, under the effect of inertial energy, and the air flow becomes larger. The pressure again drops and the reed returns to the normal position it had at the beginning, and the cycle of events repeats at the resonant frequency of the system. This action has the effect of converting a steady air stream into a saw-toothed pulsation that contains the fundamental and its harmonics.

The harmonium or the free-reed organ operates through a series of air-actuated free reeds tuned to specific notes. The air supply driving the reeds is provided by two pedal operated pedals connected to bellows that connects to a wind chest on the top of which reeds are mounted. This type of organ features a pianolike keyboard. Each key operates a valve that controls the air supply to the reed, with one key for each reed. Reed organs are usually equipped with stops for connecting banks of reeds. Thus, a number of reeds can be activated by a single key.

The accordion shown in Figure 17.27 functions by the player's arms working a bellows which provides the air supply to the reeds, alternatively creating pressure (when the bellows is compressed) and a partial vacuum (when the bellows is expanded). The air-actuated reeds, tuned to notes of a musical scale, are controlled through a keyboard. Each key connects to a valve that controls two separate reeds. One reed operates under pressure and the other under a vacuum. In some versions of the instrument, two differently tuned reeds are assigned to each key that one tone is produced with expansion of the bellows and another tone with the compression. The vibrational action of the reed is that of a free-end bar, Accordions, which are really portable versions of reed organs, are usually equipped with stops to connect

FIGURE 17.27. Two accordion versions: (a) button keyboard accordion and (b) piano keyboard accordion. (Photographs of the Hohner Corona II diatonic accordion and Hohner Atlantic IV piano keyboard accordion courtesy of Hohner, Inc.)

individual reeds so as to form banks of reeds. This enables operation of several reeds from a single key. In playing the accordion, the performer uses the left hand to move the bellows and to play the bass parts and the accompaniment and uses the right hand to play the melody.

The simple organ shown in Figure 17.27(a) contains ten melody buttons or keys and two bass keys. Each key produces two different tones, one tone upon bellows expansion and the other upon bellows compression. The piano accordion of Figure 17.27(b) is equipped with a two-octave piano-style keyboard (with black and white keys) and twelve bass and chord buttons. The same tone is produced on both compression and expansion cycles. There is no standard size for accordions. More elaborate accordions have been constructed with melody keyboards covering up to four octaves, and some of the larger accordions contain as many as 120 bass buttons arranged in six rows rather than the two rows of six buttons each as shown in the figure. The first and the second rows provide the bass notes while the other four rows produce, in the following order, the major, minor, dominant-seventh, and diminished-seventh chords. A medium-sized accordion with a piano-style keyboard can cover in its melody section a frequency range from F_3 to A_6.

The mouth organ or harmonica, which comes in a number of versions, is played by a performer's breath providing the air supply through both exhalation and inhalation. The instrument consists of a set of tuned free reeds mounted on a wood, metal, or plastic box, with channels leading from orifices to the reeds along one side of the box. Each channel connects to two differently tuned reeds, one reed operating under pressure (exhalation) and the other under a partial vacuum (inhalation). Each reed behaves as a cantilever bar in operation. To play a melody the mouth covers one or more holes, with the tip of the tongue providing some additional control. Harmonicas fall into three principal categories: the simple harmonica (a single row of 10 holes, 20 different notes), the concert harmonica (two rows, one row tuned an octave higher than the other), and the chromatic harmonica. The chromatic harmonica, while similar in appearance to the concert type, is really a set of two harmonicas, one placed above the other. The bottom instrument is tuned one semitone above the upper instrument. This instrument contains a slide which is moved by a button to the right. When the slide is in, the upper holes are closed and the lower holes are open; the converse occurs when the slide is out. A fairly typical simple 10-hole harmonica can cover $2\frac{1}{2}$ octaves from C_3 to F_5.

The reed organ pipe contains an air-actuated reed coupled to a conical pipe, with the fundamental frequencies of the reed and the pipe itself fairly matching. In some installations, a combination of a conical pipe and cylindrical pipe is used. Because the pipe and the reed are intimately coupled, the resonant frequency is established by the combination of these two components. The resonant frequency can be changed by altering the resonant frequency of either the reed or the pipe. A reed organ pipe is customarily tuned by changing the effective stiffness of the reed by moving a tuning spring that is in contact with the reed. A cross-sectional view of the organ pipe mechanism is shown in Figure 17.28. The organ may also be tuned by changing the effective length of the pipe by a metal rollback in the side of the pipe near the open end. Both the shape of the reed and that of the pipe determine the timbre. *Voicing* is the process of selecting and adjusting these components to yield the proper timbre. Reed organ pipes are generally designed to imitate orchestral instruments such as the trumpet, tuba, oboe, clarinet, the human voice, and so on. These pipes are classified as chorus and orchestral reeds.

The clarinet consists of a single mechanical reed coupled to a cylindrical tube with a flared open end. Control over the effective length of the resonating air column is provided by a number of holes which may be opened or closed by fingers, either directly or through keys. Any change in the effective length of

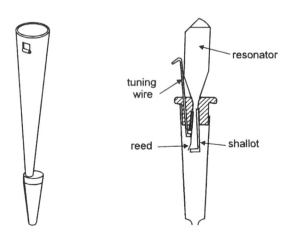

PERSPECTIVE VIEW OF AN
INSTRUMENTAL REED PIPE

MECHANISM OF A TYPICAL
VOX HUMANA REED PIPE

FIGURE 17.28. Reed organ pipe and reed mechanism.

the air column results in a change of the resonant frequency. The reed position in the bottom of the mouthpiece functions in the manner described above for Figure 17.26. The throttling action of the reed changes a steady air stream into a saw-toothed pulsation which contains the fundamental and its harmonics. When sounded alone, the reed produces a sound rich in harmonics. The quality of the tone is improved when the reed is coupled to the cylinder air column. The clarinet produces a range covering more than three octaves, from D_3 to F_6. The clarinet's overall length is about 63 cm. The bass clarinet is a larger instrument, being 94 cm in overall length, that produces tones in a lower frequency range, from D_2 to F_5. This lower frequency range is obtained by doubling the tube on itself.

The saxophone operates in a similar fashion to the clarinet. The principal difference in its construction is that the diameter of the tube at the reed end is larger, thus resulting in a lower acoustic impedance. Also, the coupling between the reed and the pipe is not as intimate as in the case of the clarinet. The buildup of the reed vibration is extremely quick, producing a sharp attack that is characteristic of the saxophone. Saxophones come in various sizes, including soprano, alto, tenor, baritone, and bass. The smallest of these, the soprano saxophone, employs a straight tube with a slightly flared open end; all the other saxophones use a curved mouthpiece and an upturned bell at the other end. Each of these types of saxophones covers a fundamental range of about two and a half octaves. The soprano produces Ab_3 to Eb_6; the alto from Db_3 to Ab_5; the tenor from Ab_2 to Eb_5; the baritone from Db_2 to Ab_4; and the bass from Ab_4 to Db_4. Their overall lengths range from 40 cm for the soprano version to nearly 100 cm for the bass saxophone.

The bagpipe of Figure 17.29, the instrument so symbolically evocative of Scotland, contains one or more combinations of air-actuated reeds coupled to a resonating pipe. A leather bag serves as a reservoir and air supply for actuating

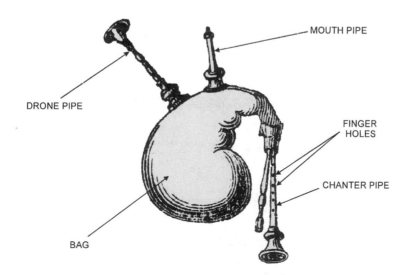

MOUTH PIPE

DRONE PIPE

FINGER HOLES

CHANTER PIPE

BAG

FIGURE 17.29. The bagpipe.

the reeds. Air is supplied to the bag by blowing with the breath. Because the reeds are supplied by a steady stream of air, the reeds sound continuously without interruption, distinguishing the bagpipe from other breath-blown instruments which do not provide such steady sounds. There are usually two or three fixed-frequency reed-pipe combinations which are called *drones*. In addition there is a variable pitch reed-pipe combination, or the *chanter*. On the chanter pipe, eight finger holes are provided so that discrete frequencies can be achieved over the range of an octave. The chanter supplies the melody, and the drones produce a harmonious steady tone. The reed and pipe mechanism is similar to that of the reed-organ pipe. However, in the modern version of the bagpipe, the reeds in the drone are of the single mechanical type and the reed in the chanter is a double mechanical type.

Double mechanical-reed instruments use two mechanical reeds for throttling a steady stream of air to produce a musical sound. All of these instruments couple the double mechanical reeds to a resonant air column. Such instruments include the oboe, English horn, oboe d'amore, bassoon, and sarrusophone.

Figure 17.30 depicts the operational principal of the double reed. When the reeds are in the normal position, that is, they are slightly apart, air is forced through the opening between the reeds. The high velocity of the air reduces the pressure between the reeds, in accordance with the effect of Bernoulli's principle. This causes the two reeds to be forced closer to each other, thereby constricting the air flow. With the air flow now reduced by the constriction, the pressure increases, causing the reeds to spring back toward their original position, but owing to their momentum they move beyond their original positions. The opening is now at its largest, and the air flow is increased accordingly. The internal pressure on these reeds are now quite small, so they return to their original position, and the cycle begins again with this sequence of events repeating at the

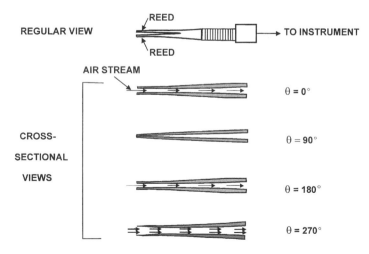

FIGURE 17.30. The action of a double reed. Air particle velocities are indicated by arrows.

rate of the resonant frequency. Thus, a steady air stream undergoes a throttling action that generates a sawtooth signal that contains the fundamental and all of its harmonics.

The oboe consists of a double mechanical reed such as that shown in Figure 17.30, coupled to a conical tube with a slight flared mouth. The effective length of the resonating air column is controlled by the number of holes that are opened or closed by the fingers either directly or through keys. The oboe covers three octaves from $B\flat$ to G_6; the overall length of this instrument is 62 cm. The English horn resembles the oboe in most respects, especially in the key and fingering system, with the principal difference being that the double mechanical reed is coupled to a tapered conical tube that terminates in a hollow spherical bulb with a relatively small mouth opening, thus producing a unique timbre. The fundamental frequency range of this 90-cm long instrument covers less than three octaves, from E_3 to $B\flat_5$. Approximately 70 cm long, the oboe d'amore is a smaller twin of the English horn, and it ranges over nearly three octaves from $G\#_3$ to $C\#_6$.

The bassoon is a noble-sounding instrument that consists of a double mechanical reed coupled to a conical tube that is doubled back on itself so that a lower frequency range is provided without compromising the portability of the instrument itself. There is no appreciable flare at the mouth. A set of holes on the side can be opened or closed by fingers either directly or through the use of keys to determine the effective length of the resonating air column. The overall length of the bassoon is about 123 cm, but the doubled conical air column is about 245 cm long. The bassoon covers roughly three octaves from $E\flat_1$ to $E\flat_3$. The contra bassoon can be considered the bigger brother of the bassoon, but its tube is folded several times to yield an air column as long as about 480 cm while keeping the overall length of the instrument to a manageable 127 cm. The fundamental frequency range runs from B_0 to F_3.

The sarrusophone, essentially a double mechanical reed coupled to a folded brass tube of conical bore with a flare at the open end, comes in several sizes covering different fundamental ranges. The effective length of the air column is varied by a number of holes that are opened and closed by cover valves operated by keys. The most common type of sarrusophone is the contrabass type, whose fundamental frequency ranges from $D\flat_1$ to B_3.

The modern organ illustrated in Figure 17.31 is really more than one type of instrument. It is considered to be a combination mechanical-reed and air-reed instrument. It consists of a large number of flue- and reed-type pipes controlled directly by manual and pedal keyboards and less directly by stops, couplers, and pistons. The organ console in Figure 17.31 is shown to contain three manuals, a pedal keyboard, tablet couplers, thumb and toe pistons, and swell pedals. Organs and organ consoles, it should be mentioned here, are not standardized. The organ may be constructed with one or more manuals and with or without a pedal keyboard. In a five-manual organ, the functions of the manuals are as follows: The first (or the lowest) controls the pipes in the choir organ; the second controls the pipes of the great organ; the third controls the pipes of the swell organ; the fourth controls the pipes of the solo organ; and the fifth controls the pipes in the echo organ. The

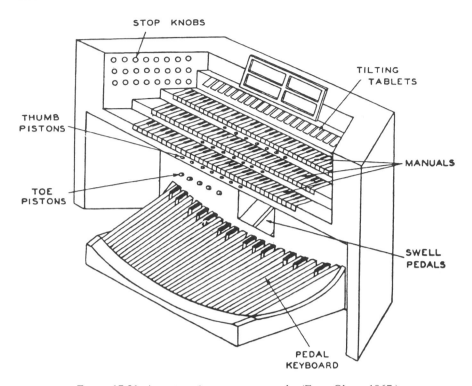

FIGURE 17.31. A contemporary organ console. (From Olson, 1967.)

pipes in the bass organ are controlled by the pedals. In a number of more elaborate installations, other instruments have been added to pipes (e.g., gongs, cymbals, bells, chimes, drums, etc.). Yet other installations contain as many as seven manuals. The typical pipe organ in use today has two to five manual keyboards and a pedal keyboard.

It can be seen from the schematic of Figure 17.32 that the organ is a combination of several organs, each organ having a different set of pipes. We have earlier described the flue organ pipe and the reed organ pipe. Pipes can be classified into basic groups: the diapasons, flutes, strings, and reeds. The more common flue and reed pipes are illustrated in Figure 17.33.

In the evolution of the organ, it was found that the tones could be improved by grouping a set of pipes to sound simultaneously by depressing a single key. Thus each key can represent any number of pipes, and each combination of pipes is called a *stop*, a term also applied to a knob along the sides of the console. Each stop knob controls a specific group of pipes.

The pedal organ features the largest flue and reed stops. The great organ, the choir organ, and the swell organ feature decreasing steps of flue and reed stops. The swell organ is placed in an enclosed space equipped with shutters arranged in a plane between the wall and the audience. The opening and closing of these

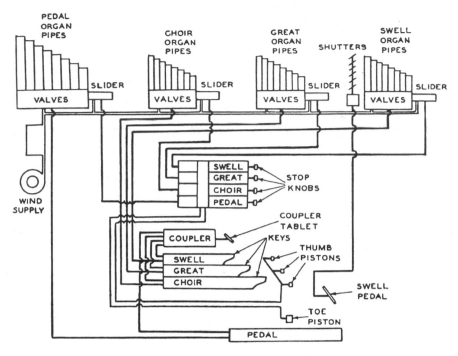

FIGURE 17.32. Schematic of the elements of the modern organ. (From Olson, 1967.)

shutters are controlled by the swell pedals in the organ console. The solo organ, if there is one, is also enclosed in a swell box.

Just above the top manual, a row of tilting tablets serve as couplers. This coupling system allows the actuation of a mechanism associated with more than one key by simply pressing one key. The different manuals can be interconnected by this

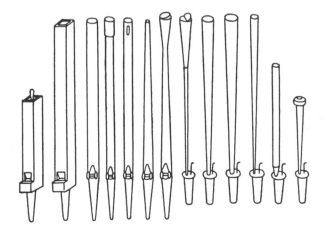

FIGURE 17.33. Different types of organ pipes. (From Olson, 1967.)

coupling arrangement; and the pedal keyboard, for example, can be tied to any manual by this coupling procedure.

In the early organs, keys were connected directly to valves that controlled the air flow to the pipes.. This required considerable strength on the organist's part to press the keys. The action was improved later by the introduction of a pneumatic system, in which the key operates a small pneumatic valve requiring a lighter force, which in turn operates a valve connected with the pipes. But this resulted in slow action owing to the slowness of the propagation rate of air impulses from the key to the valves at the organ pipes. As a result the console had to be located very near the pipes. Except for small organs, modern organs now use either electropneumatic or electric action to actuate the pipes. In the electropneumatic action, pressing down on the key energizes an electromagnet to move a set of valves in an air chamber, causing bellows to expand and move a linkage that opens up a larger valve which in turn lets in the air supply into the wind chest to activate the organ pipes. The advantages of the electropneumatic system include the following: a light force is required to press the key, the action is swift, the keys can be interconnected by merely flipping a switch, and the console can be located almost any distance from the organ and even moved about, because the wiring connecting the console to the different organs can be grouped together inside a reasonably small, flexible cable. The all-electrical action essentially constitutes a servomechanism setup, where the key activates a relay that opens the wind chest to the organ pipes.

Because of the large size and number of pipes, organs require large amounts of air for actuating the pipes. The air pressure required is quite small, ranging from about 1 kPa gauge pressure for the early hand-powered organs to as much as 12 kPa air gauge pressure in some modern units. The air supply for modern organs is provided by an electric fan or an electrically driven centrifugal pump.

Lip-Reed Instruments

In these types of instruments, the lips serve as the reeds and the air supply is provided by the lungs. The puckered upper and lower lips can be imagined as a pair of double mechanical reeds of Figure 17.30, with the operational principle being virtually the same. The frequency of the pulses, which occur as the result of the lips throttling the air flow, correspond to the resonant frequency of the lips combined with the associated instrument. Because the coupling between the lips and the instrument's mouthpiece is fairly loose, the fundamental resonant frequency of the lips must correspond to that of the instrument. However, the combined resonant frequency of the horn and the lips can be varied slightly up or down by changing the tension of the lips, thus changing the resonant frequency of the system. The modern lip-reed instruments are the bugle, trumpet, cornet, trombone. French horn, and tuba—collectively referred to as the *brasses*.

This simplest of the brass instruments is the bugle, which consists of a cupped mouthpiece attached to a coiled tube with a low rate of taper, terminating in a bell-shaped mouth. The length of the air column is fixed; there are no valves, and the number of notes that can be played is limited. The overall length of the standard bugle is about 57 cm, and the notes it can play are C_4, G_4, C_5, E_5, G_5, Bb_5, and C_6.

As it is equipped with three sets of keys that operate piston valves, the trumpet is a more versatile instrument than the bugle. The trumpet is constructed of a coiled tube almost 3 m in length with a slight taper, terminating in a bell-shaped mouth. The first third of the tube is almost circular, with the remainder slightly conical except for the last 30 cm which flares into a bell-shaped mouth. Pushing down on the keys of the piston valves adds to the length of the tube. With three valves closed and opening them singly, opening in three different combination of pairs, or with all three valves opened, it becomes possible to obtain eight different lengths of the resonating tube. A tuning slide or bit is provided so that the resonant frequencies can be matched with other instruments. The trumpet, which has an overall length of 57 cm, covers about three octaves, from E_3 to Bb_5. A mute in the form of a pear-shaped piece of metal or plastic can be inserted into the bell in order to change the quality of the tone and attenuate the output.

Only about 36 cm long, the cornet is a smaller version of the trumpet. In addition to size, the cornet has a bore that is cylindrical rather than tapered. It covers the frequency range from G_3 to Bb_5. The French horn, arguably the most beautiful brass instrument, consists of a mouthpiece coupled to a slightly tapered coiled tube of 365 cm length, which terminates in a large bell-shaped mouth. The modern version of the horn is equipped with three sets of rotary valves controlled by three keys. This valve system provides eight different lengths for the resonating tube, and hence a series of different resonant frequencies corresponding to the notes of the musical scale. The large size of the mouth of the French horn renders it possible for the player to inert his hand and raise or lower the pitch or to produce a muted effect. The French horn can sound over three octaves, from B_1 to F_5. Its overall length is nearly 58 cm.

The trombone and the bass trombone differ from the other brass instruments in that they each feature a telescoping section of the tube that can be moved by the performer to vary the length of the tube and hence the resonant frequency. The trombones are the only instruments beside the violin family that can provide a continuous glide through the musical scale. The pitch is determined by the position of the slide as well as the performer's lips and the lung pressure applied. As with the case of performers of the violin family, the player must possess an accurate sense of pitch and the ability to produce the correct note. The trombone's U-shaped tube is nearly 3 m long, with a slight conical taper that culminates in a bell-shaped mouth. It is capable of covering nearly $2\frac{1}{2}$ octaves, from E_2 to B_4. The bass trombone is larger, covering three octaves from A_1 to Gb_4.

The tuba's mouthpiece is attached to a coiled tube nearly 6 m in length that gradually increases in its cross-sectional area until nearly the end when it terminates in a large bell-shaped bell. Three piston valves are provided, each adding a different length to the tubing. Thus, eight different lengths are provided, in the same basic manner as the trumpet or cornet. Some versions of the tuba include a fourth valve, with a corresponding increase in the number of different resonant frequencies. Largest of the brass instruments, a typical tuba measures approximately 1 m. The tuba covers three octaves from F_1 to F_4. Different versions of the tuba exist, with a variety of sizes and forms. The sousaphone (named after band conductor/composer John Sousa of "March King" fame) is the largest of the tuba family.

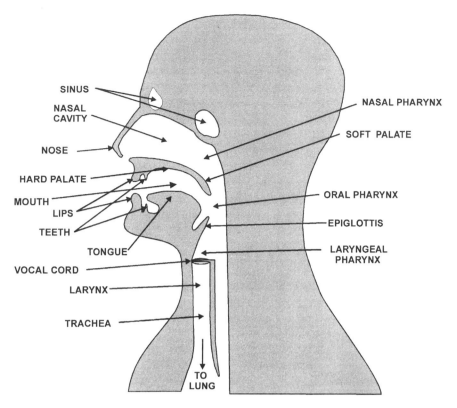

FIGURE 17.34. The human voice mechanism in a sectional view of the head.

In concluding this section on wind instruments, it might also be mentioned that the human voice also classifies as a wind instrument, that is, it is a reed instrument in which the vocal cords serve as the reed. The mechanism of the voice consists of three sections: the lungs and associated muscles to serve as the air supply, the larynx bearing the vocal cords for converting the air flow into a periodic modulation, and the vocal cavities of the pharynx, mouth, and nose which vary the tonal content of the output of the larynx.

A sectional view of the voice system in the human head is shown in Figure 17.34. The frequency of the vibration is governed by the tension of the vocal cords, the inertance, and the combined acoustical impedance of the vocal cords and the vocal cavities. The vocal mechanism is a complex one, entailing a number of acoustical elements that can be varied by the singer to yield a wide variety of tones, that differ in harmonic content, quality, loudness, duration, growth, and decay. According to Machlis,

the oldest and still the most popular of all (musical) instruments is the human voice. In no other instrument is contact between the performer and medium so intimate. In none other is expression so personal and direct. The voice is the ideal exponent of lyric melody and has consistently been the model for those who made instruments as well as for those who played them.

A human voice can range over two octaves, but variations do occur among different individuals as to the range, to say nothing of the beauty or quality of tones. It is a rare voice such as Yma Sumac's that can cover four octaves or more. An effortless coloratura such as Joan Sutherland, a powerful *heldentenor* such as Lauritz Melchior, or a sonorous basso such as Alexander Kipnis or Martti Tevala comes along once a generation. On the average, a soprano can range from C_4 to C_6; an alto, from G_3 to F_5; a tenor, from D_3 to C_5; a baritone, from A_2 to G_4; and the bass, from E_2 to D_4.

17.11   Percussion Instruments

Percussion instruments are made to sound by striking or shaking. The vibration system that is excited by an impact may be a bar, rod, plate, membrane, or a bell. Two classes of percussion instruments exist, viz. definite pitch, in which the tone has a identifiable pitch or fundamental frequency, and the indefinite pitch, in which the musical sound does not have a definite pitch or fundamental frequency. The former category includes the tuning fork, xylophone, celesta, glockenspiel, bells, and chimes. Drums, the triangle, tambourine, castanets, and cymbals fall into the latter category.

The tuning fork, discussed in Chapter 5 is used as a standard frequency sound source. The resonant frequency of a tuning fork is determined by its dimensions and its material.

The xylophone consists of a number of metal or wood bars, having specific resonant frequencies, mounted horizontally and supported on soft material at two nodal points. The vibration of each bar is that of a free bar, with the resonant frequency dependent on the dimensions of the bar. A pipe acts as a resonator by being coupled to each bar to improve the coupling between the bar and the air, thus yielding greater sound output. Felt-covered hammers are used to strike the bars. Differential models of xylophones are available to provide fundamental frequency ranges of two to four octaves, generally from C_3 to E_7. The marimba is basically an enlarged version of the xylophone, which has African and South American roots, that usually covers about five octaves, from F_2 to F_7.

The glockenspiel (also called orchestral bells) resembles a small xylophone, but small wooden hammers are used to strike the notes. The bell lyre is another version of the glockenspiel. A set of steel bars is mounted on a lyre-shaped frame. A ball-shaped hammer is used to strike the bars. The bell lyre is used in marching bands as it is small enough be held almost vertically by one hand while the other hand holds the hammer. The glockenspiel can be constructed to cover up to three octaves. The frequency range usually falls between C_3 and C_6.

The celesta contains a series of resonant steel bars with specific resonant frequencies. The bars are actuated by hammers linked to keys forming a keyboard.

Thus, the celesta resembles a small upright piano. The steel bars are suspended above wood resonating boxes, which are designed to improve coupling of the bars to the air, thus improving the sound output. Dampers are provided, and a single pedal, called a *sustaining pedal*, removes the dampers so that the energy of the vibrating bar may be dissipated over a longer period. The celesta is a four-octave instrument that ranges from C_4 to C_8.

Chimes or tubular bells are a series of resonant brass tubes suspended vertically from a wooden frame, and a damping bar is controlled by a foot pedal. The tubular bells, whose resonant frequencies correspond to the notes of the musical scale and are played by being struck by a wooden mallet, range up to two octaves, generally between G_1 and G_3. The tubular bells are intended as substitute for real bells. Bells themselves also constitute musical instruments. They are actuated through side-to-side movements by clappers which hang loosely inside the bell. The fundamental frequency of a bell depends on the geometry of the bell, the wall thickness, and the density and moduli of the metal. A *carillon* is a set of fixed bells tuned to a musical scale. A number of arrangements have been devised to play the bells: the bells may be struck directly by hammers held in the hand of the carillonneur, or they may be struck by a clapper through linkages to a keyboard which is played by a carillonneur. With a mechanical setup, fists are used with considerable force to operate the keys. In electrified carillons, the clappers are activated by solenoids activated by the keys of a keyboard. In this setup the force required to move the keys are greatly decreased to a level no greater than that for playing a piano or an organ. A carillon usually covers the range of three or more octaves.

The kettledrum or *timpani* consists of a large hemispherical bowl over which a specially treated leather skin is stretched. Several types of sticks are used to strike the membrane, according to the percussion effect desired. These sticks may have striking surfaces made of sponge, felt, rubber, or wood. The kettledrum may be tuned and changed in its fundamental frequency when so desired by varying the tension of the membrane through adjustment of head screws and by movement of a tuning pedal. The kettledrum emits a low-frequency sound of a definite pitch. The pedal provides a quick yet accurate variation of the pitch that a melody can be played on the kettledrum. Two versions of the timpani are standard, the smaller one having a diameter of $58\frac{1}{2}$ cm and the larger one, a diameter of about 76 cm. The smaller unit covers the frequency range from $B\flat_2$ to F_3, and the larger from F_3 to C_3.

An interesting development of rather recent past is that of the steel drum, which is the principal instrument of Trinidad, where it originated, and of other Caribbean nations. Steel drums developed in Trinidad in the 1940s because the use of "bamboo tamboo" sticks, which supplied rhythmic cadences at the annual Carnival festivities, were proscribed after matters got out of hand and these sticks were used as weapons in melees between rival bands and fights with the police. But these musicians were determined to continue their musical ways, and their resourcefulness turned them to buckets, garbage cans, brake drums, and whatever was available. The first steel drums were rhythmic rather than melodic. With the availability of the 55-gallon steel drums used by the petroleum industry after World War II, the continued development of tuned steel drums continued apace to the point that they must now be considered rather sophisticated examples of musical instruments and intricate workmanship. These instruments are now achieving ever greater popularity in North American and in Europe, to the point where an aspiring musician can study steel drums in a well-developed curriculum at Northern Illinois University.

A variety of steel drums or pans are available. Steel drums are generally fabricated from 55-gallon oil drums. The drums in a steel band may consist of a variety of drums that have been termed soprano, ping pong, double tenor, guitar, cello, and bass. The soprano or ping pong can have anywhere from 26 to 36 different notes, but a bass drum may have only three or four notes. Because of the relative paucity of notes on a single drum, the bass drummer is likely to play on a half a dozen drums in the same manner as a timpanist in a symphonic orchestra.

These drums are fabricated by first hammering the head of an oil barrel into the shape of a shallow basin. A pattern of grooves is cut with a nail to delineate sections of different notes. Each section is "ponged up" or shaped with a hammer. After the drum is heat-tempered, each section is tuned by the adept use of a hammer. Figure 17.35 shows the layout and the tonal segmentation of different types of steel drums.

Indefinite pitch instruments include the triangle, which is a steel rod bent into a triangular shape that is formed by a metal beater. Because the triangle undergoes complex vibration when struck, the fundamentals and the overture form an indefinite noise mixture. The bass drum, a hollow cylinder of wood or metal covered at each end by a stretched membrane or parchment or skin, is struck with a soft headed stick; its vibration is quite complex; owing to nodal contributions by the air column and the stretched membrane. Thumbscrews supply the means of adjusting the stretch of the drumskin. The bass drums are constructed in sizes ranging from 61 cm to as much as 3 m, with approximately 78 cm being the most common diameter. The military drum is a smaller rendition of the bass drum, being about 40 cm in diameter and about 30 cm in depth. At nearly 36 cm in diameter, the snare drum is even smaller than the military drum. Across its lower head, cords of catgut are stretched so these cords vibrate when struck by the membrane. This results in a sound output that is buzzing and rattling.

The tambourine consists of a hoop of wood or metal with a single membrane stretched over one end. Smaller circular metal disks are inserted in pairs in the hoop and loosely strung on wires. During performance, the tambourine is held in one hand and the membrane struck with the fingers or palms of the other hand.

A cymbal is a brass circular disk that is concave at its center. Its vibration is that of a circular plate supported at its center. Cymbals equipped with handles can be held, one in each hand, and struck together. A single cymbal can be supported in a fixed horizontal position on a stand or on some other support and struck with a drumstick. A sock cymbal consists of a pair of horizontally positioned cymbals, one being fixed and the other being moved by a pedal. Cymbals are made in various sizes, from as little as 5 cm to 51 cm in diameter.

Castanets, used to accentuate Spanish-style music, are hollow shells of hard wood. These are held in the hand and clapped together. They can also be mounted on a handle in such a manner that shaking the handle clacks the castanets.

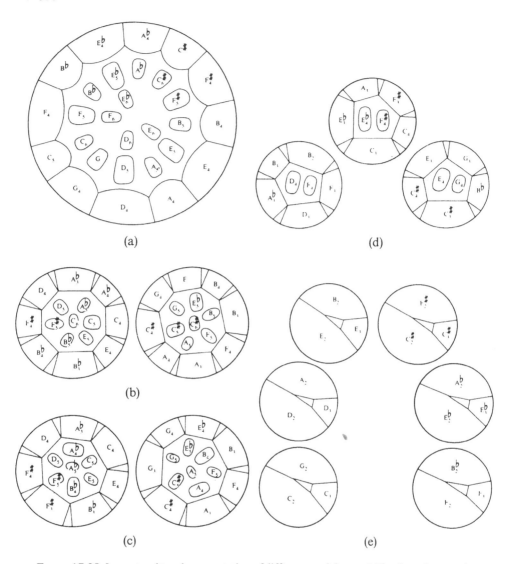

FIGURE 17.35. Layout and tonal segmentation of different steel drums: (a) lead pan (soprano), (b) double tenor (alto), (c) double second (tenor), (d) cello, and (e) bass. (From Fletcher and Rossing, *The Physics of Musical Instruments*, New York: Springer-Verlag, 1991, p. 570, with acknowledgment to Clifford Alexis.)

The gong is a round plate of hammered bronze, with the edges curved up so that the shape takes on the appearance of a shallow pan. The gong is usually suspended by strings attached to a frame. It is excited by striking with felt-covered hammer, with the sound output resembling a heavy roar. The diameter of the gong ranges from about 45 cm to 90 cm.

17.12   Electrical and Electronic Instruments

A tone can be generated by a number of electrical means. The means of producing a tone can be achieved by (a) interruption of an air stream (as with a siren which consists of a rotating perforated wheel), an electrically driven diaphragm, (as in the automotive horn) and electronic alternator and loudspeaker combination (electric organs), an electronic oscillator–amplifier–loudspeaker combination (the electric piano is an example). The electric acoustic guitar is an adaptation, where an electromechanical transducer is attached to the bridge of the instrument. The transducer converts the mechanical vibrations into a correspondingly varying electrical signal which is transmitted to an amplifier. The amplifier increases the strength of the signal and sends it on to a loudspeaker which converts the electrical signal into the corresponding acoustical output. A volume control in the amplifier or integrated into the guitar makes it possible to adjust the output sound level. The sound of the electric acoustic guitar sounds different from that of the conventional instrument.

An electric guitar (as opposed to an electric acoustic guitar), such as the Fender Stratocaster®, uses magnetic pickups, *not* located in the bridge. The strings are magnetized by permanent magnets in the pickup. When the strings are plucked, a coil of wire inside the pickup transduces the magnetic energy into an electric current which is sent through a cable to an amplifier and from thence to a loudspeaker.

Generating music by electronic means has a history older than most people realize. Early electronic instruments include the Theremin (1919), the Ondes Martenot (1928), the Trautonium (1928), and the Hammond organ (1929).[2] The early means of electrically generating music were based on the technology that prevailed in the recording and sound reproduction industry of the time. Electronics progressed rapidly during World War II and even more rapidly with the development of transistors and eventually with the advent of microcircuitry.

Because of the rising costs of traditional pipe organs and the desire then prevailing for creating sustained tones, much attention was conferred on developing and marketing the electronic organ. Vacuum-tube oscillators and, later on, solid-state oscillators were used to generate the tones in an electronic organ. The schematic of a transistor oscillator network is shown in Figure 17.36. Power from the output network is fed back to the input network. Oscillations occur when more power is developed in the output than is necessary for the loss in the input circuitry combined with appropriate phase relations between the current and voltages in the input, feedback, and output networks. Under these conditions, the reactions consists of regular surges of power at a frequency that depends on the constants of the resonant elements in the input or output networks. The resonant elements include quartz, crystal, tuning fork, inductance-capacitance, and so on. These electronic systems are capable of simulating the wave shape of almost any musical

---

[2] An interesting review of these instruments and their history may be found in texts by Rossing (1990) and Strong and Plitnik (1983).

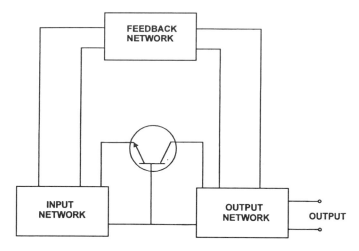

FIGURE 17.36. Schematic of a transistor oscillator network.

instrument. One type of electrical organ makes use of air-driven reeds and electrostatic pickups. Most of the electronic organs consist of two manual keyboards and a pedal keyboard and incorporate a system of stops and couplers. Loudspeakers may be housed in a separate cabinet, and the driving electronics housed in the console.

The electronic organ has been displaced in the 1980s by the growing popularity of synthesizers. Modular synthesizers made their appearance in the 1960s in both analog and digital versions. The analog synthesizer consists of a group of signal generating and signal processing modules operating on electrical voltages. The modular approach to analog synthesis is embodied in three fundamental voltage-controlled modules: (a) the voltage-controlled oscillator (VCO), (b) voltage-controlled amplifier (VCA), and (c) voltage-controlled filter (VCF).

The VCO is really a function generator that produces a periodic voltage signal and is the initial source of pitched sounds. The frequency of the periodic signal is determined by a control voltage. Usually when the control voltage increases by one volt, the frequency doubles (i.e., it goes up one octave). The control mode is thus exponential, which fits beautifully into the scheme of things in music and in electronics. Controllers can be fabricated in the form of keyboards, and musical effects of modulation (trill, vibrato, or glissando) are independent of the DC value of the control voltage. Switching keys of a melody require only the addition of a constant to the series of control voltages representing the notes of a melody. The exponential feature inherent in the voltage control is appropriate electronically, given the fact the collector current in a bipolar transistor is an exponential function of the base-emitter voltage. In addition to being stable and accurate, the VCO needs a subsequent filter to shape the waveforms for the desired tone color. A high purity sine wave output is normally needed, particularly if the VCO output is processed by a nonlinear waveform shaper, distorter or a ring modulator or an FM synthesizer. Otherwise an excessively dense spectrum that is downright unmusical can be created.

The voltage-controlled amplifier is a voltage amplifier with a gain that depends linearly upon a control voltage. An exponential control is generally available, but it is more usual for the amplifier gain to grow linearly with the control voltage with a slope of unity gain for a 5-V control input and a maximum gain of 2. Ideally the gain should be zero when the control voltage is zero or negative. It is the task of the VCA to turn on tones. Because the oscillators run continuously, the VCA bears the responsibility of shaping the amplitude envelope of the tone. Here the control voltage comes from an envelope generator. When a note is begun from a keyboard (or sequencer), the envelope generator begins a transient phase. The envelope generator sustains an appropriate level for the duration of the tone and ends the tone with an exponential decay. As with the tone control parameters on a synthesizer, the performer can set the envelope transients, the sustain level, and decay time by adjusting individual potentiometers on a modular analog instrument and through programming on a digitally controlled keyboard. Special timbres can be created by adding two sounds, one of them with a delayed onset. To achieve this the output of two VCAs can be added, where one of them is controlled by a delayed envelope. A tremolo can be added by a VCA controlled by a subsonic periodic waveform.

A VCA can be utilized to multiply two separate audio signals, producing amplitude modulation, with one signal serving as the carrier and the other serving as a modulator. The output consists of the spectral components of the carrier plus sidebands. More commonly used than amplitude modulation is the *balanced modulation* or ring modulation. While the VCA is a two-quadrant multiplier (i.e., the control voltage must be positive for a nonzero output), the balanced modulator is a four-quadrant multiplier.

As the terminology implies, the definitive frequency of a voltage-controlled filter (VCF) is controlled by a control voltage. In a low-pass VCF, for example, the cutoff frequency increases from a low value to a high value as the control voltage increases. A scaling of 1 V per octave is the standard. There is a large number of different filtering circuits available, but predominant design seems to be the state-variable filter which is constructed from two integrators and a summer. This design bears the advantage of incorporating (a) three simultaneous outputs (low-pass, bandpass, high-pass), (b) a constant value of Q independent of band frequency, (c) adjustability of Q through controlled feedback, and (d) stability of the sine oscillation mode. This type of filter features asymptotic slopes of ±12 dB/octave in the low-pass and high-pass outputs and ±6 dB/octave for the band-pass output. A four-pole low-pass filter with an asymptotic slope of ±24 dB/octave is often preferred by musicians to deal with the more subdued sounds, particularly in the lower registers.

Voltage control is applied in filters to generate dynamic effects, particularly musical attacks. As an example of this application, consider the fact that the harmonics of a brass instrument entering the attack phase are in the order of ascending frequencies (Risset and Mathews, 1965). This effect is simulated electronically by raising the cutoff frequency of a low-pass VCF with an envelope generator at the onset of the tone. On the other end of the tone, the decay of an envelope generator is utilized to decrease the cutoff frequency of the low-pass VCF to simulate the proclivity of high-frequency modes to become damped more rapidly than the low-frequency modes in a free vibrator (such as a percussion instrument).

A number of other musical effects can be generated. For example, in a delay-and-add application for a digital delay line, specifically the generation of reverberation, an input signal is passed through a delay line that is tapped at different delay intervals. Signals at these taps are fed back into the input as a weighted sum. This weight determines the reverberation time of the system. In the *flanger* use is made of comb filtering. The flanger uses a single delay-and-add. The delay time is slowly modulated so that the comb filtering changes with time, with typical delays in the order of 2 ms. *Chorusing*, which is the attempt to make a single voice sound like a group of many, functions in the same manner, except several modulated delay-and-add circuits are used to transmit to different channels, and the delays are considerably longer, typically in the order of 10 ms.

In the early years of the synthesizer, control was a cumbersome matter, involving patch cords for connecting analog synthesizers. The early digital synthesizers were

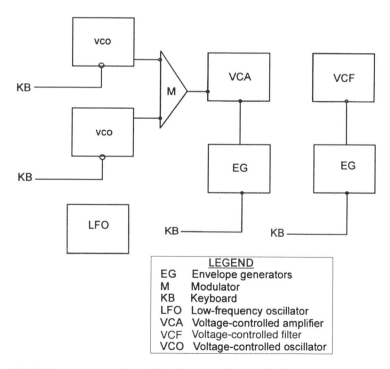

FIGURE 17.37. Arrangement of modules for an analog synthesizer. The signals from the two keyboard-manipulated voltage-controlled oscillator (VCO) are summed into a voltage-controlled amplifier (VCA) and then relayed to a voltage-controlled filter (VCF). Both VCA and VCF are controlled by the envelope generators (EG). A third oscillator operating at low frequency (LFO) is on hand for modulation of any voltage-controlled module.

even more cumbersome, none of them operating in real time. In the early 1970s, prepatched analog synthesizers were introduced, primarily for creating special effects in a live performance. The setup up of the modules is given in Figure 17.37. In the mid-1970s, the modules of Figure 17.37 were combined into a single integrated chip. One such chip was placed under each key of the keyboard, so each key essentially constituted the control for a miniature synthesizer. The PolyMoog was the first and the last instrument to be constructed this way. Subsequent designs now employ digital scanning keyboards and assign a specific sound synthesis chain to each key. The number of chains (generally 8, 12, or 16) establishes the maximum number of simultaneous notes.

The *sequencer* was introduced to provide additional automated control. The earlier sequencers generated several channels of control voltages in repetitive sequences. The analog sequencer could even replace a keyboard for generating a repetitive bass line. The sequencers have evolved into small computers which have almost unlimited musical capabilities. Thanks to computer memory, they can recall and reproduce the control sequences for multiple parameters for all the voices in the entire performance. Data entry is achieved through the use of an organ-type keyboard, and the programming allows for editing the data and displaying or printing out of the data in the traditional musical notation.

17.13   The MIDI Interface for Synthesizers and Digital Synthesis

With rapid advances in the development and manufacture of microprocessors and an equally rapid drop in the cost of these microprocessors, the digital control of synthesizer functions became inevitable. In 1983, the manufacturers of commercial synthesizers adopted the Musical Instrument Digital Interface (MIDI) standard, which expands the range of control possibilities for the musical performer by the adoption of an interface that makes it possible to transfer control data among different synthesizers or other processors as well as computers. The interface is a unidirectional asynchronous serial line operating at 31.25 kbaud (1 kbaud = 1000 bits/s). A MIDI word is defined as 10 bits long—8 data plus start and stop bits. To enable long interconnections, the interface uses a 5-mA current loop. Repeaters with each MIDI-specification instrument allow for daisy-chaining. Sixteen logical channels exist for data transmission, and instruments may be set to obey data on one channel or data on all channels.

Some musical tradition remains in the way MIDI commands are defined to emphasize the use of the keyboard in electronic instruments. Specific data words are defined for key on/off, key velocity, key aftertouch, and for encoding the position of modulation wheels (this device was originally introduced in the MiniMoog and it remains a standard). The performers are free to use the parameters as they like, and the MIDI has even been used to control stage lighting. Data of arbitrary length can be transmitted by concaternating MIDI words. Owing to its high data rate and flexible control structure, the MIDI control has proven to be very powerful and well accepted in the industry and by performers of popular music.

With the affordability and availability of digital processors, it becomes inevitable that digital oscillators can become precise enough to replace VCOs for improved flexibility and stability. Moreover, a single digital oscillator can be constructed to create several simultaneous voices.

A cost-effective digital technique is that of FM synthesis, which involves the generation of a frequency-modulated waveform given by

$$x(t) = \sin[\omega_c t + \beta \sin(\omega_m t)]$$

where ω_c denotes the carrier frequency, ω_m, the modulating frequency, and β is the modulating index which is given by $\beta = \Delta\omega/\omega_m$, with $\Delta\omega$ being the maximum frequency excursion. This FM algorithm provides for flexible addressing. A spectrum of the FM signal contains a component at the carrier frequency and a component at the sidebands displaced in frequency by $\pm n\omega_m$, where n is the sideband order. Sidebands of order n have amplitudes that are proportional to Bessel functions $J_n(\beta)$. As the modulation index is increased, the bandwidth of the signal increases. A dynamically varying spectrum is achieved by varying the modulation index as the tone is initiated. When ω_c and ω_m are commensurate, the spectrum may be harmonic; otherwise, it may be inharmonic.

The FM algorithm has proven highly successful in creating both sustained and percussive musical sounds, but the principal disadvantage of the FM technique is that the control is unintuitive. Bessel functions are not monotonic functions of their arguments, so it becomes difficult, almost impossible, for the performer sense what will happen if the modulation index is changed one way or the other. The Yamaha musical instrument company installed FM oscillators on large-scale integrated circuits in its DX series of instruments, the most popular electronic instruments ever. These instruments contain six or four dynamically controlled oscillators in a voice, together with possible feedback among the oscillators.

A more recent methodology of creating music electronically is that of *physical modeling*. A mathematical model is developed for the production of a tone by a traditional instrument or by some other mechanical device. The model may result in a set of coupled, usually nonlinear differential equations which can be solved by computerized means. The solution to these equations are played through a digital analogue computer. This is a most time-consuming process and has been applied commercially for only a few years. It is hoped with this methodology that even greater subtlety can be captured in the reproduction of actual instruments.

17.14   The Orchestra and the Band

The orchestra is a fairly full-fledged group of musicians playing on string, wind, and percussion instruments under the direction of a conductor. An orchestra differs from a band in that the main body of the music is generated by string instruments, whereas the band is generally made up of musicians playing on wind and percussion instruments. Orchestras differ according to the demands of the music, as the requirements differ for playing symphonies and overtures, or providing accompaniment for operas, musicals, and oratories, or incidental music for theatrical performances, or ballet or dance music.

Four groups usually constitute a symphony orchestra, namely, the strings, woodwinds, brass, and percussion. Violins, violas, violoncellos, contrabasses, harps and (usually one) piano comprise the strings section. Woodwinds include the flutes, piccolos, oboes, bassoons, English horns, contra bassoons, clarinets, and bass clarinets. Brass instruments include trumpets, trombones, French horns, and tubas. In the percussion group there may be included timpani, bass, military and snare drums, tambourines, gongs, celestas, glockenspiel, tubular chimes, xylophones, castanets, and triangles. In order to achieve sufficient sound output for suitable artistic effects in large halls, the orchestra may have to consist of 80–120 players. An organ may be added to the performance where required. Some performances may require fewer players, and in some situations, a limited-size orchestra can achieve a dramatic effect through the use of sound reinforcement systems. Figure 17.38(a) shows a plan view of the arrangement of the instruments of a symphony orchestra and the placement of its conductor. Figure 17.38(b) illustrates the layout of a smaller orchestra; and 17.38(c) that of a small dance band. An orchestra for performances of popular and dance music may number anywhere from 5 to 25.

A band consists of musicians playing wind and percussion instruments. They are generally more suited for outdoor performances, because the main body of the tone is produced by the brass and woodwind sections which can produce considerable acoustic power in free space. Bands help to sustain unison in marching groups and can whip up enthusiasm in athletic meets. A standard band generally features a complete range of woodwind, brass, and percussion instruments. The number of performers can vary from 25 to 50 or even more. The military band resembles the standard band except that it may include a fife, drum, or bugle corps. A concert band may include contrabasses, timpani, harps, and other instruments that lack the portability for marching. There is virtually no difference between the dance band and the dance orchestra. Table 17.4 shows the components of a typical symphony orchestra and those of a standard band.

The frequency ranges of the fundamental frequencies of various musical instruments, including the human singing voices, are displayed in Figure 17.39.

17.15   Recording Equipment

Recording of sound occurs in different ways and it can be defined as the storing of information of how the amplitude of sound varies with time. The stored information can be retrieved by playback at a later time through control of a source of sound waves. The twentieth century has seen the evolution of the phonograph cylinder into the 78 rpm recordings which in turn gave way to the 45 rpm and $33\frac{1}{3}$ rpm vinyl disks. These mechanical types of recordings use wavy grooves on a surface

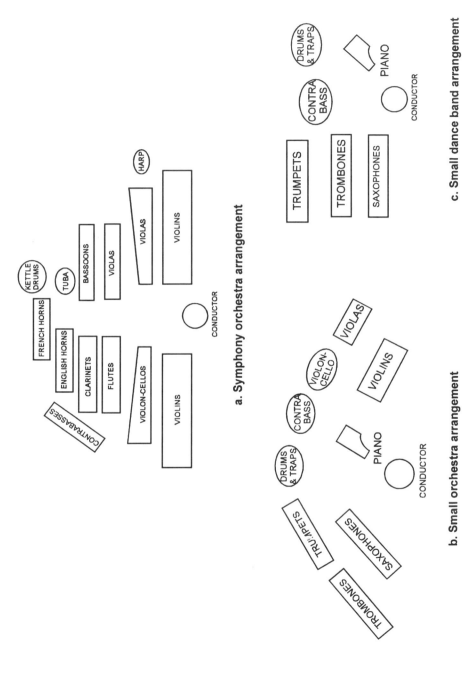

a. Symphony orchestra arrangement

b. Small orchestra arrangement

c. Small dance band arrangement

FIGURE 17.38. Arrangement of players for (a) symphony orchestra, (b) smaller orchestra, and (c) small dance band.

TABLE 17.4. Elements of a Typical Orchestra and of a Typical Band.

Orchestra		Standard Band	
Instruments	Number of Players	Instruments	Number of Players
First violins	10–20	Flutes	4
Second violins	14–18	Piccolo	1
Violas	10–14	Clarinets	14
Violoncellos	8–12	Oboes	2
Contrabasses	8–10	Bassoons	2
Flutes	2–3	Sarrusophones	2
Piccolos	1–2	Saxophones	4
Oboes	3	Cornets	4
English horn	1	Trumpets	2
Bassoon	3	French horns	4
Contra bassoon	1	Trombones	4
Clarinets	3	Tubas	6
Bass Clarinet	1	Snare drum	1
Trumpets	4	Bass drum	1
Trombones	4	Percussion	1–5
French horns	4–12		
Tuba	1		
Timpani	1		
Harp	1		
Percussion	1–5		

Percussion includes triangles, bells, cymbals, castanets, and xylophones.

In addition, other instruments such as a piano or an organ may be included. Percussion includes bass, snare and military drums, gongs, cymbals, tambourines, celestas, glockenspiels, tubular chimes, xylophones, castanets, and triangles.

that guide a stylus to replicate the signals that were recorded. Reel-to-reel tape recorders have been used since the 1940s to make recordings, and many great performances have been archived on this medium. These master tapes are used to make vinyl disks and eventually CD disks. While tapes carry the advantage of being relatively easy to edit, they do lack the random accessibility of any type of disk. Tapes solely for playback are now principally in the compact cassette format. The advent of sound in cinema was made possible by photographic tracks that vary in transparency or by reflection of light (laser disc or compact disk).

Magnetic Recording

Figure 17.40 shows a fundamental record–play system using magnetic recording. A closeup of the interior construction of the magnetic head is also included in the figure. Tape D is made of a smooth, durable plastic such as polyester (Mylar) and is coated on one side with a magnetizable material. Most commonly a dried

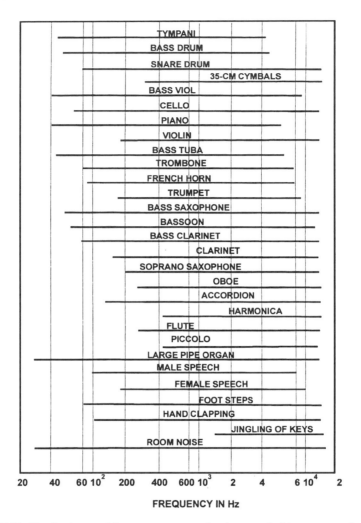

FIGURE 17.39. The fundamental frequency ranges of various musical instruments, including the human voice.

suspension of acicular gamma ferrite particles about 0.25 μm long in a lacquer binder.[3] The thickness of the magnetic layer is of the order of 12 μm or less. The tape unreels from supply reel A and threads over a magnetic head B that supplies

---

[3] After the tape has been coated and while the suspension is still damp, it is subjected to a magnetic field that aligns the magnetic particles in order to increase recording signal strength and minimize noise. Other alternative magnetic materials include chromium dioxide and cobalt-coat gamma ferrite. It is obvious that tapes have to be manufactured under stringent clean-room conditions because the slightest traces of contaminants cause impaired recordings.

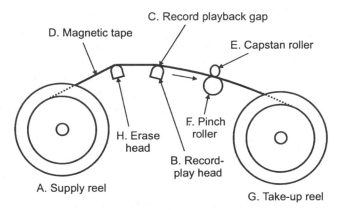

MAGNETIC RECORD/PLAYBACK SYSTEM

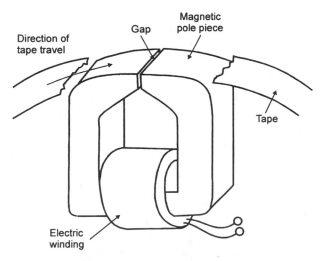

CLOSE-UP SHOWING MAGNETIC HEAD CONSTRUCTION

FIGURE 17.40. The elements of a magnetic tape recording and playback system, with details of the interior construction of a magnetic head.

a magnetic field at recording gap C. The magnetic field varies in its strength according to the sound being recorded. The now magnetized tape is pulled past the head by capstan roller E and pinch roller F and spooled onto a takeup reel G. In rewinding the tape, the recording head is deenergized. For playback, the tape leaves the supply reel, as before, but the winding on head B is connected to an amplifier which, in turn, connects to a loudspeaker. The magnetized elements induce a voltage in the head winding, which translates into a replication of the gap field variations during recording. The recording may be erased by an additional head H that produces a strong, steady AC field. It is not necessary to erase the tape if it is being recorded over, because the erase head is automatically energized when recording, thus removing previously recorded material.

A magnetic head is designed to concentrate the magnetic field in the smallest possible region of tape, thus giving a high recording density of maximum number of wavelengths possible. To achieve these high densities, the head gap must be made very small, in the order of 1 or 2 μm, which yields low output voltage and requires very smooth tape surface and good contact between the head and the tape. The head cores are usually fabricated of Permalloy (80% Ni, 20% Fe) for high permeability and low magnetic retention. In more recent times ceramic-like magnets have been used for better wear and they do not require lamination.

Tape speeds have been standardized at 30, 15, 7.5, 3.75, and 1.875 in./s for analog recording. The highest speed is used for master recordings, in situations where the highest quality is demanded in covering the 20–20 kHz range. Cassettes run at 1.875 in./ speed. All of these speeds are much lower than that required for digital recording, which necessitates bandwidths in the megahertz range. Digital audio recorders make use of the video recording techniques that entail rotating heads. These heads scan the tape at 228 in./s while the tape itself moves at only 0.66 in./s.

Digital Recording

The capabilities of analog recording pale in comparison with the advantages of digital recording. Zero wow and flutter, more than 90 dB signal-to-noise ratio, and incredibly low distortion levels are the most outstanding attributes of digital recording. To achieve digital recording, the analog signal is sampled at regular intervals. Sampling of the amplitude is done with almost absolute accuracy (one part per 10^9 in a 96 dB signal-to-noise ratio system). The amplitude is recorded as a number, say 59,959,498. The sampling should be done at more than twice the Nyquist number (the highest frequency in the audio signal). For a signal with a maximum of 20,000 Hz, a rate of 44,000 samplings per second is quite practicable ($20,000 \times 2 + 10\%$). Because the series of digital samples occur in the megahertz range, they are recorded and played back on videotape recorders. The audio program in digital form can be processed with digital features, error correctors, and other techniques that cannot be achieved with analog means. The numbers comprising the signal can be stored successively in buffers and read out in perfect crystal-controlled time intervals, even if there exist erratic mechanical fluctuations in the tape drive. Thus, wow and flutter are eliminated. The digital signals are restored to analog format through digital-to-analog (D/A) converters.

Experiments on the digitization of sound were conducted during the late 1950s and early 1960s for the purpose of computer analysis, speech synthesis, simulation of music, and simulation of reverberation, mostly at the Bell Telephone Laboratories in Murray Hill, NJ (Mathews, Miller, and David, 1961). In the early 1970s, commercial digital recordings were commonly used by recording studios for master recordings, but the results were distributed by analog means. A significant breakthrough occurred in the early 1980s, when Sony Corporation of Japan and Philips BV of Netherlands came out with the compact disc (CD) that quickly replaced the analog long-play (LP) record as the most common medium for play-back of recorded music. Later on, machines for digital recording on cassette video tape and then the rotary-head digital audiotape (R-DAT) came on the market. In the past few years, the digital compact cassette (DCC) developed by Philips-Matsushita and the MiniDisc™ (MD) by Sony arrived on the recording scene. The DVD disk is a multimedia device for playback through video monitors as well as through audio channels. A DVD-disk player can also function as a audio-only playback unit for compact disks, but CD players cannot play the audio portion of DVD disks.

17.16   Playback Audio Equipment

Figure 17.41 shows a schematic of a fairly complete playback system for multi-channel sound reproduction. Whether the system is monophonic (single-channel), stereo (two-channel), or multichannel (more than two channels), there are three stages to sound reproduction: the first stage consists of the source or sources of program material; the second stage is the preamplification and amplification. The third stage consists of converting the electrical signal into acoustic signals, and this occurs in loudspeakers or headphones.

Two-channel (stereo) setups reproduce the spatiality of the original program. When properly recorded and reproduced, the strings located on one side of the

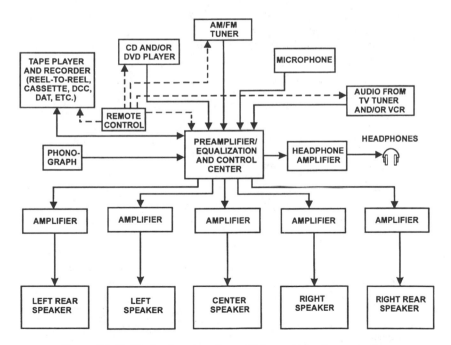

FIGURE 17.41. Playback system for multichannel sound reproduction.

orchestra become reproduced more strongly in the same side loudspeaker than in the loudspeaker on the other side, giving the listener the illusion of the strings' location. Multichanneling is intended to reproduce not only the spatiality of the program but also provide some degree of the acoustic ambience of the hall where the program was recorded. The acoustics of the listening room, of course, affect the reproduced sound, which may even assume additional characteristics that may not be so desirable.

The sources constituting the first stage can include one or more of the following: microphone, AM/FM (radio) tuner, phonograph (consisting of a turntable, arm, and cartridge), reel-to-reel or cassette player, compact disk (CD) player, digital versatile disk (DVD) player, and the audio output of a television receiver. The second stage consists of a preamplifier that provides the means to select signal sources, the proper equalization for program sources, and means of controlling the volume of the program material. After the initial preamplification stage, the program material is amplified in the amplifier. An amplifier may be multichanneled in that there is a separate amplifier for each channel, and a number of these amplifiers may be mounted on a single chassis.[4] An *integrated amplifier* combines on a single chassis the preamplifier (which acts as a controller) and the amplifier. A stand-alone amplifier is referred to as a *power amplifier* and it functions in conjunction with a separate preamplifier. A chassis that combines a radio tuner with a preamplifier and a power amplifier is called a *receiver*. This type of construction makes for more economical production, with a single power supply serving all of these subcomponents, and conservation of rack or shelf space; its principal disadvantage is potentially increased heat output from the electronics and the lack of flexibility in upgrading individual subcomponents.

The third stage consists of loudspeakers and or headphones. A tremendous variety of loudspeakers are available on the market. The most common type is the electrodynamic speaker, which can range from a single-cone type to more elaborate multidriver units consisting of woofers to reproduce low-frequency sounds, midrange-drivers to handle the frequencies between the low frequencies produced by the woofers, and the high frequencies produced by tweeters. A two-way loudspeaker system divides its program material between a woofer and a tweeter. A three-way loudspeaker system incorporates mid-range drivers, and four- and five-way units have also been constructed. In order that the individual drivers receive the proper frequencies to the exclusion of the program contents outside of their respective optimal operating ranges, special types of bandpass filters, or *crossover networks*, are employed to separate out the high frequencies from the signals being

---

[4] In extremely elaborate (and expensive!) systems, an electronic active crossover system may be inserted between the preamplifier and the amplifier, and the customary passive crossover system built inside the loudspeaker system is bypassed. Each channel's input into the electronic crossover separates into appropriate frequency bands which are fed into separate amplifiers. Each of the amplifiers links directly to the individual drivers. A single channel would require three amplifiers, one to feed the woofer, another to feed the midrange, and the third to feed the tweeter.

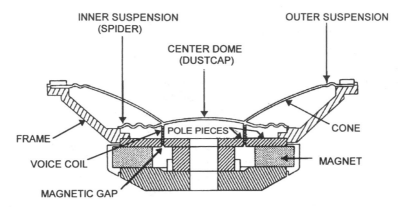

FIGURE 17.42. Elements of an electrodynamic speaker. (Courtesy of JBL Professional.)

fed into the woofers, to band-pass the mid-frequencies into the midrange drives, and to channel only the high-frequency portion of the signals into the tweeters. Because much of the acoustical energy is contained within the low-frequency portion of the signals, passage of unfiltered signals into the tweeters can destroy these drivers.

Figure 17.42 shows the construction of a electrodynamic driver. Three separate but interrelated subsystems constitute the driver. The motor system consists of the magnet, pole piece, front plate, and voice coil. The diaphragm, generally a cone and a dust cap or a one-piece dome, constitutes the second subsystem; and the suspension system, which includes the spider and the surround, constitutes the remaining subsystem.

In the motor assembly, the back and front plates and the pole pieces are made from a highly permeable material, such as iron, which provides a path for the magnetic field of the ring-shaped magnet, which is usually of ceramic-ferrite material. The magnetic circuit is completed at the gap, with a strong magnetic field existing in the air space between the pole piece and the front plate. The coil, which is connected to a pair of input terminals, is wound around a thin cylinder that is attached to the speaker cone which, in turn, is mounted at its outer edge through a flexible surround to the frame. A spider, essentially a movable membrane also attached to the frame, positions centrally the diaphragm at its inner edge and the voice coil. The signal containing the program material feeds into the voice coil, which generates a change in the magnetic field imposed by the permanent magnet surrounding the coil. If an alternating current is fed into the coil, the flow of the current in one direction will cause the voice coil to move in one direction, and the reverse flow will cause the coil to move in the opposite direction. The cone under the impetus of the moving voice coil acts as a piston in moving the air in front of it.

The ideal cone would act as a perfectly rigid piston pushing against the air. The transfer of the motion from the piston to the air is bound in terms of frequency by resonance frequency of the cone at the low end (here the ability to transfer energy to the air is limited by mechanical constraints) and by the radiation impedance at the upper limit. This upper frequency limit occurs from the fact that it is a function of both the nature of radiation impedance of the air and the radius of the radiating surface. Smaller radiating surfaces can reproduce higher frequencies more effectively than larger surfaces, which accounts for the smaller sizes of tweeters. Real-world cones, however, are not perfectly rigid and will flex depending on the traits of the materials they are constructed from. Cone flexure has a critical effect on the high frequency efficiency, the sound pressure level output, and driver polar response. Driver materials may differ in degrees of stiffness and transmit vibrations at different speeds internally, but they tend to produce the same sort of flexures or modes. These modes have been discussed for circular membranes fixed at their outer edges in Chapter 6.

The most important function of the speaker enclosure is to control transmission of the driver's rear radiated sound energy in order to avoid its mutual cancellation with forward-radiated energy at low frequencies. The enclosure also acoustically "loads" the driver by providing a suitable acoustic impedance to match the characteristics of the driver and the requirements of the speaker system with regard to low-frequency and large-signal performance. The enclosure must be made sufficiently rigid so that the cabinet vibrations and resonances do not add appreciably to the program material. Two major classes of electrodynamic-driver enclosures are (1) the infinite baffle/closed box in which the sole radiation source is the driver diaphragm and (2) the bass reflex which is vented to augment the driver's radiation at low frequencies. The bass reflex may be a vented system which incorporates a tuned aperture of a specified cross-sectional area and port length in which the enclosed air mass resonates with the enclosure's air spring. This causes a woofer's back wave to communicate with the external acoustic space with the enclosure effectively acting as a Helmholtz resonator.

Another version of the bass reflex system is the port/passive radiator (PR), in which the vent is replaced with an acoustically driven diaphragm (aptly called a *drone cone*). Another type of enclosure is the transmission line (TL) which effectively functions as a low-pass filter with a 90° phase shift, absorbing all of the rear wave energy of the woofer except for frequencies below 75 Hz. The TL enclosure provides a folded path or a labyrinth equal in length to the 25% of wave length at or just above the resonance of the woofer. Damping material such as Dacron®, fiber wool, or fiberglass fills much of the labyrinth. The TL enclosures generally provide excellent, clean bass response from relatively compact floor-standing cabinets. The principal disadvantage of TL is the complexity of the cabinet structure.

Other types of speaker systems are available on the market. Among these are the electrostatic speakers, the ribbon-type speaker system, and there are others that consist of hybrids, for example, a system that combines the electrodynamic drivers with an electrostatic or ribbon tweeter. A ribbon type of drive essentially consists of a voice coil patterned (like the conductive pattern on a printed circuit board) onto a magnetizable diaphragm (i.e., the "ribbon") made of a strong, flexible materials such as Mylar. The diaphragm is suspended in the front of a flat magnet that matches the area of the diaphragm. A varying signal sent into the coil causes a change in the magnetic field that causes the ribbon to move with respect to the magnet, thus causing sound to be radiated. If the magnet is perforated in one way or the other, the ribbon acts as a dipole sound generator.

In the electrostatic speaker (which is similar operationally to the condenser microphone but functioning in reverse), a voltage is maintained between a thin diaphragm and another surface. In effect the two thinly separated surfaces are acting as a large condenser, with the varying voltages fed to one of the elements, causing the diaphragm to move. Because of the small excursions of the diaphragm, the electrostatic driver does not function well in the low-frequency range, but it is capable of providing excellent quality high-frequency signals. The highly capacitive loading of the electrostatic speaker may be problematic for some power amplifiers which are usually designed to handle primarily resistive impedances.

Headphones are essentially miniaturized speakers mounted in earcups, which are in turn attached to headbands that serve the purpose of holding these drivers against the ears. There are also a number of headphone models that use electrostatic elements either by themselves or as supplemental components to electrodynamic elements.

References

Ågren, C.-H., and K. A. Stetson. 1972. Measuring the resonances of treble viol plates by hologram interferometry and designing an improved instrument. *Journal of the Acoustical Society of America* 51: 1971–1983.

Computer Music Journal 16:4. 1992. (A special issue on physical modeling. Of interest are articles by Woodhouse, Keefe, and Smith.)

Cremer, Lothar. 1984. *The Physics of the Violin*. Translated from the German by J. S. Allen. Cambridge, MA: MIT Press. (Originally published as *Physik der Geige*, Stuttgart: S. Hirzel Verlag, 1981.)

Fletcher, Neville H., and Thomas D. Rossing. 1991. *The Physics of Musical Instruments*. New York: Springer-Verlag. (A modern classic in its own right, it is probably the best scientific text devoted to the physics of the entire gamut of musical instruments, with some interesting insights on musical history.)

Hartmann, William Morris. 1997. Electronic and computer music. *Encyclopedia of Acoustics* (Malcolm J. Crocker, ed.), vol. 4. Ch. 138, pp. 1679–1685.

Helmholtz, Herman L. F. 1954. *On the Sensations of Tone*, 4th ed. Translated by A. J. Ellis. New York: Dover.

Hutchins, Carleen M. (ed.). 1967. Founding a family of fiddles. *Physics Today* 20: 23–27.

Hutchins, Carleen M. (ed.). 1975. *Musical Acoustics, Part I. Violin Family Components*. Stroudsburg, PA: Dowden, Hutchinson and Ross.

Hutchins, Carleen M. 1976. *Musical Acoustics, Part II. Violin Family Functions*. Stroudsburg, PA: Dowden, Hutchinson and Ross.

Hutchins, Carleen M. 1981. The acoustics of violin plates. *Scientific America* 245(4):170–186.

Hutchins, Carleen M. 1983. A history of violin research. *Journal of the Acoustical Society of America* 73: 1421–1440.

Hutchins, C. M., K. A. Stetson, and P. A. Taylor. Clarification of "free plate tap tones" by holographic interferometry. *Journal of the Catgut Society* 16: 12–23.

International MIDI Association. 1983. *MIDI Musical Instrument Digital Interface Specification 1.0.* Sun Valley, CA.

Jeans, James. 1968. *Science and Music.* New York: Dover P. (Originally published by Cambridge University Press in 1937. A clear exposition on the fundamentals of musical acoustics and music itself.)

Machlis, Joseph, and Kristine Forney (contributor). 1995. *The Enjoyment of Music: An Introduction to Perceptive Listening,* 7th ed. New York: Norton and Company. (*The* standard text in music appreciation courses in universities and colleges worldwide, it contains a most fascinating history of the evolution of music to the modern times and an overview of the technical aspects of music and musical instruments.)

Mathews, Max V., J. V. Miller, and E. E. David, Jr. 1961. Pitch synchronous analysis of voiced sounds. *Journal of the Acoustical Society of America* 33: 1725–1736. (The pioneering paper on digitization of audio signals.)

Mathews, Max V. 1969. *The Technology of Computer Music.* Cambridge, MA: MIT Press. (Written by a pioneer in the field.)

Meyer, Jürgen. 1985. The tonal quality of violins. *Proceedings SMAC 83.* Stockholm: Royal Swedish Academy of Music.

Meyer, Jürgen. 1975. Akustische untersuchungen zur klangqualität von geigen. *Instrument-bau* 29(2):2–8.

Moog, R. V. 1965. *Journal of the Audio Engineering Society* 13: 200–206. (By the man who started it all—the father of the Moog synthesizer.)

Olson, Harry F. 1967. *Music, Physics, and Engineering,* 2nd ed. New York: Dover P. (A classic in the field of musical acoustics written by a major researcher in the field of sound reproduction.)

Raman, C. V. 1918. On the mechanical theory of the vibrations of bowed strings and of musical instruments of the violin family, with experimental verification of the results. *Indian Association for the Cultivation of Science Bulletin* 15: 1–18. (Excerpted 1975 in *Musical Acoustics, Part II. Violin Family Functions.* C. M. Hutchins (ed.). Stroudsburg, PA: Dowden, Hutchinson and Ross.

Risset. J. C., and M. V. Mathews. 1965. *Journal of the Audio Engineering Society* 13: 200–206.

Roederer, Juan G. 1995. *The Physics and Psychophysics of Music: An Introduction,* 3rd ed. New York: Springer-Verlag. (An interesting combined approach to the physical processes of producing music and human perception of music. While intended for music theorists, composers, performers, music psychologists, and therapists, this text includes some materials on recent findings that may prove useful to acousticians, psychoacousticians, audiologists, and neuropsychologists.)

Rossing, T. D. 1990. *The Science of Sound,* New York: Addison Wesley; Chs. 27–29.

Savart, F. *Des instruments de musique* (1840). Translated by D. H. Fletcher. In: Hutchins (1976), pp. 15–18.

Strong, W. J., and G. R. Plitnik. 1983. *Music, Speech, and High Fidelity.* Salt Lake City: Soundprint.

Young, Robert W. 1939. *Journal of the Acoustical Society of America* 11(1): 134.

18
Vibration and Vibration Control

18.1 Introduction

Noise often results from vibration. Many sources of vibration exist and they include impact processes, such as blasting, pile driving, hammering, and die stamping; machinery such as motors, engines, fans, blowers, and pumps; turbulence in fluid systems; and transportation vehicles. Attenuation of vibration generally cuts down on the noise levels and in many cases lengthens the service life of the machinery itself. Damping, correction of imbalances, and configuration of flow paths constitute the principal measures of cutting down on the deleterious effects of vibration.

18.2 Modeling Vibration Systems

We commence with a basic one-degree-of-freedom system of Figure 18.1, which consists of a mass, a spring, and a damper (also referred to as a *dashpot*). The system has only one degree of freedom because it is constrained to move only in the x-direction. Summing up all the forces acting on the mass and applying Newton's second law, we obtain

$$m\frac{d^2x}{dt^2} + C\frac{dx}{dt} + kx = f(t) \tag{18.1}$$

where

$$m = \text{mass}$$
$$t = \text{time}$$
$$C = \text{coefficient of damping}$$
$$k = \text{linear elastic constant}$$

Damping occurs from energy dissipation due to hysteresis, sliding friction, fluid viscosity, and other causes. The damping force may be proportional to velocity (as stated in the above equation) or it may even be proportional to some other

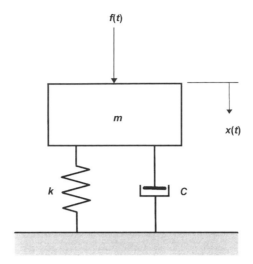

FIGURE 18.1. Model of a vibrating system with one degree of freedom.

power of velocity. Sliding friction is often represented as a constant force in a direction opposing the velocity. Viscous damping proportional to and opposite in direction to velocity serves as a reasonable model in many situations. Moreover this assumption is quite amenable to mathematical treatment. When $f(t)$ is set to zero, equation (18.1) describes the free, viscous-damped, one-degree-of-freedom system,

$$m\frac{d^2x}{dt^2} + C\frac{dx}{dt} + kx = 0 \tag{18.2}$$

We shall employ Laplace transforms to treat equation (18.2). Dividing (18.2) by mass m and assuming for the moment that the initial conditions are zero, we can write equation (18.2) using the Laplace transform variable s, as

$$s^2 X(s) + \frac{C}{m}sX(s) + \frac{k}{m}X(s) = sx(0) + \dot{x}(0) + \frac{C}{m}x(0)$$

or

$$\left(s^2 + \frac{C}{m}s + \frac{k}{m}\right)X(s) = \left(s + \frac{C}{m}\right)x(0) + \dot{x}(0) \tag{18.3}$$

Here $\dot{x} \equiv dx/dt$. Setting the parenthesized term on the left-hand side of equation (18.3) equal to zero gives

$$s^2 + \frac{C}{m}s + \frac{k}{m} = 0$$

which can be rewritten in the form

$$s^2 + 2\xi\omega_n s + \omega_n^2 = 0 \tag{18.4}$$

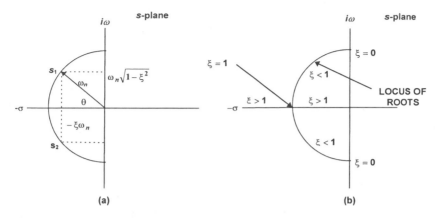

FIGURE 18.2. Root locations and nomenclature for a second-order dynamic system. The s-plane to the right shows the domain of the roots as a function of the damping ratio ξ.

where

$$2\xi\omega_n = \frac{C}{m}, \qquad \omega_n^2 = \frac{k}{m} \tag{18.5}$$

Equation (18.4) is the characteristic equation of which roots essentially determine the response of the system [i.e., the position of the mass as a function of time or $x(t)$]. Solving for the roots of equation (18.4) yields

$$s_1, s_2 = \frac{-2\xi\omega_n \pm \sqrt{(2\xi^2\omega_n)^2 - 4\omega_n^2}}{2} \tag{18.6}$$

These roots can assume complex values and so can be plotted in the s-plane, where $s = \sigma + i\omega$ and $\omega = 2\pi f$. The general plot is shown in the complex plane of Figure 18.2(a). For this plot it is assumed that the system is stable and the value of ξ falls between zero and unity. The term ξ is the *damping ratio* and ω_n is the *undamped natural frequency*. In Figure 18.2(a) it is noted that $\xi = \cos\theta$. Both ξ and ω_n constitute the key factors that determine the roots of the characteristic equation and hence the response of the system. Four cases of damping are of interest:

Case 1: $\xi < 1$ (underdamped system)
  The roots are $s_1, s_2 = -\xi\omega_n \pm i\omega\sqrt{1 - \xi^2}$ and system response is given by

$$x(t) = Ae^{-\xi\omega_n t} \cos\left(\omega_n\sqrt{1 - \xi^2}t + \theta\right)$$

Case 2: $\xi > 1$ (overdamped system)
  The roots are $s_1, s_2 = -\xi\omega_n \pm i\omega\sqrt{\xi^2 - 1}$ and system response is given by

$$x(t) = Ae^{-\left(\xi\omega_n + \omega_n\sqrt{\xi^2 - 1}\right)t} + Be^{-\left(\xi\omega_n - \omega_n\sqrt{\xi^2 - 1}\right)t}$$

Case 3: $\zeta = 1$ (critically damped system)

The roots are $s_1, s_2 = -\omega_n$ and system response is given by

$$x(t) = Ate^{-\omega_n t} + Be^{-\omega_n t}$$

Case 4: $\xi = 0$ (undamped system)

The roots are $s_1, s_2 = \pm i\omega_n$ and system response is given by

$$x(t) = A\cos(\omega_n t + \theta)$$

Figure 18.2(b) shows the domains of the roots in the s plane as a function of the damping ration ξ for a constant value of ω_n.

In order for us to appreciate the influence of system damping, consider the system of Figure 18.1 which is subjected to an initial displacement x_0 and which has no external forcing function acting on it. The dependency of the response of the system on the root locations is shown in Figure 18.3 for ω_n held constant. In Figure 18.3(a) the system is very sluggish and returns to the equilibrium position very slowly. In 18.3(b), the system response is rapid, but it overshoots the equilibrium position and eventually the oscillation dies out due to the effect of damping. Figure 18.3(c) represents the critically damped system which promptly returns to equilibrium position with no overshoot. This represents the dividing line between the overdamped system (Case 2) and the underdamped system (Case 1). The roots for the underdamped system occur in complex-conjugate pairs, the roots for the overdamped system are real and unequal, and the roots for the critically damped system are real and equal. Figures 18.3(b) and 18.3(d) shows, respectively, the difference between a lightly damped system and a heavily damped system. The response of the more heavily damped system dies out more quickly.

If the roots lie on the $i\omega$ axis as shown in Figure 18.3(e), no damping occurs and the system will oscillate forever. If any roots appear on the right half of the s plane, then the response will increase monotonically with time and the system will become unstable, as shown in Figure 18.3(f). Thus the $i\omega$ axis represents the line of demarcation between stability and instability.

From equation (18.4) the natural frequency of the system is

$$\omega_n = \sqrt{\frac{k}{m}} \qquad (18.7)$$

and the damping ratio is

$$\xi = \frac{C}{2\sqrt{km}} \qquad (18.8)$$

When $\xi = 1$, critical damping occurs. We set $C = C_c$ for this value of $\xi = 1$. Then

$$\xi = 1 = \frac{C_c}{2\sqrt{km}}$$

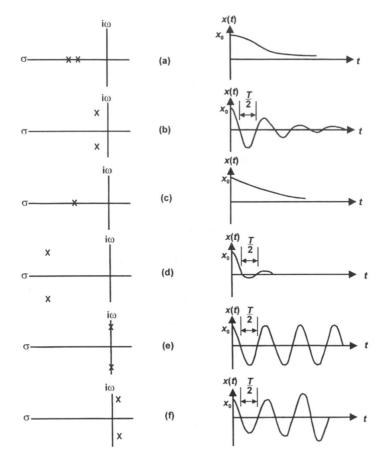

FIGURE 18.3. System responses showing the influence of root location in a second-order dynamic system.

from which we obtain

$$C_c = 2\sqrt{km} \tag{18.9}$$

as the critical damping factor. The damping ratio is often written as

$$\xi = \frac{C}{C_c} \tag{18.10}$$

Example Problem 1:

The system of Figure 18.1 has the following parameters: the weight W of mass m is 28.5 N, $C = 0.0650$ N-s/cm, and $k = 0.422$ N/cm. Determine the undamped natural frequency of the system, its damping ratio, and the type of response the system would have if the mass was to be initially displaced and released.

Solution

The mass is found from

$$m = \frac{W}{g} = \frac{28.5 \text{ N}}{9.807 \text{ m/s}^2} = 2.91 \text{ kg}$$

Then from equation (18.7)

$$\omega_n = \sqrt{\frac{42.2 \text{ N/m}}{2.91 \text{ kg}}} = 3.81 \text{ rad/s}$$

From equation (18.8)

$$\xi = \frac{C}{2\sqrt{km}} = \frac{6.50 \text{ N-s/m}}{2\sqrt{(42.2 \text{ N/m})(2.91 \text{ kg})}} = 0.293$$

Because the damping ratio is less than unity, the response will be underdamped. The system response is that of the form shown in Figure 18.3(b).

Example Problem 2.

For the system of Example Problem 1, find the amount of additional damping needed to have the system become critically damped.

Solution

From equation (18.9), the condition for a critically damped system is

$$C_c = 2\sqrt{km} = 2\sqrt{(42.2 \text{ N/m})(2.91 \text{ kg})} = 22.16 \text{ N-s/m} = 0.2216 \text{ N-s/cm}$$

The additional damping required is:

$$\Delta C = 0.2216 - 0.0650 = 0.1566 \text{ N-s/cm}$$

With this additional damping, equation (18.2) becomes

$$\ddot{x}(t) + \frac{22.16}{2.91}\dot{x}(t) + \frac{42.2}{2.91}x(t) = 0$$

The characteristic equation in the Laplace transform variable form becomes:

$$s^2 + \frac{22.16}{2.91}s + \frac{42.2}{2.91} = 0$$

or

$$(s + 3.81)^2 = 0$$

which demonstrates that Case 3 of Figure 18.3 is valid.

18.3   General Solution for the One-Degree Model of Simple System

An initial displacement is introduced to the model system of Figure 18.1, and it is desired to determine the response of this system. The differential equation in standard mathematical shorthand notation is

$$\ddot{x}(t) + 2\xi\omega_n\dot{x}(t) + \omega_n^2 x(t) = 0 \tag{18.11}$$

which undergoes a Laplacean transformation into

$$s^2 X(s) - sx(0) - \dot{x}(0) + 2\xi\omega_n[sX(s) - x(0)] + \omega_n^2 X(s) = 0$$

Let the initial displacement be represented by $x(0) = x_0$ and the initial velocity $\dot{x}(0) = 0$. The preceding equation reduces to

$$X(s) = x_0 \frac{s + 2\xi\omega_n}{s^2 + 2\xi\omega_n s + \omega_n^2} \tag{18.12}$$

The graphical residue technique, described by example problem 4 in Appendix C, can be used to find the inverse transform for equation (18.12) and hence the solution of equation (18.11):

$$x(t) = \frac{x_0}{\sqrt{1-\xi^2}} e^{-\xi\omega_n t} \cos\left(\omega_n\sqrt{1-\xi^2}t - \frac{\pi}{2} + \cos^{-1}\xi\right)$$

which reduces to

$$x(t) = \frac{x_0}{\sqrt{1-\xi^2}} e^{-\xi\omega_n t} \sin\left(\omega_n\sqrt{1-\xi^2}t + \theta\right) \tag{18.13}$$

from the trigonometric relationship $\cos(\varphi - \pi/2) = \sin\varphi$ and by setting

$$\theta = \cos^{-1}\xi \tag{18.14}$$

Example Problem 3

Consider the system of Figure 18.1 which is initially displaced and then suddenly released. Its resultant motion is described by

$$x(t) = 3.0e^{-3.43t} \sin(11.4t + 60°)$$

Find the system's damping ratio, natural frequency, and the initial displacement.

Solution

Applying equations (18.13) and (18.14),

$$\theta = 60° = \cos^{-1}\xi, \quad \xi = 0.500, \quad \text{and } \xi\omega_n = 3.43$$

Hence $\omega_n = 6.86$. The initial condition is established from

$$3.0 = \frac{x_0}{\sqrt{1 - \xi^2}} = \frac{x_0}{\sqrt{1 - (0.500)^2}} = 1.155 x_0$$

$$x_0 = 2.598$$

Because $\xi < 1$, this system is underdamped, and details of the response are shown in Figure 18.4. It is noted in the figure that because $x_1 > x_0$, that is, a slope of x exists initially, an initial velocity is present. The actual damping is $\xi \omega_n$, and the damped period is defined as

$$\tau_d = \frac{2\pi}{\omega_n \sqrt{1 - \xi^2}} \tag{18.15}$$

In Figure 18.4, it is evident that the decay rate $e^{-\xi \omega_n t}$ of the free oscillation depends on the system damping. The greater the damping, the faster is the rate of decay.

Let us now establish the relationship between the rate of decay and damping. First, the time response at two distinct points, each of which is a quarter period from the crossover points in the first two lobes, must be determined. These points are the points where the sine function equals unity and not the peak points of the damped response in Figure 18.4. From equation (18.4) and equation (18.13), it is noted that

$$x_1 = \frac{x_0}{\sqrt{1 - \xi^2}} e^{-\xi \omega_n t_1} \sin\left(\omega_n \sqrt{1 - \xi^2} t_1 + \theta\right) \tag{18.16}$$

and

$$x_2 = \frac{x_0}{\sqrt{1 - \xi^2}} e^{-\xi \omega_n (t_1 + \tau_d)} \sin\left(\omega_n \sqrt{1 - \xi^2}(t_1 + \tau_d) + \theta\right) \tag{18.17}$$

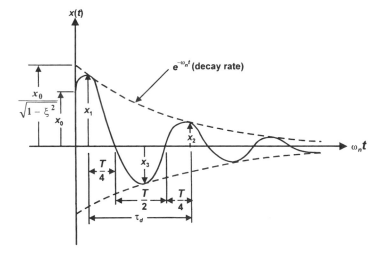

FIGURE 18.4. Response of a second-order system that is underdamped.

Taking the ratio of the two amplitudes represented by equations (18.16) and (18.17) and noting that in this case that the sine functions equals unity, we obtain

$$\frac{x_1}{x_2} = \frac{e^{-\xi \omega_n t_1}}{e^{-\xi \omega_n (t_1 + \tau_d)}} = e^{\xi \omega_n \tau_d}$$

The logarithmic decrement δ is defined at the natural logarithm of the ratio of two points such as x_1 and x_2. Invoking equation (18.14) and this definition for logarithmic decrement, we can derive

$$\delta = \ln\left(\frac{x_1}{x_2}\right) = \frac{2\pi \xi}{\sqrt{1 - \xi^2}} \tag{18.18}$$

Thus, the logarithmic decrement δ is expressed in terms of the damping ration ξ of the system.

A negative value of the response function can also be used to find the logarithmic decrement, but in this case τ_d should be replaced by $\tau_d/2$, because a half period is used in this evaluation. For this situation

$$\delta = 2 \ln\left(\frac{x_1}{x_3}\right)$$

The values of the response function at any two points can be used to find the two unknowns ξ and ω_n. It is also important to realize that the ratio of the points in the response curve, e.g. x_1/x_2, x_2/x_3, or x_3/x_4, will be identical only in the presence of viscous damping.

18.4 Forced Vibration

Forced vibration occurs when $f(t) \neq 0$ in equation (18.1). The forcing function can be a harmonic excitation, an example of which is the imbalance in rotating machinery such as a motor. Or it can be an impulse type of excitation, such as that produced by a hammer or it can be simply the weight of the moving part itself.

Harmonic Excitation

A harmonic forcing function can be represented by $F_0 \sin \omega t$. The differential equation for the model of Figure 18.1 assumes the form

$$m\ddot{x} + C\dot{x} + kx = F_0 \sin \omega t \tag{18.19}$$

With the assumption of zero initial conditions, the Laplacian transform of equation (18.19) yields

$$(ms^2 + Cs + k)X(s) = F_0\left(\frac{\omega}{s^2 + \omega^2}\right) \tag{18.20}$$

and solving for $X(s)$

$$X(s) = \frac{F_0}{ms^2 + Cs + k}\left(\frac{\omega}{s^2 + \omega^2}\right) \tag{18.21}$$

As a rule, the solution $x(t)$ will consist of two parts, viz. the complementary solution and a particular solution. The former corresponds to the transient part of the total solution and the latter to the steady-state part. The transient portion of the solution is in the form of equation (18.13) and it is determined by the residues of the complex poles, wherein the poles constitute the solution of

$$ms^2 + Cs + k = 0$$

The steady-state solution of equation (18.19) is a sinusoidal oscillation expressed as

$$x(t) = x(\omega)\sin(\omega t - \phi) \tag{18.22}$$

and this solution is determined by the residues at the complex poles $s = \pm i\omega$. While the exact solution can be derived by using the residue method, the solution of equation (18.19) can be readily obtained through the standard differential methods. The advantage of the Laplace transform method that it facilitates finding the frequency and stability information.

Through equations (18.21) and (18.22), the magnitude of the steady-state oscillation can be determined from

$$X(s) = \frac{F_0}{ms^2 + Cs + k} \tag{18.23}$$

From equation (18.23), the magnitude of the oscillation can be determined as a function of frequency. Because $s = \sigma + i\omega$ and $\omega = 2\pi f$, substituting $s = i\omega$ in equation (18.23) yields

$$x(i\omega) = \frac{F_0}{-m\omega^2 + i\omega C + k}$$

The denominator is a complex number, so that the magnitude of this number is equal to

$$\sqrt{(k - m\omega^2)^2 + (\omega C)^2}$$

The magnitude of the oscillation as a function of frequency is

$$x(\omega) = \frac{F_0}{\sqrt{(k - m\omega^2)^2 + (C\omega)^2}} = \frac{F_0/k}{\sqrt{[1 - (m/k)\omega^2]^2 + (C\omega/k)^2}} \tag{18.24}$$

From the definitions of equations (18.7)–(18.10), equation (18.24) can be expressed as

$$\frac{X(\omega)}{F_0/k} = \frac{1}{\sqrt{[1 - (\omega/\omega_n)^2]^2 + [2\xi(\omega/\omega_n)]^2}} \tag{18.25}$$

TABLE 18.1. System Response as a Function of Frequency.

Frequency	Response	Controlling Parameter
$\omega^2 \ll \omega_n^2$	$x(\omega) = \frac{F_0}{k}$	Stiffness controlled
$\omega^2 \gg \omega_n^2$	$x(\omega) = \frac{F_0}{m\omega^2}$	Mass controlled
$\omega^2 = \omega_n^2$	$x(\omega) = \frac{F_0}{c\omega}$	Damping controlled

The magnitude is now a function of only two quantities, the ratio (ω/ω_n) and the damping ration ξ. The phase angle ϕ is also a function of these two parameters and it is given by

$$\tan \phi = \frac{2\xi(\omega/\omega_n)}{1 - (\omega/\omega_n)^2} \tag{18.26}$$

The system undergoes resonance when the excitation frequency $f = \omega/2\pi$ equals the natural frequency of the dynamic system $f_n = \omega_n/2\pi$. The damping ration ξ affects the magnitude of the oscillation peak at resonance and the sharpness of this resonant peak.

In considering the system response as a function of frequency, we observe that the response varies with the frequency as shown in Table 18.1. The relationships listed in the table show that each parameter listed—the stiffness, the mass and damping—effectively controls the response only within a limited region. For example, the damping is primarily effective at resonance. The selection of any vibratory corrective measure depends on whether the excitation frequency is less than, greater than, or equal to the resonant frequency of the system.

The effect on the amplitude by stiffness, mass, or damping is exemplified by the *magnification factor* M.F., defined as

$$\text{M.F.} = \frac{x(\omega)}{F_0/k} = \frac{1}{\sqrt{[1 - (\omega/\omega_n)^2]^2 + [2\xi(\omega/w_n)]^2}} \tag{18.27}$$

At resonance $\omega = \omega_n$, and therefore

$$(\text{M.F.})_{\text{resonance}} = \frac{1}{2\xi} \tag{18.28}$$

A measure of the shape of the resonance peak is given by the bandwidth at the half-power points, as shown in Figure 18.5. These points are the two points, one to the right and one to the left of the peak, which have a magnitude equal to $(1/\sqrt{2})$ of the value of the peak. The square root occurs because power is proportional to the square of the magnitude.

Let us set $h \equiv \omega_n/\omega$. At half power points, equation (18.27) becomes

$$\frac{1}{2\sqrt{2\xi}} = \frac{1}{\sqrt{(1 - h^2)^2 + (2\xi h)^2}}$$

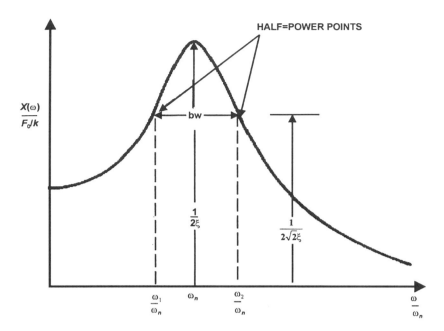

FIGURE 18.5. Principal features of a resonance peak.

Solving the preceding equation algebraically for h^2 results in

$$h^2 = 1 - 2\xi \pm 2\xi\sqrt{1 + \xi^2}$$

We also assume small values of damping (i.e., $\xi \ll 1$) and neglect second-order terms. Then the following result occurs:

$$h^2 = 1 \pm 2\xi$$

Then

$$h_1 = \left(\frac{\omega}{\omega_n}\right)^2 = 1 - 2\xi, \qquad h_2 = \left(\frac{\omega}{\omega_n}\right)^2 = 1 + 2\xi$$

Hence

$$h_2 - h_1 = 4\xi$$

We approximate

$$\frac{\omega_2^2 - \omega_1^2}{\omega_n^2} \approx \frac{2(\omega_2 - \omega_1)}{\omega_n}$$

The bandwidth bw is given by

$$\text{bw} = \frac{(\omega_2 - \omega_1)}{\omega_n} = \frac{\Delta\omega}{\omega} = 2\xi \qquad (18.29)$$

The reciprocal of the bandwidth is the quality fact Q, expressed as

$$Q = \frac{\omega_n}{\Delta \omega} = \frac{1}{2\xi} \tag{18.30}$$

When a frequency response of system is plotted in the format of Figure 18.5, this figure in conjunction with equation (18.30) can provide the basis for finding the equivalent viscous damping of the system.

Excitation by Impulse

Impulse occurs very commonly as a cause of vibration excitation in the industrial environment. An impulse force is one that acts for a very short time. A hammer is an example that provides an impulse force. Such a force can be represented by the Dirac delta function or distribution which has the following properties: consider a pulse that starts at time $t = \varepsilon$ and has width a and height $1/a$. Let a approach zero, so that the pulse essentially approaches zero width and infinite height so that

$$\int_0^\infty \delta(t - \varepsilon)\, dt = 1, \qquad 0 < \varepsilon < \alpha \tag{18.31}$$

Here $\delta(t - \varepsilon)$ is the unit impulse function at $t = \varepsilon$. Equation (18.31) states that the area or strength of the impulse is unity. If we specify a pulse of width a and height A/a, then the impulse function $A\delta(t - \varepsilon)$ is of strength A at $t = \varepsilon$.

A system subjected to impulse force of strength A has a response determined by the following equation:

$$m\ddot{x} + C\dot{x} + kx = A\delta(t) \tag{18.32}$$

The Laplacian $\mathcal{L}[A\delta(t - \varepsilon)]$ is equal to A; and so if the initial conditions are zero, then equation (18.32) transforms to

$$X(s) = \frac{A}{ms^2 + Cs + k}$$

If the system is damped slightly, then through the use of techniques described in Appendix C, the response can be calculated to yield

$$x(t) = \frac{A}{k\omega_n\sqrt{1 - \xi^2}} e^{-\xi\omega_n t} \sin \omega_n \sqrt{1 - \xi^2} t$$

where ω_n and ξ are the natural frequency and damping ratio, respectively, as defined previously.

Example Problem 4

Consider the system of Figure 18.1 in which the weight $W = 11.5\,\text{N}, k = 9.45\,\text{N/cm}$, and $C = 0.092\,\text{N/cm/s}$. The maximum value of the harmonic force exciting the system is 14.25 N (or $F = 14.25 \sin \omega t$). Find the amplitude at resonance (which represents the maximum amplitude of the steady-state motion of the mass for small damping ratios).

Solution

From equation (18.8)

$$\xi = \frac{C}{2\sqrt{km}} = \frac{9.2 \text{ N/m/s}}{2\sqrt{(945 \text{ N/m})(11.5 \text{ N/9.807 m/s}^2)}} = 0.138$$

From equation (18.27)

$$\frac{x}{F_0/k} = \frac{1}{2\xi}$$

Hence at resonance, the amplitude at resonance is

$$x_{res} = \frac{14.25 \text{ N/945 N/m}}{0.276} = 0.0546 \text{ m} = 5.46 \text{ cm}$$

Static Deflection

In Figure 18.1 the deadweight W constitutes a static load that causes a deflection δ_{st} that is given by

$$\delta_{st} = \frac{W}{k} = \frac{mg}{k} \tag{18.33}$$

where g represents the gravitational constant. Because the natural frequency of the system is found from

$$f_n = \frac{1}{2\pi}\sqrt{\frac{k}{m}} \tag{18.34}$$

we can combine equations (18.33) and (18.34) to obtain the well-known relationship between static deflection and natural frequency:

$$f_n = \frac{1}{2\pi}\sqrt{\frac{g}{\delta_{st}}} \tag{18.35}$$

Example Problem 5

A large machine weighs 875 N and the static deflection of the springs supporting the machine is 0.83 cm. Find the undamped natural frequency.

Solution

From equation (18.35)

$$f_n = \frac{1}{2\pi}\sqrt{\frac{980.7 \text{ cm/s}^2}{0.83 \text{ cm}}} = 5.47 \text{ Hz}$$

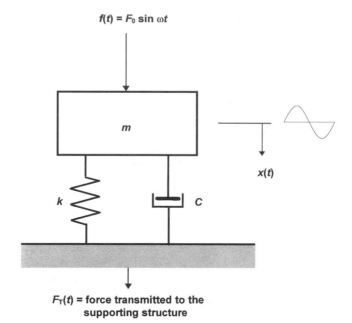

FIGURE 18.6. A spring-damper system that is subjected to force excitation.

18.5   Vibration Control

Transmissibility

In the case of a forcing function being harmonic in nature, two cases of vibration transmission can occur. One case occurs when force is transmitted to the supporting structure, and the opposite case occurs when the motion of the supporting structure is transmitted to the machine. In Figure 18.6 $f(t) = F_0 \sin \omega t$ represents the harmonic force imparted to the system and $f_T(t)$ is the force transmitted through the spring/damper system to the supporting structure. The force sent through the spring and damper to the supporting structure is given by

$$f_T(t) = kx + C\dot{x}$$

The magnitude of this force is a function of the frequency:

$$F_T = \sqrt{[kx(\omega)]^2 + [C\omega X(\omega)]^2} = kx(\omega)\sqrt{1 + \left(\frac{c\omega}{k}\right)^2} \qquad (18.36)$$

Inserting equation (18.24) into (18.36) results in

$$F_T = \frac{F_0\sqrt{1 + \left(\frac{C\omega}{k}\right)^2}}{\sqrt{\left(1 - \frac{m\omega^2}{k}\right)^2 + \left(\frac{C\omega}{k}\right)^2}} \tag{18.37}$$

Applying the definitions for ω_n and ζ, and defining *transmissibility* T as the ratio of the amplitude of the force transmitted to the supporting structure to that of the exciting force, we now have

$$T = \frac{F_T}{F_0} = \frac{\sqrt{1 + \left(\frac{2\xi\omega}{\omega_n}\right)^2}}{\sqrt{\left[1 - \left(\frac{\omega}{\omega_n}\right)^2\right]^2 + \left(2\xi\frac{\omega}{\omega_n}\right)^2}} \tag{18.38}$$

Motion Excitation

As the corollary to the force excitation model of Figure 18.6, the system for motion excitation is given in Figure 18.7. Variable x represents the motion of the dynamic system and variable y represents the harmonic displacement of the supporting base. The dynamics of this system is characterized by the following equation:

$$m\ddot{x} + C(\dot{x} - \dot{y}) + k(x - y) = 0 \tag{18.39}$$

The ratio of the magnitudes of the two displacements as a function of frequency denotes the transmissibility given by

$$T = \frac{x}{y} = \frac{\sqrt{k^2 + (C\omega)^2}}{\sqrt{(k - m\omega^2)^2 + (C\omega^2)^2}} = \frac{\sqrt{1 + \left(2\xi\left(\frac{\omega}{\omega_n}\right)\right)^2}}{\sqrt{\left[1 - \left(\frac{\omega}{\omega_n}\right)^2\right]^2 + \left(2\xi\frac{\omega}{\omega_n}\right)^2}} \tag{18.40}$$

The right-hand side of equation (18.40), which was expressed in terms of ω_n and ξ, is identical to the transmissibility of equation (18.38). This equality indicates that the methodologies employed to protect the supporting structure under force excitation are also applicable to insulating the dynamic system from motion excitation.

Equations (18.38) and (18.40) are used to plot Figure 18.8 in order to illustrate the interrelation between the damping ratio ξ, the ratio of disturbing frequency to the natural frequency ω/ω_n, and transmissibility T. If the ratio $\omega/\omega_n < \sqrt{2}$, then transmissibility $T > 1$, which means that the input disturbance is amplified, not decreased. In the region $\omega/\omega_n > \sqrt{2}$, T decreases with increasing ξ. At resonance condition $\omega/\omega_n, = 1$, transmissibility can be quite large. The region $\omega/\omega_n > 2$ is the only domain where isolation is possible, and there T (which is less than unity) decreases with decreasing ξ, and this indicates that better isolation can be achieved with very little or no damping.

The curves of Figure 18.8 thus demonstrate the effectiveness of an isolator in mitigating vibration. It is also apparent that isolators should be selected to avoid

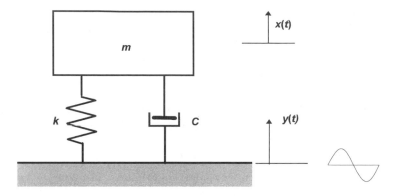

FIGURE 18.7. A spring-damper system that is subjected to motion excitation.

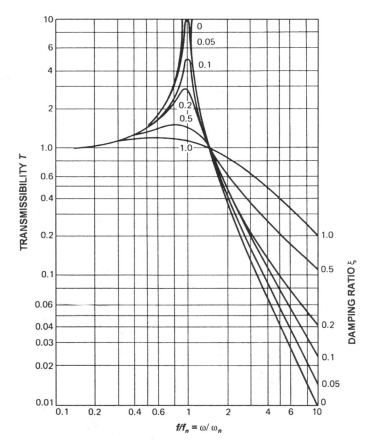

FIGURE 18.8. Plot of transmissibility versus frequency ration f/f_n as a function of the damping ratio of a linear system with one degree of freedom.

exciting the natural frequencies of the system, and that damping is important in the range of resonance, when the system is operating near resonance or merely passing through resonance during startup. From scrutiny of the isolation region it is noted that the larger the ratio ω/ω_n (or the smaller the value of ω_n), the smaller the transmissibility will be. From the relation of equation (18.7), ω_n can be made quite small by selecting soft springs, which, in conjunction with light damping, provides good isolation. The natural frequency can also be reduced by increasing the mass but this also increases the dead weight of the dynamic system, thus imposing a greater load on the supporting structure.

A measure of the effectiveness of isolation as a function of frequency can be gained by the parameter *percent isolation* (%I) which is defined by:

$$\%I = 100(1 - T)$$

where T is the transmissibility which is ≤ 1. The isolation efficiency is usually plotted with the disturbing frequency and natural frequencies as shown in Figure 18.9 for a damping ratio ξ of zero.

Example Problem 6

A machine weighs 208.5 N and is supported on a spring having a spring constant of 935.23 N/cm. A rotating mass within the machine generates a disturbing force

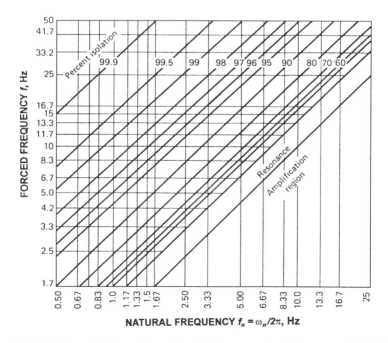

FIGURE 18.9. Isolation efficiency $[(1 - T) \times 100]$ versus natural frequency f_n of a linear single-degree-of-freedom system with zero damping ratio ξ.

of 51.23 N at 4800 rpm, owing to imbalance of the rotating mass. Determine the force transmitted to the mounting base for $\xi = 0.12$.

Solution

The static deflection of the spring is first obtained,

$$\delta_{st} = \frac{W}{k} = \frac{208.5 \text{ N}}{935.23 \text{ N/cm}} = 0.223 \text{ cm}$$

which, according to equation (18.35), gives us the natural frequency of the system

$$f_n = \frac{1}{2\pi} \sqrt{\frac{980.7 \text{ cm/s}^2}{0.223 \text{ cm}}} = 10.6 \text{ Hz}$$

The disturbing frequency generated by the imbalance is $4800/60 = 80$ Hz. It follows that

$$\frac{\omega}{\omega_n} = \frac{f}{f_n} = \frac{80}{10.6} = 7.58$$

From equation (18.40) or Figure 18.8 we find

$$T = \frac{\sqrt{1 + \left(2\xi\left(\frac{\omega}{\omega_n}\right)\right)^2}}{\sqrt{\left[1 - \left(\frac{\omega}{\omega_n}\right)^2\right]^2 + \left(2\xi\frac{\omega}{\omega_n}\right)^2}} = \frac{\sqrt{1 + (2 \times 0.12 \times 7.58)^2}}{\sqrt{[1 - (7.58)^2]^2 + (2 \times 0.12 \times 7.58)^2}}$$

$$= \frac{2.076}{\sqrt{3187.33 + 3.509}} = \frac{2.076}{56.49} = 0.0367$$

The force transmitted to the mounting structure is

$$F_T = (0.0367)(51.23 \text{ N}) = 1.88 \text{ N}$$

18.6   Techniques for Vibration Control

Modification of Source

It is often possible to mitigate vibration by modifying the source of the vibration. For example, the vibrating structure may be made more rigid, closer tolerances of machining can eliminate or lessen imbalance, or the system mass and stiffness can be adjusted so that the resonant frequencies of the system do not coincide with the forcing frequency (this procedure is called *detuning*), or it may be possible to cut down on the number of coupled resonators between the vibrational source and the component of interest (this procedure is called *decoupling*).

Isolation

There are three principal types of isolators, viz. (a) metal springs, (b) elastomeric pads, and (c) resilient pads. Metal springs are generally the best for low-frequency isolation, and they possess the advantage of being impervious to the effects of temperature, humidity, corrosion, presence of solvents and they allow maximum deflections. But springs have almost no damping and hence can lead to very high transmissibility at resonance. But dampers can be included in parallel with springs to minimize resonance effects.

Elastomeric mounts are generally constructed of either natural rubber or synthetic rubber materials such as neoprene. Synthetic rubber is generally far more impervious to environmental effects than natural rubber. These types of mounts are generally used to isolate relatively small electrical and mechanical devices from relatively high forcing frequencies. Elastomeric compounds inherently contain damping. Rubber can be used in either tension, compression, or shear modes, but it is generally used in compression or shear and hardly ever used in tension mode.

Isolation or resilient pads include a variety of materials such as cork, felt, fiberglass, and special plastics. These items can be purchased in sheets and cut into gaskets to fit the particular application. Also, they can be stacked to provide different degrees of isolation. Cork is available in squares 1–2.5 cm thick. Cork can withstand corrosion and solvents, but felts being made of organic materials cannot be utilized in an environment where solvents are present. Cork and felt have damping ratios that typically range from $\xi = 0.05$ to 0.06.

Inertia Blocks

Certain very large machines such as reciprocating compressors generate large inertia forces that result in unacceptably large motion which can cause the machine to function improperly. One means of limiting this motion is to mount the equipment on an inertia base, which consists of a heavy steel or concrete mass. This mass limits the motion by the dint of its heaviness overcoming the inertia forces produced by the mounted equipment. Also, low natural frequency isolation needs a large deflection isolator such as a soft spring. But the use of soft springs can lead to rocking motions that cannot be tolerated. Hence, an inertia block mounted on the appropriate isolators can serve to effectively limit the motion as well as provide the required isolation. Inertia blocks also help to lower the center of gravity and thus provide an additional degree of stability. The introduction of additional mass through inertia blocks decreases vibrational amplitudes and minimizes rocking. Alignment errors can be minimized because of the greater stiffness of the base, and moreover, these blocks can serve as a noise barrier between the floor on which they are mounted and the mounted equipment itself.

Vibration Absorbers

In order to better understand its operation, let us consider the vibration absorber model of Figure 18.10. The components of the vibration absorber are m_2 and k_2.

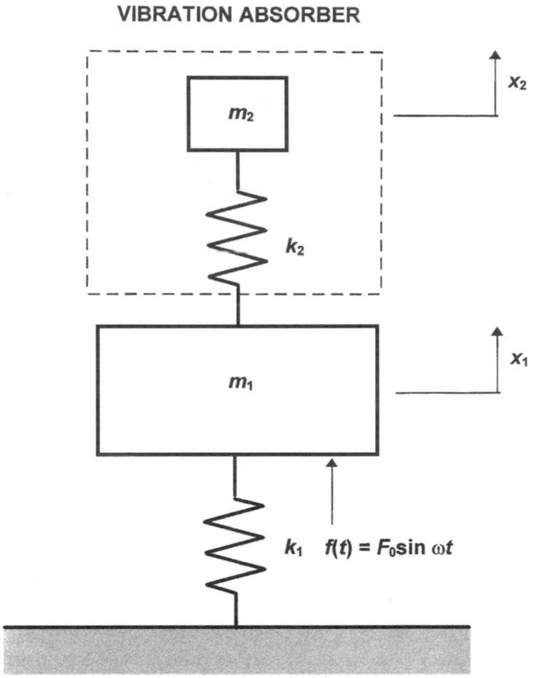

VIBRATION ABSORBER

FIGURE 18.10. The analytical model for a vibration absorber.

The applicable equations of motion for the system are

$$m_1\ddot{x}_1 + k_1 x_1 + k_2(x_1 - x_2) = f$$
$$m_2\ddot{x}_2 + k_2(x_2 - x_1) = 0$$

The frequency response can be gotten from the following transformed equations:

$$-m_1\omega^2 X_1 + (k_1 + k_2)X_1 - k_2 X_2 = F_0$$
$$-k_2 X_1 + (-m_2\omega^2 + k_2)X_2 = 0$$

Rearranging these last two equations:

$$(-m_1\omega^2 + k_1 + k_2)X_1 - k_2 X_2 = 0 \qquad (18.41)$$

$$-k_2 X_1 + (-m_2\omega^2 + k_2)X_2 = 0 \qquad (18.42)$$

The two simultaneous equations may now be written in the matrix format as follows:

$$\begin{bmatrix} -m_1\omega^2 + k_1 + k_2 & -k_2 \\ -k_2 & -m_2\omega^2 + k_2 \end{bmatrix} \begin{bmatrix} X_1 \\ X_2 \end{bmatrix} = \begin{bmatrix} F_0 \\ 0 \end{bmatrix} \qquad (18.43)$$

Now consider what happens to equations (18.41)–(18.43) when the forcing frequency ω equals the natural frequency $\omega_2 = (k_2/m_2)^{1/2}$ of the vibration absorber.

This condition leads to

$$X_1 = 0$$
$$X_2 = -\frac{F_0}{k_2} \tag{18.44}$$

From equations (18.44) it can be perceived that when the natural frequency of the vibrating absorber is tuned to the vibrational forcing frequency, the motion of the principal mass m_1 is ideally zero and the spring force of the absorber is at all times equal and opposite to the applied force F_0. It follows that no net force is transmitted to the supporting structure. Because the natural frequency must be tuned to the vibration forcing frequency, the absorber is customarily used for constant speed machinery. Even though this isolation feature is useful over a broad band of frequencies, absorption is useful principally for very narrow-band or single-frequency control.

Active Systems

Active isolation and absorption systems normally incorporate a feedback control, which may be electromagnetic, electronic, fluidic, pneumatic, mechanical, or a combination thereof. Because the cost of such systems is quite high, they are used mainly in precision instruments such as electronic microscopes, lasers, stabilized platforms, and other devices where a high degree of isolation is required. Active control is also applied where a high degree of isolation is required, for example, in the case of an air-conditioning system that is mounted on the top floor of a building above work areas and offices. Because the cost of computerization is continuously falling, active control is becoming a more viable option in many applications (including automotive suspensions), particularly for isolating very low-frequency vibration (<1 Hz) and also for tracking changes in natural frequencies due to shifting modes of operation.

Damping

Of the elements of vibration models, damping constitutes the most difficult of parameters to deal with, but it does provide an important means of cutting down vibration. Damping reduces the transmission of vibration through a structure, decreases amplitudes of resonance, and attenuates free vibration and vibrations caused by impact. The reduction of vibrational energy in structures as the result of damping lessens the amount of noise radiated by the structural areas.

A measure of damping is that of the *system loss factor*, denoted by η_s and defined as the ratio of damping energy loss per radian to the peak potential energy. Let us assume an oscillation represented by

$$x(t) = A \sin(\omega t + \phi)$$

Then the velocity is

$$\dot{x}(t) = A\omega \cos(\omega t + \phi)$$

The damping force is given by $C\dot{x}(t)$; and the amount of energy dissipated through a spatial increment dx is $C\dot{x}(t)dx$. Because $\dot{x} = dx/dt$, we write

$$C\dot{x}(t)\,dt = C\dot{x}^2(t)\,dt = \frac{1}{\omega}C\dot{x}^2(t)\,d(\omega t)$$

Then the energy dissipated in a cycle is obtained from

$$\frac{C}{\omega}\int_0^{2\pi}\dot{x}^2(t)\,d(\omega t) = CA^2\omega\int_0^{2\pi}\cos^2(\omega t + \phi)\,d(\omega t) = \pi CA^2\omega$$

The energy dissipated per radian is $CA^2\omega/2$, and the peak potential energy is $kA^2/2$. Therefore, the system loss factor is

$$\eta_s = \frac{CA^2\omega/2}{kA^2/2} = \frac{C\omega}{k} \tag{18.45}$$

Introducing equations (18.8) and (18.9), equation (18.45) becomes

$$\eta_s = 2\xi\frac{\omega}{\omega_n} \tag{18.46}$$

At resonance, η_s assumes the value of 2ξ. This indicates that systems with large loss factors are highly damped. For small values of damping, applying equations (18.18), (18.28), and (18.29):

$$\eta_s = 2\xi = \frac{\delta}{\pi} = \frac{1}{(\text{M.F.})_{\text{resonance}}} = \text{bw}$$

In the analysis of structural damping, complex stiffness and complex moduli are principal parameters. We recall that the steady-state oscillation is described by the following expression,

$$(-\omega^2 m + i\omega C + k)X = F_0$$

which can be rewritten as

$$(-\omega^2 m + k^*)X = F_0$$

where

$$k^* = k + i\omega C$$

The imaginary stiffness is the damping. Employing equation (18.45), we can express the complex stiffness as

$$k^* = k(1 + i\eta_s)$$

In an analogous manner, the complex Young's modulus E for damping material can expressed as

$$E^* = E(1 + i\eta_M)$$

where E is the real part of the Young's modulus of the damping material and η_M is the loss factor of the damping material.[1]

The loss factors and complex moduli vary with frequency and temperature. It has been assumed that the damping force is proportional to velocity and independent of amplitude, but there are situations where the damping of the materials depends mainly on vibration amplitude, so the assumption of viscous damping must be applied judiciously.

Damping converts mechanical energy into thermal energy; and while there are a number of mechanisms for such energy conversion, we describe here those that are the most useful ones. Damping materials also reduce sound transmission. When a sound wave strikes a structure, causing its surface to vibrate, the vibrating surface produces a reflected wave and a transmitted wave [Figure 18.11(a)]. The transmission loss through the structure varies with frequency as shown in Figure 18.11(b) for a given temperature. The region of damping control nestles between the low-frequency region where stiffness reigns as the controlling parameter and the higher-frequency region where mass predominates as the controlling

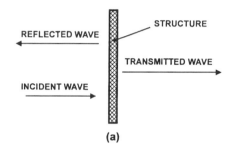

(a)

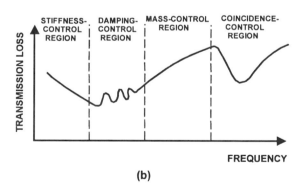

(b)

FIGURE 18.11. Sound transmission characteristics: (a) incident wave resolving into a reflected wave and a transmitted wave and (b) transmission loss in a structure as a function of frequency.

---

[1] η_S is the loss factor of the system structure *and* damping material.

parameter. Between these two regions, many natural vibratory modes of the structure exist, and this region is the only one where transmission loss depends greatly upon resonance conditions. Here structural damping is the controlling parameter.

Internal Damping

A material with very high damping internal damping properties could be utilized to eliminate noise emanating from a structure. Ferromagnetic materials and certain magnesium and cobalt alloys exhibit such properties but these materials are generally too costly to use as structural materials and they may not meet the strength criteria.

Damping Mechanisms

Figure 18.12 shows three types of damping mechanisms, namely, the tuned damper, free viscoelastic layer, and constrained viscoelastic layer.

The tuned damper of Figure 18.12(a) consists of a mass attached to a point of vibration through a spring and dashpot or a viscoelastic spring. This device is not very useful because it functions well at only a single frequency or in a very

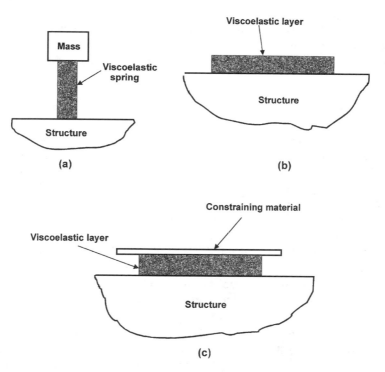

FIGURE 18.12. Different types of damping mechanisms: (a) tuned damper, (b) free viscoelastic layer, and (c) constrained viscoelastic layer.

narrow frequency band. Moreover, any change in temperature is likely to change the tuning frequency of the damper.

The other two mechanisms of Figure 18.12 entail adding viscoelastic layers to a structure that is to be damped. When a structure consisting of different layers of materials undergoes bending, the layers will extend or deform in shear. The resultant deformation causes energy dissipation, a phenomenon that constitutes the basis for the two viscoelastic mechanisms.

In the mechanism of Figure 18.12(b), a layer of free (i.e., uncovered) viscoelastic material is bonded to the main structure. The damping material thickness should be about one-third the thickness of the structure or wall thickness. One thick layer on one side of the structure is generally more effective than two lesser layers on either side of the structure. This technique, which entails extensional damping, results in an economical, highly damped structure that is easy to fabricate.

The mechanism of Figure 18.12(c) is essentially a sandwich arrangement, where a viscoelastic layer is added to the structure. The covered (hence, constrained) viscoelastic layer provides high extensional damping, but the entire structure becomes harder and more expensive to assemble together.

In order to be effective, the damping action must store a major portion of the energy present in the entire system. Damping is best applied to points where stretching or bending is the maximum, because these are the locations of maximal energy storage.

Viscoelastic layer techniques can be employed in a large variety of applications ranging from walls, enclosures, barriers, conveyers, chutes, racks, and hoppers to the most specialized, technologically sophisticated electronic instruments.

18.7   Finite Element Analysis

We have used relatively simple mathematical models to deal with vibration, but the analysis of plates, shells, and other continuous systems can be difficult to analyze without the aid of computers. Experimental modal analysis can provide needed information, but this requires that a real structure be constructed and instrumented to yield the desired data. Finite-element analysis (FEA), however, allows the problem to be represented with some detail and permits the designer to optimize a design by investigating the effects of minor changes in the model upon the static and dynamic states of the structure being evaluated. FEA is also used not only to determine the statics and dynamics of beams, plates, shells, trusses, and other solid bodies, but also to treat problems involving fluids, including airborne noise propagation. A number of FEA programs can predict stresses and strains, temperature distribution stemming from heat and mass transfer as well as vibratory states. Transient states are amenable to treatment by FEA. Because FEA is a powerful analytical tool, it no longer became necessary to fabricate a series of actual structures before freezing a design. The most prominent programs containing general codes for vibrational analysis include NASTRAN, developed by the U.S. National Aeronautics and Space Administration and now in public domain; MSC-NASTRAN®, a proprietary code developed from NASTRAN coding available from MacNeal-Schwindler Corp; and ANSYS®, another proprietary code that is available from Swanson Analysis System, Inc.

A structure undergoing analysis is modeled by subdividing it into various types of finite elements from the finite-element library. The library may include more than 100 element types, including beam elements, triangular and rectangular plate and shell elements, conical shell elements, and mass, damping, and stiffness elements. The preprocessing stage in FEA entails the creation of a finite-element mesh to depict the structure being evaluated. The elements may be sized automatically by proportional spacing so that the mesh is denser in areas of greater concern (particularly where sudden changes or discontinuities in geometry can lead to higher values of stresses or steeper temperature gradients). Constraints and forcing functions are specified. The input data is converted into matrix format by the FEA code. The form of matrix equation is

$$\mathbf{ma} + \mathbf{Cv} + \mathbf{kx} = \mathbf{F}$$

where

$\mathbf{m}$, $\mathbf{C}$, and $\mathbf{k}$ = the mass, damping, and stiffness matrices, respectively, for the finite element representation of the structure being analyzed

$\mathbf{a}$, $\mathbf{v}$, and $\mathbf{x}$ = the acceleration, velocity, and displacement vectors, respectively

$\mathbf{F}$ = the forcing function vector

The matrices are manipulated to determine natural frequencies, frequencies, mode shapes, response amplitudes, and/or dynamic stresses due to harmonic, random, or transient forcing functions. In the postprocessing phase, the FEA program arranges the output in a convenient format for engineering evaluation. The deformed structure can be portrayed graphically as an overlay over the undeformed mesh diagram. In order to show the difference, mode shapes or displacement amplitudes are exaggerated in the display. In almost all situations, a great deal of engineering time must be spent in defining even a simple problem by modeling it and setting up a mesh. But once a solution is obtained, redesign to improve vibrational characteristics can be done quickly and effectively by modifying the stored data. FEA software can be installed and used on personal computers, but for more elaborate designs it may be necessary to use minicomputers or mainframes in order to deal with much larger matrices (cf. Brooks, 1986; Hughes, 1987; Grandin, 1986; and Huebner and Thornton, 1982).

18.8   Vibration Measurements

Vibration displacement and vibrational velocity can be measured but the most common measurements of vibration are those of acceleration. The basic transducer used to measure vibration and shock is the *accelerometer*. Most of the other components in vibration measurement systems are similar to those used to measure airborne sound (cf. Chapter 9). Many instruments such as the fast Fourier transform (FFT) analyzers are designed to be used for both acoustic and vibration applications.

Accelerometers

Accelerometers are usually mounted directly on a vibrating body, using a threaded stud, adhesive, or wax. Other versions mount the accelerometer in a probe that is held against a vibrating body. A typical accelerometer houses one or more piezoelectric elements against which rests a mass that is preloaded by a stiff spring. When an accelerometer becomes subject to acceleration imparted by the vibration being measured, the mass exerts a force on the piezoelectric element that is proportional to the acceleration. The charged developed in the piezoelectric element is, in turn, proportional to the force. The accelerometer may be designed so that the piezoelectric element is stressed in compression or in shear. The piezoelectric elements are usually quartz crystals or specially processed ceramic materials.

There is more than one type of sensitivity of interest in accelerometer specifications. The *charge sensitivity*, measured in picocoulombs/g, and *voltage sensitivity*, measured in terms of mV/g, are important, depending whether the accelerometer is used with charge measuring or voltage measuring equipment. Also, the *transverse sensitivity*, which is the sensitivity to acceleration in a plane normal to the principal accelerometer axis should be a low value, preferably less than 3% of the main axis sensitivity at low frequencies.

The preamplifier, which constitutes the second stage in signal processing, serves two purposes: one is to amplify the vibration pickup signal, which is generally quite weak, and the other is to serve as an impedance transformer between the accelerometer and the subsequent chain of equipment. A preamplifier may be designed to function as a voltage amplifier in which case the output voltage is directly proportional to the input voltage or it may function as a charge amplifier in which case the output voltage is proportional to the input charge Each type of amplifier has its own advantages and disadvantages. When a charge amplifier is used, changes in cable length (which modify the cable capacitance) have negligible effects on the measurements. When a voltage amplifier is employed, the system will be extremely sensitive to changes in cable capacitance. Because the input resistance of a voltage amplifier cannot be disregarded, the extremely low-frequency response of the system may be affected. But voltage amplifiers are usually less expensive and may be more reliable because they contain fewer components.

18.9   Random Vibrations

Most of the vibration problems discussed earlier in this chapter are *deterministic*, that is, the forcing function can be described as a function of time. An example of a deterministic problem is the imbalance of a shaft operating at a known speed.

Random vibrations, on the other hand, result from excitation which can be only described statistically. Structural excitation of an aircraft fuselage due to jet engine noise or turbulent flow is considered to be a *random process*. Both frequency and amplitude vary and they do not establish a deterministic pattern.

Probability Density

Because the amplitude or acceleration of random vibration cannot be determined as a function of time, it is described in terms of its probability density. One model widely used is the Gaussian distribution or normal probability density curve expressed as follows in normalized form:

$$p(x) = \frac{1}{\sigma\sqrt{2\pi}} e^{-0.5x^2/\sigma^2} \tag{18.47}$$

and the probability of the value of x falling between a and b is

$$P(a < x < b) = \int_a^b p(x)\,dx \tag{18.48}$$

where

$p(x) =$ the probability density of the function

$x =$ the amount the function differs from the mean

$\sigma =$ the standard deviation

$P =$ the probability of x falling within a particular range

It is apparent that the total area under the probability density curve must be unity:

$$P(-\infty < x < \infty) = \int_{-\infty}^{+\infty} p(x)\,dx = 1$$

Mean-Square Value and Autocorrelation

The temporal mean square of a function is defined by

$$x_{rms} = \sqrt{\lim_{T\to\infty}\left(\frac{1}{T}\int_0^T x^2(t)\,dt\right)} \tag{18.49}$$

where

$x_{rms} =$ the root mean square of function x

$T =$ the time interval

The temporal autocorrelation function describes, on the average, the way in which the instantaneous value of a function depends on previous values. It is given by

$$R(\tau) = \lim_{T\to\infty)}\left(\frac{1}{T}\int_0^T x(t)x(t+\tau)\,dt\right) \tag{18.50}$$

where

$$R(\tau) = \text{autocorrelation fuction}$$

$$\tau = \text{the time interval between measurements}$$

$$t = \text{time}$$

Ergodic Processes

A function may vary in such a manner that there is no narrow time interval which can be truly representative of the function. However, if the time interval is sufficiently long and the probability distribution functions are independent of the time interval during which they were measured, the function then represents a stationary process. This means that the mean square value measured during one interval should be equal to the mean square value measured during a later interval. The autocorrelation function will also be unaffected by a time shift. If mean-square values and the autocorrelation functions of a number of ensembles of data are equal to the temporal values, then the process may be considered *ergodic* as well as stationary.

Spectral Density

The power spectral density, also known as the mean-square spectral density, can be determined from the autocorrelation function as follows:

$$S(\omega) = \frac{\int_{-\infty}^{+\infty} R(\tau)e^{-\omega\tau}\,d\tau}{2\pi} \tag{18.51}$$

where

$$S(\omega) = \text{mean-square spectral density, in } g^2/(\text{rad/s}) \text{ or } (\text{m/s}^2)^2/(\text{rad/s})$$

$$\omega = \text{radial frequency, rad/s}$$

Equation (18.51) is used in analytical studies. The inverse relationship is

$$R(\tau) = \int_{-\infty}^{+\infty} S(\omega)e^{i\omega\tau}\,d\omega$$

Vibration measurements obtained from an FFT analyzer or another spectral instrument may be expressed in dB re 1 g or dB re 1 m/s^2, within each frequency band. These values may be converted to mean square spectral density $W(f)$ (usually expressed in g^2/Hz or some other engineering units), where $W(f)$ and $S(\omega)$ are related by

$$W(f) = 4\pi S(\omega)$$

where

$$W(f) = \text{spectral density, units}^2/\text{Hz, defined for positive frequencies only}$$

f = frequency, Hz

$S(\omega)$ = spectral density, units^2/(rad/s), defined for both positive and
negative frequencies (a mathematical artifice)

If the vibration measurements are sufficiently representative and the process is
stationary, the mean square value is then given by

$$x_{rms}^2 = \lim_{T \to \infty} \left(\frac{1}{T} \int_0^T x^2(t)\, dt \right) = \int_0^\infty W(f)\, df = R(0)$$

where

$$x_{rms}^2 = \text{mean-square value, } g^2$$
$$R(0) = \text{the autocorrelation function for } \tau = 0$$

White Noise

White noise is a random signal which has a constant mean-square spectral density
for all frequencies from zero to infinity, that is,

$$W_{\text{white}}(f) = W_0 \qquad \text{for } 0 < f < \infty$$

This idealization cannot be achieved physically, because it would amount to requiring an infinite amount of power. More realistically, band-limited white noise
can be achieved and it is a random signal having a constant spectral density over
a specified range:

$$W_{\text{white}}(f) = W_0 \qquad \text{for } f_1 < f < f_2$$

White noise generators are designed to generate signals with random vibration
in amplitude and frequency, with relatively constant spectral density over various
frequency ranges.

References

Beranek, Leo L. (ed.). 1971. *Noise and Vibration Control*. New York: McGraw-Hill.

Broch, J. T. 1973. *Mechanical and Shock Measurements*. Nærum, Denmark: Brüel & Kjær
Instruments Company.

Brooks, P. 1986. Solving vibration problems on a PC. *Sound and Vibration* 20(11): 26–32.

Crandall, S. H., and W. D. Mark. 1963. *Random Vibration in Mechanical Systems*.
New York: Academic Press.

Crede, C. E. 1965. *Shock and Vibration Concepts in Engineering Design*. Englewood Cliffs,
NJ: Prentice-Hall.

Grandin, H. 1986. *Fundamentals of the Finite Element Method*. New York: Macmillan.

Harris, Cyril M. (ed.). 1991. *Handbook of Acoustical Measurements and Noise Control*,
3rd ed. New York: McGraw-Hill; Chs. 6–10.

Harris, C. M. (ed.). 1988. *Shock and Vibration Handbook*, 3rd ed. New York: McGraw-Hill.

Huebner, K. H., and E. A. Thornton. 1982. *The Finite Element Method for Engineers*, 2nd ed. New York: Wiley-Interscience.

Hughes, T. J. R. 1987. *The Finite Element Method: Linear Static and Dynamic Finite Element Analysis*. Englewood Cliffs, NJ: Prentice-Hall.

Problems for Chapter 18

1. In the spring-dashpot system of Figure 18.1, the weight $W = 50$ N, the spring constant $k = 0.30$ N/cm, and damping ratio $\xi = 0.35$. Find the natural frequency and the viscous damping of the system.

2. Establish the characteristic equation for the spring-dashpot system of Problem 1 and determine the additional viscous damping needed to yield a critically damped system.

3. In the system of Problem 1, the mass is displaced 15 cm and then suddenly released. Set up the equation for the system response, which describes the position of the mass as a function of time.

4. Consider a spring-dashpot system which has the following parameters: $W = 40$ N, $C = 0$, and $k = 0.39$ N/cm. If the mass is initially displaced and released, how will the system response?

5. In a spring-dashpot system, $W = 40$ N, $C = 0.10$ N/cm/s and a forcing function $4.0 \sin \omega_n t$ drives the system. What kind of system response will occur as a function of time?

6. In a spring-damper system of Figure 18.1, $W = 250$ N, $C = 0.13$ N/cm/s, and $k = 32$ N/cm. How much does the spring deflect under the dead weight load of the mass? What is the natural frequency of the system? If the system undergoes a forced harmonic oscillation described by $F = F_0 \sin 12t$, which one of the system parameters effectively controls the response of the system?

7. A mass is supported by four identical springs, each having a spring constant of 2.6 N/cm. If the spring deflects 1.5 cm, what is the weight of the mass being supported?

8. Consider the system described in the last problem. It is forced with a sinusoidal signal with a frequency twice that of its natural frequency. Find the degree of vibration isolation that will be obtained.

9. A mechanical system is being designed for installation in a plant. It bears the following parameters: $W = 2000$ N and $k = 3.65$ N/cm. If the forcing frequency $= 12$ rad/s, how much damping can the systems tolerate if the isolation must exceed 90%?

10. In a spacecraft, delicate electronic sensors have to be isolated from a panel that vibrates at 50 rad/s. It is required that at least 90% vibration isolation can be achieved by using springs to protect the equipment. Assume that the damping ratio $\xi = 0$. What static deflection is required?

Appendix A
Physical Properties of Matter

A. Solids

Solid	Density, ρ (kg/m^3)	Young's Modulus, E (Gpa)	Shear Modulus, G (Gpa)	Poisson's Ratio, ν	Speed, c (m/s)	Characteristic Impedance, $\rho_0 c$ $10^6 \times$ Pa·s/m
Aluminum	2,700	71	24	0.33	5150	13.9
Brass	8,500	104	38	0.37	3500	29.8
Copper	8,900	122	44	0.35	3700	33.0
Iron (cast)	7,700	105	44	0.28	3700	28.5
Lead	11,300	16.5	5.5	0.44	1200	13.6
Nickel	8,800	210	80	0.31	4900	43.0
Silver	11,300	78	28	0.37	2700	28.4
Steel	7,700	195	83	0.28	5050	39.0
Glass (Pyrex)	2,300	62	25	0.24	5200	12.0
Quartz (X-cut)	2,650	79	39	0.33	5450	14.5
Lucite	1,200	4	1.4	0.4	1800	2.15
Concrete	2,600	—	—	—	—	—
Ice	920	—	—	—	—	—
Cork	240	—	—	—	—	—
Wood, oak	720	—	—	—	—	—
Wood, pine	450	—	1	0.4	1450	1.6
Hard rubber	1,100	2.3	—	0.5	70	0.065
Soft rubber	950	0.005	—	—	—	—
Rubber, ρ-c	1,000	—	—	—	—	—

B. Liquids

Liquid	Temperature (°C)	Density, ρ_0 (kg/m^3)	Bulk Modulus, B (Gpa)	Ratio of Specific Heats, γ	Speed (m/s)	Characteristic Impedance, ρ_0/c 10^6(Pa · s/m)	Coefficient of Shear Viscosity, η (Pa · s)
Alcohol (ethyl)	20	790	—	—	1150	0.91	0.0012
Castor oil	20	950	—	—	1540	1.45	0.96
Glycerine	20	1,260	—	—	1980	2.5	1.2
Mercury	20	13,600	25.3	1.13	1450	19.7	0.0016
Seawater	13	1,026	2.28	1.01	1500	1.54	0.001
Turpentine	20	870	1.07	1.27	1250	1.11	0.0015
Water (fresh)	20	998	2.18	1.004	1481	1.48	0.001
Fluidlike sea bottoms							
Red clay	—	1,340	—	—	1460	1.96	—
Calcareous ooze	—	1,570	—	—	1470	2.31	—
Coarse silt	—	1,790	—	—	1540	2.76	—
Quartz sand	—	2,070	—	—	1730	3.58	—

C. Gases (at one atmosphere pressure, 101.3 kPa)

Gas	Temperature (°C)	Density, ρ_0 (kg/m^3)	Ratio of Specific Heats, γ	Speed (m/s)	Characteristic Impedance, $\rho_0 c$ (Pa·s/m)	Coefficient of Shear Viscosity, η (Pa·s)
Air	0	1.293	1.402	331.6	428	0.000017
Air	20	1.21	1.402	343	415	0.0000181
Hydrogen	0	0.09	1.41	1269.5	114	0.0000088
CO_2 (low-frequency)	0	1.98	1.304	258	512	0.0000145
CO_2 (high-frequency)	0	1.98	1.40	268.6	532	0.0000145
Oxygen	0	1.43	1.40	317.2	453	0.00002
Steam	100	0.6	1.324	404.8	242	0.000013

D. Conversion Factors

To Convert	Into	Multiply By	Conversely, Multiply By
atm (atmosphere)	mm Hg at 0°C	760	1.316×10^{-3}
	lb/in.^{2}	14.70	6.805×10^{-2}
	N/m^2 (Pa)	1.0132×10^5	9.872×10^{-6}
	kg/m^2	1.033×10^4	9.681×10^{-5}
°C (Celsius)	°F (fahrenheit)	$[(°C \times 9)/5] + 32$	$(°F - 32) \times 5/9$
cm (centimeter)	in. (inch)	0.3937	2.540
	ft (foot)	3.281×10^{-2}	30.48
	m (meter)	10^{-2}	10^2
cm^2 (square centimeter)	in.^{2}	0.1550	6.452
	ft^2	1.0764×10^{-3}	929
	m^2	10^{-4}	10^4
cm^3 (cubic centimeter)	in.^{3}	0.06102	16.387
	ft^3	3.531×10^{-5}	2.832×10^4
	m^3	10^{-6}	10^6
dyne	lb (force)	2.248×10^{-6}	4.448×10^5
	N (newton)	10^{-5}	10^5
dynes/cm^2	lb/ft^2 (force)	2.090×10^{-3}	478.5
	N/m^2 (Pa)	10^{-1}	10
ft (foot)	in. (inch)	12	0.08333
	cm (centimeter)	30.48	3.281×10^{-2}
	m (meter)	0.3048	3.281
ft^2 (square foot)	in.^{2}	144	6.945×10^{-3}
	cm^2	9.290×10^2	0.010764
	m^2	9.290×10^{-2}	10.764
ft^3 (cubic foot)	in.^{2}	1728	5.787×10^{-4}
	cm^3	2.832×10^4	3.531×10^{-5}
	m^3	2.832×10^{-2}	35.31
hp (horsepower)	W (watt)	745.7	1.341×10^{-3}
in. (inch)	ft (foot)	0.0833	12
	cm (centimeter)	2.540	0.3937
	m (meter)	0.0254	39.37

(Continued)

D. Conversion Factors (*continued*)

To Convert	Into	Multiply By	Conversely, Multiply By
in.^{2} (square inch)	ft^2	6.945×10^{-3}	144
	cm^2	6.452	0.1550
	m^2	6.452×10^{-4}	1550
in.^{3} (cubic inch)	ft^3	5.787×10^{-4}	1.728×10^3
	cm^3	16.387	6.102×10^{-2}
	m^3	1.639×10^{-5}	6.102×10^4
kg (kilogram)	lb (weight)	2.2046	0.4536
	slug	0.06852	14.594
	g (gram)	10^3	10^{-3}
kg/m^2	lb/in.^{2} (weight)	1.422×10^{-3}	703.0
	lb/ft^2 (weight)	0.2048	4.882
	g/cm^2	10^{-1}	10
m (meter)	in. (inch)	39.371	2.540×10^{-2}
	ft (foot)	3.2808	0.30481
	cm (centimeter)	10^2	10^{-2}
m^2 (square meter)	in.^{2}	1550	6.452×10^{-4}
	ft^2	10.764	9.290×10^{-2}
	cm^2	10^4	10^{-4}
m^3 (cubic meter)	in.^{3}	6.102×10^4	1.639×10^{-5}
	ft^3	35.31	2.832×10^{-2}
	cm^3	10^6	10^{-6}
microbar (dynes/cm^2)	lb/in.^{2}	1.4513×10^{-5}	6.890×10^4
	lb/ft^2	2.090×10^{-3}	478.5
	N/m^2 (Pa)	10^{-1}	10
Np (neper)	dB (decibel)	8.686	0.1151
N (newton)	lb (force)	0.2248	4.448
	dynes	10^5	10^{-5}
N/m^2 (pascal, Pa)	lb/in.^{2} (force)	1.4513×10^{-2}	6.890×10^3
	lb/ft^2 (force)	2.090×10^{-2}	47.85
	dynes/cm^2	10	10^{-1}
lb (force) (pound)	N (newton)	4.448	0.2248
lb (weight) (pound)	slug	0.03108	32.17
	kg (kilogram)	0.4536	2.2046
lb/in.^{2} (weight)	lb/ft^2 (weight)	144	6.945×10^{-3}
	kg/m^2	703	1.422×10^{-3}
lb/in.^{2} (force)	lb/ft^2 (force)	144	6.945×10^{-3}
	N/m^2 (Pa)	6894	1.4506×10^{-4}
lb/ft^2 (weight)	lb/in.^{2} (weight)	6.945×10^{-3}	144
	g/cm^2	0.4882	2.0482
	kg/m^2	4.882	0.2048
lb/ft^2 (force)	lb/in.^{2} (force)	6.945×10^{-3}	144
	N/m^2 (Pa)	47.85	2.090×10^{-2}
slugs	lb (weight)	32.17	3.108×10^{-2}
	kg (kilogram)	14.594	6.852×10^{-2}
W (watt)	hp (horsepower)	1.341×10^{-3}	745.7

Appendix B
Bessel Functions

B.1 The Bessel Differential Equation

The Bessel differential equation of order n, expressed as

$$\left[x^2 \frac{d^2}{dx^2} + x \frac{d}{dx} + (x^2 - n^2) \right] f(x) = 0$$

carries the solutions consisting of (1) the Bessel functions of the first kind $J_n(x)$ for all x, (2) the Bessel functions of the second kind $Y_n(x)$ (also called Neuman functions), and (3) the Bessel functions of the third kind $H_n^{(1)}(x)$ and $H_n^{(2)}(x)$ for all x greater than zero.

B.2 Relationships Between Solutions

$$H_n^{(1)} = J_n + i Y_n$$
$$H_n^{(2)} = J_n - i Y_n$$
$$J_{-n} = (-1)^n J_n$$
$$Y_{-n} = (-1)^n Y_n$$

B.3 Series Expansions for J_0 and J_1

$$J_0 = 1 - \frac{x^2}{2^2} + \frac{x^4}{2^2 \cdot 4^2} - \frac{x^6}{2^2 \cdot 4^2 \cdot 6^2} + \cdots$$
$$J_1 = \frac{x}{2} - \frac{2x^3}{2 \cdot 4^2} + \frac{3x^5}{2 \cdot 4^2 \cdot 6^2} - \cdots$$

B.4 Approximations for Small Argument, $x < 1$

$$J_0 \rightarrow 1 - \frac{x^2}{4}, \qquad J_1 \rightarrow \frac{x}{2} - \frac{x^3}{16},$$

$$Y_0 \rightarrow \frac{2}{\pi} \ln x, \qquad Y_1 \rightarrow -\frac{2}{\pi} \frac{1}{x}$$

B.5 Approximations for Large Arguments, $x > 2\pi$

$$J_n \rightarrow \sqrt{\frac{2}{\pi x}} \cos\left(x - \frac{n\pi}{2} - \frac{\pi}{4}\right), \qquad Y_n \rightarrow \sqrt{\frac{2}{\pi x}} \sin\left(x - \frac{n\pi}{2} - \frac{\pi}{4}\right)$$

$$H_n^{(1)} \rightarrow \sqrt{\frac{2}{\pi x}} e^{i(x - \frac{n\pi}{2} - \frac{\pi}{4})}, \qquad H_n^{(2)} \rightarrow \sqrt{\frac{2}{\pi x}} e^{-i(x - \frac{n\pi}{2} - \frac{\pi}{4})}$$

B.6 Recursion Relations

The relations listed below hold true for C being any of the Bessel functions of the first, second, or third kind or for linear combinations of these functions.

$$C_{n+1} + C_{n-1} = \frac{2n}{x} C_n$$

$$\frac{dC_0}{dx} = -C_1$$

$$\frac{dC_n}{dx} = \frac{1}{2}(C_{n-1} - C_{n+1})$$

$$\frac{d}{dx}(x^n C_n) = x^n C_{n-1}$$

$$\frac{d}{dx}\left(\frac{1}{x^n} C_n\right) = -\frac{1}{x^n} C_{n+1}$$

B.7 Modified Bessel Functions

$$I_n(x) = i^{-n} J_n(ix)$$

$$I_0(x) = J_0(ix) = 1 + \frac{x^2}{2^2} + \frac{x^4}{2^2 \cdot 4^2} + \frac{x^6}{2^2 \cdot 4^2 \cdot 6^2} + \cdots$$

$$\int x I_0(x)\, dx = x I_1(x)$$

$$\int I_1(x)\, dx = I_0(x)$$

$$I_0(x) - I_2(x) = \frac{2}{x} I_1(x)$$

B.8 Tables of Bessel Functions, Zeros, and Inflection Points

A. Bessel Functions or Orders 0, 1, and 2

x	$J_0(x)$	$J_1(x)$	$J_2(x)$	$I_0(x)$	$I_1(x)$	$I_2(x)$
0.0	1.0000	0.0000	0.0000	1.0000	0.0000	0.0000
0.2	0.9900	0.0995	0.0050	1.0100	1.1005	0.0050
0.4	0.9604	0.1960	0.0197	1.0404	0.2040	0.0203
0.6	0.9120	0.2867	0.0437	1.0921	0.3137	0.0464
0.8	0.8463	0.3688	0.0758	1.1665	0.4329	0.0843
1.0	0.7652	0.4401	0.1149	1.2661	0.5652	0.1358
1.2	0.6711	0.4983	0.1593	1.3937	0.7147	0.2026
1.4	0.5669	0.5419	0.2074	1.5534	0.8861	0.2876
1.6	0.4554	0.5699	0.2570	1.7500	1.0848	0.3940
1.8	0.3400	0.5815	0.3061	1.9895	1.3172	0.5260
2.0	0.2239	0.5767	0.3528	2.2796	1.5906	0.6890
2.2	0.1104	0.5560	0.3951	2.6292	1.9141	0.8891
2.4	+0.0025	0.5202	0.4310	3.0492	2.2981	1.1111
2.6	−0.0968	0.4708	0.4590	3.5532	2.7554	1.4338
2.8	−0.1850	0.4097	0.4777	4.1575	3.3011	1.7994
3.0	−0.2601	0.3391	0.4861	4.8808	3.9534	2.2452
3.2	−0.3202	0.2613	0.4835	5.7472	4.7343	2.7884
3.4	−0.3643	0.1792	0.4697	6.7848	5.6701	3.4495
3.6	−0.3918	0.0955	0.4448	8.0278	6.7926	4.2538
3.8	−0.4026	+0.0128	0.4093	9.5169	8.1405	5.2323
4.0	−0.3971	−0.0660	0.3641	11.302	9.7594	6.4224
4.2	−0.3766	−0.1386	0.3105	13.443	11.705	7.8683
4.4	−0.3423	−0.2028	0.2501	16.010	14.046	9.6259
4.6	−0.2961	−0.2566	0.1846	19.097	16.863	11.761
4.8	−0.2404	−0.2985	0.1161	22.794	20.253	14.355
5.0	−0.1776	−0.3276	+0.0466	27.240	24.335	17.505
5.2	−0.1103	−0.3432	−0.0217	32.584	29.254	21.332
5.4	−0.0412	−0.3453	−0.0867	39.010	35.181	25.980
5.6	+0.0270	−0.3343	−0.1464	46.738	42.327	31.621
5.8	0.0917	−0.3110	−0.1989	56.039	50.945	38.472
6.0	0.1507	−0.2767	−0.2429	67.235	61.341	46.788
6.2	0.2017	−0.2329	−0.2769	80.717	73.888	56.882
6.4	0.2433	−0.1816	−0.3001	96.963	89.025	69.143
6.6	0.2740	−0.1250	−0.3119	116.54	107.31	84.021
6.8	0.2931	−0.0652	−0.3123	140.14	129.38	102.08
7.0	0.3001	−0.0047	−0.3014	168.59	156.04	124.01
7.2	0.2951	+0.0543	−0.2800	202.92	188.25	150.63
7.4	0.2786	0.1096	−0.2487	244.34	227.17	182.94
7.6	0.2516	0.1592	−0.2097	294.33	274.22	222.17
7.8	0.2154	0.2014	−0.1638	354.68	331.10	269.79
8.0	0.1716	0.2346	−0.1130	427.57	399.87	327.60

B. Zeros of J_m: $J_m(j_{mn}) = 0$.

$$j_{mn}$$

m \ n	0	1	2	3	4	5
0	—	2.40	5.52	8.65	11.79	14.93
1	0	3.83	7.02	10.17	13.32	16.47
2	0	5.14	8.42	11.62	14.80	17.96
3	0	6.38	9.76	13.02	16.22	19.41
4	0	7.59	11.06	14.37	17.62	20.83
5	0	8.77	12.34	15.70	18.98	22.22

C. Inflection Points of J_m: $(dJ_m/dx)_{j'_{mn}} = 0$.

$$j'_{mn}$$

m \ n	1	2	3	4	5
0	0	3.83	7.02	10.17	13.32
1	1.84	5.33	8.54	11.71	14.86
2	3.05	6.71	9.97	13.17	16.35
3	4.20	8.02	11.35	14.59	17.79
4	5.32	9.28	12.68	15.96	19.20
5	6.41	10.52	13.99	17.31	20.58

Appendix C
Using Laplace Transforms to Solve Differential Equations

C.1 Introduction

Not all aspects of the of Laplace transforms will be presented here, but enough of the characteristics will be discussed to help the reader understand and appreciate the elegance and powerfulness of the procedures that can be effectively applied to solve linear differential equations. The essential idea of Laplace transforms is that they are used to convert a differential equation into an algebraic one in order to solve the differential equation.

Let $f(t)$ be a function of t for $t > 0$. Its Laplace transform is defined as:

$$F(t) = \int_0^\infty f(t)e^{-st} = \mathcal{L}[f(t)] \tag{C.1}$$

where s is a complex variable which can be expressed as:

$$s = \alpha + i\beta \tag{C.2}$$

The function $f(t)$ is Laplace transformable for $\alpha > 0$ if

$$\lim_{T \to 0} \int_0^T |f(t)|e^{-\alpha t}\, dt < \infty$$

If $f(t) = A$ (a constant), for $t > 0$, the Laplace transform of the constant A is:

$$\mathcal{L}[A] = \int_0^\infty Ae^{-st}\, dt = -\frac{1}{s}e^{-st}\Big|_0^\infty = \frac{A}{s} = F(s)$$

We therefore have the Laplace transform pair

$$f(t) = A \Leftrightarrow F(s) = \frac{A}{s}$$

If $f(t) = e^{-at}$ for $t > 0$, then the Laplace transform of this exponential function is

$$\mathcal{L}e^{-at} = \int_0^\infty e^{-at}e^{-st}\, dt = \int_0^\infty e^{-(s+a)t} = \frac{A}{s}$$

and hence

$$f(t) = e^{-at} \Leftrightarrow F(s) = \frac{1}{s+a}$$

C.2   Laplace Transforms of Derivatives

Consider a function $f(t)$ and its derivative $df(t)/dt$ that are both Laplace transformable. If the function $f(t)$ has the Laplace transform $F(s)$, then

$$\mathcal{L}\frac{df(t)}{dt} = sF(s) - f(0+) \qquad (C.3)$$

Equation (C.3) is readily proven by using integration by parts, that is,

$$F(s) = \int_0^\infty f(t)e^{-st}\, dt$$

Set

$$u = f(t), \qquad du = \frac{df(t)}{dt}\, dt$$

$$dv = e^{-st}\, dt, \qquad v = -\frac{1}{s}e^{-st}$$

It then follows that

$$F(s) = uv\big|_0^\infty - \int_0^\infty v\, du = -\frac{1}{s}f(t)e^{-st}\bigg|_0^\infty + \frac{1}{s}\int_0^\infty \frac{df(t)}{dt}e^{-st}\, dt$$

$$= \frac{f(0+)}{s} + \frac{1}{s}\int_0^\infty \frac{df(t)}{dt}e^{-st}\, dt$$

or

$$\int_0^\infty \frac{df(t)}{dt}e^{-st}\, dt = sF(s) - f(0+)$$

The term $f(0+)$ represents the value of $f(t)$ as $t \to 0$ from the positive side. In a similar manner, it can be demonstrated that

$$\mathcal{L}\frac{d^2 f(t)}{dt^2} = s^2 F(s) - sf(0+) - f(0+) \qquad (C.4)$$

C.3   Solving Differential Equations

We discuss here a differential equation that is of greatest interest in treatment of vibration problems:

$$m\ddot{x}(t) + C\dot{x}(t) + kx(t) = f(t) \qquad (C.5)$$

which also can be expressed as

$$\ddot{x} + 2\xi\omega_n\dot{x}(t) + \omega_n^2 x(t) = f_\alpha(t) \tag{C.6}$$

We now apply equations (C.3) and (C.4)–(C.6) to obtain

$$s^2X(s) - sx(0+) - \dot{x}(0+) + 2\xi\omega_n[sX(s) - x(0+)] + \omega_n^2 X(s) = F_\alpha(s)$$

Solving for $X(s)$:

$$X(s) = \frac{F_\alpha(s) + x(0+)(s + 2\xi\omega_n) + \dot{x}(0+)}{s^2 + 2\xi\omega_n s + \omega_n^2} \tag{C.7}$$

Equation (C.7) can be written in the form

$$X(s) = \frac{A(s)}{B(s)} \tag{C.8}$$

Equation (C.7) and its variation (C.6) have been derived from a second-order equation, and the techniques we describe below are valid for any order function.
   For equations with simple roots, we can write equation (C.8) in the form

$$X(s) = \frac{A(s)}{B(s)} = \frac{A(s)}{(s + a_1)(s + a_2)\cdots(s + a_n)} \tag{C.9}$$

It is assumed that $B(s)$ is of higher order than $A(s)$ throughout the discussion below. We can expand equation (C.9) through the use of partial fraction expansions, which will result in:

$$X(s) = \frac{C_1}{s + a_1} + \frac{C_2}{s + a_2} + \cdots + \frac{C_n}{s + a_n} \tag{C.10}$$

where

$$C_k = \lim_{s \to -a_k} (s + a_k)\frac{A(s)}{B(s)} \tag{C.11}$$

Example Problem 1

Find the inverse Laplace tranform of the function

$$X(s) = \frac{13(s + 3)}{s(s + 1)((s + 2)}$$

Solution

The roots of the numerator (for example, $s = -3$) are called *zeros* of the function. The roots of the denominator (viz., 0, −1, −2) are called *poles*. A partial fraction expansion of the poles is written as follows:

$$\frac{13(s+3)}{s(s+1)(s+2)} = 13\left(\frac{C_1}{s} + \frac{C_2}{s+1} + \frac{C_3}{s+2}\right)$$

The *residues* C1, C2, and C3 are obtained from (C.11) as follows:

$$C_1 = \lim_{s \to 0} \frac{s+3}{(s+1)(s+2)} = \frac{3}{(1)(2)} = \frac{3}{2}$$

$$C_2 = \lim_{s \to -1} \frac{s+3}{s(s+2)} = \frac{2}{(-1)(1)} = -2$$

$$C_3 = \lim_{s \to -2} \frac{s+3}{s(s+1)} = \frac{1}{(-2)(-1)} = \frac{1}{2}$$

Hence

$$X(s) = 13\left(\frac{3/2}{s} + \frac{-2}{s+1} + \frac{1/2}{s+2}\right)$$

Applying the transform pairs we derived in Section C.2:

$$x(t) = 13\left(\frac{3}{2} - 2e^{-t} + \frac{1}{2}e^{-2t}\right)$$

When we know the form of the inverse transform above we can derive the residue graphically by inspection. The method for doing this is simply divide the product of all vectors from the zeros to the pole whose residue is being determined by the product of the vectors from all the other poles to the pole under consideration. This technique is shown in Figure C.1. In Figure C.1a, C_1 is derived as

$$C_1 = \frac{3\angle 0°}{(2\angle 0°)(1\angle 0°)} = 3/2$$

In Figure C.1b, we get

$$C_2 = \frac{2\angle 0°}{(1\angle 0°)(1\angle 180°)} = \frac{2}{(1)(-1)} = -2$$

and from Figure C.1c:

$$C_3 = \frac{1\angle 0°}{(1\angle 180°)(2\angle 180°)} = \frac{1}{(-1)(-2)} = \frac{1}{2}$$

Note that the numbers obtained from the use of the graphical residue technique are the same as those derived from equation (C.11). The sign of the residue can be readily established by simply counting the number of poles and zeros to the right in the s plane of the pole whose residue is being determined. If the number is even, the sign of the residue is positive; and if the number is odd, the sign is negative.

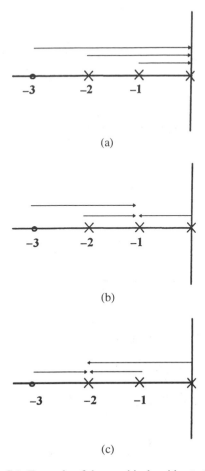

(a)

(b)

(c)

FIGURE C.1. Example of the graphical residue technique.

Example Problem 2

Consider the function

$$X(s) = \frac{8(s+2)(s+4)}{s(s+1)(s+3)}$$

Find the inverse Laplace transform.

Solution

Use the pole-zero pattern of Figure C.2 to apply the graphical technique. The inverse transform can be written down by inspection as

$$x(t) = 8\left[\frac{(2)(4)}{(1)(3)} - \frac{(1)(3)}{(1)(2)}e^{-t} - \frac{(1)(1)}{(2)(3)}e^{-3t}\right] = 8\left[\frac{8}{3} - \frac{3}{2}e^{-t} - \frac{1}{6}e^{-3t}\right]$$

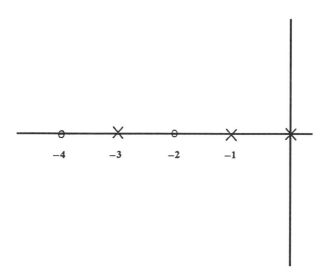

FIGURE C.2. Pole-zero pattern for Example Problem 2.

C.4   Equations with Multiple-Order Roots

Consider a function $X(s)$ which has multiple-order roots of the form

$$X(s) = \frac{A(s)}{B(s)} = \frac{A(s)}{(s + a_1)^m (s + a_2) \cdots (s + a_n)}$$

The partial fraction expansion of the function must be of the format

$$\frac{A(s)}{B(s)} = \frac{C_{11}}{(s + a_1)^m} + \frac{C_{12}}{(s + a_1)^{m-1}} + \cdots + \frac{C_{1m}}{s + a_1} + \frac{C_2}{s + a_2} + \cdots + \frac{C_n}{s + a_n}$$

The residues $C_2, C_3, \ldots, C_n$ are determined in the same manner as was described in the preceding section, but the coefficients $C_{11}, C_{12}, \ldots, C_{1m}$ require special treatment. They are evaluated from the following expression:

$$C_{1j} = \frac{1}{(j-1)!} \frac{d^{j-1}}{ds^{j-1}} (s + a_1)^m \frac{A(s)}{B(s)} \tag{C.12}$$

which is evaluated at $s = -a_1$. Then the inverse Laplace is obtained from the transform relation:

$$\mathcal{L}^{-1}\left[\frac{1}{(s+a)^j}\right] = \frac{t^{j-1}}{(j-1)!} e^{-at} \tag{C.13}$$

The following example problem will demonstrate the procedure.

Example Problem 2

Find the inverse Laplace transform for

$$X(s) = \frac{s+1}{(s+2)^2(s+3)} = \frac{C_{11}}{(s+2)^2} + \frac{C_{12}}{s+2} + \frac{C_2}{s+3}$$

Solution

Use equation (C.11) to obtain

$$C_{11} = \left.\frac{s+1}{s+3}\right|_{s=-2} = -1, \qquad C_2 = \left.\frac{s+1}{(s+2)^2}\right|_{s=-3} = -2$$

Equation (C.12) is now applied to obtain the coefficient C_{12}:

$$C_{12} = \frac{1}{(2-1)!}\frac{d^{2-1}}{ds^{2-1}}\left(\frac{s+1}{s+3}\right) = \frac{d}{ds}\left(\frac{s+1}{s+3}\right) = \left.\left(\frac{1}{s+3} - \frac{s+1}{(s+3)^2}\right)\right|_{s=-2} = 2$$

which now results in

$$X(s) = \frac{-1}{(s+2)^2} + \frac{2}{s+2} - \frac{2}{s+3}$$

The inverse Laplace transform now becomes

$$x(t) = -te^{-2t} + 2e^{-2t} - 2e^{-3t}$$

C.5   Equations with Complex Roots

Let us consider the case of finding the inverse Laplace transform of a function of the form

$$X(s) = \frac{(s+b_1)(s+b_2)}{[(s+\alpha)^2 + \beta^2](s+a_1)(s+a_2)(s+a_3)}$$

We could evaluate the inverse transform using a partial fraction expansion, but this can be quite a complicated procedure especially when complex roots are entailed. In order to deal with this problem, it is preferable to use the relation between the transform pair

$$\frac{as+b}{(s+\alpha)^2 + \beta^2} \Leftrightarrow 2Me^{-\alpha t}\cos(\beta t - \phi) \tag{C.13}$$

where M and ϕ are unknowns that depend on a and b.

In order to demonstrate that M and ϕ can be evaluated by the graphical approach, suppose that the above function has the pole-zero pattern of Figure C.3a. Only the

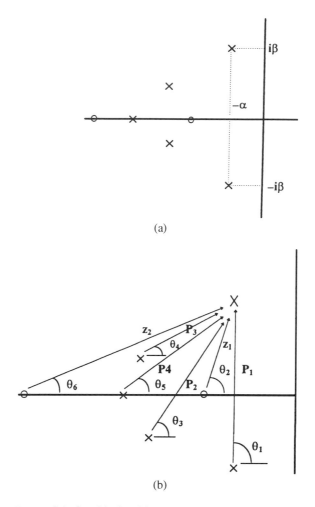

FIGURE C.3. Graphical residue approach for complex roots.

residues of the complex poles at $-\alpha \pm i\beta$ will be determined. Using the graphical approach for simple roots, we get from using Figure C.3b:

$$X(s) = \frac{\dfrac{(z_1 e^{i\theta_2})(z_2 e^{i\theta_6})}{(P_1 e^{i\theta_1})(P_2 e^{i\theta_3})(P_3 e^{i\theta_4})(P_4 e^{i\theta_5})}}{s + \alpha - i\beta} + \frac{\dfrac{(z_1 e^{-i\theta_2})(z_2 e^{-i\theta_6})}{(P_1 e^{-i\theta_1})(P_2 e^{-i\theta_3})(P_3 e^{-i\theta_4})(P_4 e^{-i\theta_5})}}{s + \alpha + i\beta}$$

$$+ \text{ additional terms} \tag{C.14}$$

We observe that the coefficient of the second term, which represents the residue at $s = -\alpha - i\beta$ is the complex conjugate of the coefficient of the first term which is the residue at $s = -\alpha + i\beta$. Equation (C.14) can be expressed as

$$X(s) = \frac{Me^{i\phi}}{s + \alpha - i\beta} + \frac{Me^{-i\phi}}{s + \alpha + i\beta} + \text{additional terms}$$

where

$$M = \frac{Z_1 Z_2}{P_1 P_2 P_3 P_4}, \qquad \phi = \theta_2 + \theta_6 - (\theta_1 + \theta_3 + \theta_4 + \theta_5)$$

Then

$$
\begin{aligned}
x(t) &= M(e^{-\alpha t + i\beta t + i\phi} + e^{-i\alpha t - i\beta t - i\phi}) + \text{additional terms} \\
&= Me^{-\alpha t}\left(e^{i(\beta t + i\phi)} + e^{-i(\beta t + \phi)}\right) + \text{additional terms} \\
&= 2Me^{-\alpha t}\frac{\left(e^{i(\beta t + i\phi)} + e^{-i(\beta t + \phi)}\right)}{2} \text{additional terms} \\
&= 2Me^{-\alpha t}\cos(\beta t + \phi) + \text{additional terms}
\end{aligned}
$$

Here M and ϕ can be determined graphically using Figure C.3. The use of the graphical residue technique for equations with complex roots is extremely powerful because the inverse transform can be determined by inspection.

Example Problem 3

Given

$$X(s) = \frac{s + 2}{s(s^2 + 4s + 5)}$$

find the inverse Laplace transform.

Solution

Figure C.4 shows the function. The residue at the pole at the origin is found by using Figure C.4b, and the complex poles are derived by using Figure C.4c. The ensuing function is

$$x(t) = \frac{2}{\sqrt{5}\sqrt{5}} + 2 \times \frac{1}{2\sqrt{5}}e^{-2t}\cos(t + \phi)$$

wherein

$$\phi = (90°) - (90° + \theta)$$
$$\theta = \left(\pi - \tan^{-1}\tfrac{1}{2}\right)$$

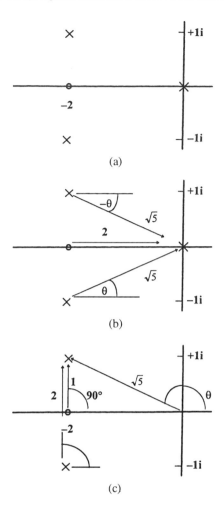

FIGURE C.4. Pole-zero diagrams for Example Problem 3.

Example Problem 4

From vibration theory $X(s)$ [cf. equation (18.12)] is of the form

$$X(s) = x_0 \frac{s + 2\xi\omega_n}{s^2 + 2\xi\omega_n s + \omega_n^2}$$

$$= \frac{x_0(s + 2\xi\omega_n)}{\left(s + \xi\omega_n + i\omega_n\sqrt{1 - \xi^2}\right)\left(s + \xi\omega_n - i\omega_n\sqrt{1 - \xi^2}\right)}$$

Determine the inverse transform.

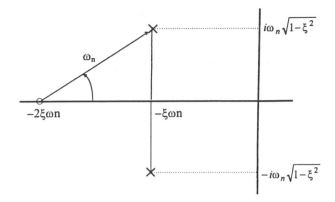

FIGURE C.5. Pole-zero diagram for Example Problem 4.

Solution

From Figure C.5 the inverse transform can be evaluated to yield

$$x(t) = x_0 \left[\frac{\omega_n}{\omega_n \sqrt{1 - \xi^2}} e^{-\zeta \omega_n t} \cos \left(\omega_n \sqrt{1 - \xi^2}\, t + \phi \right) \right]$$

where

$$\phi = \theta - \pi/2$$

$$\theta = \cos^{-1} \xi$$

and we now have

$$x(t) = \frac{x_0}{\sqrt{1 - \xi^2}} e^{-2\omega_n t} \cos \left(\omega_n \sqrt{1 - \xi^2}\, t - \pi/2 + \cos^{-1} \xi \right)$$

Index